Preparing for the AP® Statistics Examination

to accompany

The Practice of Statistics

SEVENTH EDITION

Daren Starnes | Josh Tabor

Erica Chauvet

Jeff Eicher, Jr.

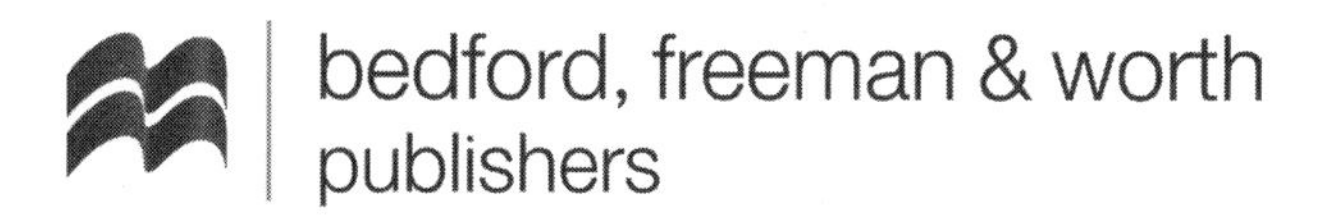

ISBN-13: 978-1-319-53546-9

Manufactured in the United States of America.

1 2 3 4 5 6   29 28 27 26 25 24

W. H. Freeman and Company
Bedford, Freeman & Worth Publishers
120 Broadway, New York, NY 10271
https://www.bfwpub.com/catalog

Table of Contents

Preface

Strive for a 5: Preparing for the AP® Statistics Examination is designed for use with *The Practice of Statistics, 7th Edition* by Daren Starnes and Josh Tabor, which aligns closely to the College Board AP Statistics Course Exam and Description (CED). *Strive for a 5* is intended to help you prepare for the AP® Statistics Exam. This book is divided into two sections: a study guide and practice tests.

SECTION 1: STUDY GUIDE

The study guide is designed for you to use while you take AP® Statistics. As each unit is covered in class, use the study guide to strengthen your understanding of important statistical concepts and terms. For each section, this guide provides learning targets, checks for understanding, practice problems, and hints to help you master the material before moving on to the next section.

This guide is designed to help you clearly identify what you need to learn, determine whether or not you have learned it, and help to close gaps in your understanding. That is, it will help you answer the questions:

"What do I need to understand?"
"What do I currently understand?"
"How can I close the gap?"

For each unit, the study guide is organized as follows.

- **Overview**: The unit content is summarized to provide a quick overview of the material. The material is broken down into the specific learning targets to help you answer the question "What do I need to understand?"

- **Unit Sections:** Each section includes a presentation of the important content to summarize the main ideas. The section content includes the following.
  - **Learning Targets**: These "I can…" statements provide a guide for you to keep track of your learning. Check these off as you become confident with each concept.
  - **Vocabulary**: Space for you to record and study definitions of important terms.
  - **Concept Explanations**: Important concepts are explained further to aid you in your study of the key points of each section.
  - **Checks for Understanding**: Mini assessments for each learning target help you answer the question "What do I currently understand?"

- **Unit Summary**: The key concepts of the unit are briefly reviewed, and a summary of the learning targets is provided to summarize what you have learned.

- **Multiple-Choice Questions**: The multiple-choice questions focus on key concepts you should grasp after reading the unit. They are designed for quick exam preparation. After checking your answers, you should reflect on the big ideas from the unit and identify the concepts you may need to study more.

- **FRAPPY!:** A "Free-Response AP Problem, Yay!" is provided after each unit. Each FRAPPY is modeled after actual AP Statistics Exam questions and provide an opportunity for you to practice your communication skills to maximize your score. These problems are meant to help you determine what further work is needed to address the question "How can I close the gap?"

- **Vocabulary Crossword Puzzle**: These puzzles provide a fun way to check your understanding of key definitions from the unit.

Answers to all problems in the study guide are available in the back of the book. Checking your answers will help you determine whether or not you need additional work on specific learning targets.

SECTION 2: PRACTICE TESTS

The "Practice Tests" section of this guide is meant to help you better understand how the exam is constructed and scored and to help you study for the exam. Make note of what you understand and what you need more work on as you complete the practice exams. Two full-length exams and answers modeled after the actual Advanced Placement Exam (40 multiple-choice questions, 5 free response questions, and an investigative task) are included to help you get a feel for the structure of the test. Try to simulate actual exam conditions as much as possible when taking these practice tests, including timing yourself so you can practice pacing.

****

We hope that your use of Strive for a 5: Preparing for the AP® Statistics Examination *will assist you in your study of statistical concepts, help you earn as high a score as possible on the exam, and provide you with an interest and desire to further your study of statistics.*

About the Authors

Erica Chauvet
Waynesburg University, PA

Erica has more than 20 years of experience in teaching AP® Statistics, AP® Calculus, and college statistics. She has been an AP® Statistics Reader and Table Leader for the past 15 years where she also serves as the social director and famously hosts the annual 1.96-mile Prediction Fun Run. Since its inception in 2017, Erica has assisted as co-editor of the weekly New York Times feature: "What's Going On In This Graph?" Erica is the solution manual author, writes the prepared tests and quizzes, creates the test banks, and manages the Achieve online homework content creation coding team for *The Practice of Statistics* as well as Statistics and Probability with Applications textbooks. Erica is also the solution manual author for a new college-level statistics text: *Introductory Statistics: A Student-Centered Approach*. Erica has worked as a writer, consultant, and reviewer of statistics and calculus material for the past 15 years. When she is not writing statistics material, she enjoys competing in a USTA tennis league and spending time with her husband and twin daughters.

Jeff Eicher, Jr.
Eastern University, St. Davids, PA

Jeff has taught mathematics, statistics, and data science to high school, undergraduate, and graduate students. He taught AP® Statistics for 11 years and has served as an AP® Reader for 8 years. He participated on The College Board's Instructional Design Team during the revision of the Course and Exam Description and the creation of AP® Classroom resources. He has contributed to a variety of textbook projects as well as Stats Medic blog posts and content creation. When he's not doing statistics, he's hanging out with his wife and daughter.

Jeff wants to thank several individuals who have been influential along his statistical journey: Mrs. Kopas and Mrs. Rosie for encouraging me to consider a career in teaching; Daren and Josh for amazing textbooks and quick email replies; Luke and Lindsey for sharing countless student-centered resources; Andrew, Robert and Jackie for inspiring lessons and great conversations; Jared and Joel for many conversations about AP Statistics; Eleanor for introducing me to "building thinking classrooms"; Greg and Jamie for welcoming me to the team; and Mike, whose bumper sticker threw me into the world of statistics in the first place.

Unit 1, Part I: Exploring One-Variable Data

"Statistical thinking will one day be as necessary for efficient citizenship as the ability to read and write." H.G. Wells

UNIT 1 PART I OVERVIEW

The world we live in is driven by statistics. We use them every day whether we realize it or not. We use statistical thinking when we complain about the price of gas or when we celebrate over a better-than-expected test score. We begin our study of statistics by mastering the art of displaying and describing data. In this unit, you will learn how to use a variety of graphical tools to display data and will also learn how to summarize data numerically.

By the end of this unit, you should understand the difference between **categorical** and **quantitative** data as well as how to display and describe these two types of data. The distinction between categorical data (ex: yes/no questions, favorite holiday) and quantitative data (ex: age – in years, GPA) is the foundation for much of what you will learn in this course. Use this guide to ensure you have a firm grasp of these concepts!

Sections in Unit 1, Part I
Section 1A: Statistics: The Language of Variation
Section 1B: Displaying and Describing Categorical Data
Section 1C: Displaying Quantitative Data with Graphs
Section 1D: Describing Quantitative Data with Numbers

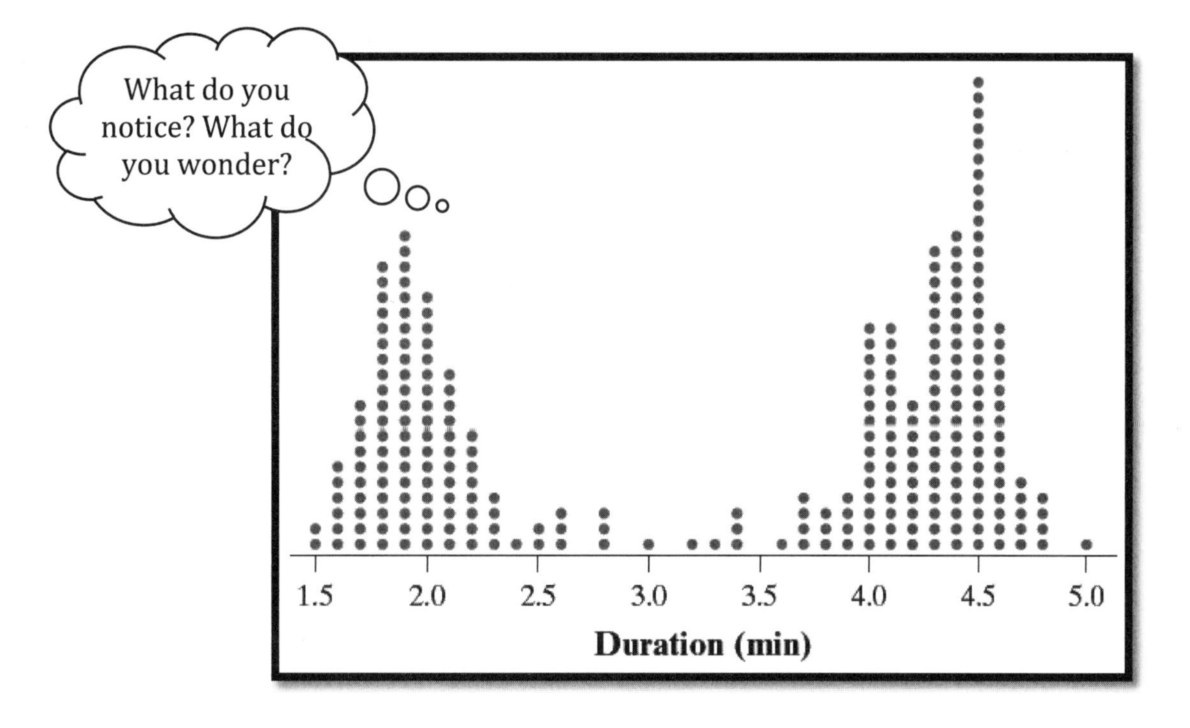

The dotplot shows the durations (in minutes) of 263 eruptions of the Old Faithful geyser.

PRACTICE FOR MASTERY

Use the following *suggested* guide to the pages and exercises in your text to practice for mastery!
Note: your teacher may assign different problems. Be sure to follow their instructions!

Day	Topics	Read	Do
1	• Introduction	p. 2-3	
2	• Individuals and Variables • Summarizing Data with Tables	p. 3-7	**1A**: 1, 3, 5, 7, 8, 9
3	• Displaying Categorical Data: Bar Graphs • Comparing Distributions of Categorical Data • Misleading Graphs	p. 10-15	**1B**: 1, 3, 5, 7, 11, 13, 18, 19
4	• Displaying Quantitative Data: Dotplots • Describing Shape • Describing Distributions of Quantitative Data	p. 21-27	**1C**: 3, 5, 7, 9, 11, 42, 45
5	• Comparing Distributions of Quantitative Data • Displaying Quantitative Data: Stemplots	p. 28-31	**1C**: 13, 15, 17, 19, 23, 40
6	• Displaying Quantitative Data: Histograms	p. 32-37	**1C**: 25, 29, 31, 33, 37, 41, 43, 44
7	Option: AP Classroom Topic Questions (Topics 1.1 – 1.6)		**1B**: 9, 15, 20 **1C**: 1, 21, 27, 46
8	• Measuring Center: The Median • Measuring Center: The Mean	p. 48-53	**1D**: 1, 3, 5, 7, 9, 43
9	• Measuring Variability: The Range • Measuring Variability: Standard Deviation • Measuring Variability: Interquartile Range • Choosing Summary Statistics	p. 54-62	**1D**: 11, 13, 15, 17, 19, 21, 23, 44
10	• Identifying Outliers • Displaying Summary Statistics: Boxplots • Comparing Distributions with Boxplots and Summary Statistics	p. 63-70	**1D**: 27, 29, 31, 33, 35, 37, 45, 46
11	Unit 1, Part I AP® Statistics Practice Test (p. 84)		Unit 1, Part I Review Exercises (p. 81)
12	Unit 1, Part I Test: Celebration of learning!	Note: Since "Test" can sound daunting, think of this as a way to show off all you have learned. It's your own celebration of learning!	

Introduction – Smelling Parkinson's Disease

Watch this video, which explains Joy's amazing ability to smell Parkinson's disease:

tinyurl.com/4synrm34

Can you smell Parkinson's Disease? See for yourself by using the Smelling Parkinson's applet at

stapplet.com/parkinsons.html

Write down your guesses.

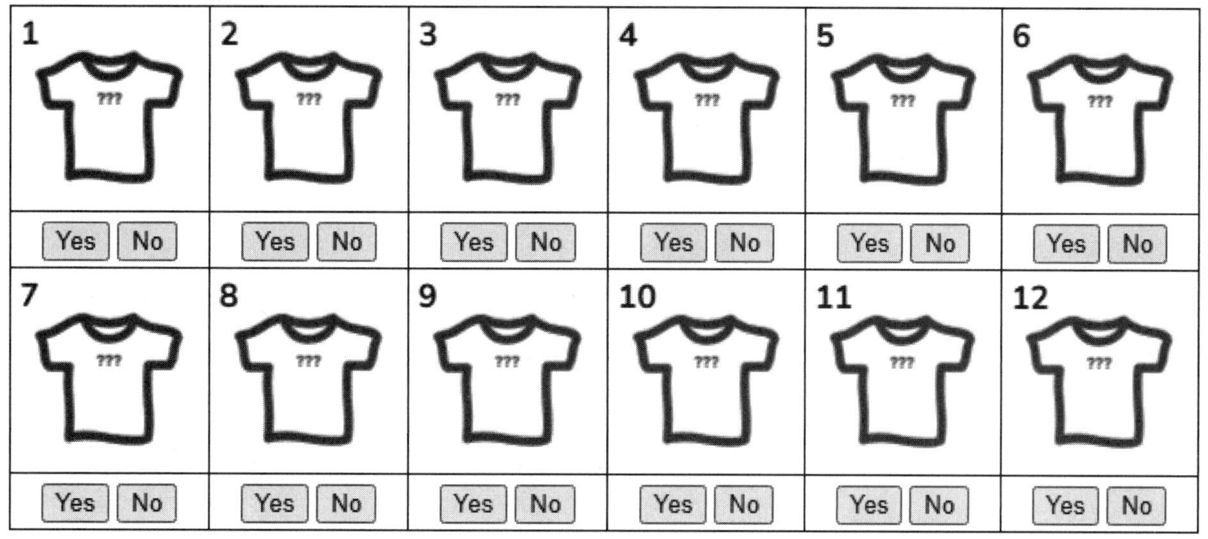

How many did you get correct? ____________________

Repeat the process several times. Record your results on the dotplot.

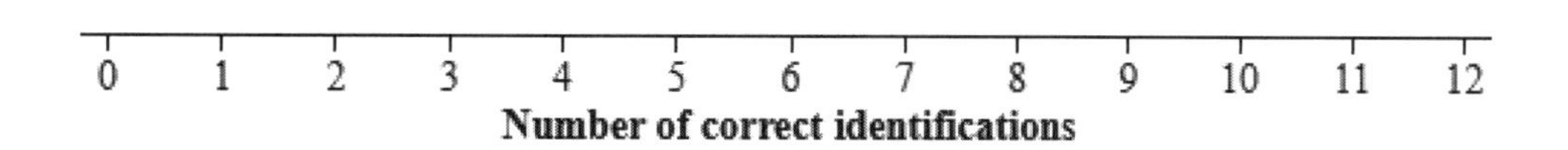

How often were you able to guess 11 or more shirts correctly by chance alone? __________

In what percentage of your guesses did you get 11 or more shirts correct by chance alone? _____

Do you think Joy is a lucky guesser or do you think Joy really can smell Parkinson's?

Section 1A – Statistics: The Language of Variation

Section Summary

In this section, you will learn the basic terms and definitions necessary for the study of statistics. You will also be introduced to the idea of moving from data analysis to inference. One of our goals in statistics is to use data from a representative sample to make an inference about the population from which the sample was selected. In order to do this, we must be able to identify the type of data we are dealing with, as our choice of statistical procedures will depend upon the type of data we have.

Learning Targets:

____ I can identify individuals and variables in a set of data and classify the variables as categorical or quantitative.

____ I can make and interpret a frequency table or a relative frequency table for a distribution of data.

Read: Vocabulary and Key Concepts

- An _____________ is an object described in a set of data. Individuals can be people, animals, or things.
- A ___________ is a characteristic that can take different values for different individuals.
- A____________________ takes values that are labels, which place each individual into a particular group, called a category.
- A ____________________ takes number values that are quantities — counts or measurements.
- Is zip code a categorical or quantitative variable? ____________________
- Is "year" a categorical or quantitative variable? ____
- The _________________ of a variable tells us what values the variable takes and how often it takes each value.
- A _________________________ shows the number of individuals having each value.
- A ___________________________ shows the proportion or percentage of individuals having each value.

Watch: Go to bfwpub.com/TPS7eStrive. Select Unit 1, Part I and watch the following Example videos:

Video	Topic	Done
Unit 1A, Census at school (p. 4)	*Individuals and variables*	
Unit 1A, Call me, maybe? (p. 6)	*Summarizing data with tables*	

Do: Check for Understanding

Concept 1: Individuals and Variables, Types of Variables

Sets of data contain information about individuals. The characteristics of the individuals are called variables. Variables can take on different values for different individuals. It is these values and the variation in them that we will learn about in this course.

Variables can fall into one of two categories: categorical or quantitative. When the characteristic we measure places an individual into one of several groups, we call it a categorical variable. When the characteristic we measure results in numerical values for which it makes sense to find an average, we call it a quantitative variable. This distinction is important, as the methods we use to describe and analyze data will depend upon the type of variable we are studying. In the rest of this unit, we will learn how to display and describe the distribution of categorical and quantitative variables.

Check for Understanding – Learning Target 1

____ *I can identify individuals and variables in a set of data and classify the variables as categorical or quantitative.*

Mr. Bush gathered some information on 6 students in his statistics class and organized it in a table.

Student	Class	ACT score	Favorite subject	Height (in.)
James	Junior	34	Statistics	70
Jen	Junior	35	Statistics	51
DeAnna	Sophomore	32	History	67
Jonathan	Freshman	28	History	69
Doug	Senior	33	Art	71
Sharon	Senior	30	Spanish	68.5

1. Who are the individuals in this dataset?

2. What variables were measured? Identify each as categorical or quantitative.

3. Three of the 6 students who were surveyed said Statistics is their favorite subject. Can we infer that 50% of all students at this school would say that Statistics is their favorite subject?

Concept 2: Frequency Table, Relative Frequency Table
A frequency table displays the count of individuals that fall into each category. A relative frequency table displays the proportion (which essentially means, "leave your answer as a decimal.") or a percent of individuals that fall into each category.

Check for Understanding – Learning Target 2
____ *I can make and interpret a frequency table or a relative frequency table for a distribution of data.*

An elementary school student surveyed 25 classmates to see if they believe in Santa. Here are the results:

Yes	No	Yes	Yes	Not sure	Yes	Yes	No	Yes
No	Yes	No	Not sure	No	Yes	Yes	Yes	Yes
No	No	Not sure	Not sure	No	Yes	Yes		

1. Make a frequency table and relative frequency table to summarize the distribution of response to the survey question.

2. How many students surveyed believe in Santa? What table did you use to answer this question?

3. What percentage of students surveyed believe in Santa? What table did you use to answer this question?

Section 1B: Displaying and Describing Categorical Data

Section Summary

In this section, you will learn how to display, describe, and compare distributions of categorical data. Bar graphs and pie charts are typically used to display categorical data. A bar graph is useful for displaying frequencies or relative frequencies, such as what percentage of students in your class drive to school, ride the bus, walk, or use other transportation. A pie chart is useful for displaying parts of the whole. A pictograph is an artistic form of a bar graph but be on guard! Pictographs can sometimes be misleading.

Learning Targets:

____ I can make and interpret bar graphs of categorical data.
____ I can compare distributions of categorical data.
____ I can identify what makes some graphs of categorical data misleading.

Read: Vocabulary and Key Concepts

- A ____________________ shows each category as a bar. The heights of the bars show the category frequencies or relative frequencies.
- A ____________________ shows each category as a slice of the "pie." The areas of the slices are proportional to the category frequencies or relative frequencies.
- A ______________________________ displays the distribution of one categorical variable in each of two or more groups.

Watch: Go to bfwpub.com/TPS7eStrive. Select Unit 1, Part I and watch the following videos:

Video	Topic	Done
Unit 1B, What's on the radio? (p. 12)	*Displaying categorical data: Bar graphs*	
Unit 1B, Screen time for teens and tweens (p. 13)	*Comparing distributions of categorical data*	
Unit 1B, Who buys iMacs? (p. 15)	*Misleading graphs*	

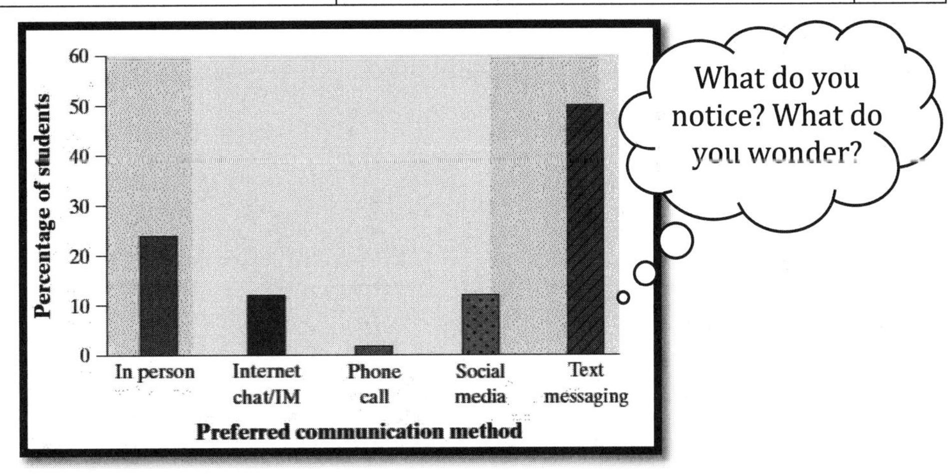

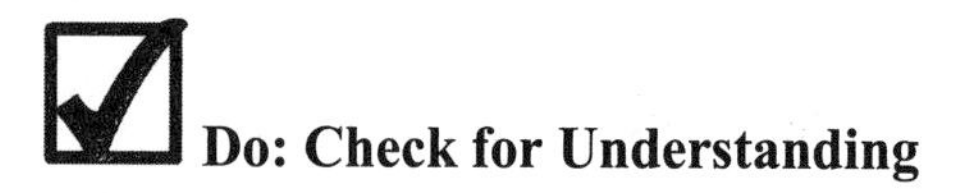

Do: Check for Understanding

Concept 1: Displaying Categorical Data

A frequency table (or a relative frequency table) displays the counts (or percentages) of individuals that take on each value of a variable. However, tables are sometimes difficult to read, and they don't always highlight important features of a distribution. Graphical displays of data are much easier to read and often reveal interesting patterns and departures from patterns in the distribution of data. We can use bar graphs and pie charts to display the distribution of categorical variables. Note, when constructing graphical displays, we must be careful not to distort the quantities. Beware of pictographs and watch the scales when displaying or reading graphs!

Check for Understanding – Learning Target 1

____ *I can make and interpret bar graphs of categorical data.*

A local business owner was interested in knowing the coffee-shop preferences of her town's residents. According to her survey of 250 residents, 75 preferred "Goodbye Blue Monday", 50 liked "The Ugly Mug", 38 chose "Morning Joe's", 50 said "One Mean Bean", 25 brewed their own coffee, and 12 preferred the national chain.

1. Construct a bar graph to display the data. Describe what you see.

2. Construct a pie chart to display the data. How does this display differ from the bar graph?

Concept 2: Comparing Distributions of Categorical Data

When comparing distributions of categorical data, first look at the frequencies or relative frequencies across the categories, then look at the frequencies or relative frequencies within the categories. When the categories (groups) are of different sizes it is better to compare relative frequencies than to compare frequencies because the comparison can be muddied when the groups are of different sizes.

Check for Understanding – Learning Target 2

____ *I can compare distributions of categorical data.*

A random sample of 174 students were asked if they prefer in-person or online learning. The side-by-side bar graph shows the responses by class.

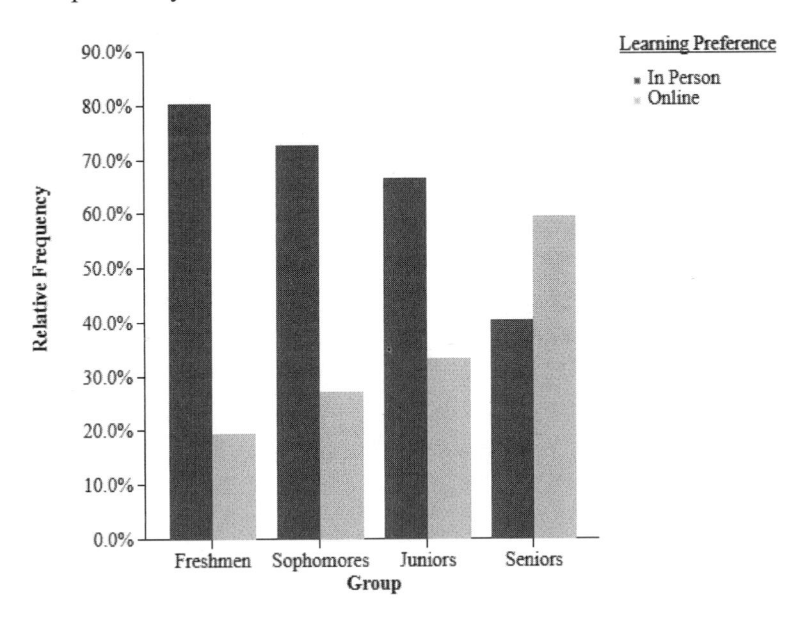

Compare the distributions of learning preference for these classes.

Concept 3: Misleading graphs

Sometimes people try to get creative with the way they set up bar graphs. Avoid the temptation! The bars should be of equal widths with an equal amount of space between the bars. Even more risky is when people try to replace the bars with pictures. Pictures rarely have the same width, so the area of the "bars" varies according to the width of the pictures. This makes the graph misleading! Also, pay attention to the vertical axis scale. It should always start at 0 and consist of equal intervals.

Check for Understanding – Learning Target 3

____ *I can identify what makes some graphs of categorical data misleading.*

1. A student observed the type of roofing materials that were used on all of the homes in their small town. The student created the following graph to display the results. How is this pictograph misleading?

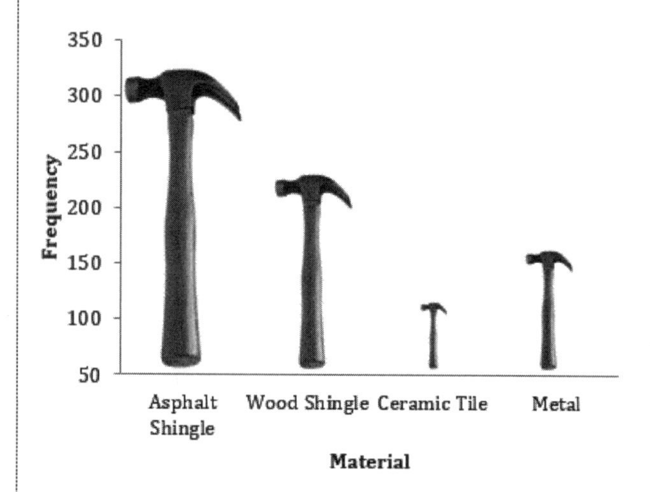

2. The bar graph displays the military spending for the top 5 countries with the largest military budgets. Is the United States' military budget over 3 times the size of China's military budget? Explain why this graph could be considered deceptive.

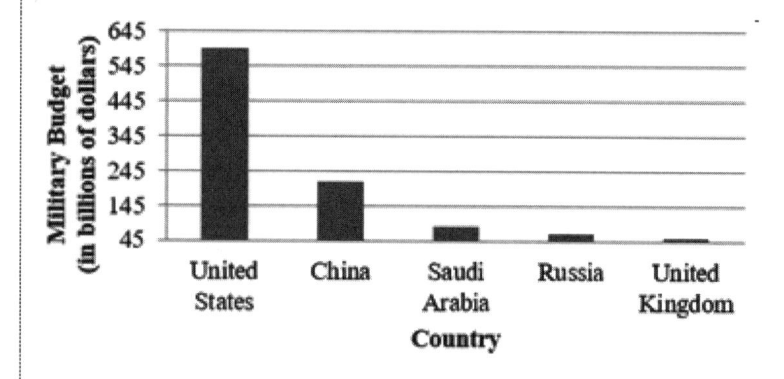

Section 1C: Displaying Quantitative Data with Graphs

Section Summary

Now, you will learn how to display quantitative data, such as the average gas mileage for a random sample of cars, or the hourly wage earned by students in your class who are employed. Dotplots, stemplots, and histograms are helpful for revealing the shape, center, and variability of a distribution of quantitative data.

Learning Targets:

____ I can make and interpret dotplots of quantitative data.
____ I can describe the shape of a distribution of quantitative data.
____ I can describe the distribution of a quantitative variable.
____ I can compare distributions of quantitative data.
____ I can make and interpret stemplots of quantitative data.
____ I can make and interpret histograms of quantitative data.

Read: Vocabulary and Key Concepts

- A quantitative variable that takes a countable set of possible values with gaps between them on the number line is a ____________ variable.
- A quantitative variable that can take any value in an interval on the number line is a ________________ variable.
- A ___________ shows each data value as a dot above its location on a number line.
- A distribution is ___________ if the right side of the graph (containing the half of the observations with the largest values) is approximately a mirror image of the left side.
- A distribution is ______________________ if the left side of the graph is much longer than the right side.
- A distribution is ______________________ if the right side of the graph is much longer than the left side.
- A distribution in which the frequency (relative frequency) of each possible value is about the same is ______________________.
- A _____________ shows each data value separated into two parts: a *stem,* which consists of the leftmost digits, and a *leaf,* consisting of the final digit. The stems are ordered from least to greatest and arranged in a vertical column. The leaves are arranged in increasing order out from the appropriate stems.
- A _______________ shows each interval as a bar. The heights of the bars show the frequencies or relative frequencies of values in each interval.

Watch: Go to bfwpub.com/TPS7eStrive. Select Unit 1, Part I and watch the following Example videos:

Video	Topic	Done
Unit 1C, Eating healthy (p. 23)	*Displaying quantitative data: Dotplots*	
Unit 1C, Quiz scores and die rolls (p. 25)	*Describing shape*	
Unit 1C, Eating healthy (p. 26)	*Describing distributions of quantitative data*	
Unit 1C, Household size in South Africa and the United Kingdom (p. 28)	*Comparing distributions of quantitative data*	
Unit 1C, Preventing concussions (p. 30)	*Displaying quantitative data: Stemplots*	
Unit 1C, Foreign-born residents (p.33)	*Displaying quantitative data: Histograms*	

 **Do: Check for Understanding**

Concept 1: Dotplots

A dotplot is one of the easiest graphs to construct, especially if you have a small data set. Dotplots are helpful for describing distributions because they keep the data intact. That is, you can identify the individual data values directly from the plot. When constructing a dotplot, always label the horizontal axis and remember to include units!

Concept 2: Describing shape

The shape of a distribution of quantitative data gives a description of the overall pattern of the data values. A distribution of quantitative data can be roughly symmetric, skewed to the left, skewed to the right, or approximately uniform. When describing shape also remember to state whether there is a single-peak, multiple peaks, or no clear peak. Lastly, notice whether there are any obvious gaps in the distribution. Only use the description "symmetric" if the distribution is perfectly symmetric. If the halves are not mirror images, but are similar, use the phrase "roughly symmetric."

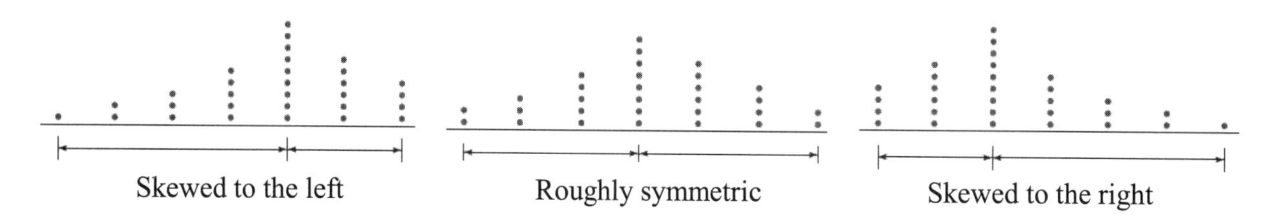

Check for Understanding – Learning Targets 1 and 2

____ *I can make and interpret dotplots of quantitative data.*

____ *I can describe the shape of a distribution of quantitative data.*

A recent study by the Environmental Protection Agency (EPA) measured the gas mileage (miles per gallon) for 30 models of cars. The results are below:

EPA-Measured MPG for 30 Cars

36.3	32.7	40.5	36.2	38.5	36.3	41.0	37.0	37.1	39.9
41.0	37.3	36.5	37.9	39.0	36.8	31.8	37.2	40.3	36.9
36.7	33.6	34.2	35.1	39.7	39.3	35.8	34.5	39.5	36.9

1. Construct a dotplot to display this data

2. Describe the shape of the distribution.

Concept 3: Describing a distribution of quantitative data - SOCV!

We construct graphs of quantitative data so we can better understand the data. Graphs help us understand what values of the variable are typical (common) and what values are unusual (rare). When describing the distribution of a quantitative variable, follow the acronym: SOCV – Shape, Outliers, Center, Variability.

- **Shape**: Is the distribution fairly symmetric? Is it skewed? How many peaks does it have?
- **Outliers**: Are there any outliers, which are values that are unusually large or unusually small?
- **Center**: What is the center? The center can be described by the mean or the median.
- **Variability**: How variable are the data? Are the values bunched up around the center, or are they spread out? What is the smallest value in the data set? The largest?

When asked to "describe a distribution" of quantitative data, be sure to address all 4 of these features.

Concept 4: Comparing distributions of quantitative data.

When comparing distributions of quantitative data, we answer the same 4 questions as when we describe the distribution of a single quantitative variable, but there is one very important difference. When comparing distributions we must <u>*compare the centers*</u> and we must <u>*compare the variabilities*</u> of the distributions. This means that a comparison phrase (like "is less than" or "is greater than") is required. We do not use a comparison phrase for shape and outliers, because those descriptions are verbal (ex. skewed right, no outliers) rather than numeric (ex. median, range).

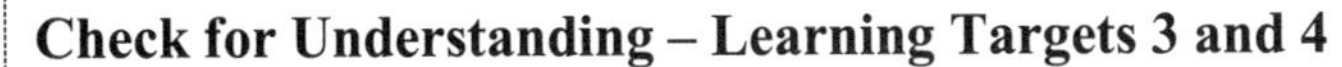

Check for Understanding – Learning Targets 3 and 4

_____ *I can describe the distribution of a quantitative variable.*

_____ *I can compare distributions of quantitative data.*

1. The dotplot below displays the Unit 1 exam scores of 34 students taking Statistics at a local university.

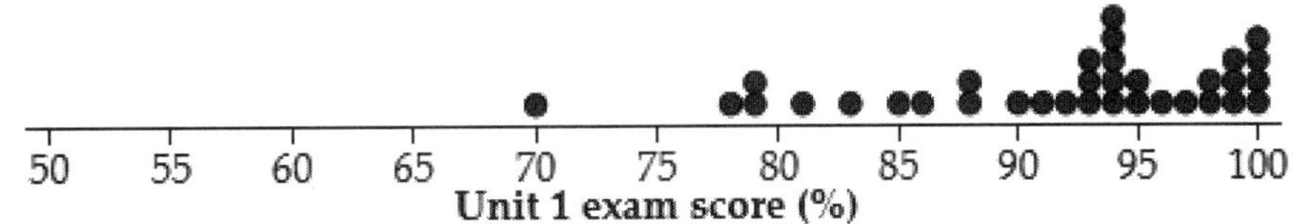

Describe the distribution.

2. The professor made another dotplot at the end of the semester to show the distribution of grade at the end of the semester for the same students.

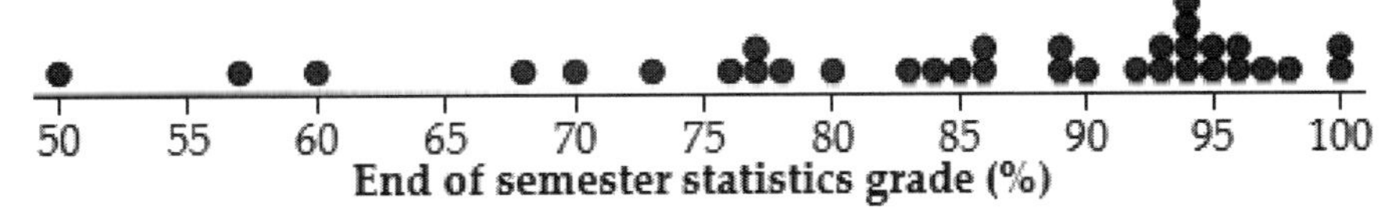

Compare the distribution of grade at the end of the semester to the distribution of Unit 1 exam score.

Concept 5: Stemplots

Like dotplots, stemplots are useful for displaying all values in a data set. Both of these displays are equally useful to display the distribution of a small quantitative data set. An advantage of stemplots is that the graph consists of the actual data, so you do not have to estimate the values in the data set.

Concept 6: Histograms

When dealing with larger sets of data, dotplots and stemplots can be a bit cumbersome and time-consuming to construct. When the data set is large, it may be easier to construct a histogram. Instead of plotting each value in the data set, a histogram displays the frequency of values that fall within equal-width classes. Don't confuse histograms with bar graphs. They look similar, but bar graphs display categorical data and histograms display quantitative data! Consider using your calculator to construct histograms. This helps to prevent errors, as you only have to correctly enter the data and then let your calculator do the rest! Be certain to use the trace feature to identify the lower and upper bounds and the height of each bin (class) accurately.

Check for Understanding – Learning Targets 5 and 6

____ *I can make and interpret a stemplot of quantitative data.*

____ *I can make and interpret a histogram of quantitative data.*

The heights (in inches) of a random sample of 24 NBA players are given below.

84	83	80	75	78	82	72	78	71	78	78	75
84	76	78	86	78	73	75	82	81	80	75	76

1. Create a split stemplot to display the distribution of height. Remember to include a key. What percentage of the players in this sample are taller than 82 inches?

The Environmental Protection Agency (EPA) expanded their study of gas mileage to 50 cars. Here are the data. (*Need Tech Help?* View the video **Tech Corner 1: Making Histograms** at bfwpub.com/TPS7eStrive)

EPA-Measured MPG for 50 Cars

36.3	32.7	40.5	36.2	38.5	36.3	41.0	37.0	37.1	39.9
41.0	37.3	36.5	37.9	39.0	36.8	31.8	37.2	40.3	36.9
36.7	33.6	34.2	35.1	39.7	39.3	35.8	34.5	39.5	36.9
36.9	41.2	37.6	36.0	35.5	32.5	37.3	40.7	36.7	32.9
42.1	37.5	40.0	35.6	38.8	38.4	39.0	36.7	34.8	38.1

2. Construct a frequency histogram of these data using your calculator. Use a bin width of 2. What percentage of these cars get at least 40 mpg?

Section 1D: Describing Quantitative Data with Numbers

Section Summary

Now you are ready to learn how to use numerical summaries to describe the center and variability of a distribution of quantitative data. You will also learn how to identify outliers in a distribution. By the end of this section, you should be comfortable choosing appropriate measures of center and variability. You will calculate the center and variability when given data, a dotplot, or a stemplot. You will estimate the center and variability when the data is presented in a histogram. You will also learn how to make a new graph…the boxplot!

Learning Targets:

____ I can find the median of a distribution of quantitative data.
____ I can calculate the mean of a distribution of quantitative data.
____ I can find the range of a distribution of quantitative data.
____ I can calculate and interpret the standard deviation of a distribution of quantitative data.
____ I can find the interquartile range (*IQR*) of a distribution of quantitative data.
____ I can choose appropriate measures of center and variability to summarize a distribution of quantitative data.
____ I can identify outliers in a distribution of quantitative data.
____ I can make and interpret boxplots of quantitative data.
____ I can use boxplots and summary statistics to compare distributions of quantitative data.

Read: Vocabulary and Key Concepts

- The ____________ is the midpoint of a distribution — the number such that about half the observations are smaller and about half are larger. To find the ________________, arrange the data values from smallest to largest.
  - If the number n of data values is odd, the median is the ____________________ in the ordered list.
  - If the number n of data values is even, use the _________________________________ in the ordered list as the median.
- The ___________ of a distribution of quantitative data is the average of all the individual data values. To find the ____________, add all the values and divide by the total number of data values.
- If a distribution of quantitative data is roughly symmetric and has no outliers, the mean and the median will be ____________________.
- For a right-skewed distribution, we expect the mean to be ________________________ the median.
- For a left-skewed distribution, we expect the mean to be ________________________ the median.
- A _____________ is a number that describes some characteristic of a sample.
- A _____________ is a number that describes some characteristic of a population.
- A statistical measure is _______________ if it is not affected much by extreme data values.
- The ____________ is resistant to outliers but the ______________ isn't.
- The ____________ of a distribution is the distance between the minimum value and the maximum value. That is, _____________ = maximum – minimum

- The ____________________ measures the typical distance of the values in a distribution from the mean. It is calculated by finding an "average" of the squared deviations of the individual data values from the mean, and then taking the square root.
- The value of the standard deviation is always ________________.
- The standard deviation is not a __________________ measure of variability.
- Standard deviation should be used to describe variability when the _________ is chosen as the measure of center.
- The ___________ of a distribution divide an ordered data set into four groups having roughly the same number of values. To find the _______________, arrange the data values left to right from smallest to largest and find the median.
- The _____________________________ is the median of the data values that are to the left of the median in the ordered list.
- The _____________________________ is the median of the data values that are to the right of the median in the ordered list.
- The __________________________ (_______) is the distance between the first and third quartiles of a distribution. In symbols: _______ $= Q_3 - Q_1$
- The __________________________ of a distribution of quantitative data consists of the minimum, the first quartile $Q1$, the median, the third quartile Q_3, and the maximum.
- A ________________ is a visual representation of the five-number summary.

Watch: Go to bfwpub.com/TPS7eStrive. Select Unit 1, Part I and watch the following Example videos:

Video	Topic	Done
Unit 1D, More chips please (p. 49)	*Measuring center: The median*	
Unit 1D, More chips please (p. 51)	*Measuring center: The mean*	
Unit 1D, More chips please (p. 55)	*Measuring variability: The range*	
Unit 1D, How many friends? (p. 56)	*Measuring variability: The standard deviation*	
Unit 1D, More chips please (p. 60)	*Measuring variability: The interquartile range (IQR)*	
Unit 1D, Lead in the water (p.61)	*Choosing summary statistics*	
Unit 1D, An NBA legend, (p. 64)	*Identifying outliers*	
Unit 1D, How big are the large fries? (p. 66)	*Displaying summary statistics: Boxplots*	
Unit 1D, Which company makes better tablets? (p. 68)	*Comparing distributions with boxplots and summary statistics*	

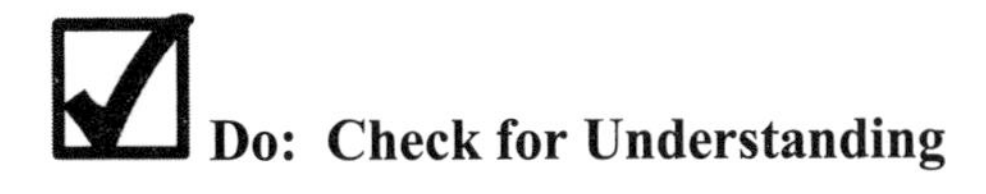

Do: Check for Understanding

Concept 1: The median – a measure of center

The median is one way to describe the center of a distribution of quantitative data. Choose the median when a distribution is skewed or has outliers because the median is a resistant measure of center, meaning it is not greatly affected by outliers. The median is also a good choice of center when the data are presented in a boxplot. When given data, remember to place them in order before finding the median!

Concept 2: The mean – a measure of center

The mean and median measure the center of a set of data in different ways. While the mean, or average, is the most common measure of center, it is not always the most appropriate. Extreme values can "pull" the mean towards them. Choose the mean as the measure of center when the distribution is roughly symmetric. In symmetric distributions, the mean and median will be approximately equal, while in a skewed distribution the mean tends to be closer to the "tail" than the median. Always consider the shape of the distribution when deciding which measure to use to describe your data!

Check for Understanding – Learning Targets 1 and 2

____ *I can find the median of a distribution of quantitative data.*

____ *I can calculate the mean of a distribution of quantitative data.*

Consider the following stemplot of the lengths of time (in seconds) it took students to complete a logic puzzle. Use it to answer the following questions. Note: 2 | 2 = 22 seconds.

Stem	Leaf
1	58
2	23
2	677778888999
3	2344
3	68
4	011223
4	6
5	00

1. Without doing any calculations, how does the mean compare to the median? How do you know?

2. Calculate and interpret the mean.

3. Calculate and interpret the median.

Technology: Use your calculator to find the mean and median for this set of data. Verify that you get the same results.

(*Need Tech Help?* View the video **Tech Corner 2: Calculating Summary Statistics** at bfwpub.com/TPS7eStrive)

Concept 3: Range – a measure of variability

Like measures of center, we have several different ways to measure variability for distributions of quantitative data. The easiest way to describe the variability of a distribution is to calculate the range (maximum – minimum). However, extreme values can cause this measure to be much greater than the variability of the majority of values.

Concept 4: Standard Deviation – a measure of variability

Another measure of variability that we will use to describe data is the standard deviation. The standard deviation measures roughly the average distance of the observations from their mean. The calculation can be quite time-consuming to do by hand, so we'll rely on technology to provide the standard deviation for us. However, be sure you understand how it is calculated!

Concept 5: Interquartile Range (IQR) – a measure of variability

A measure of variability that is resistant to the effect of outliers is the interquartile range (*IQR*). To find the *IQR*, arrange the observations from smallest to largest and determine the median. Then find the median of the lower half of the data. This number is the first quartile. The median of the upper half of the data is the third quartile. The distance between the first quartile and third quartile is the interquartile range.

Check for Understanding – Learning Targets 3, 4, and 5

_____ *I can find the range of a distribution of quantitative data.*

_____ *I can calculate and interpret the standard deviation of a distribution of quantitative data.*

_____ *I can find the interquartile range (IQR) of a distribution of quantitative data.*

The population of a random sample of 7 countries in Europe are, in millions: 84.6, 68.2, 10.5, 0.6, 5.9, 3.7, and 8.9

1. Find the range of the distribution.

2. Calculate and interpret the standard deviation for this sample of 7 countries.

3. Find the interquartile range.

Technology: Use your calculator to find the standard deviation and interquartile range for this set of data. Verify that you get the same results.

Concept 6: Choosing appropriate measures of center and variability

When using the mean to describe the center of a distribution, use standard deviation to describe the variability of the distribution. Be careful though! Both the mean and standard deviation are affected by outliers! It is best to use these measures when a distribution is roughly symmetric and does not have any outliers. When a distribution is skewed or has outliers, use the median as the measure of center and *IQR* as the measure of variability.

Check for Understanding – Learning Target 6

____ *I can choose appropriate measures of center and variability to summarize a distribution of quantitative data.*

The histogram displays the distribution of price for 35 homes recently sold in a suburban area.

1. Which measures of center and variability would you choose to describe the distribution? Explain your answer.

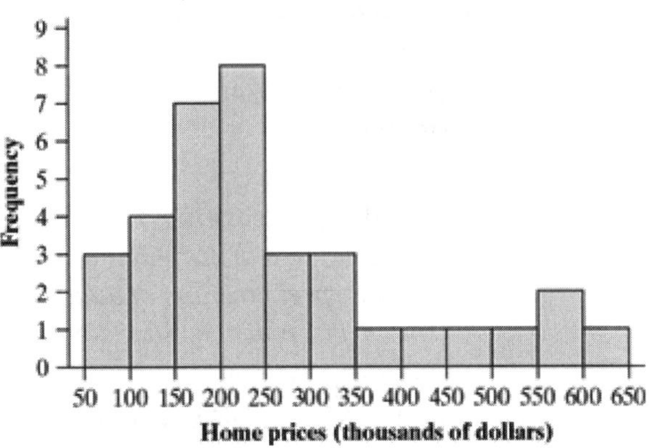

2. In what interval does the median home price fall? How does the mean compare to the median home price?

A specific drone is advertised to have eight minutes of flight time on a single battery charge. A student collected data to determine if the advertised time was accurate. A stemplot shows the observed flight times in minutes.

```
6 | 8
7 | 1 4 4 4 4
7 | 5 5 5 6 7 8 9 9
8 | 0 1 1 2
8 | 7
```

Key: 6 | 8 represents a flight that lasted 6.8 minutes.

3. Which measures of center and variability would you choose to describe the distribution? Explain your answer.

Concept 7: Outliers

Not only does the *IQR* provide a measure of variability, but it also provides a way to identify outliers. According to the 1.5 × *IQR* rule, any value that falls more than 1.5 × *IQR* above the third quartile or below the first quartile is considered an outlier.

Concept 8: Boxplots

The minimum, maximum, median, and quartiles make up the "five-number summary." These 5 numbers lead to a useful display – the boxplot. One advantage that the boxplot has over all other quantitative data displays is that it explicitly shows outliers using dots, or asterisks, that are separated from the rest of the boxplot. The boxplot also clearly shows the value of the median and makes calculating the *IQR* a breeze! (*Need Tech Help?* View the video **Tech Corner 3: Making Boxplots** at bfwpub.com/TPS7eStrive)

Concept 9: Using boxplots and summary statistics to compare distributions of quantitative data

Boxplots are especially effective for comparing the distribution of a quantitative variable in two or more groups. Remember to discuss shape, outliers, center, and variability as you did with comparative dotplots, stemplots, and histograms.

Check for Understanding – Learning Targets 7, 8, and 9

____ *I can identify outliers in a distribution of quantitative data.*
____ *I can make and interpret boxplots of quantitative data.*
____ *I can use boxplots and summary statistics to compare distributions of quantitative data.*

The length (in pages) of Mr. Molesky's favorite books are noted below

242	346	314	330	340	322	284	342
368	170	344	318	318	375	332	

1. Complete the following summary statistics.

n	mean	SD	min	Q_1	med	Q_3	max
15	316.267	51.63					

2. Identify any outliers in the distribution.

3. Mrs. Molesky reads age-appropriate novels to the Molesky children. A boxplot of the number of pages in the most recent 10 books she read to the children is given below. Above Mrs. Molesky's boxplot, make a boxplot showing the distribution of book length for Mr. Molesky.

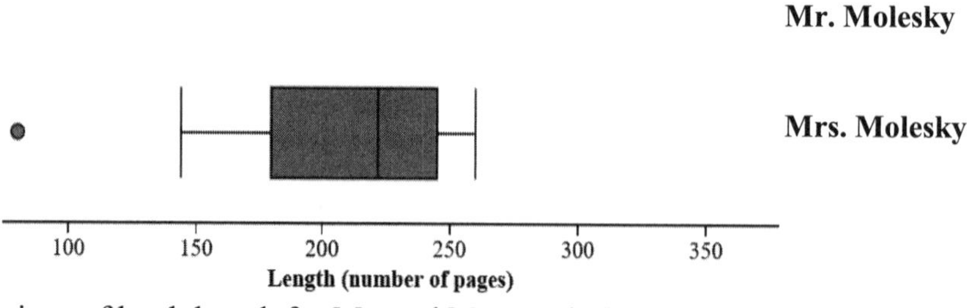

4. Compare the distributions of book length for Mr. and Mrs. Molesky.

Unit 1 Part I Summary: Exploring One-Variable Data

In this unit, we learned that statistics is the art and science of data. When working with data, it is important to know whether a variable is categorical or quantitative, as this will determine the most appropriate display for the distribution. A bar graph or pie chart is typically used to display categorical data. Quantitative data are typically displayed using a dotplot, stemplot, histogram, or boxplot. When examining quantitative data, a graph will help us describe the shape of the distribution and suggest the most appropriate numeric measures of center and variability. When exploring quantitative data, describe the shape, outliers, center, and variability in context. Look for an overall pattern and note any striking departures from that pattern.

As you study, be sure to focus on *understanding*, not just mechanics. While it may be easy to "plug the data" into your calculator, simply making graphs and calculating values is not the point of statistics. Rather, focus on being able to explain *how* a graph is constructed and *why* you would choose a certain display or numeric summary. Get in this habit early...your calculator is a powerful tool, but it cannot replace your thinking and communication skills!

How well do you understand each of the learning targets?

Learning Target	Got It!	Almost There	Needs Work
I can identify individuals and variables in a set of data and classify the variables as categorical or quantitative.			
I can make and interpret a frequency table or a relative frequency table for a distribution of data.			
I can make and interpret bar graphs of categorical data.			
I can compare distributions of categorical data.			
I can identify what makes some graphs of categorical data misleading.			
I can make and interpret dotplots of quantitative data.			
I can describe the shape of a distribution of quantitative data.			
I can describe the distribution of a quantitative variable.			
I can compare distributions of quantitative data.			
I can make and interpret a stemplot of quantitative data.			
I can make and interpret a histogram of quantitative data.			
I can find the median of a distribution of quantitative data.			
I can calculate the mean of a distribution of quantitative data.			
I can find the range of a distribution of quantitative data.			
I can calculate and interpret the standard deviation of a distribution of quantitative data.			
I can find the interquartile range (*IQR*) of a distribution of quantitative data.			
I can choose appropriate measures of center and variability to summarize a distribution of quantitative data			
I can identify outliers in a distribution of quantitative data.			
I can make and interpret boxplots of quantitative data.			
I can use boxplots and summary statistics to compare distributions of quantitative data.			

Unit 1, Part I Multiple Choice Practice

Directions. Identify the choice that best completes the statement or answers the question. Check your answers and note your performance when you are finished.

1. You measure the age (years), weight (pounds), and breed (beagle, golden retriever, pug, or terrier) of 200 dogs. How many variables did you measure?
   (A) 1
   (B) 2
   (C) 3
   (D) 200
   (E) 203

2. You open a package of Lucky Charms cereal and count how many there are of each marshmallow shape. The <u>distribution</u> of the variable "marshmallow" is:
   (A) The shape: star, heart, moon, clover, diamond, horseshoe, balloon.
   (B) The total number of marshmallows in the package.
   (C) Seven—the number of different shapes that are in the package.
   (D) The seven different shapes and how many there are of each.
   (E) Because "shape" is a categorical variable, it doesn't have a distribution.

3. Two friends compared their score distributions from several games of bowling. Here are dotplots of their scores. Which of the following is incorrect?

   (A) Both distributions are skewed to the right.
   (B) Bowler 1 tended to have lower scores than Bowler 2.
   (C) The variability in scores for Bowler 1 is Similar to the variability in scores for Bowler 2.
   (D) Bowler 1 appears to have an outlier at 206.
   (E) Bowler 1 may sometimes lose to Bowler 2.

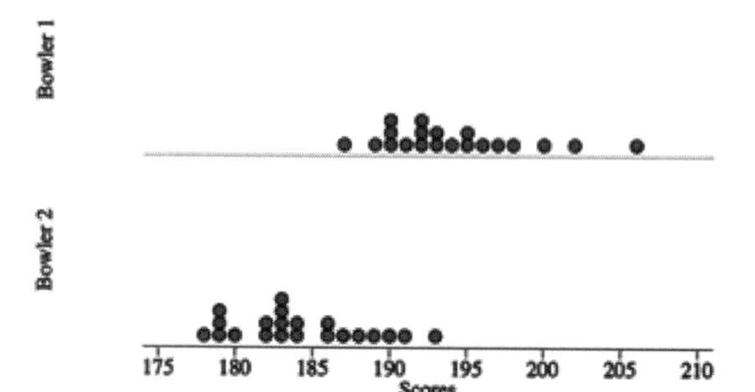

4. For a physics course containing 10 students, the maximum possible point total for the quarter was 200. The point totals for the 10 students are given in the stemplot below.

Stem	Leaves
11	6 8
12	1 4 8
13	3 7
14	2 6
15	
16	
17	9

   Which statement about this distribution is true?
   (A) The distribution of point totals is roughly symmetric.
   (B) 50% of the students earned more than 137 points.
   (C) The standard deviation of point total is greater than the interquartile range of point total.
   (D) The mean point total is greater than the median point total.
   (E) There are no possible outliers in the distribution of point total.

5. When drawing a histogram, it is important to
   (A) have a separate class interval for each observation to get the most informative plot.
   (B) make sure the heights of the bars exceed the widths of the class intervals so that the bars are true rectangles.
   (C) label the vertical axis so the reader can determine the counts or percent in each class interval.
   (D) leave large gaps between bars. This allows room for comments.
   (E) scale the vertical axis according to the variable whose distribution you are displaying.

6. A set of data has a mean that is much larger than the median. Which of the following statements is most consistent with this information?
   (A) The distribution is symmetric.
   (B) The distribution is skewed left.
   (C) The distribution is skewed right.
   (D) The distribution is bimodal.
   (E) The data set probably has a few low outliers.

7. The following is a boxplot of the birth weights (in ounces) of a sample of 160 infants born in a local hospital.

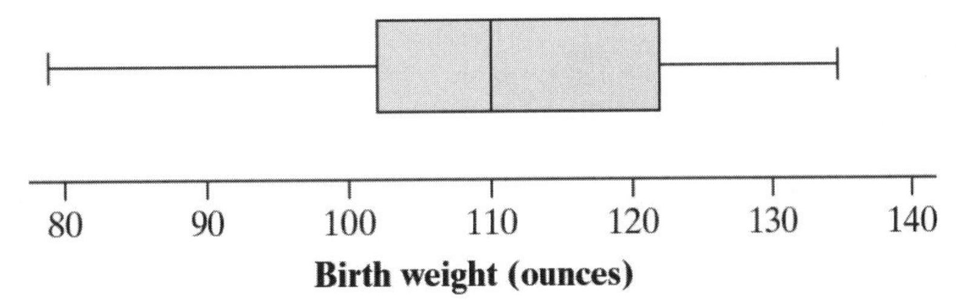

   About 40 of the birth weights were below what value?
   (A) 92 ounces
   (B) 102 ounces
   (C) 112 ounces
   (D) 122 ounces
   (E) 132 ounces

8. A sample of production records for an automobile manufacturer shows the following figures for production per shift:

   705   700   690   705

   What is the standard deviation of this sample?
   (A) 6.12
   (B) 7.07
   (C) 8.66
   (D) 50
   (E) 20

9. You catch 10 cockroaches in your bedroom and measure their lengths in centimeters. Which of these sets of numerical descriptions are *all* measured in centimeters?
   (A) median length, variance of lengths, largest length
   (B) median length, first and third quartiles of lengths
   (C) mean length, standard deviation of lengths, median length
   (D) mean length, median length, variance of lengths.
   (E) both (B) and (C)

10. A policeman records the speeds of cars on a certain section of roadway with a radar gun. The histogram below shows the distribution of speeds for 251 cars.

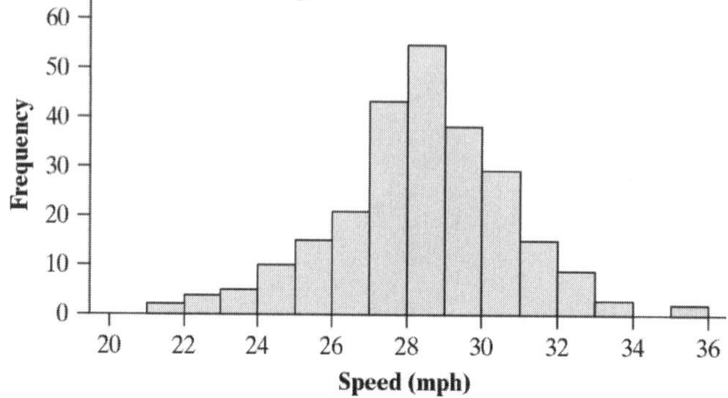

Which of the following measures of center and spread would be the best ones to use when summarizing these data?
   (A) Mean and interquartile range.
   (B) Mean and standard deviation.
   (C) Median and range.
   (D) Median and standard deviation.
   (E) Median and interquartile range.

Check your answers below. If you got a question wrong, check to see if you made a simple mistake or if you need to study that concept more. After you check your work, identify the concepts you feel very confident about and note what you will do to learn the concepts in need of more study.

#	Answer	Concept	Right	Wrong	Simple Mistake?	Need to Study More
1	C	Variables				
2	D	Categorical variables				
3	B	Comparing distributions				
4	D	Distribution basics				
5	C	Constructing histograms				
6	C	Skewed distributions				
7	B	Interpreting boxplots				
8	B	Calculating standard deviation				
9	E	Summary statistics units				
10	B	Choosing statistics				

FRAPPY! Free Response AP® Problem, Yay!

The following problem is modeled after actual Advanced Placement Statistics free response questions. Your task is to generate a complete, concise response in 15 minutes. After you generate your response, read over the solution and scoring guide. Score your response and note what, if anything, you would do differently to increase your own score.

SugarBitz candies are packaged in 15 oz. snack-size bags. The back-to-back stemplot below displays the distribution of weight (in ounces) of two samples of SugarBitz bags filled by different filling machines. The minimum weight overall was 14.1 oz., and the maximum weight overall was 15.9 oz.

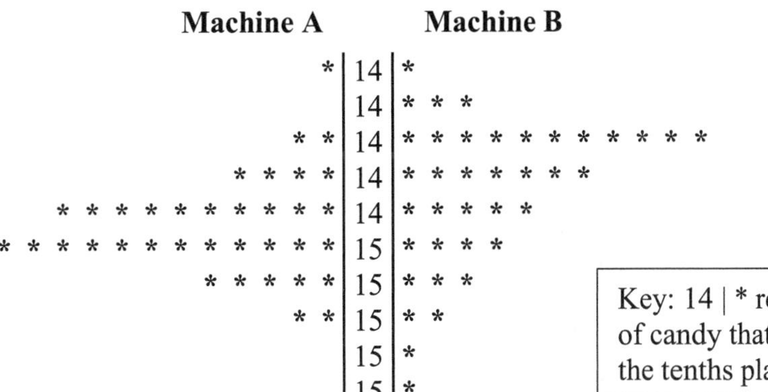

Key: 14 | * represents a snack-sized bag of candy that weighs 14.* ounces, where the tenths place is not explicitly revealed.

(a) Compare the distributions of weight for bags packaged by each machine.

(b) The company wishes to be as consistent as possible when packing its snack bags. Which machine would be *least* likely to produce snack bags of SugarBitz that have a consistent weight? Explain.

(c) Suppose the company filled its bags using the machine you chose in part (b). Which measure of center, the mean or the median, would be closer to the advertised 15 oz.? Justify your answer.

FRAPPY! Scoring Rubric

Use the following rubric to score your response. Each part receives a score of "Essentially Correct," "Partially Correct," or "Incorrect." When you have scored your response, reflect on your understanding of the concepts addressed in this problem. If necessary, note what you would do differently on future questions like this to increase your score.

Intent of the Question

The goals of this question are (1) to determine your ability to use graphical displays to compare and contrast two distributions and (2) to evaluate your ability to use statistical information to make a decision.

Solution

(a) Both distributions are single-peaked. However, Machine A's distribution is roughly symmetric while Machine B's is skewed to the right. The center of the weights for Machine A (median A = about 15 or 15.1 oz.) is slightly greater than that of Machine B (median B = about 14.6 or 14.7 oz.). There is more variability in the weights produced by Machine B. Machine A has one low value (14.1 oz.) that does not fall with the majority of weights. However, it does not appear to be extreme enough to be considered an outlier. Machine B does not have any obvious outliers.

(b) Both machines produce bags of varying weight. However, Machine B has a greater variability as evidenced by a greater overall range. Machine B would be least likely to produce a consistent weight for the snack bags.

(c) The mean would be closer to the advertised 15 oz. weight. This is because in a skewed distribution, the mean is pulled away from the median in the direction of the tail. In Machine B's distribution, the median is at about 14.6 to 14.7 oz. so we would expect the mean to be slightly greater than the median, making it closer to 15 oz.

Scoring:

Parts (a), (b), and (c) are scored as essentially correct (E), partially correct (P), or incorrect (I).

Part (a) is essentially correct if you correctly identify similarities and differences in the shape, center, and spread for the two distributions and do so in context.
Part (a) is partially correct if you correctly identify similarities and differences in two of the three characteristics for the two distributions or identify all 3 characteristics but exclude context.
Part (a) is incorrect if you only identify one similarity or difference of the three characteristics for the two distributions or name two but exclude context.

Part (b) is essentially correct if Machine B is chosen using rationale based on its measure of spread of the packaged weights.
Part (b) is partially correct if B is chosen, but the relevant explanation does not refer to the variability in the weights.
Part (b) is incorrect if B is chosen but no explanation or an irrelevant explanation is provided OR if A is chosen.

Part (c) is essentially correct if the mean is chosen, and the explanation addresses the fact that the mean will be greater than the median in a skewed right distribution.
Part (c) is partially correct if the mean is chosen, but the explanation is incomplete or incorrect.
Part (c) is incorrect if the mean is chosen, but no relevant explanation is given OR if the median is chosen.

NOTE: If Machine A was chosen in part (b) and the solution to part (c) indicates either the mean or median would be appropriate due to the fact that they will be approximately equal in a symmetric, mound-shaped distribution, part (c) should be scored as essentially correct.

4 Complete Response
All three parts essentially correct

3 Substantial Response
Two parts essentially correct and one part partially correct

2 Developing Response
Two parts essentially correct and no parts partially correct
One part essentially correct and two or one parts partially correct
Three parts partially correct

1 Minimal Response
One part essentially correct and no parts partially correct
No parts essentially correct and two parts partially correct

My Score:
What I did well:
What I could improve:
What I should remember if I see a problem like this on the AP Exam:

Unit 1, Part I Formula Sheet

Create a one-page summary of important concepts and formulas found in this unit. This will be a valuable resource as you prepare for the AP Exam.

Unit 1 Part I Crossword Puzzle

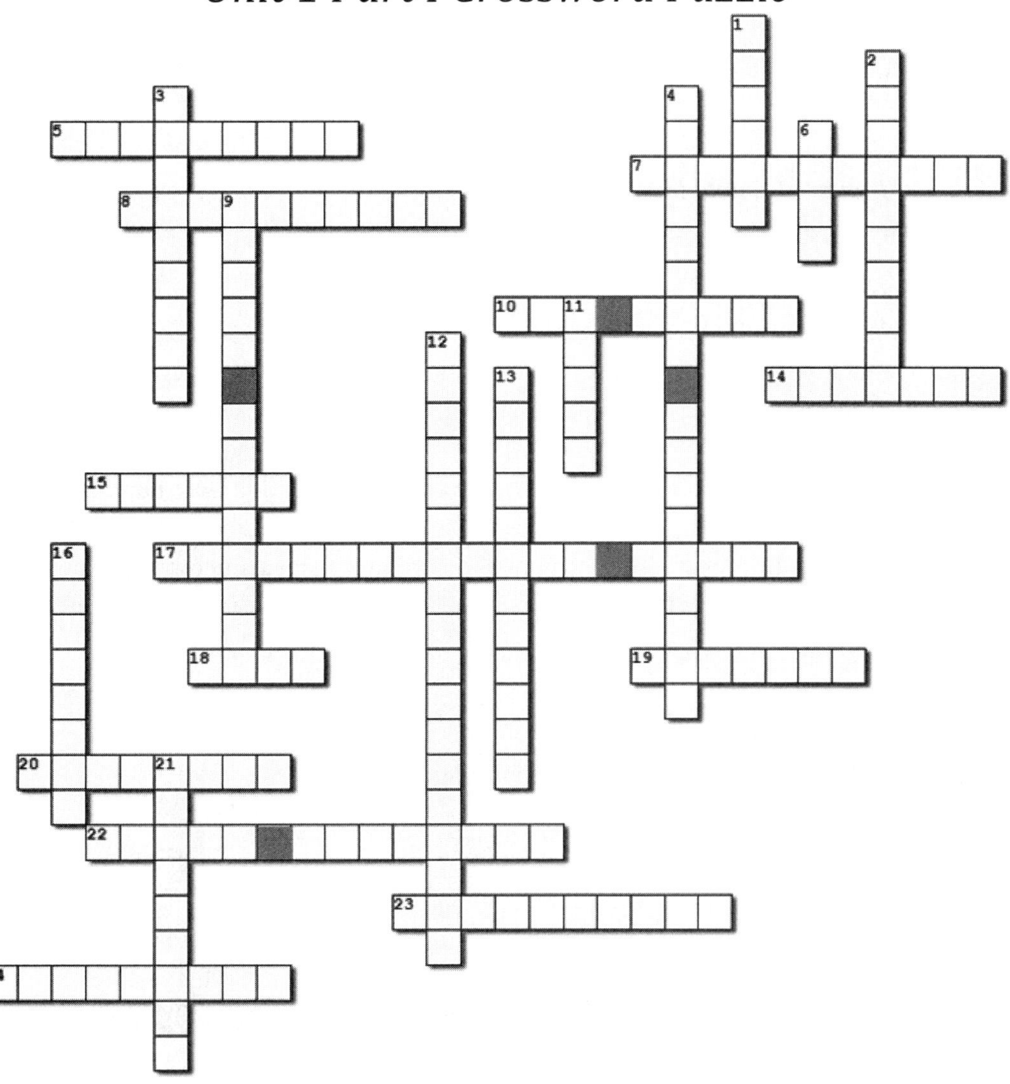

Across:

5. A graphical display of quantitative data that shows the frequency of values in intervals using bars.

7. Type of variable that takes values that are labels, which place each individual into a particular group.

8. The science and art of collecting, analyzing, and drawing conclusions from data.

10. Graph that displays the counts or percents of categories of a categorical variable.

14. A value that falls more than 1.5*IQR* above Q_3 or below Q_1.

15. This measure of center is resistant.

17. A measure of variability that describes the middle 50% of a distribution.

18. A measure of center. Also called the average.

19. A graphical display of quantitative data that shows outliers as isolated dots.

20. A graphical display of quantitative data that shows the actual data values.

22. This value is the middle of the lower half of a data set.

23. A quantitative variable that can take any value in an interval on a number line.

24. The shape of a distribution in which the left and right sides are exact mirror images of each other.

Down:

1. Shape of a distribution in which one side of the graph is much longer than the other.

2. An object described in a set of data. Can be people, animals, or things.

3. A number that describes a sample.

4. When a distribution is roughly symmetric, which measure of variability should you use to describe the distribution?

6. When describing a distribution, use this acronym.

9. This value is the middle of the upper half of a data set.

11. A measure of variability found by subtracting the minimum value from the maximum value.

12. The typical distance of observations from their mean.

13. Type of variables that takes number values that are quantities – counts or measurements.

16. A quantitative variable that takes a countable set of possible values with gaps between them.

21. A number that describes the population.

Unit 1, Part II: Exploring One-Variable Data

"It is not knowledge, but the act of learning, not possession but the act of getting there, which grants the greatest enjoyment." Carl Friedrich Gauss

UNIT 1 PART II OVERVIEW

In Unit 1, Part I, we filled our toolbox with graphical and numerical tools for describing distributions of categorical and quantitative data. We now have a clear strategy for exploring data. Unit 1, Part II covers a key concept in statistics – describing the location of an observation within a distribution. You will learn how to measure position using percentiles as well as by using a standardized measure based on the mean and standard deviation. We'll discover that sometimes the overall pattern of a large number of observations is so regular that it can be described by a smooth curve. The normal distribution will be explored in Section 1F. You will not only learn the properties of the normal distributions, but also how to perform a number of calculations with them. The concepts introduced in this section will be revisited throughout the course, so be sure to master them!

Sections in Unit 1, Part II
Section 1E: Describing Position and Transforming Data **Section 1F**: Normal Distributions

PRACTICE FOR MASTERY

Use the following *suggested* guide to the pages and exercises in your text to practice for mastery!
Note: your teacher may assign different problems. Be sure to follow their instructions!

Day	Topics	Read	Do
1	◆ Measuring Position: Percentiles ◆ Measuring Position: Standardized Scores (*z*-scores) ◆ Comparing Relative Positions in Distributions of Quantitative Data	p. 88-93	**1E**: 1, 3, 5, 7, 9, 11, 13, 29, 32
2	◆ Cumulative Relative Frequency Graphs ◆ Transforming Data	p. 94-100	**1E**: 15, 17, 19, 23, 25, 30, 31, 33, 34
3	◆ Normal Curves ◆ The Empirical Rule ◆ Assessing Normality	p. 107-116	**1F**: 1, 3, 5, 7, 9, 11, 13, 40, 41, 42
4	◆ Finding Areas in a Normal Distribution	p. 117-123	**1F**: 15, 17, 19, 21, 23, 43
5	◆ Finding Percentiles in a Normal Distribution ◆ Calculating the Mean or Standard Deviation of a Normal Distribution	p. 124-128	**1F**: 25, 27, 31, 33, 35, 44
6	Unit 1, Part II AP® Statistics Practice Test (p. 138)		Unit 1, Part II Review Exercises (p. 137)
7	Unit 1, Part II Test: Celebration of learning!		

Section 1E: Describing Position and Transforming Data

Section Summary

Do you ever wonder where you fit in? What is your class rank? How tall are you compared to your classmates? This section is all about describing the location of an individual value in a distribution. You have probably encountered measures of location before through the concept of percentiles. In this section, you will learn how to calculate and interpret percentiles as well as how to identify percentiles through a cumulative relative frequency graph. You will also learn a new way to describe location using the mean and standard deviation. Standardized scores, or *z*-scores, will be introduced as a way to describe location within a distribution and to compare observations from different distributions. Finally, you will learn how to adjust the measures of center and variability when each of the individual data values are adjusted. This is a huge time saver because if you already know the mean and standard deviation of the height of every student in your high school (in inches) but need to know the mean and standard deviation in centimeters, you do not have to convert the height of each student and recalculate the mean and standard deviation from scratch. Whew!

Learning Targets:

____ I can calculate and interpret a percentile in a distribution of quantitative data.

____ I can calculate and interpret a standardized score (*z*-score) in a distribution of quantitative data.

____ I can use percentiles or standardized scores (*z*-scores) to compare the relative positions of individual values in distributions of quantitative data.

____ I can use a cumulative relative frequency graph to estimate percentiles and individual values in a distribution of quantitative data.

____ I can describe the effect of adding, subtracting, multiplying by, or dividing by a constant on the shape, center, and variability of a distribution of data.

Read: Vocabulary and Key Concepts

- The *p*th ____________________ of a distribution is the value with *p*% of observations less than or equal to it.
- Should we say an observation is "in" the 75th percentile or "at" the 75th percentile? _____________
- The median is roughly at the ____ percentile. Q_1 is at about the ____ percentile and Q_3 is at about the ______ percentile.
- The ____________________________ for an individual value in a distribution tells us how many standard deviations from the mean the value falls, and in what direction.
- To find the standardized score (*z*-score), use this formula: _________________________.
- A ______________________________________ plots a point corresponding to the percentile of a given value in a distribution of quantitative data. Consecutive points are then connected with a line segment to form the graph.

- When we ***add or subtract*** the same positive number (a) to/from each data value:
  - ❖ How do we recalculate the new measures of center? ___________________________
  - ❖ Does adding/subtracting the same number to/from each data value affect the variability of the distribution? __________ Range, *IQR*, and standard deviation will stay the same!
  - ❖ Does adding/subtracting the same number to/from each data value affect the shape of the distribution? __________ The shape of the distribution will stay the same!
- When we ***multiply or divide*** each data value by the sane positive number (b):
  - ❖ How do we recalculate the new measures of center? ___________________________
  - ❖ How do we recalculate the new measures of variability? ___________________________
  - ❖ Does multiplying/dividing each data value by the same positive number affect the shape of the distribution? __________ The shape of the distribution will stay the same!

Transformation Example: The dotplot displays the price of a gallon of gas in 5 cities.

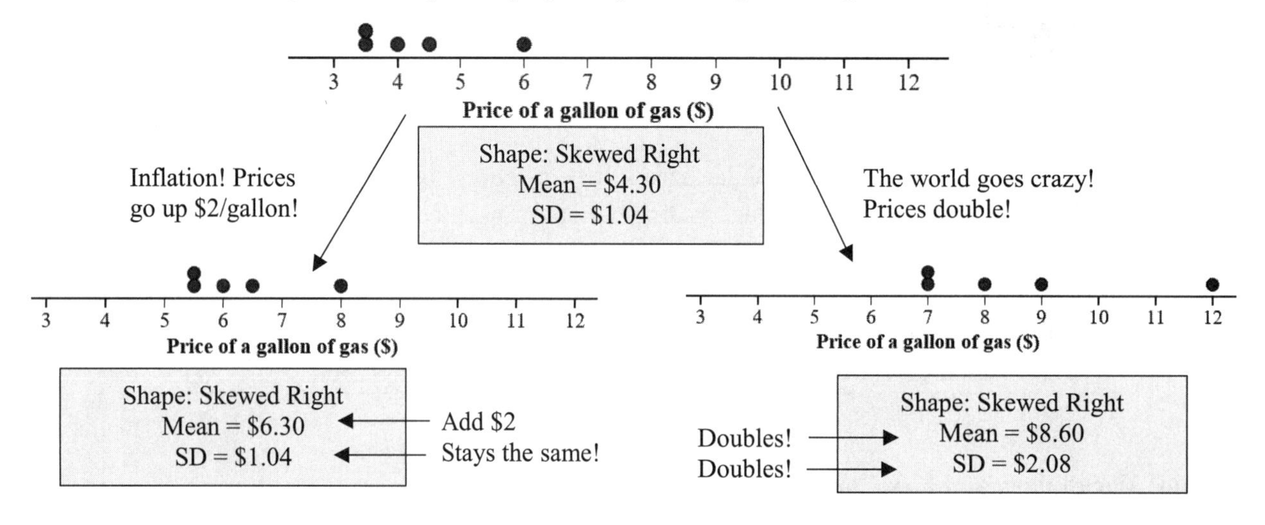

Watch: Go to bfwpub.com/TPS7eStrive. Select Unit 1, Part II and watch the following Example videos:

Video	Topic	Done
Unit 1E, How much lead is in the water? (p. 90)	*Measuring position: Percentiles*	
Unit 1E, How much lead is in the water? (p. 91)	*Measuring position: Standardized scores (z-scores)*	
Unit 1E, Growing like a beanstalk (p. 93)	*Comparing relative position in distributions of quantitative data*	
Unit 1E, Ages of U.S. presidents (p. 95)	*Cumulative relative frequency graphs*	
Unit 1E, Too cool at the cabin? (p. 98)	*Transforming data*	

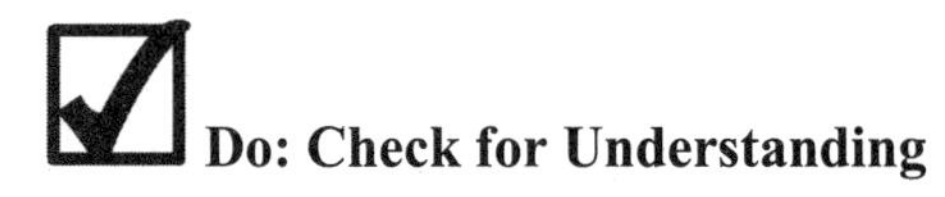

Do: Check for Understanding

Concept 1: Measuring Position - Percentiles

A common way to measure position within a distribution is to tell what percent of observations fall at or below the value in question. Percentiles are relatively easy to find – especially when the data is presented in order or in a dotplot or stemplot. To find a percentile, simply calculate the percent of values that fall at or below the value of interest.

Check for Understanding – Learning Target 1

____ *I can calculate and interpret a percentile in a distribution of quantitative data.*

Every 4th of July, Nathan's Famous hot dog eating contest is held. The dotplot shows the number of hot dogs eaten by the male contestants.

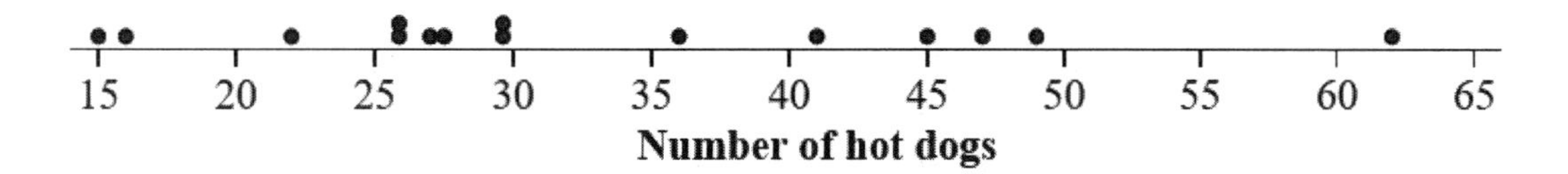

1. Calculate and interpret his percentile of the contestant who ate 29.75 hot dogs.

2. How many hot dogs did the contestant at the 80th percentile eat?

Here are the results from the female division during the same year.

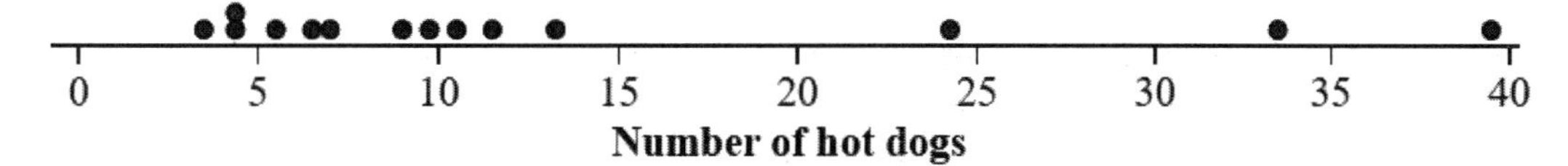

3. Who did better in their division: Mele, who ate 13.25 hot dogs in the female division, or Oji who ate 36 hot dogs in the male division? Calculate and compare their percentiles.

Concepts 2 and 3: Measuring and Comparing Position – Standardized Scores

When describing distributions, we learned that measures of center and variability are both very important characteristics. It follows, then, that measuring the position of an observation in a distribution should consider both the center and the variability of the distribution. After all, saying a particular observation falls 5 points above the average doesn't mean much unless you know how varied the observations are. If the distribution has little variability, an observation that is 5 points above the average might be an extreme value. However, if the distribution has a lot of variability, being 5 points above average might not be a big deal. We *must* consider center and spread when describing location. The standardized value, or z-score, of an observation does just that. The z-score tells us how many standard deviations a particular observation falls above or below the mean. This method of describing location not only allows us to describe individuals within a distribution, but also allows us to compare the positions of individuals in different distributions. We will standardize values A LOT in this course. Master the concept now!

Check for Understanding – Learning Targets 2 and 3

____ *I can calculate and interpret a standardized score (z-score) in a distribution of data.*

____ *I can use percentiles or standardized scores (z-scores) to compare the relative positions of individual values in distributions of quantitative data.*

Do you watch the Super Bowl? The table gives summary statistics on the number of people who watched the Super Bowl, in millions, for the past 20 years.

n	Mean	SD	Min	Q_1	Med	Q_3	Max
20	103	8.9	86	96	103	112	115

1. Calculate and interpret the z-score for 2023, in which 113 million people watched the Super Bowl.

2. The Covid pandemic was just beginning in 2020. The number of viewers that year had a standardized score of –0.112. How many viewers watched the Super Bowl that year?

3. In 2023, 1.5 billion people globally watched the World Cup Final, which has a mean viewership of 1 billion viewers globally with a standard deviation of 0.251 billion viewers. Which event: World Cup Final or the Super Bowl had the more impressive viewership in 2023?

Concepts 4: Cumulative Relative Frequency Graphs

Cumulative relative frequency graphs (also known as ogives or percentile plots) provide a graphical tool to find percentiles in a distribution of quantitative data. The vertical axis displays the cumulative relative frequency, which is a fancy way of saying "percentile." The vertical change over each interval gives the percentage of individuals in each interval. This means that segments that are "steeper" contain a greater number of individuals in the corresponding interval than segments that are less steep. Let's explore this more by digging deeper into the example provided in the text.

Check for Understanding – Learning Target 4

____ *I can use a cumulative relative frequency graph to estimate percentiles and individual values in a distribution of quantitative data.*

The cumulative relative frequency graph describes the distribution of age when each of the first 46 U.S. presidents first took office.

1. Joe Biden took office at age 78 years and 61 days (about 78.167 years old). Use the cumulative relative frequency graph to estimate Joe Biden's percentile.

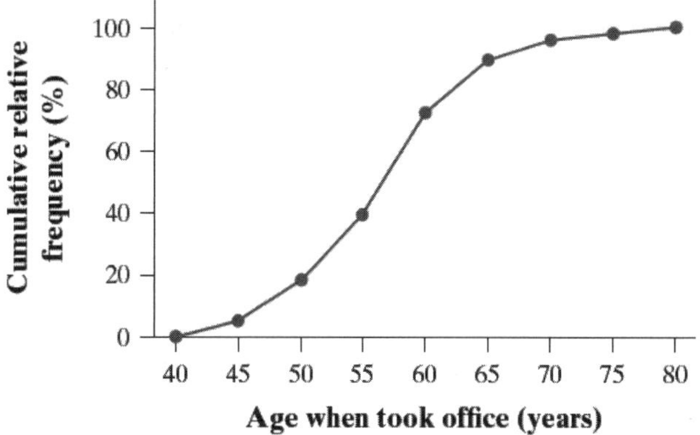

2. Joe Biden is the oldest individual to become president of the United States. What is Joe Biden's exact percentile in the distribution of age for U.S. presidents? Compare this to your answer to Question 1.

3. Estimate the median of the distribution of age when taking office.

4. Here is a table which summarizes the age at which each of the first 46 U.S. presidents first took office. In what interval is the median age? How do you know? Compare this to your answer to Question 3.

Age	Frequency	Relative frequency	Cumulative frequency	Cumulative relative frequency
40 to <45	2	2/46 = 0.0435 = 4.35%	2	2/46 = 0.0435 = 4.35%
45 to <50	6	6/46 = 0.1304 = 13.04%	8	8/46 = 0.1739 = 17.39%
50 to <55	10	10/46 = 0.2174 = 21.74%	18	18/46 = 0.3913 = 39.13%
55 to <60	15	15/46 = 0.3261 = 32.61%	33	33/46 = 0.7174 = 71.74%
60 to <65	8	8/46 = 0.1739 = 17.39%	41	41/46 = 0.8913 = 89.13%
65 to <70	3	3/46 = 0.0652 = 6.52%	44	44/46 = 0.9565 = 95.65%
70 to <75	1	1/46 = 0.0217 = 2.17%	45	45/46 = 0.9783 = 97.83%
75 to <80	1	1/46 = 0.0217 = 2.17%	46	46/46 = 1.00 = 100%

Concept 5: Transforming Data

When we find z-scores, we are transforming our data to a standardized scale. That is, we subtract the mean and divide by the standard deviation, converting the observation from its original units to a standardized scale. Sometimes we transform data to switch between measurement units (inches to centimeters, Fahrenheit to Celsius, etc.). When we do this, it is important to know what happens to the center and spread of the transformed distribution. Adding (or subtracting) a constant to each observation in a set of data will have an effect on the center of the distribution, but not on the spread or the shape. Multiplying (or dividing) each of the observations by a positive constant will change the center and the spread of the distribution, but not its shape. Most importantly, while transforming data in these ways may change the center and spread of a distribution, the locations of individual observations remain unchanged! So, if you had a z-score of 1.5 on a quiz and your teacher decided to double everyone's score and give an additional 5 points, your z-score would STILL be 1.5!

Check for Understanding – Learning Target 5

____ *I can analyze the effect of adding, subtracting, multiplying by, or dividing by a constant on the shape, center, and variability of a distribution of quantitative data.*

Here are the number of points earned by 23 students on a statistics quiz.

0	0 1 2
1	2 2 4 8
2	1 1 3 4 5 9 9
3	0 0 0 3 6
4	4 5 7
5	0

Key: 5 | 0 = 50 points

1. Describe the distribution of quiz scores.

2. Suppose the teacher adds 5 points to each student's score. How would the distribution of adjusted quiz scores compare to the original distribution of quiz scores?

3. Suppose the teacher decided to double the original scores instead of adding points. How would the distribution of doubled scores compare to the original distribution of quiz scores?

Section 1F: Normal Distributions

Section Summary

Sometimes the overall pattern of distribution is so regular that it can described by a smooth curve. This section is devoted to studying the normal curve. You will learn the basic properties of normal distributions and how to use normal distributions to perform a variety of calculations. You will also use graphical and numerical evidence to determine whether or not a distribution of quantitative data can be described as approximately normal. For distributions that can be described by a normal curve, you will learn how to determine the proportion of observations that fall into given intervals. You will also learn how to find the value corresponding to a specified percentile in a normal distribution. A variety of tools and methods will be presented to help you perform normal calculations by hand and on your calculator. Remember to always interpret the results in the context of the situation! Make sure you are comfortable with not only performing the calculations in this section but also in describing what the results mean!

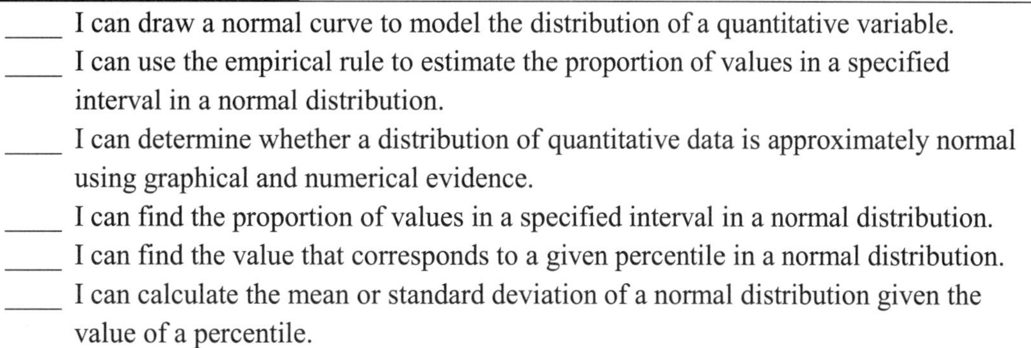

Learning Targets:

____ I can draw a normal curve to model the distribution of a quantitative variable.

____ I can use the empirical rule to estimate the proportion of values in a specified interval in a normal distribution.

____ I can determine whether a distribution of quantitative data is approximately normal using graphical and numerical evidence.

____ I can find the proportion of values in a specified interval in a normal distribution.

____ I can find the value that corresponds to a given percentile in a normal distribution.

____ I can calculate the mean or standard deviation of a normal distribution given the value of a percentile.

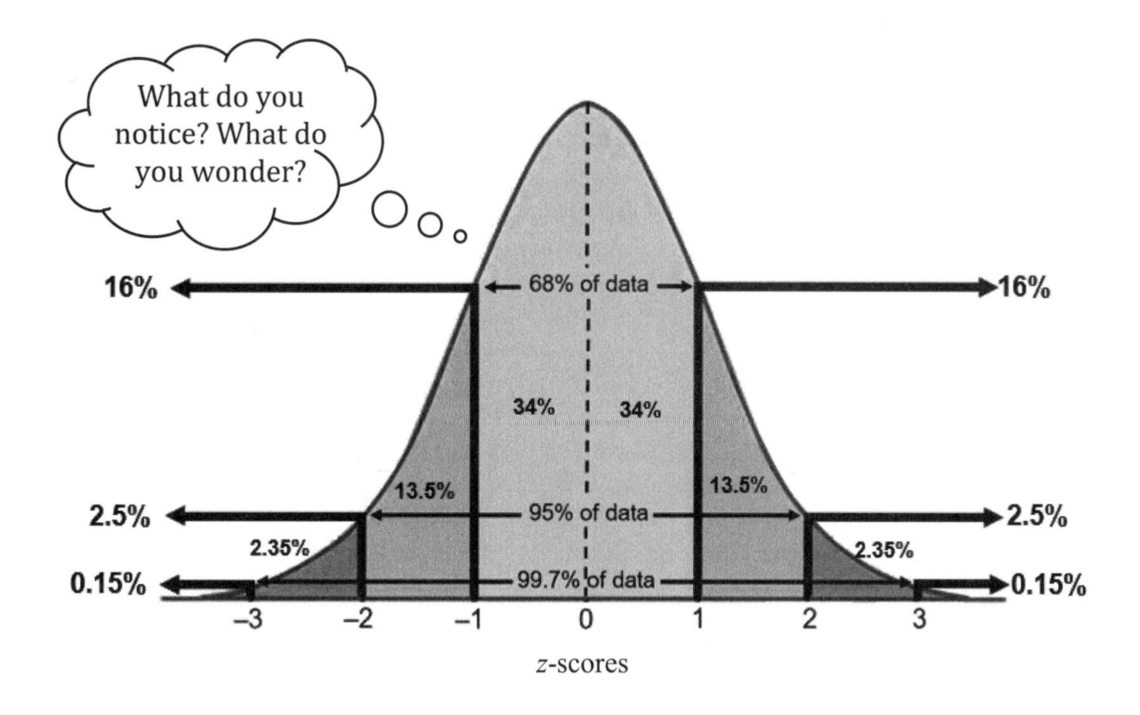

Read: Vocabulary and Key Concepts

- A _____________________ is described by a symmetric, single-peaked, mound-shaped curve called a normal curve. Any normal distribution is completely specified by two parameters: its mean ____ and standard deviation _____.
- In a normal distribution with mean μ and standard deviation σ, the empirical rule states that
  - About ______% of the values fall within 1σ of the mean μ.
  - About ______% of the values fall within 2σ of the mean μ.
  - About ______% of the values fall within 3σ of the mean μ.
- When assessing if a distribution of quantitative data is approximately normal, start with a ___________ and then add ______________ summaries. If a graph of the data is clearly skewed, has multiple peaks, or isn't bell-shaped, that's evidence the distribution is ____________________.
- Even if a graph of the data looks roughly symmetric and bell-shaped, we shouldn't assume that the distribution is approximately normal. Use the _________________________ to gain additional evidence in favor of, or against normality.
- A standard normal distribution is the normal distribution with mean _____ and standard deviation ____.
- Which has a greater area in a normal distribution with mean, 155 and standard deviation, 3: the area < 160 or the area ≤ 160? ____

Watch: Go to bfwpub.com/TPS7eStrive. Select Unit 1, Part II and watch the following Example videos:

Video	Topic	Done
Unit 1F, Stop the car! (p. 110)	*Normal curves*	
Unit 1F, Stop the car! (p. 113)	*The empirical rule*	
Unit 1F, Is the amount of cream in Oreo cookies normally distributed? (p. 115)	*Assessing normality*	
Unit 1F, Stop the car! (p. 120)	*Finding area to the left in a normal distribution*	
Unit 1F, Can Nelly clear the trees? (p. 121)	*Finding area to the right in a normal distribution*	
Unit 1F, Can Nelly hit the green? (p. 123)	*Finding areas between two values in a normal distribution*	
Unit 1F, How tall are 3-year-old girls? (p. 126)	*Finding percentiles in a normal distribution*	
Unit 1F, Get off your phone! (p. 127)	*Calculating the mean or SD of a normal distribution*	

Do: Check for Understanding

Concepts 1 and 2: Normal Curves and The Empirical Rule

You have learned that exploring quantitative data requires making a graph, describing the overall shape, and providing a numerical summary of the center and spread. Many distributions of real data and chance outcomes are symmetric, single-peaked, and bell-shaped and can be described by normal curves. A normal curve is an idealized model for the population distribution of a quantitative variable where the center is indicated by the mean (μ) and the variability is indicated by the standard deviation (σ). When labeling a normal curve, begin by placing the value of the mean at the center of the curve and then label 1, 2, and 3 standard deviations from the mean.

The empirical rule was discovered through observation and experience. The empirical rule states that in any normal distribution, approximately 68% of the observations will fall within one standard deviation of the mean, approximately 95% of observations will fall within two standard deviations, and approximately 99.7% of observations will fall within three standard deviations of the mean. This fact allows us to perform calculations about the approximate proportion of observations in a distribution that fall into certain intervals without even knowing all of the individual observations!

Check for Understanding – Learning Targets 1 and 2

____ *I can draw a normal curve to model the distribution of a quantitative variable.*

____ *I can use the empirical rule to estimate the proportion of values in a specified interval in a normal distribution.*

According to the Environmental Protection Agency (EPA), U.S. residents generate, on average, 4.48 pounds of trash per day with a standard deviation of 1.32 pounds.

1. Draw a normal curve. Label the mean and the points that are 1, 2, and 3 standard deviations from the mean.

2. About what percentage of U.S. residents make more than 5.8 pounds of trash per day?

3. About what percentage of U.S. residents make between 1.84 and 3.16 pounds of trash per day?

Concept 3: Assessing Normality

Normal distributions are handy models for some distributions of data. However, when we are given data, how can we know if the distribution is approximately normal? A good first step is to create a histogram, dotplot, or stemplot of the data. If the distribution is not roughly symmetric, single-peaked, and bell-shaped then the distribution is not approximately normal. However, it is important to note that even if a distribution is roughly symmetric, single-peaked, and bell-shaped, it is *not* necessarily normal. If the distribution passes the "graphical check," then we should compare the distribution of the data to the empirical rule. Specifically, does about 68% of the data fall within 1SD of the mean? Does about 95% of the data fall within 2SD of the mean? Does about 99.7% of the data fall within 3SD of the mean? If a distribution passes the graphical check and the numerical check, then we can confidently say the distribution is approximately normal.

Check for Understanding – Learning Target 3

____ *I can determine whether a distribution of quantitative data is approximately normal using graphical and numerical evidence.*

Here are the number of push-ups that the 23 students on the JV and High School basketball teams did in one minute:

15, 19, 20, 22, 23, 26, 27, 27, 28, 28, 28, 28, 31, 32, 34, 35, 38, 39, 41, 42, 44, 45, 50

Use graphical and numerical methods to determine whether these data are approximately normally distributed.

Concepts 4 and 5: Normal Distribution Calculations

Not all observations of interest will fall one, two, or three standard deviations from the mean. That is, we might be interested in knowing what proportion of observations are more than 2.1 standard deviations above the mean. In cases like this, the empirical rule can help us estimate the proportion, but we'll want to be more exact. Because all normal distributions share the same properties, the standard normal table and normal calculations using technology allow us to perform calculations for *any* observation in an approximately normal distribution.

To perform a normal calculation, draw and label a normal curve and shade the area of interest. Perform calculations by standardizing the boundary value(s) and then using Table A or use your calculator to find the desired area under the normal curve without standardizing. Be sure to write your conclusion in context! (*Need Tech Help?* View the videos **Tech Corner 4: Finding Areas in a Normal Distribution** and **Tech Corner 5: Finding Percentiles in a Normal Distribution** at bfwpub.com/TPS7eStrive)

Check for Understanding – Learning Targets 4 and 5

____ *I can find the proportion of values in a specified interval in a normal distribution.*

____ *I can find the value that corresponds to a given percentile in a normal distribution.*

The Bureau of Labor Statistics found that many U.S. adults hardly spend any time reading for personal interest. The normal curve shows the distribution of number of minutes spent reading for personal interest, per day, for U.S. adults.

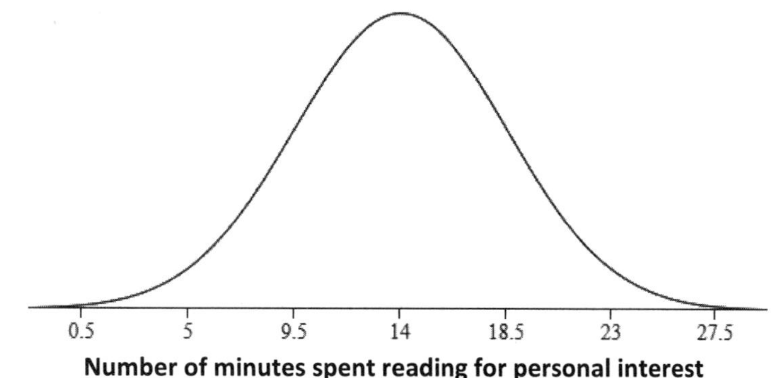

1. What are the mean and standard deviation of this distribution?

2. What percentage of U.S. adults read for personal interest more than 25 minutes per day?

3. What percentage of U.S. adults read for personal interest less than 12 minutes per day?

4. What proportion of U.S. adults read for personal interest between 5 and 10 minutes per day?

5. What is the 75th percentile of this distribution?

Concept 6: Calculating the Mean of SD of a Normal Distribution

All good math problems have a "forward" version and a "backwards" version. You have learned to find the proportion of values in a normal distribution when given the mean and standard deviation. Now we will find the mean or standard deviation, given the proportion of values in a normal distribution. The thought process is the same, but if you were only using technology in previous problems, watch out! You'll need to use the z-score formula this time around.

Check for Understanding – Learning Target 6

____ *I can calculate the mean or standard deviation of a normal distribution given the value of a percentile.*

1. According to a study conducted by the U.S. Department of Agriculture Economic Research, the amount of ice cream that U.S. residents eat in a year is normally distributed with a mean of 13 pounds. Additionally, the 25th percentile of the distribution is 10.471 pounds. What is the standard deviation of this distribution?

2. The Federal Communications Commission carried out a study to estimate the number of robocalls made to U.S. cellphone owners in a period of one month. The standard deviation of the distribution is 2.45 calls. The report also revealed that 60% of U.S. cell phone owners received more than 7.38 calls during the one-month study. What is the mean number of robocalls made to U.S. cellphone owners that month?

Unit 1, Part II Summary: Exploring One-Variable Data

In this unit, we expanded our toolbox for working with quantitative data. We learned how to describe the location of an individual within a distribution by determining its percentile or by calculating a standardized score (z-score) based on the mean and standard deviation of the distribution. We also learned that the normal curve is a helpful model for describing many quantitative variables. It is important to know how to use graphical and numerical evidence to assess the approximate normality of a distribution. If a distribution is normal, you can use normal distribution calculations to answer a number of questions about observations within the set of data. The empirical rule and the standard normal table are useful tools when performing calculations about observations in normal distributions.

The concepts introduced in this unit will form the basis of much of our study of inference later in the course. Standardizing data, justifying normality, and performing normal calculations are critical skills for statistical inference. Be sure to practice them because you will be using these skills a lot!

How well do you understand each of the learning targets?

Learning Target	Got It!	Almost There	Needs Work
I can calculate and interpret a percentile in a distribution of quantitative data.			
I can calculate and interpret a standardized score (z-score) in a distribution of quantitative data.			
I can use percentiles or standardized scores (z-scores) to compare the relative positions of individual values in distributions of quantitative data.			
I can use a cumulative relative frequency graph to estimate percentiles and individual values in a distribution of quantitative data.			
I can analyze the effect of adding, subtracting, multiplying by, or dividing by a constant on the shape, center, and variability of a distribution of quantitative data.			
I can draw a normal curve to model the distribution of a quantitative variable.			
I can use the empirical rule to estimate the proportion of values in a specified interval in a normal distribution.			
I can determine whether a distribution of quantitative data is approximately normal using graphical and numerical evidence.			
I can find the proportion of values in a specified interval in a normal distribution.			
I can find the value that corresponds to a given percentile in a normal distribution.			
I can calculate the mean or standard deviation of a normal distribution given the value of a percentile.			

Unit 1, Part II Multiple Choice Practice

Directions. Identify the choice that best completes the statement or answers the question. Check your answers and note your performance when you are finished.

1. The 16th percentile of a normally distributed variable has a value of 25 and the 97.5th percentile has a value of 40. Which of the following is the best estimate of the mean and standard deviation of the variable?
   (A) Mean ≈ 32.5; Standard deviation ≈ 2.5
   (B) Mean ≈ 32.5; Standard deviation ≈ 5
   (C) Mean ≈ 32.5; Standard deviation ≈ 10
   (D) Mean ≈ 30; Standard deviation ≈ 2.5
   (E) Mean ≈ 30; Standard deviation ≈ 5

2. What portion of observations from a standard normal distribution take values larger than 0.75?
   (A) 0.2266
   (B) 0.2500
   (C) 0.7704
   (D) 0.7764
   (E) 0.8023

3. A study was conducted to determine how long students spend washing their hands in the restroom. A student stood to the side, pretending to be using her phone, and discretely timed 10 students. Here are the number of seconds each student washed their hands: 3, 8, 15, 2, 3, 7, 20, 6, 9, 5. What is the percentile of the student who washed their hands for 5 seconds?
   (A) 10%
   (B) 20%
   (C) 30%
   (D) 40%
   (E) 50%

4. The distribution of the heights of students in a large class is roughly normal. The average height is 68 inches, and approximately 99.7% of the heights are between 62 and 74 inches. Thus, the standard deviation of the height distribution is approximately equal to
   (A) 2
   (B) 3
   (C) 4
   (D) 6
   (E) 9

5. The mean age (at inauguration) of all U.S. Presidents is approximately normally distributed with a mean of 54.6. Barack Obama was 47 when he was inaugurated, which is the 11th percentile of the distribution. George Washington was 57. Approximately what percentile is George Washington at?
   (A) 6th
   (B) 35th
   (C) 39th
   (D) 63rd
   (E) 65th

6. Which of the following statements are false?
   I. The standard normal table can be used with z-scores from any distribution
   II. The mean is always equal to the median for any normal distribution.
   III. Every symmetric, bell-shaped distribution is approximately normal
   IV. The area under a normal curve is always 1, regardless of the mean and standard deviation.

   (A) I and II
   (B) I and III
   (C) II and III
   (D) III and IV
   (E) None of the above gives the correct set of false statements.

7. High school textbooks don't last forever. The lifespan of all high school statistics textbooks is approximately normally distributed with a mean of 9 years and a standard deviation of 2.5 years. What percentage of the books last more than 10 years?
   (A) 11.5%
   (B) 34.5%
   (C) 65.5%
   (D) 69.0%
   (E) 84.5%

8. The distribution of the time it takes for different people to solve their Strive for a Five crossword puzzle is strongly skewed to the right, with a mean of 10 minutes and a standard deviation of 2 minutes. The distribution of z-scores for those times is

   (A) normally distributed, with mean 10 and standard deviation 2
   (B) skewed to the right, with mean 10 and standard deviation 2
   (C) normally distributed, with mean 0 and standard deviation 1
   (D) skewed to the right, with mean 0 and standard deviation 1
   (E) Skewed to the right, but the mean and standard deviation cannot be determined without more information.

9. The cumulative relative frequency graph below shows the distribution of lengths (in centimeters) of ears for a sample of adults. The third quartile for this distribution is approximately:

   (A) 5.5 cm
   (B) 6.0 cm
   (C) 6.7 cm
   (D) 7.0 cm
   (E) 7.5 cm

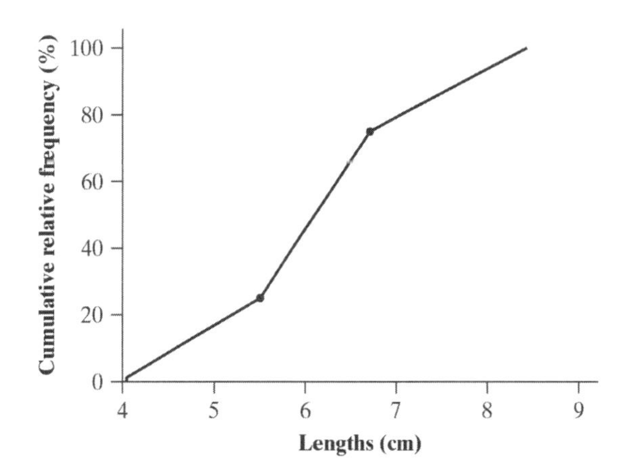

10. The plot shown below is a dotplot of the distribution of number of siblings for a random sample of students. Can the distribution of number of siblings be described as approximately normal?

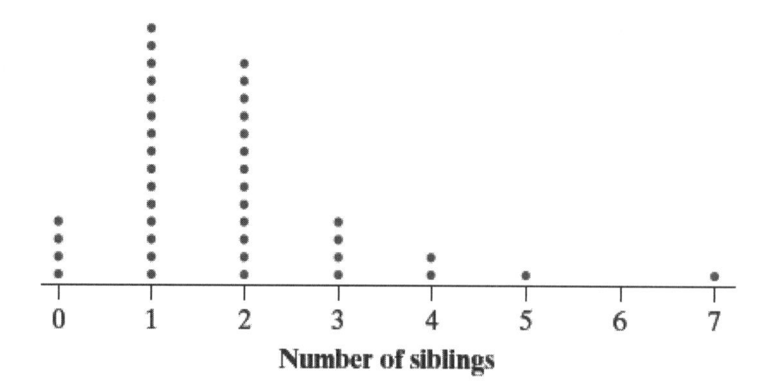

(A) Yes, the distribution of number of siblings is approximately normal because this distribution is as you would expect: most people have few siblings, and a few people have many siblings.
(B) Yes, the distribution of number of siblings is approximately normal because the percentage of values that fall within 1, 2, and 3 SD of the mean closely follow the empirical rule.
(C) No, the distribution of number of siblings cannot be described as approximately normal because the graphical evidence indicates that the distribution is skewed to the right.
(D) No, the distribution of number of siblings cannot be described as approximately normal because the graphical evidence indicates that the distribution is uniformly distributed.
(E) No, the distribution of number of siblings cannot be described as approximately normal because there is a potential high outlier, and a normal distribution never has outliers.

Check your answers below. If you got a question wrong, check to see if you made a simple mistake or if you need to study that concept more. After you check your work, identify the concepts you feel very confident about and note what you will do to learn the concepts in need of more study.

#	Answer	Concept	Right	Wrong	Simple Mistake?	Need to Study More
1	E	Empirical Rule				
2	A	Standard Normal Table				
3	D	Percentile				
4	A	Empirical Rule				
5	E	Standard Normal Calculations				
6	B	Properties of Normal Distributions				
7	B	Standard Normal Calculations				
8	D	Standardized Scores				
9	C	Cumulative Relative Frequency Graph				
10	C	Assessing Normality				

FRAPPY! Free Response AP® Problem, Yay!

The following problem is modeled after actual Advanced Placement Statistics free response questions. Your task is to generate a complete, concise response in 15 minutes. After you generate your response, read over the solution and scoring guide. Score your response and note what, if anything, you would do differently to increase your own score.

Final exam grades are determined by the percent correct on the exam. A teacher's records indicate that performance on the exam is normally distributed with mean 82 and standard deviation 5. Grades on the exam are assigned using the scale below.

Grade	Percent Correct
A	$94 \leq$ percent ≤ 100
B	$85 \leq$ percent < 94
C	$76 \leq$ percent < 85
D	$65 \leq$ percent < 76
F	$0 \leq$ percent < 65

(a) Sketch a normal distribution to illustrate the proportion of students who would earn a B. Calculate this proportion.

(b) Students who earn a B, C, or D are considered to "meet standards." Based on this grading scale, what percent of students will receive a score that places them in a category other than "meets standards"?

(c) What grade would the student who scored at the 25th percentile earn on this exam? Justify your answer.

FRAPPY! Scoring Rubric

Use the following rubric to score your response. Each part receives a score of "Essentially Correct," "Partially Correct," or "Incorrect." When you have scored your response, reflect on your understanding of the concepts addressed in this problem. If necessary, note what you would do differently on future questions like this to increase your score.

Intent of the Question

The goal of this question is to determine your ability to perform and interpret normal calculations.

Solution

(a) $P(\text{grade} = \text{B})$

$= P(85 \leq \text{percent} < 94)$

$= P\left(\frac{85-82}{5} \leq z \leq \frac{94-82}{5}\right)$

$= P(0.6 \leq z < 2.4)$

$= 0.9918 - 0.7257$

$= 0.2661$

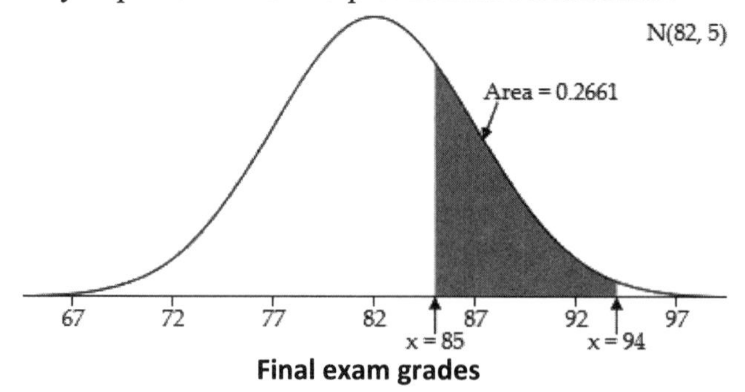

Using technology: normalcdf(lower: 85, upper: 94, mean: 82, SD: 5) = 0.2661

(b) $P(\text{A or F}) = P(\text{A}) + P(\text{F})$

$= P\left(z < \frac{65-82}{5}\right) + P\left(z \geq \frac{94-82}{5}\right)$

$= P(z < 3.40) + P(z \geq 2.40)$

$= 0.0003 + 0.0082$

$= 0.0085$

Using technology: $P(\text{A or F})$ = normalcdf(lower: –1000, upper: 65, mean: 82, SD: 5)
+ normalcdf(lower: 94, upper: 1000, mean: 82, SD: 5)
= 0.0085

(c) A z-score of –0.6745 corresponds to the 25th percentile.

$$-0.6745 = \frac{x-82}{5}$$

$$(-0.6475)(5) = x - 82$$

$$-3.2375 = x - 82$$

$$x = 78.63$$

Using technology: Invnorm(area: 0.25, mean: 82, SD: 5) = 78.628.

Scoring:

Parts (a), (b), and (c) are scored as essentially correct (E), partially correct (P), or incorrect (I).

Part (a) is essentially correct if (1) the appropriate probability is illustrated using a labeled normal curve and (2) the proportion is correctly computed.
Part (a) is partially correct if only one of the above elements is correct

Part (b) is essentially correct if the response (1) recognizes the need to look at grades of A and F and (2) correctly computes the tail probabilities and adds them together.

Part (b) is partially correct if the response considers only an A or an F and calculates the corresponding tail area correctly OR recognizes the need to look at A and F but only calculates one of the tail areas correctly OR approximates the probabilities using the empirical rule OR computes the proportion that will "meet standards" OR states the correct answer without supporting work.

Part (c) is essentially correct if (1) the correct *z*-score is identified for the 25th percentile and (2) the correct corresponding score is calculated with correct supporting work.
Part (c) is partially correct if only one of the above elements is correct.

NOTE: If the student makes an error in part (b) and correctly uses that probability in part (c) to compute a reasonable probability, part (c) is essentially correct.

4 Complete Response

All three parts essentially correct

3 Substantial Response

Two parts essentially correct and one part partially correct

2 Developing Response

Two parts essentially correct and no parts partially correct

One part essentially correct and one or two parts partially correct

Three parts partially correct

1 Minimal Response

One part essentially correct and no parts partially correct

No parts essentially correct and two parts partially correct

My Score:
What I did well:
What I could improve:
What I should remember if I see a problem like this on the AP Exam:

Unit 1, Part II Formula Sheet

Create a one-page summary of important concepts and formulas found in this unit. This will serve as a valuable resource as you prepare for the AP Exam.

Unit 1, Part II Crossword Puzzle

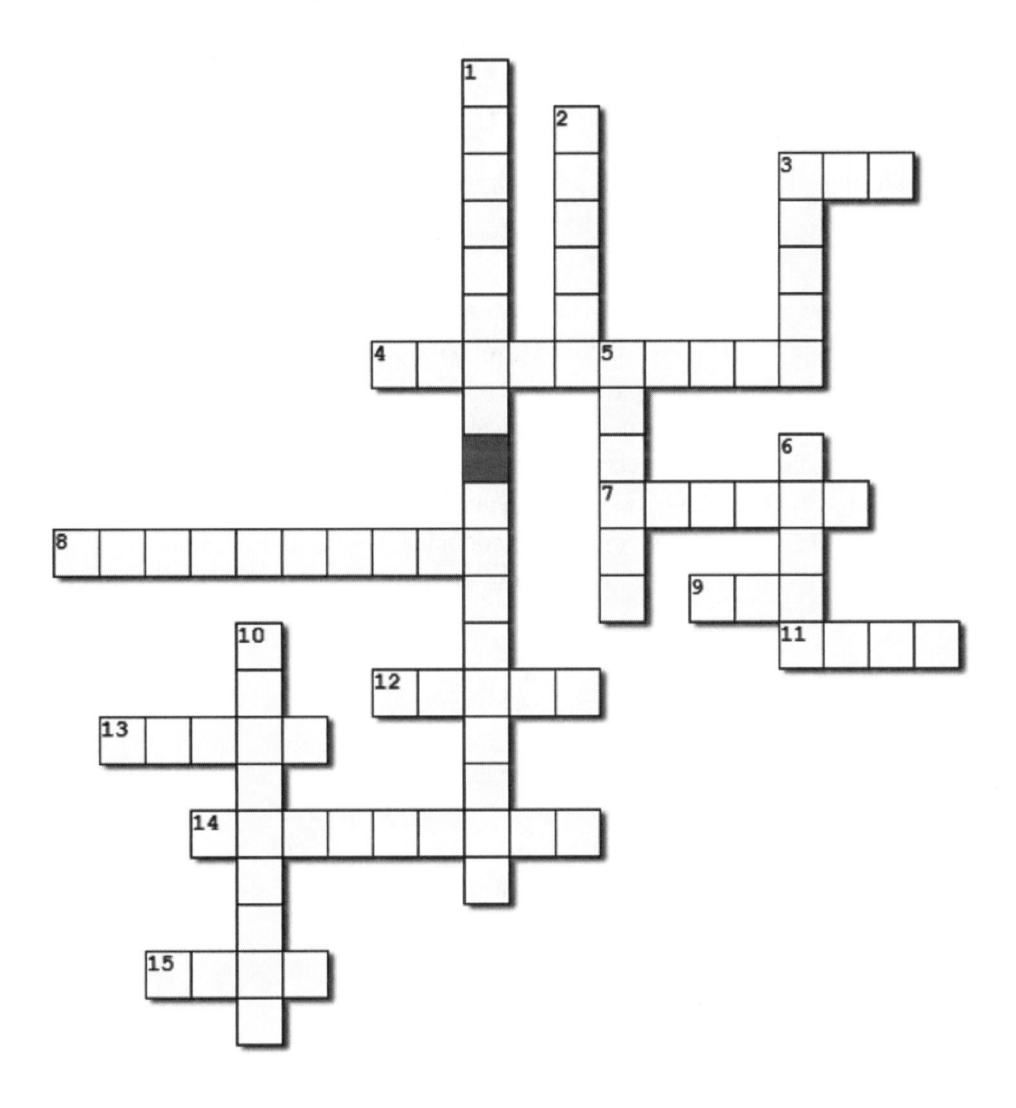

Across:

3. In a normal distribution, about 95% of the observations fall within ____ SD of the mean.
4. The value of a distribution with p% of observations less than or equal to it.
7. In a normal distribution, the mean has the same value as the ____.
8. A ____ relative frequency graph plots a point corresponding to the percentile of a given value in a distribution.
9. In a normal distribution, about 68% of the observations fall within ____ SD of the mean.
11. The standard normal table tells us the area under the standard normal curve to the ____ of z.
12. Multiplying every value in a data set by a positive constant does not change the ____ of the distribution.
13. To assess normality of a distribution, first make a ____ of the data, such as a dotplot.
14. We ____ data when we change each value by adding, subtracting, multiplying, or dividing by a constant.
15. The center of a normal distribution.

Down:

1. In a normal distribution, the ____ is the distance from the mean to the change-of-curvature points on either side of the mean.
2. The standardized score for an individual in a distribution is called a ____.
3. In a normal distribution, about 99.7% of the observations fall within ____ SD of the mean.
5. The standard ____ curve has a mean of 0 and standard deviation of 1.
6. When using a calculator to perform normal calculations, be sure to ___ the inputs.
10. The ____ rule tells us the percent of observations that fall within 1, 2, and 3 standard deviations of the mean in a normal distribution.

Unit 2: Exploring Two-Variable Data

"You can only predict things after they've happened." Eugene Ionesco

UNIT 2 OVERVIEW

Our statistics toolbox now contains a variety of ways to explore a single quantitative variable (graphs, numerical summaries, normal distributions). Further, we have learned ways to explore a single categorical variable (bar graph, pie chart, calculating percentages). Now you are ready to explore and describe the *relationship* between two categorical or two quantitative variables. In this unit, we will learn how to analyze patterns in "bivariate" relationships by creating graphs and calculating summary statistics. Also, we will learn how to describe the relationship between two quantitative variables with mathematical models that can be used to make predictions based on the relationship between the variables. Investigating the relationship between two variables is a key component of statistical study and is the final skill necessary for inclusion in our data exploration toolbox.

Sections in Unit 2

What do you notice? What do you wonder?

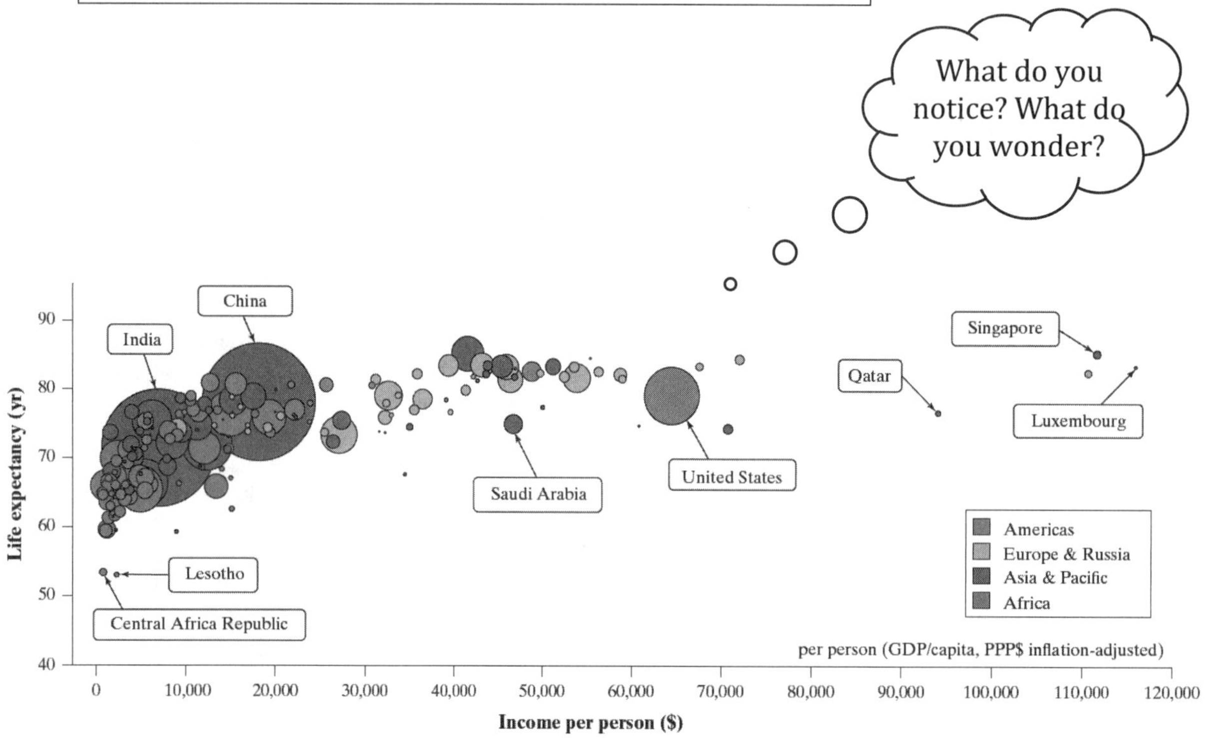

PRACTICE FOR MASTERY

Use the following *suggested* guide to the pages and exercises in your text to practice for mastery!
Note: your teacher may assign different problems. Be sure to follow their instructions!

Day	Topics	Read	Do
1	◆ Explanatory and Response Variables ◆ Summarizing Data on Two Categorical Variables ◆ Displaying the Relationship Between Two Categorical Variables ◆ Describing the Relationship Between Two Categorical Variables	p. 142-153	**2A**: 1, 3, 5, 7, 9, 11, 13, 20 – 23
2	◆ Displaying the Relationship Between Two Quantitative Variables ◆ Describing a Scatterplot	p. 158-164	**2B**: 1, 3, 5, 7, 30, 33, 34
3	◆ Interpreting Correlation ◆ Cautions about Correlation ◆ Calculating Correlation	p. 165-172	**2B**: 9, 11, 15, 19, 23, 31, 32
4	◆ Prediction ◆ Residuals ◆ Interpreting the Slope and *y* Intercept	p. 180-187	**2C**: 1, 3, 7, 9, 11, 41
5	◆ The Least-Squares Regression Line ◆ Using Technology to Calculate a Least-Squares Regression Line ◆ Determining Whether a Linear Model is Appropriate: Residual Plots ◆ How Well the Line Fits the data: Interpreting r^2 and s.	p. 187-200	**2C**: 13, 17, 21, 23, 27, 31, 37, 39, 43
6	Option: AP Classroom Topic Questions (Topics 2.1-2.8)		**2B**: 13, 17, 21 **2C**: 5, 15, 25, 38, 40, 42
7	◆ Influential Points	p. 210-215	**2D**: 1, 3, 5, 20, 21
8	◆ Transformations to Achieve Linearity ◆ Choosing the Most Appropriate Model	p. 215-222	**2D**: 7, 9, 11, 13, 15, 18, 19
9	Unit 2 AP® Statistics Practice Test (p. 235)		Unit 2 Review Exercises (p. 232)
10	Unit 2 Test: Celebration of learning!	Note: Since "Test" can sound daunting, think of this as a way to show off all you have learned. It's your own celebration of learning!	

Section 2A: Relationships Between Two Categorical Variables

Section Summary

Many statistical studies examine more than one variable. So far, we have learned methods to graph and describe relationships between categorical variables. In this unit, we'll learn that the approach to data analysis that we learned for a single quantitative variable can also be applied to explore the relationship between two quantitative variables. That is, we'll learn how to plot our data and add numerical summaries. We'll then learn how to describe the overall patterns and departures from patterns that we see. Finally, we'll learn how to create a mathematical model to describe the overall pattern. This section will focus primarily on displaying the relationship between two quantitative variables and describing their form, direction, and strength. Like the previous units, you will find that technology can be used to do most of the difficult calculations. However, be sure you understand *how* the calculator is determining its results and *what* those results mean!

Learning Targets:

____ I can identify the explanatory and response variables in a given setting.
____ I can calculate statistics for two categorical variables.
____ I can display the relationship between two categorical variables.
____ I can describe the relationship between two categorical variables.

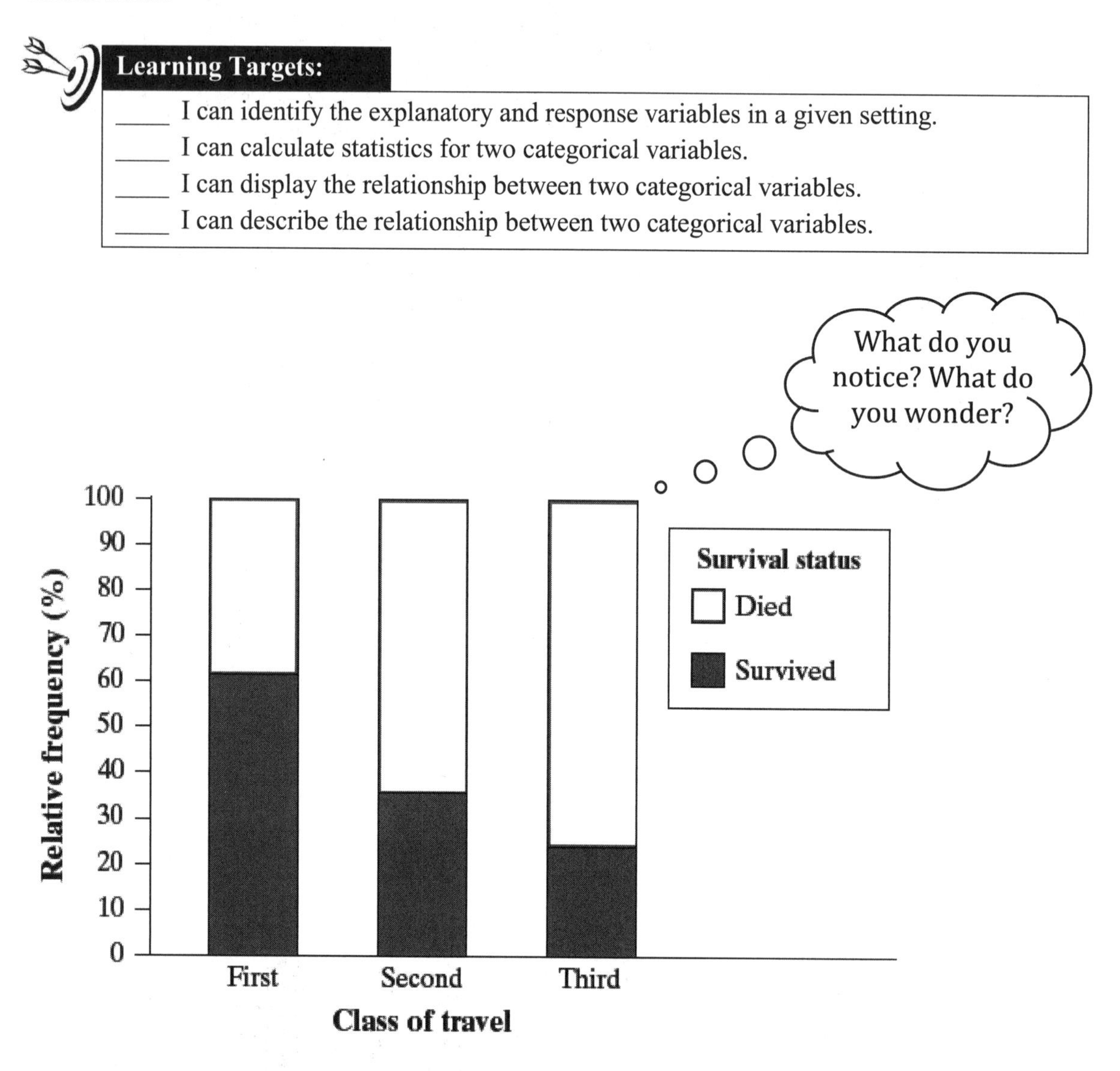

In 1912, on its first voyage across the Atlantic, the luxury liner *Titanic* struck an iceberg and sank. Some passengers made it off the ship in lifeboats, but many died. Here is a segmented bar graph of class of travel and survival status for adult passengers on the *Titanic.*

Read: Vocabulary and Key Concepts

- A ____________________________ measures an outcome of a study.
- An ___________________________ may help predict or explain changes in a response variable.
- A _____________________________ is a table of counts or relative frequencies that summarizes data on the relationship between two categorical variables for some group of individuals.
- A _______________________________ gives the percentage or proportion of individuals in a two-way table that have a specific value for one categorical variable. A marginal relative frequency is calculated by dividing a row or column total by the total for the entire two-way table.
- A _______________________________ gives the percentage or proportion of individuals in a two-way table that have a specific value for one categorical variable and a specific value for another categorical variable. A joint relative frequency is calculated by dividing the value in one cell by the total for the entire two-way table.
- A _____________________________ gives the percentage or proportion of individuals that have a specific value for one categorical variable among a group of individuals that share the same value of another categorical variable (the condition). A conditional relative frequency is calculated by dividing the value in one cell of a two-way table by the total for the appropriate row or column.
- A ____________________________ displays the distribution of a categorical variable as segments of a bar, with the area of each segment proportional to the number of individuals in the corresponding category.
- A ____________________ is a modified segmented bar graph in which the width of each bar is proportional to the number of individuals in the corresponding category.
- There is an ______________________ between two variables if knowing the value of one variable helps us predict the value of the other. If knowing the value of one variable does not help us predict the value of the other, there is no association between the variables.
- What would a segmented bar graph look like if there were no association between the variables?

  ____
- Association does not imply __________________!

Watch: Go to bfwpub.com/TPS7eStrive. Select Unit 2 and watch the following Example videos:

Video	Topic	Done
Unit 2A, Diamonds and color (p. 143)	*Explanatory and response variables*	
Unit 2A, A Titanic disaster, part 1 (p. 146)	*Summarizing data on two categorical variables*	
Unit 2A, A Titanic disaster, part 2 (p. 150)	*Displaying the relationship between two categorical variables*	
Unit 2A, A Titanic disaster, part 3 (p. 150)	*Describing the relationship between two categorical variables*	

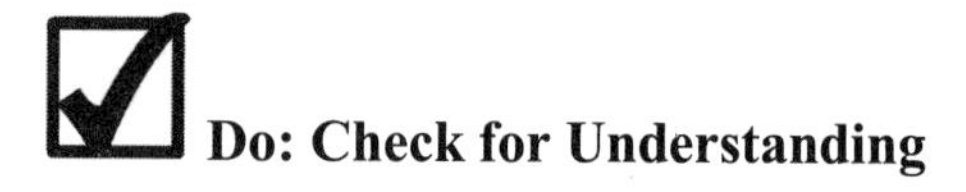

Concept 1: Explanatory and Response Variables

The purpose of many studies involving two variables is to use one variable to make a prediction about the other. For this reason, it is necessary to identify which variable in a situation is *explanatory* and which is the *response.* The explanatory variable is the one we think explains the relationship or "predicts" changes in the response variable. It is important to know the difference as identifying the explanatory and response variable will determine how we display the data. Note: you may have learned about *independent* and *dependent* variables in an earlier math or science class. Those are just different names for explanatory and response variables. We'll avoid using independent and dependent here, though, because those terms have a different meaning later in the course.

Check for Understanding – Learning Target 1

____ *I can identify the explanatory and response variables in a given setting.*

Identify the explanatory and response variable in the following situations:

1. How does test anxiety affect test performance? Researchers measured students' test anxiety and recorded their subsequent performance on a standardized test.

2. Is brain size related to memory? A recent study measured the volume of each subject's hippocampus and then administered a verbal retention assessment.

Concept 2: Calculating Statistics Using a Two-Way Table

Data on two categorical variables is best summarized in a two-way table. The column variables often (but not always) display the explanatory variable and the rows often (but not always) display the response variable. When calculating relative frequencies from a two-way table pay attention to *what* you are asked to find: the proportion or percentage "of." The "of" in these problems indicates the denominator of the desired fraction. Sometimes you will find the percentage "of people in the sample" who ...[characteristic]. In this case, the denominator would be the table total, and you would identify the correct row (or column) that is described by the characteristic. The "of" is a giveaway! Look for it!

Check for Understanding – Learning Target 2

____ *I can calculate statistics for two categorical variables.*

A famous comedian once said, "If you don't vote, you lose the right to complain." A random sample of U.S. adults was selected. The two-way table classifies the respondents by age group and whether or not they voted in the most recent midterm election.

		Age group			
		18 to 29	30 to 59	60+	Total
Voted	Yes	17	34	54	105
	No	61	88	40	189
	Total	78	122	94	294

1. What percentage of people in the sample voted in the most recent midterm election? (Note: This is a marginal relative frequency.)

2. What percentage of people in the sample are 30 to 59-years-old and did not vote in the most recent midterm election? (Note: This is a joint relative frequency.)

3. What percentage of 18 to 29-year-olds in the sample voted in the most recent midterm election? What percentage of 30 to 59-year-olds voted? 60+ year-olds? (Note: These are conditional relative frequencies.)

Concepts 3 and 4: Displaying and Describing the Relationship Between Two Categorical Variables

Two-way tables can be cumbersome to read, and it can be difficult to grasp the relationship between the variables when looking at numbers in a table. Once again, graphical displays come to the rescue! Data presented in two-way tables can be displayed in a segmented bar graph, which has one bar for each category of the explanatory variable broken into segments according to the proportions of the response variables. Another useful graph is a mosaic plot, which is just like a segmented bar graph except that the width of the bars are proportional to the number of individuals in each category of the explanatory variable. These graphs help answer the question: What story do the data tell?

Check for Understanding – Learning Targets 3 and 4

____ *I can display the relationship between two categorical variables.*

____ *I can describe the relationship between two categorical variables.*

Let's explore the relationship between age group and voting status further. Here are the data.

		Age group			
		18 to 29	30 to 59	60+	Total
Voted	Yes	17	34	54	105
	No	61	88	40	189
	Total	78	122	94	294

1. In the previous Check for Understanding, you calculated the conditional relative frequencies of individuals who voted by age group. Use those calculations and the grids below to create a segmented bar chart and a mosaic plot to display the relationship between age group and voting status.

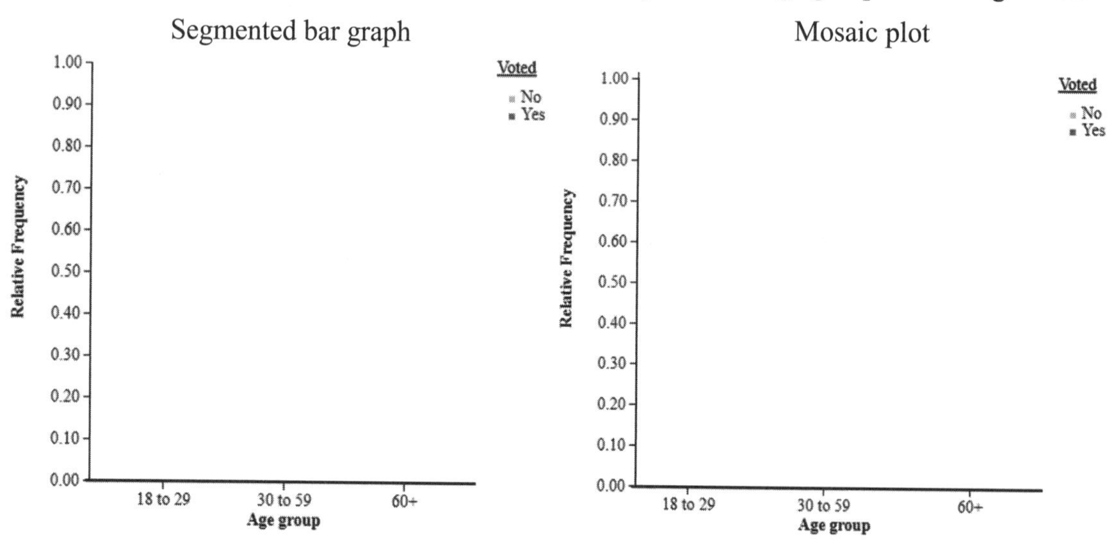

2. Describe what the graphs reveal about the relationship between age group and voting status for the U.S. adults in the sample.

Section 2B: Relationships Between Two Quantitative Variables

Section Summary

We have been working systematically through the concept of exploring data. In Unit 1, we explored one-variable *categorical* data graphically using bar charts and numerically by calculating percentages. Then we explored one-variable *quantitative* data graphically using dotplots, stemplots, histograms, boxplots, and normal curves. To complete the "graphical, then numerical" pattern, we explored one-variable quantitative data numerically by calculating measures of center (mean, median) and variability (range, *IQR*, SD).

In Unit 2, we turned our focus to two-variable data. As in Unit 1, we began with categorical data. We took a graphical (segmented bar graph, mosaic plot) and numerical (percentages) approach to investigate the relationship between two categorical variables. Now, we are ready for the final step! We will explore two-variable quantitative data graphically and numerically. Welcome to the wonderful world of scatterplots!

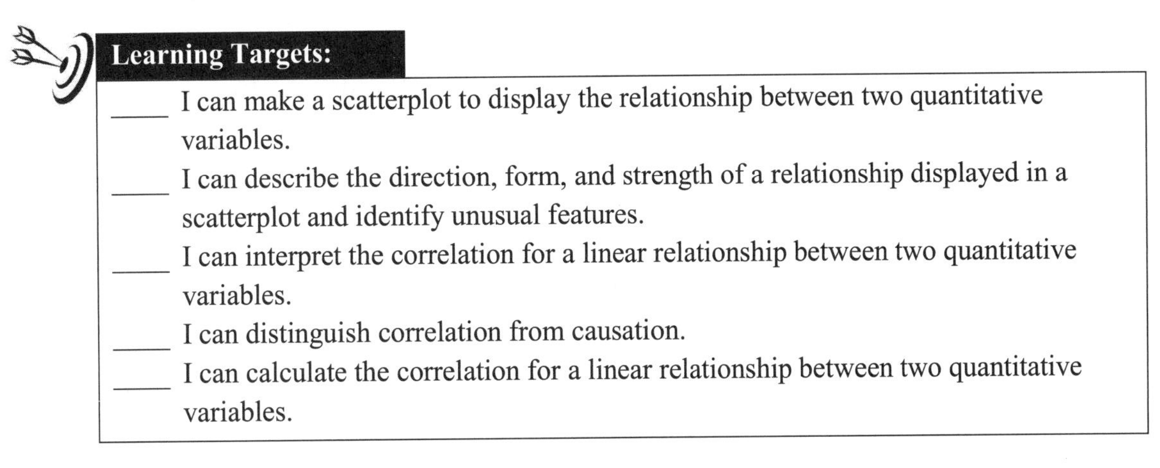

Learning Targets:

____ I can make a scatterplot to display the relationship between two quantitative variables.

____ I can describe the direction, form, and strength of a relationship displayed in a scatterplot and identify unusual features.

____ I can interpret the correlation for a linear relationship between two quantitative variables.

____ I can distinguish correlation from causation.

____ I can calculate the correlation for a linear relationship between two quantitative variables.

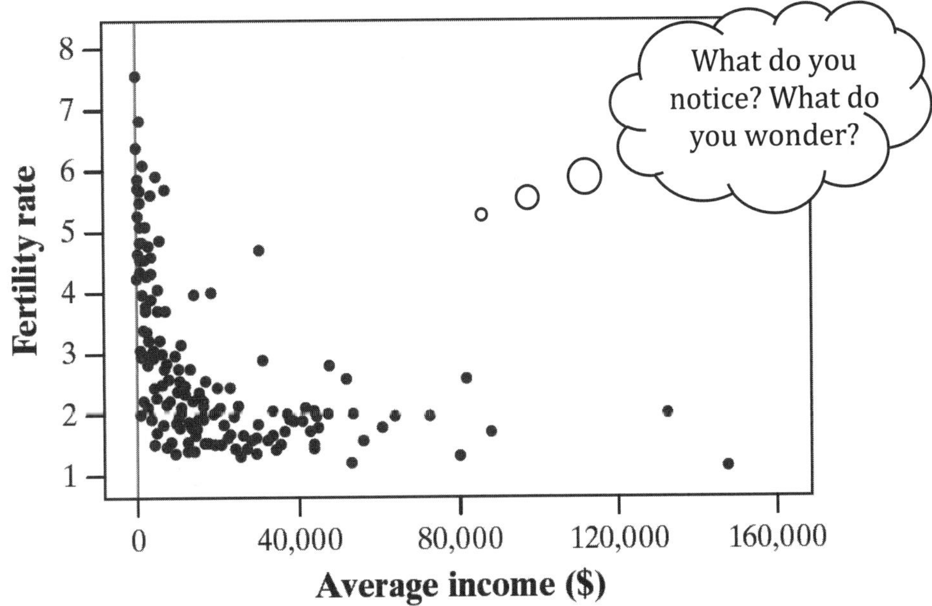

The scatterplot shows the relationship between the average income (gross domestic product per person, in dollars) and fertility rate (number of children per woman) in 187 countries.

Read: Vocabulary and Key Concepts

- A ________________________ shows the relationship between two quantitative variables measured for the same individuals. The values of one variable appear on the horizontal axis, and the values of the other variable appear on the vertical axis. Each individual in the data set appears as a point in the graph.
- When labeling the axes of a scatterplot, the name of the _______________ variable should be placed under the horizontal axis and the name of the ________________ variable should be placed to the left of the vertical axis.
- Two variables have a __________________ association when the values of one variable tend to increase as the values of the other variable increase.
- Two variables have a __________________ association when the values of one variable tend to decrease as the values of the other variable increase.
- There is ______ association between two variables if knowing the value of one variable does not help us predict the value of the other variable.
- The ____________________ r gives the direction and measures the strength of the linear association between two quantitative variables.
- The correlation is always a value between _____ and _____.
- When $r > 0$, the direction of the relationship is _______________ and when $r < 0$, the direction of the relationship is ____________.
- Correlation has no ____________ of measurement.
- Correlation does not measure form and should only be used to describe ___________ relationships.

Watch: Go to bfwpub.com/TPS7eStrive. Select Unit 2 and watch the following Example videos:

Video	Topic	Done
Unit 2B, Height and free-throws (p. 160)	*Displaying the relationship between two quantitative variables*	
Unit 2B, Old Faithful and fertility rate (p. 163)	*Describing a scatterplot*	
Unit 2B, Height and free-throws, part 2 (p. 166)	*Interpreting correlation*	
Unit 2B, Nobel chocolate (p. 168)	*Cautions about correlation*	
Unit 2B, Grab that candy (p. 170)	*Calculating correlation*	

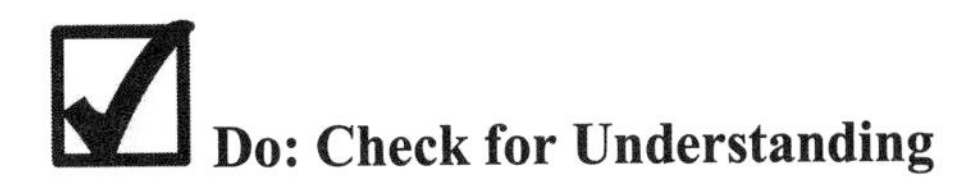

Do: Check for Understanding

Concepts 1 and 2: Making and Describing Scatterplots

As we learned in Unit 1, plotting data should always be our first step in data exploration. The most useful graph for exploring relationships between two-variable quantitative data is the scatterplot. Making a scatterplot is pretty easy. 1) Determine which variable goes on which axis. (Hint: The eXplanatory variable goes on the x-axis!) 2) Label and scale the axes. 3) Plot individual data values. Once the scatterplot is constructed, describe what you see. What is the form of the relationship? Is it linear? Nonlinear? What direction does the relationship take? Is it positive? Negative? How strong is the relationship? Do the points follow the pattern closely, or are they widely scattered? Finally, are there any unusual observations?

Check for Understanding – Learning Targets 1 and 2

____ *I can make a scatterplot to display the relationship between two quantitative variables.*

____ *I can describe the direction, form, and strength of a relationship displayed in a scatterplot and identify unusual features.*

The table shows the growth of Google Chrome and the decline of smoking in the United States since 2009. Create a scatterplot to display this relationship, then describe the relationship.

Year	**2009**	**2012**	**2015**	**2018**	**2021**	**2024**
Google Chrome market share (%)	2	22	42	55	62	65
Percent of U.S. adults who smoke	20.6	18.1	15.1	13.7	11.5	11.1

(*Need Tech Help?* View the video **Tech Corner 6: Making Scatterplots** at bfwpub.com/TPS7eStrive)

Concepts 3, 4, and 5: Correlation

Scatterplots are great tools for displaying the direction, form, and strength of the relationship between two quantitative variables. Often, we want to know whether or not the relationship is linear and, if so, how strong the linear relationship is. However, our eyes aren't the most accurate judges of the strength of linear relationships. Thankfully, the correlation r measures the strength of the linear relationship!

Some key points to remember about r:

- r is always between -1 and 1. A correlation of $r = 1$ or $r = -1$ indicates a perfect linear relationship.
- Positive relationships have a positive correlation and negative relationships have a negative correlation.
- Correlation has no units of measurement. It would be incorrect to say that the correlation between age and height is $r = 0.9$ inches/year.
- Finally, the closer r is to 1, the stronger the linear relationship between the quantitative variables.

Here is a summary, showing some values of the correlation coefficient, and appropriate descriptions based on those values:

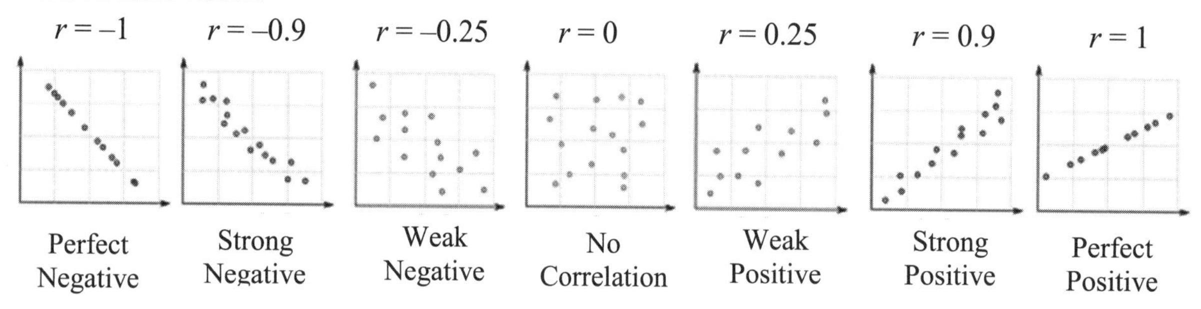

Check for Understanding – Learning Targets 3, 4, and 5

____ *I can interpret the correlation for a linear relationship between two quantitative variables.*

____ *I can distinguish correlation from causation.*

____ *I can calculate the correlation for a linear relationship between two quantitative variables.*

1. Using the data from the previous Check for Understanding, calculate and interpret the correlation coefficient. What does this value tell you about the relationship? (*Need Tech Help?* View the video **Tech Corner 7: Calculating Correlation** at bfwpub.com/TPS7eStrive)

2. A reporter observes the scatterplot of the relationship between Google Chrome market share and the percent of U.S. adults who smoke, from the last CYU and writes an article with the headline: "Google Chrome Drives the Decline of Smoking!" Is this headline justified? Explain your answer.

Section 2C: Linear Regression Models

Section Summary

In the previous section, we learned that we can display the relationship between two quantitative variables using a scatterplot. Further, we can use a scatterplot to describe the direction, form, and strength of the relationship. The correlation coefficient r allows us to describe the situation by telling us how strong the *linear* relationship between the variables is. In this section, we'll learn how to summarize the overall pattern of a linear relationship by finding the equation of the least-squares regression line. This line can be used not only to model the linear relationship, but also to make predictions based on the overall pattern. As with correlation, our calculator will do most of the work for us. Your job is to be able to interpret and apply the results!

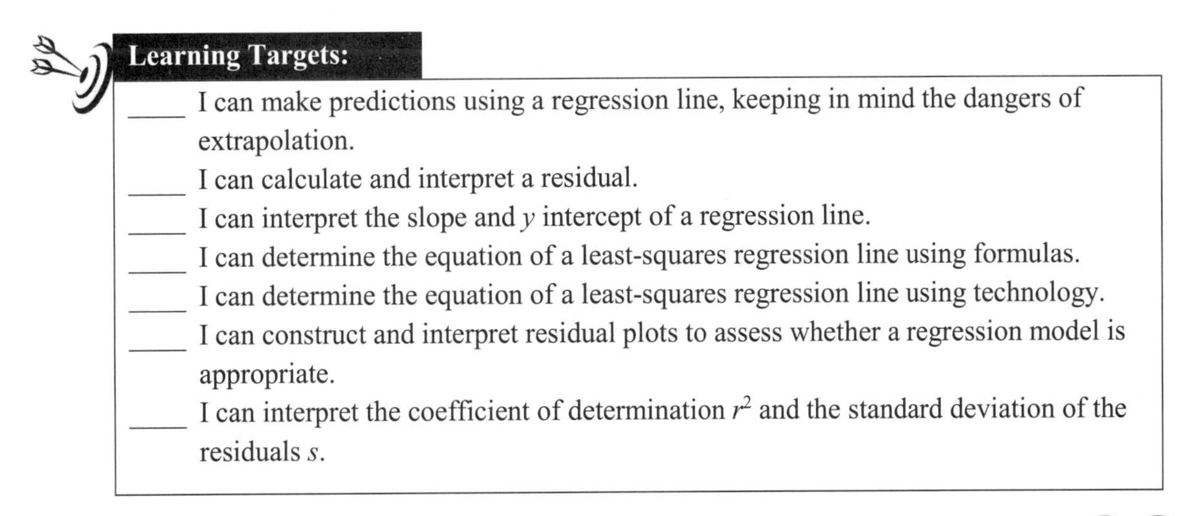

Learning Targets:

- ____ I can make predictions using a regression line, keeping in mind the dangers of extrapolation.
- ____ I can calculate and interpret a residual.
- ____ I can interpret the slope and y intercept of a regression line.
- ____ I can determine the equation of a least-squares regression line using formulas.
- ____ I can determine the equation of a least-squares regression line using technology.
- ____ I can construct and interpret residual plots to assess whether a regression model is appropriate.
- ____ I can interpret the coefficient of determination r^2 and the standard deviation of the residuals s.

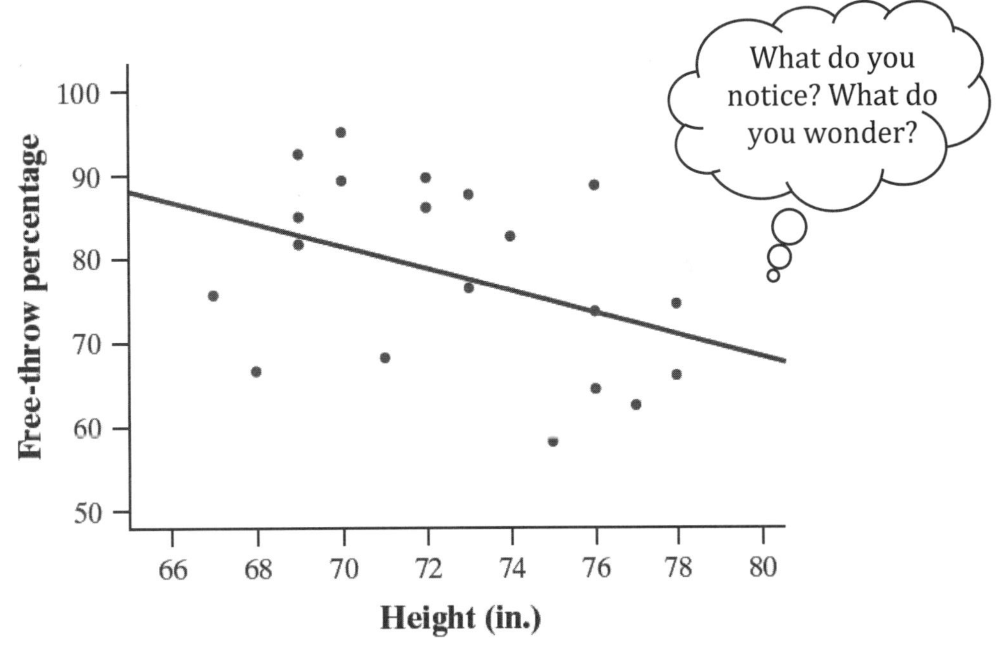

The scatterplot shows the relationship between height (in inches) and free throw percentage for a random sample of WNBA players.

Read: Vocabulary and Key Concepts

- A ____________________ is a line that models how a response variable y changes as an explanatory variable x changes. Regression lines are expressed in the form $\hat{y} = a + bx$, where $\hat{y}$ is the predicted value of y for a given value of x.
- ____________________ is the use of a regression line for prediction outside the interval of x-values used to obtain the line. The further we extrapolate, the less reliable the predictions become.
- A ________________ is the difference between the actual value of y and the value of y predicted by the regression line. That is, residual = actual y – predicted y = $y - \hat{y}$.
- In the regression equation $\hat{y} = a + bx$, a is the ____________, the predicted value of y when $x = 0$ and b is the ____________, the amount by which the predicted value of y changes when x increases by 1 unit.
- The ______________________________ is the line that makes the sum of the squared residuals as small as possible.
- A ____________________ is a scatterplot that displays the residuals on the vertical axis and the values of the explanatory variable (or the predicted values) on the horizontal axis.
- The ______________________________ r^2 measures the percent reduction in the sum of squared residuals when using the least-squares regression line to make predictions, rather than the mean value of y. In other words, r^2 measures the proportion (or percentage) of the variation in the response variable that is accounted for (or explained) by the explanatory variable in the linear model.
- The ____________________________________ s measures the size of a typical residual. That is, s measures the typical distance between the actual y-values and the predicted y-values.

Watch: Go to bfwpub.com/TPS7eStrive. Select Unit 2 and watch the following Example videos:

Video	Topic	Done
Unit 2C, How much candy can you grab? (p. 183)	*Prediction*	
Unit 2C, Can you grab more than expected? (p. 184)	*Residuals*	
Unit 2C, Grabbing more candy (p. 186)	*Interpreting the slope and y intercept*	
Unit 2C, Foot length and height (p. 189)	*The least-squares regression line*	
Unit 2C, Foot length and height, part 2 (p. 190)	*Using technology to calculate a least-squares regression line*	
Unit 2C, Pricing diamonds (p. 194)	*Residual plots*	
Unit 2C, Grabbing candy, again (p. 199)	*Interpreting r^2 and s*	

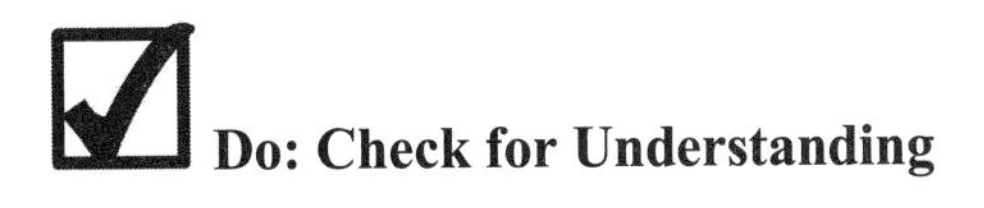

Do: Check for Understanding

Concept 1: Regression Lines - Prediction and Extrapolation

A regression line is used to model linear relationships between two quantitative variables where the explanatory variable is represented by x and the response variable is represented by y. It is important to distinguish y from $\hat{y}$. The values of (x, y) are given in a data table or scatterplot. We use the symbol $\hat{y}$ to represent the value we predict that y will be for a specific value of x. All values of $\hat{y}$ are on the regression line $\hat{y} = a + bx$. Use caution not to extrapolate when making predictions, though, as we do not know if the relationship between the variables extends far beyond the observed values of x!

Concept 2: Calculate and Interpret a Residual

When we use a regression line to make a prediction, the prediction often is not perfect. A "residual" is a fancy word for prediction error. The difference between an actual value of y and the y value that is predicted by the regression line is called a residual. Be sure to note whether the actual value of y is greater than or less than the value of y that is predicted by the regression line!

Concept 3: Interpret the Slope and y intercept

In Algebra class, you likely heard the phrase "rise over run" as a way of describing the slope (b), which is the coefficient of x in the equation $\hat{y} = a + bx$. The phrase "rise over run" applies here too. However, when we describe slope, we say how much the predicted value of y "rises" (or decreases) for each 1 unit increase in "run." For example, if the slope is 0.5, we would say that the predicted value of y increases by 0.5 for each 1 unit increase in x. The y intercept (a) is the constant in the equation $\hat{y} = a + bx$. The y intercept is always on the y-axis, where $x = 0$. This helps us interpret the y intercept, which is the predicted value of y when $x = 0$. But be careful: the y intercept does not always have a practical meaning!

Check for Understanding – Learning Targets 1, 2, and 3

____ *I can make predictions using a regression line, keeping in mind the dangers of extrapolation.*

____ *I can calculate and interpret a residual.*

____ *I can interpret the slope and y intercept of a regression line.*

The regression equation $\hat{y} = 1.5 + 0.25x$ describes the relationship between x = age (in years) and y = height (in feet) for a random sample of children.

1. Use the regression equation to predict the height of a 12-year-old child. Is the prediction reasonable?

2. Now, predict the height of a 40-year-old individual. Is the prediction reasonable?

3. Alyse is a 12-year-old child who is 4.25 feet tall. Calculate and interpret the residual for Alyse.

4. Interpret the slope and the y intercept of the regression line.

Concept 4: Determine the Equation of the Least-Squares Regression Line Given Summary Statistics

The equation of the least-squares regression line can be calculated by hand, if you know the mean and standard deviation of the variables and the correlation r. First, calculate the slope by multiplying the correlation by the standard deviation of the response variable and dividing by the standard deviation of the explanatory variable. Then use the fact that every least-squares regression line passes through the point ($\bar{x}$, $\bar{y}$) to find the y intercept. Use the equation $\bar{y} = a + b\bar{x}$. You know the value of $\bar{y}$, b, and $\bar{x}$. Substitute those values and solve for a.

Concept 5: Determine the Equation of the Least-Squares Regression Line Using Technology

You should only calculate the equation of the least-squares regression line using summary statistics if you have to. If you are given data, use your calculator to find the equation of the least-squares regression line. If you are given computer output, your work is even easier! The value of the y intercept and slope are in the "Coef" column in order: y intercept, then slope.

Check for Understanding – Learning Targets 4 and 5

____ *I can determine the equation of a least-squares regression line using formulas.*

____ *I can determine the equation of a least-squares regression line using technology.*

Stores love to sell gift cards because they know a portion of all gift cards purchased go unused, resulting in more profit for the store. Let's focus on $25 gift cards. A random sample of stores recorded the number of $25 gift cards sold in the 2-month period leading up to Christmas. They also recorded the number of those gift cards that were redeemed within 2 years. The mean number of gift cards sold is 355 with a standard deviation of 48.5. The mean number of gift cards redeemed within 2 years is 294 with a standard deviation of 48. The correlation between number of gift cards sold and number of cards redeemed is 0.92.

1. What is the equation of the least-squares regression line relating number of gift cards sold to number of gift cards redeemed?

2. Interpret the slope in context. For each additional one hundred $25 cards purchased, how many are predicted to be redeemed within the next 2 years?

Does anxiety affect test performance? Researchers measured students' test anxiety and recorded their performance on a standardized test. Note: Higher anxiety scores indicate higher levels of test anxiety. Here are the data:

Anxiety	23	14	14	0	7	20	20	15	21	4
Exam Score	43	59	48	77	50	52	46	60	51	70

3. Use technology to determine the least-squares regression line to predict exam score from anxiety.

(*Need Tech Help?* View the video **Tech Corner 8: Calculating Least-Squares Regression Lines** at bfwpub.com/TPS7eStrive)

Concept 6: Residual Plots

Our eyes aren't always the best judge of linear relationships. While correlation r gives us a better understanding of the strength of the linear relationship, we still need to assess how well the least-squares regression line fits the observed data. If it fits well, it may be a useful prediction tool. If it doesn't fit well, we may want to search for a model that fits it better. One way to assess how well the least-squares regression line fits our data is to make a residual plot. Plotting the residuals gives us more information about the relationship between quantitative variables and helps us assess how well a linear model fits the data. If the residual plot displays a curved pattern, a better (perhaps nonlinear) model might exist.

Concept 7: Interpreting r^2 and s

We can also assess the fit of the least-squares regression line by interpreting the coefficient of determination r^2. r^2 is a measure of how well the regression model explains the response. Specifically, it is interpreted as the fraction of variation in the values of y that is explained by the least-squares regression line of y on x. For example, if $r^2 = 0.82$, we can say that 82% of the variation in y is due to the linear relationship between y and x. 18% is due to factors other than x. The standard deviation of the residuals is symbolized by s. This measures the typical distance between the actual y-values and the predicted y-values.

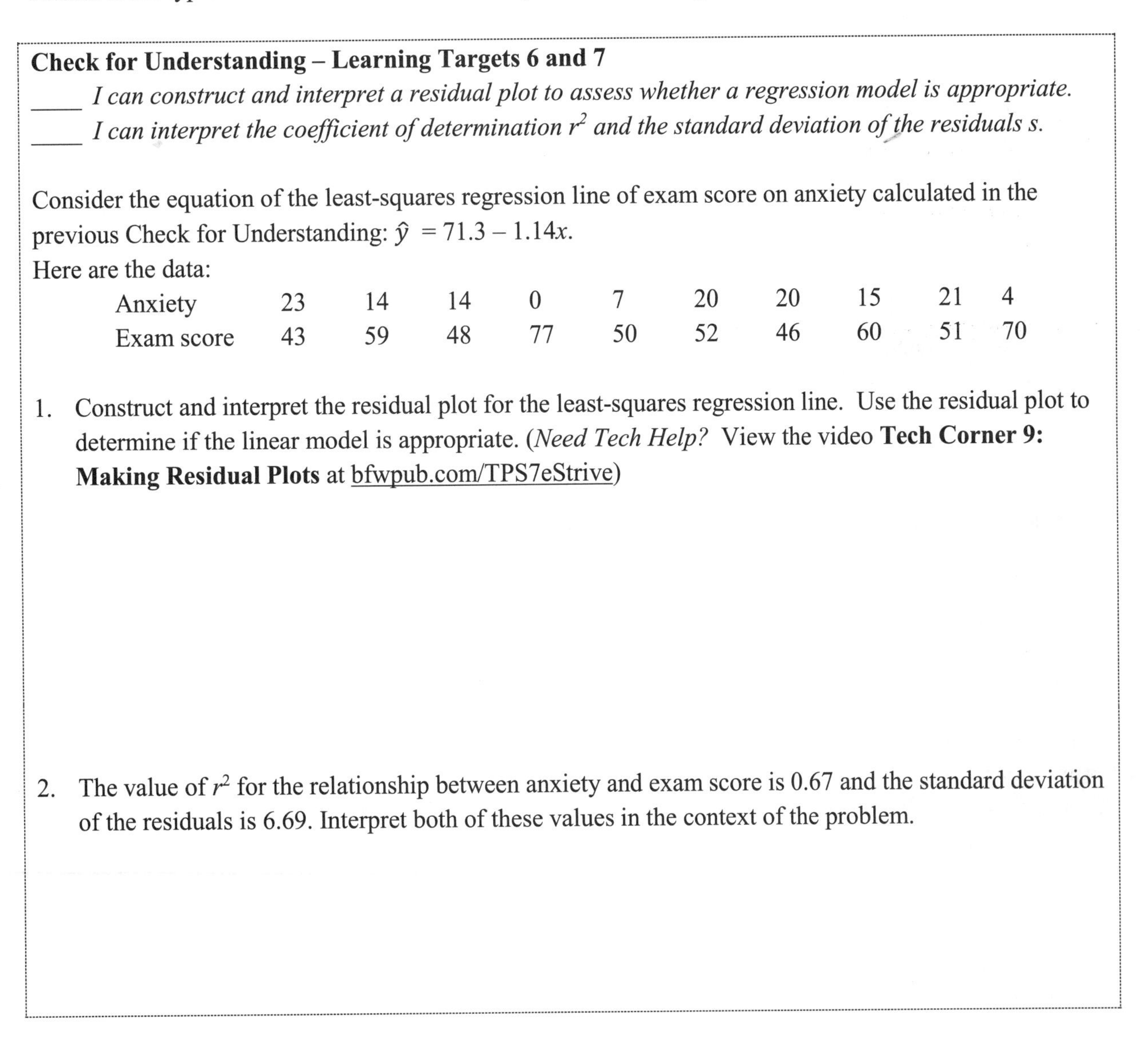

Check for Understanding – Learning Targets 6 and 7

____ *I can construct and interpret a residual plot to assess whether a regression model is appropriate.*

____ *I can interpret the coefficient of determination r^2 and the standard deviation of the residuals s.*

Consider the equation of the least-squares regression line of exam score on anxiety calculated in the previous Check for Understanding: $\hat{y} = 71.3 - 1.14x$.

Here are the data:

Anxiety	23	14	14	0	7	20	20	15	21	4
Exam score	43	59	48	77	50	52	46	60	51	70

1. Construct and interpret the residual plot for the least-squares regression line. Use the residual plot to determine if the linear model is appropriate. (*Need Tech Help?* View the video **Tech Corner 9: Making Residual Plots** at bfwpub.com/TPS7eStrive)

2. The value of r^2 for the relationship between anxiety and exam score is 0.67 and the standard deviation of the residuals is 6.69. Interpret both of these values in the context of the problem.

Section 2D: Analyzing Departures from Linearity

Section Summary

In this section, you will learn how to model nonlinear relationships. Since you already know how to model a linear relationship, you will learn how to transform nonlinear data so that a linear model would be appropriate. That is, you will apply mathematical transformations to one or both variables to "straighten" out the scatterplot. By finding a linear model for the transformed data, you can make predictions involving the original data. The better the fit of your model, the better your prediction!

Learning Targets:

____ I can describe how unusual points influence the least-squares regression line, the correlation, r^2, and the standard deviation of the residuals.

____ I can calculate predicted values from linear models using variables that have been transformed with logarithms or powers.

____ I can determine which of several models does a better job of describing the relationship between two quantitative variables.

Read: Vocabulary and Key Concepts

- Points with ____________________ in regression have much larger or much smaller x-values than the other points in the data set.
- An ________________ in regression is a point that does not follow the pattern of the data and has a large residual.
- An ____________________________ in regression is any point that, if removed, substantially changes the slope, y intercept, correlation, coefficient of determination, or standard deviation of the residuals.
- What function "undoes" $\ln(x)$? In other words, what is the inverse of $y = \ln x$? __________
- What function "undoes" $\log(x)$? In other words, what is the inverse of $y = \log x$? __________
- When choosing between different models to describe a relationship between two quantitative variables, choose the model whose residual plot has the most ___________________________.

  If more than one model produced randomly scattered residual plot, choose the model with the largest

  ____.

Watch: Go to bfwpub.com/TPS7eStrive. Select Unit 2 and watch the following Example videos:

Video	Topic	Done
Unit 2D, Forests and trees (p. 214)	*Influential points*	
Unit 2D, Go fish! (p. 218)	*Transformations to achieve linearity*	
Unit 2D, Stop that car! (p. 220)	*Choosing the most appropriate model*	

Do: Check for Understanding

Concept 1: Influential Points

Influential points have a substantial impact on the slope, the y intercept, correlation, r^2, or the standard deviation of the residuals. When an unusual observation is extreme in the x-direction, the point may be a high leverage point. If a point does not follow the pattern formed by the other points in the data set and has a large residual, it is a potential outlier.

Check for Understanding – Learning Target 1

_____ *I can describe how unusual points influence the least-squares regression line, the correlation, r^2, and the standard deviation of the residuals.*

1. The Rosenberg self-esteem scale (RSES) is commonly used in psychological research to measure a person's self-esteem. The RSES was administered to a random sample of 15 business professionals. Their height was also measured. The scatterplot shows the data.

Describe how the point at (64, 38) affects:

- The equation of the least-squares regression line.
- r
- r^2
- s

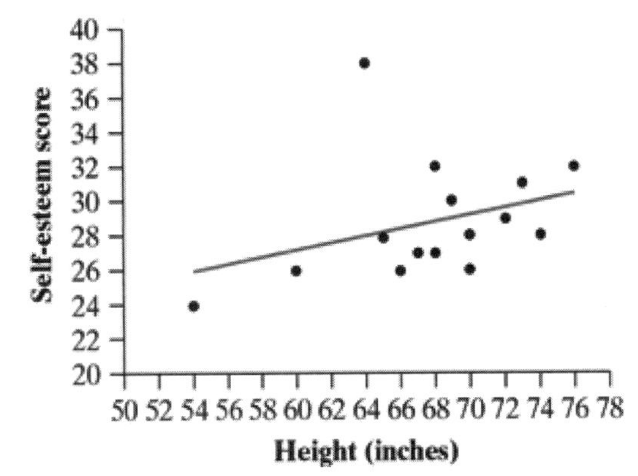

Is this point an outlier, a high-leverage point, or both? Explain your reasoning.

2. The scatterplot shows the relationship between height and weight for a random sample of NFL quarterbacks.

Describe how the point at (77, 210) affects:

- The equation of the least-squares regression line.
- r
- r^2
- s

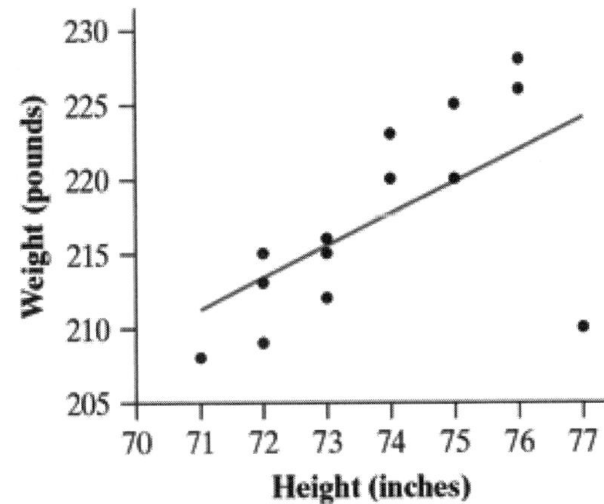

Concept 2: Making Predictions using Linear Models based on Transformed Data

To write the least-squares regression line for transformed data, the trickiest step is to correctly identify the (potentially transformed) explanatory and response variables. This information is provided in the problem statement, but if you are not certain here are two other tricks. (1) Are you given a scatterplot? If so, then simply look at the horizontal and vertical axis labels. Use those in the regression equation in place of $\hat{y}$ and x. (2) Are you given computer output? The transformation of the explanatory variable is given below "Constant" in the first column.

Check for Understanding – Learning Target 2

____ *I can calculate predicted values from linear models using variables that have been transformed with logarithms or powers.*

The following data represent the lengths (mm) and diameters (mm) of the humerus bones of the *Moleskius Primateum* species of monkeys once thought to inhabit Northern Minnesota.

Diameter	17.6	26	31.9	38.9	45.8	51.2	58.1	64.7	66.7	80.8	82.9
Length	159.9	206.9	236.8	269.9	300.6	323.6	351.7	377.6	384.1	437.2	444.7

Here is a residual plot for the untransformed data:

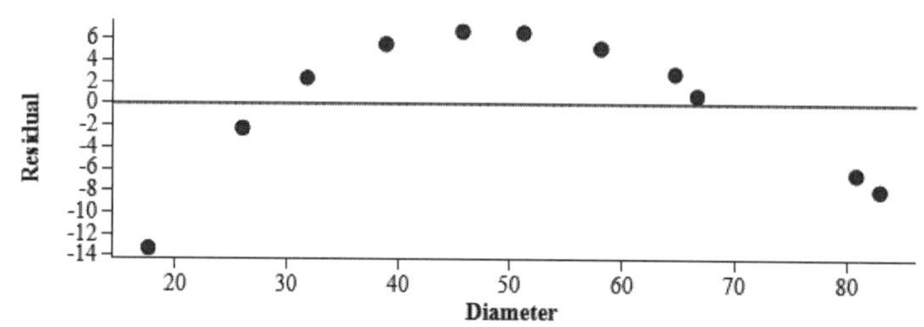

Studies suggest the diameter and length are related by a model of the form *length* = *a*(*diameter*)$^{0.7}$. Here is a scatterplot showing the relationship between diameter$^{0.7}$ and length, along with computer output from a linear regression analysis of the transformed data.

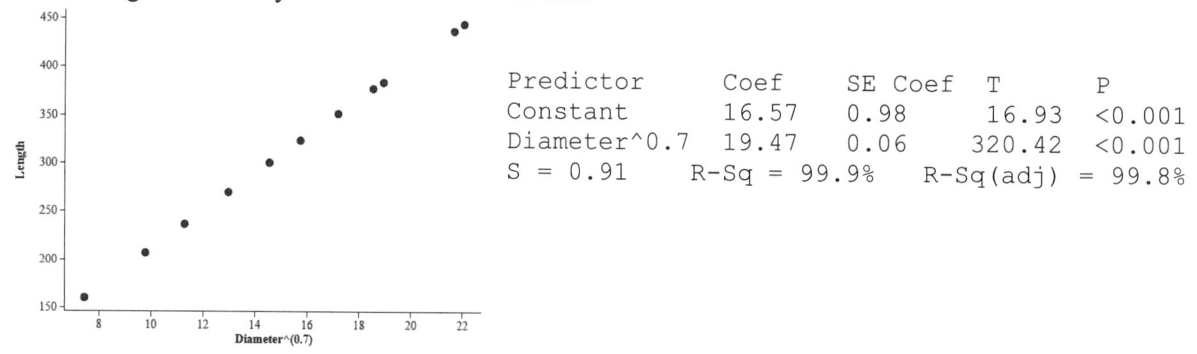


```
Predictor        Coef     SE Coef    T          P
Constant         16.57    0.98       16.93    <0.001
Diameter^0.7     19.47    0.06      320.42    <0.001
S = 0.91     R-Sq = 99.9%     R-Sq(adj) = 99.8%
```

You discover a portion of a *Moleskius Primateum* humerus bone with a diameter of 47mm. Use your model to predict how long the entire bone was.

Concept 3: Choosing the most appropriate model

When transforming nonlinear data, sometimes more than one type of transformation can produce data that appears roughly linear. How can we know which model is best? Here are two tips:

1) Choose the model for which the residual plot has the most random scatter.
2) If it is unclear which residual plot has the most random scatter, choose the model with the greatest value of r^2.

It is not recommended to compare the standard deviation of the residuals, because the standard deviation of the residual depends upon the units of the data, which are unlikely to be the same across various models.

Check for Understanding – Learning Target 3

____ *I can determine which of several models does a better job of describing the relationship between two quantitative variables.*

The following data describe the number of police officers (thousands) and the violent crime rate (per 100,000 population) in a sample of states.

Police	86.2	9.2	45	39.9	6	11.8	2.9	14.6	30.5	12.3	46.2	15.2	10.9
Crime	1090	559	1184	1039	303	951	132	763	635	726	840	373	523

Here is linear regression output from three different models, along with a residual plot for each model. Model 1 is based on the original data. Models 2 and 3 involve transformations of the original data.

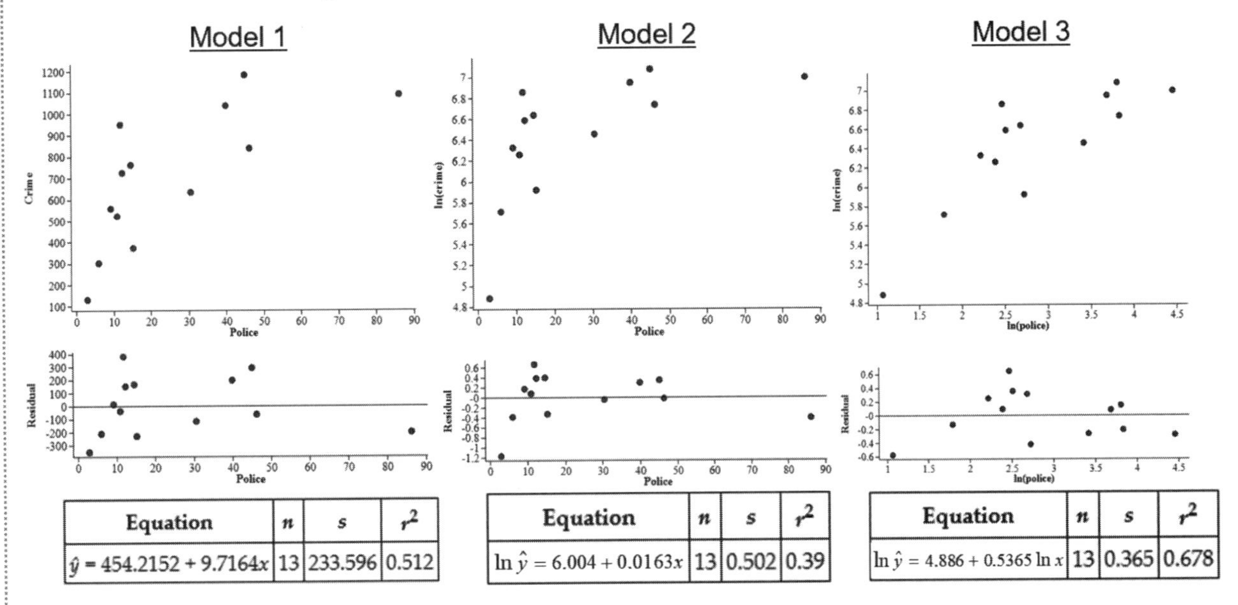

Equation	n	s	r^2
$\hat{y} = 454.2152 + 9.7164x$	13	233.596	0.512

Equation	n	s	r^2
$\ln \hat{y} = 6.004 + 0.0163x$	13	0.502	0.39

Equation	n	s	r^2
$\ln \hat{y} = 4.886 + 0.5365 \ln x$	13	0.365	0.678

1. Which model does the best job of summarizing the relationship between number of police and violent crime rate?

2. Use your model to predict the violent crime rate for a state with 25,400 police officers.

Unit 2 Summary: Exploring Two-Variable Quantitative Data

In this unit, we expanded our toolbox for working with categorical and quantitative data. We learned how to display and describe the relationship between two categorical variables (segmented bar chart, percentages). We also learned how to analyze and describe the relationship between two quantitative variables (scatterplots, linear regression). Using scatterplots, we can display the relationship and describe the direction, strength, and form of the overall pattern between two variables. Correlation provides a numerical summary of the strength of the linear relationship between the variables and the equation of the least-squares regression line provides a model that can be used to make predictions. Residual plots, the standard deviation of the residuals, and the coefficient of determination help us assess the fit of the least-squares regression line and may suggest whether or not a linear model is appropriate. Finally, we learned that outliers and influential points can affect the least-squares regression line, r, r^2, and s. And always remember: Association does not imply causation!

When the relationship between two variables is curved, you can use transformations to "straighten" the scatterplot. You can find the least-squares regression line for the transformed data and make predictions for the original relationship. The most common methods for transforming data involve taking powers, roots, or logarithms of one or both variables. To determine which transformation does the best job of "straightening" the relationship, examine residual plots and compare values of r^2.

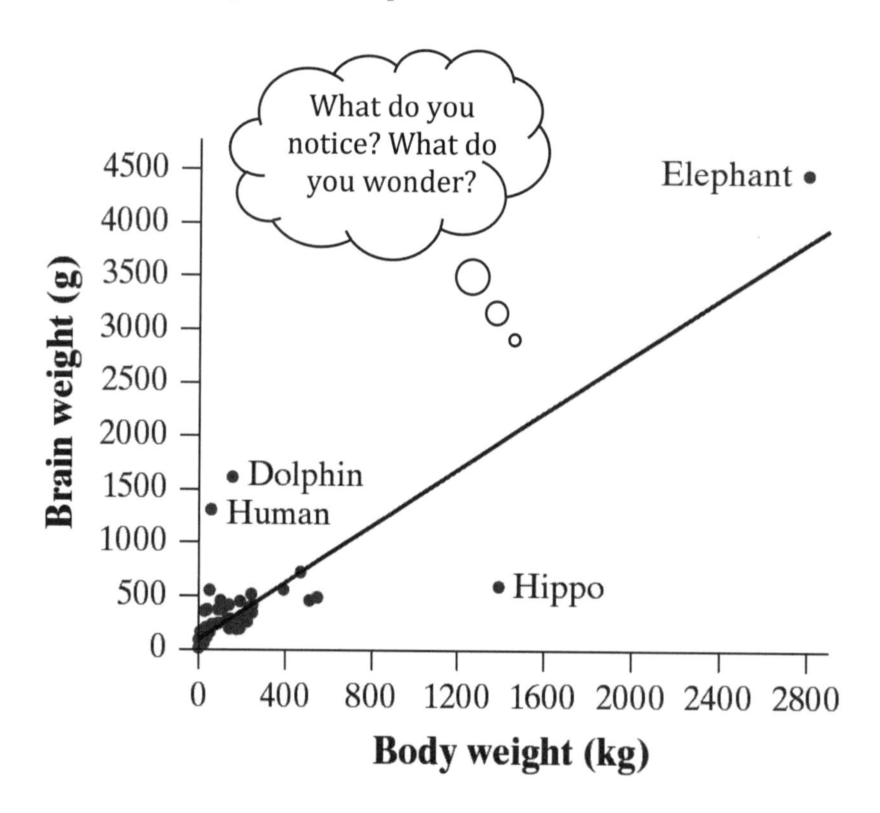

The scatterplot shows the average brain weight (in grams) versus average body weight (in kilograms) for 96 species of mammals.

How well do you understand each of the learning targets?

Learning Target	Got It!	Almost There	Needs Work
I can identify the explanatory and response variables in a given setting.			
I can calculate statistics for two categorical variables.			
I can display the relationship between two categorical variables.			
I can describe the relationship between two categorical variables.			
I can make a scatterplot to display the relationship between two quantitative variables.			
I can describe the direction, form, and strength of a relationship displayed in a scatterplot and identify unusual features.			
I can interpret the correlation for a linear relationship between two quantitative variables.			
I can distinguish correlation from causation.			
I can calculate the correlation for a linear relationship between two quantitative variables.			
I can make predictions using a regression line, keeping in mind the dangers of extrapolation.			
I can calculate and interpret a residual.			
I can interpret the slope and y intercept of a regression line.			
I can determine the equation of a least-squares regression line using formulas.			
I can determine the equation of a least-squares regression line using technology.			
I can construct and interpret residual plots to assess whether a regression model is appropriate.			
I can interpret the coefficient of determination r^2 and the standard deviation of the residuals s.			
I can describe how unusual points influence the least-squares regression line, the correlation, r^2, and the standard deviation of the residuals.			
I can calculate predicted values from linear models using variables that have been transformed with logarithms or powers.			
I can determine which of several models does a better job of describing the relationship between two quantitative variables.			

Unit 2 Multiple Choice Practice

Directions. Identify the choice that best completes the statement or answers the question. Check your answers and note your performance when you are finished.

1. A study is conducted to determine if one can predict the academic performance of a first-year college student based on their high school grade point average. The explanatory variable in this study is
   (A) academic performance of the first-year student.
   (B) high school grade point average.
   (C) the experimenter.
   (D) number of credits the student is taking.
   (E) the college.

2. Consider the following scatterplot which describes the relationship between stopping distance (in feet) and air temperature (in degrees Centigrade) for a certain 2,000-pound car travelling 40 mph.

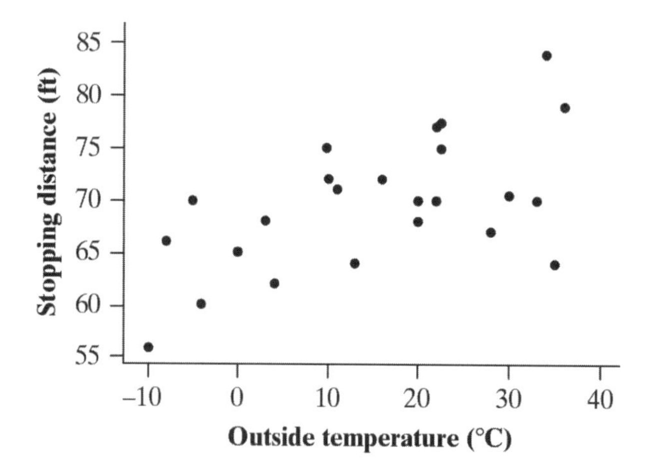

   Do these data provide strong evidence that warmer temperatures actually *cause* a greater stopping distance?
   (A) Yes. The strong straight-line association in the plot shows that temperature has a strong effect on stopping distance.
   (B) No. $r \neq +1$
   (C) No. We can't be sure the temperature is responsible for the difference in stopping distances.
   (D) No. The plot shows that differences among stopping distances are not large enough to be important.
   (E) No. The plot shows that stopping distances go down as temperature increases

3. If another data point were added with an air temperature of 0° C and a stopping distance of 80 feet, the correlation would
   (A) decrease, since this new point is an outlier that does not follow the pattern in the data.
   (B) increase, since this new point is an outlier that does not follow the pattern in the data.
   (C) stay nearly the same since correlation is resistant to outliers.
   (D) increase, since there would be more data points.
   (E) Whether this data point causes an increase or decrease cannot be determined without recalculating the correlation.

4. Consider the following scatterplot of amounts of CO (carbon monoxide) and NOX (nitrogen oxide) in grams per mile driven in the exhausts of cars. The least-squares regression line has been drawn in the plot.

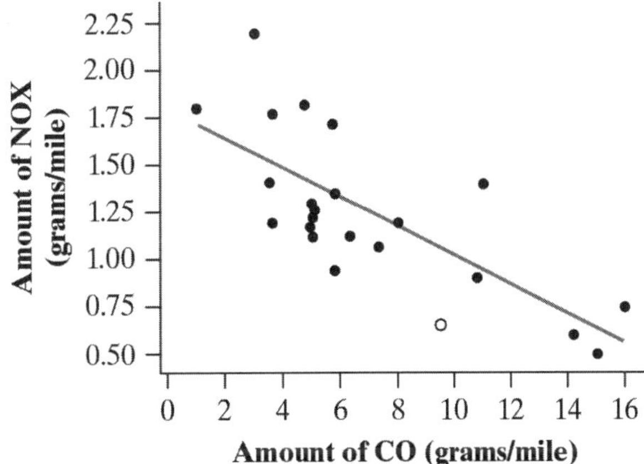

Based on the scatterplot, the least-squares line would predict that a car that emits 2 grams of CO per mile driven would emit approximately how many grams of NOX per mile driven?

(A) 4.0
(B) 1.25
(C) 2.0
(D) 1.7
(E) 0.7

5. In the scatterplot shown in problem 4, the point indicated by the open circle

(A) has a negative value for the residual.
(B) has a positive value for the residual.
(C) has a zero value for the residual.
(D) has a zero value for the correlation.
(E) is an outlier.

6. A study of the effects of television measured how many hours of television each of 125 grade school children watched per week during a school year and their reading scores. The study found that children who watch more television tend to have lower reading scores than children who watch fewer hours of television. The study report says that "Hours of television watched explained 25% of the observed variation in the reading scores of the 125 subjects." The correlation between hours of TV and reading score must be

(A) $r = 0.25$.
(B) $r = -0.25$.
(C) $r = -0.5$.
(D) $r = 0.5$.
(E) Can't tell from the information given.

7. A fishery's biologist studying whitefish in a Canadian lake collected data on the length (in centimeters) and egg production for 25 female fish. A scatterplot of the results and computer regression analysis of egg production versus fish length are given below. Note that number of eggs is given in thousands (i.e., "40" means 40,000 eggs).

Predictor	Coef	SE Coef	T	P
Constant	-142.74	25.55	-5.59	0.000
Fish length	39.250	5.392	7.28	0.000

S = 6.75133   R-Sq = 69.7%   R-Sq(adj) = 68.4%

Which of the following statements is a correct interpretation of the slope of the regression line?

(A) For each 1-cm increase in the fish length, the predicted number of eggs increases by 39.25 cm.
(B) For each 1-cm increase in the fish length, the predicted number of eggs decreases by 142.74 cm.
(C) For each 1-unit increase in the number of eggs, the predicted fish length increases by 39.25 cm.
(D) For each 1-unit increase in the number of eggs, the predicted fish length decreases by 142.74 cm.
(E) For each 1-cm increase in the fish length, the predicted number of eggs increases by 39,250.

8. Use of the internet has increased steadily since 1990. A scatterplot of this growth shows a strongly non-linear pattern. However, a scatterplot of ln(internet users) *versus* year is much closer to linear. Below is a computer regression analysis of the transformed data (note that natural logarithms are used).

Predictor	Coef	SE Coef	T	P
Constant	-951.10	43.45	-21.89	0.000
Year	0.4785	0.02176	21.99	0.000

S = 0.2516   R-Sq = 98.2%   R-Sq(adj) = 98.0%

Which of the following best describes the model that is given by this computer printout?

(A) $\widehat{\text{internet users}} = -951.10 + 0.4785(\text{year})$
(B) $\widehat{\text{internet users}} = e^{-951.10}(\text{year})^{0.4785}$
(C) $\widehat{\text{internet users}} = 10^{-95110}(\text{year})^{0.4785}$
(D) $\ln(\widehat{\text{internet users}}) = -951.10 + 0.4785(\text{year})$
(E) $\ln(\widehat{\text{internet users}}) = -951.10 + 0.4785 \ln(\text{year})$

9. Suppose the relationship between a response variable y and an explanatory variable x is modeled well by the equation $y = 3.6(0.32)^x$. Which of the following plots is most likely to be roughly linear?

(A) A plot of y against x
(B) A plot of y against $\log x$
(C) A plot of $\log y$ against x
(D) A plot of 10^y against x
(E) A plot of $\log y$ against $\log x$

10. Like most animals, small marine crustaceans are not able to digest all the food they eat. Moreover, the percentage of food eaten that is assimilated (that is, digested) decreases as the amount of food eaten increases. A residual plot for the regression of Assimilation rate (as a percentage of food intake) on Food Intake (μg/day) is shown here.

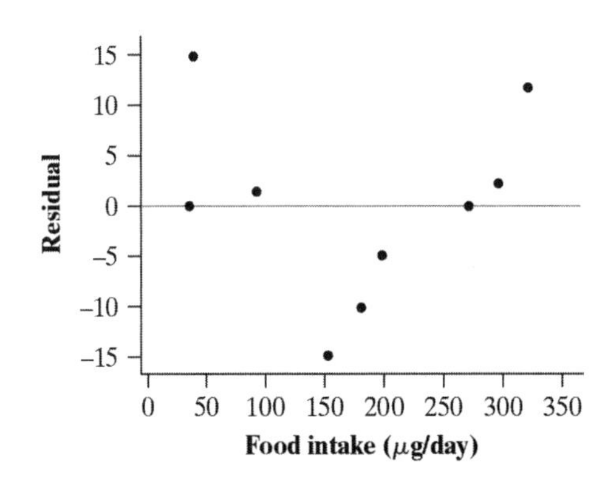

A scatterplot of ln(Assimilation) versus ln(Food Intake) is strongly linear, suggesting that a linear regression of these transformed variables may be more appropriate. Below is a computer regression analysis of the transformed data (note that natural logarithms are used).

```
Predictor            Coef     SE Coef        T          P
Constant           6.3324      0.5218    12.14      0.000
ln Food Intake    -0.6513      0.1047    -6.22      0.000

S = 0.247460       R-Sq = 84.7%       R-Sq(adj) = 82.5%
```

When food intake is 250 g/day, what is the predicted assimilation rate from this model?

(A) 2.7%
(B) 15.4%
(C) 27.4%
(D) 34.3%
(E) 54.4%

Check your answers below. If you got a question wrong, check to see if you made a simple mistake or if you need to study that concept more. After you check your work, identify the concepts you feel very confident about and note what you will do to learn the concepts in need of more study.

#	Answer	Concept	Right	Wrong	Simple Mistake?	Need to Study More
1	B	Explanatory vs. Response				
2	C	Association vs. Causation				
3	A	Correlation				
4	D	Predicting with the LSRL				
5	A	Residuals				
6	C	Coefficient of Determination				
7	E	Slope of the LSRL				
8	D	Log *y* Transformation				
9	C	Exponential Transformation				
10	B	Prediction from ln Transformation				

FRAPPY! Free Response AP® Problem, Yay!

The following problem is modeled after actual Advanced Placement Statistics free response questions. Your task is to generate a complete, concise response in 15 minutes. After you generate your response, read over the solution and scoring guide. Score your response and note what, if anything, you would do differently to increase your own score.

A recent study was interested in determining the optimal location for fire stations in a suburban city. Ideally, fire stations should be placed so the distance between the station and residences is minimized. One component of the study examined the linear relationship between the amount of fire damage y (in thousands of dollars) and the distance between the fire station and the residence x (in miles). The results of the regression analysis are below.

Predictor	Coef	SE Coef	T	P
Constant	10.28	1.42	7.237	0.000
Distance	4.92	0.39	12.525	0.000

S = 2.232   R-Sq = 0.9235   R-Sq(adj) = 0.9176

(a) Write the equation of the least-squares regression line. Define any variables used. Interpret the slope of the equation in context.

(b) A home located 3 miles from the fire station received $22,300 in damage. Use your equation in part (a) to calculate and interpret the residual for this observation.

(c) Identify and interpret the correlation coefficient.

FRAPPY! Scoring Rubric

Use the following rubric to score your response. Each part receives a score of "Essentially Correct," "Partially Correct," or "Incorrect." When you have scored your response, reflect on your understanding of the concepts addressed in this problem. If necessary, note what you would do differently on future questions like this to increase your score.

Intent of the Question

The goal of this question is to determine your ability to interpret computer regression output and explain key concepts of linear regression.

Solution

(a) $\text{fire }\widehat{\text{damage}} = 10.28 + 4.92(\text{distance})$ OR $\hat{y} = 10.28 + 4.92x$ where x = distance and y = damage.

For each additional mile between the fire station and residence, the predicted additional amount in damage increases by about $4,920.

(b) $\text{fire }\widehat{\text{damage}} = 4.92(3) + 10.28 = 25.04$
residual = 22.3 – 25.04 = –2.74.
The actual amount of damage is $2,740 less than the amount of damage predicted by the model.

(c) Because $r^2 = 0.9325$, $r = 0.96$. The linear relationship between a residence's damage from a fire and its distance from a fire station is strong because r is close to 1 and positive because r takes the same sign as the slope of the least-squares regression line.

Scoring

Parts (a), (b), and (c) are scored as essentially correct (E), partially correct (P), or incorrect (I).

Part (a) is essentially correct if the response (1) correctly identifies the least-squares regression equation in context or with variables defined and (2) correctly interprets the slope.
Part (a) is partially correct if the response fails to define the variables in context or reverses the coefficients OR if the slope is not correctly defined in context (ex. predicts 4.92 dollars instead of $4920). Incorrect, otherwise.

Part (b) is essentially correct if (1) the correct residual is calculated and (2) the interpretation is correct.
Part (b) is partially correct if only one of the above elements is correct. Incorrect, otherwise.

Part (c) is essentially correct if the correlation coefficient is correctly identified and interpreted correctly in context. (The linear relationship between damage and distance is strong and positive).
Part (c) is partially correct if one of the elements (linear relationship, strong, positive) is missing OR if r is incorrectly calculated. Incorrect, otherwise.

4 Complete Response
All three parts essentially correct

3 Substantial Response
Two parts essentially correct and one part partially correct

2 Developing Response
Two parts essentially correct and no parts partially correct
One part essentially correct and one or two parts partially correct
Three parts partially correct

1 Minimal Response
One part essentially correct and no parts partially correct
No parts essentially correct and two parts partially correct

My Score:
What I did well:
What I could improve:
What I should remember if I see a problem like this on the AP Exam:

Unit 2 Formula Sheet

Create a one-page summary of important concepts and formulas found in this unit. This will serve as a valuable resource as you prepare for the AP Exam.

Unit 2 Crossword Puzzle

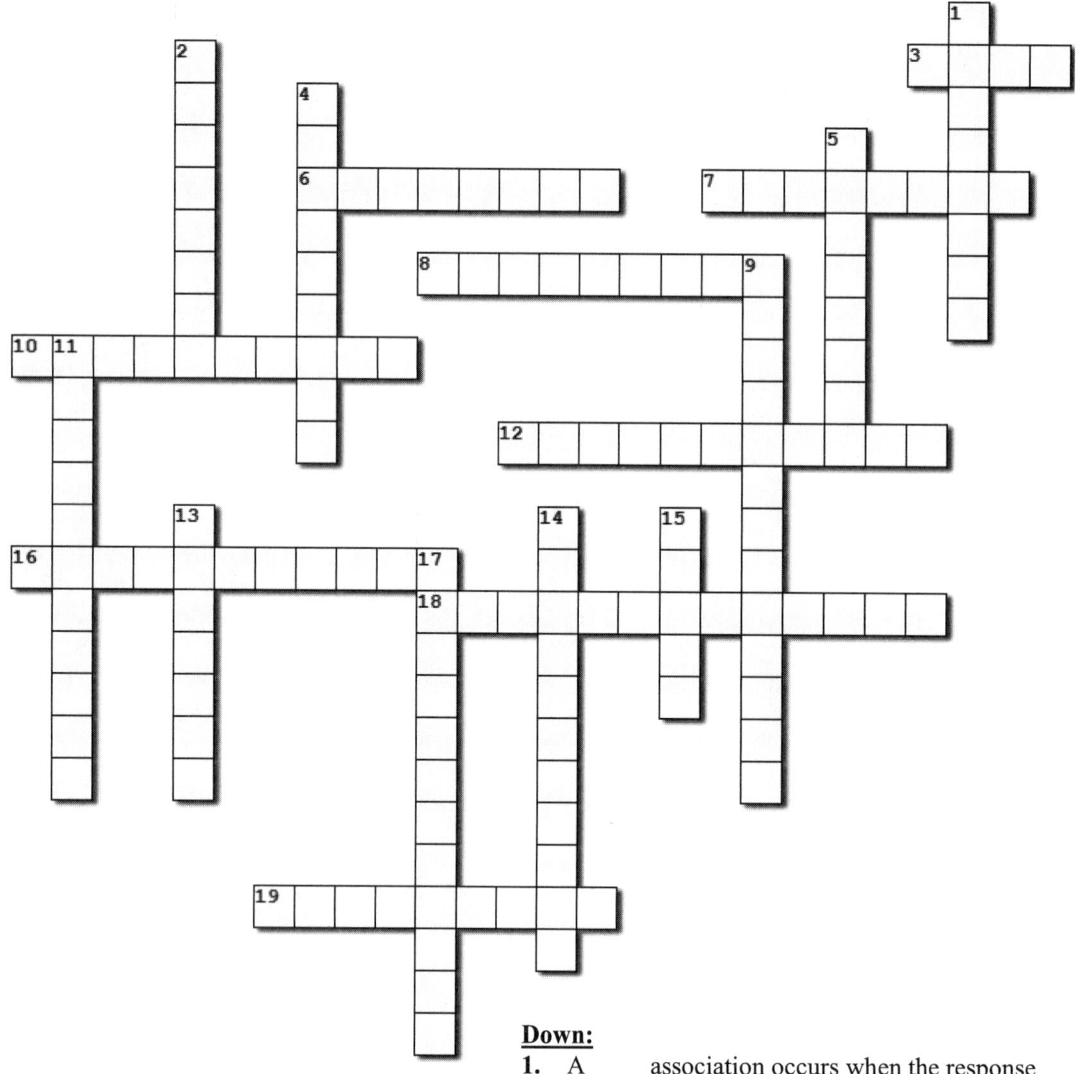

Across:

3. The ____ of a scatterplot is usually linear or nonlinear.
6. This variable measures the outcome of a study.
7. The ____ of a relationship is determined by how closely the points follow a clear form.
8. $\hat{y}$ is the ___ value of the y variable for a given value of x.
10. A line that describes the relationship between two quantitative variables.
12. A graphical display of the relationship between two quantitative variables.
16. Individual points that substantially change the correlation, slope, or y intercept of a least-squares regression line.
18. The use of a regression line to make a prediction far outside the observed x values.
19. Association does not imply ____!

Down:

1. A ____ association occurs when the response variable tends to increase as the explanatory variable increases.
2. A ____ association occurs when the response variable tends to decrease as the explanatory variable increases.
4. The ____ of a scatterplot indicates a positive or negative association between the variables.
5. The difference between an observed value of the response variable and the value predicted by the regression line.
9. The coefficient of ____ describes the fraction of the variability in the values of y that is explained by the least-squares regression line.
11. This variable may help explain or predict changes in another variable.
13. An individual value that falls outside the overall pattern of the relationship.
14. A value that measures the strength of the linear relationship between two quantitative variables.
15. The amount by which y is predicted to change when x increases by 1 unit.

Unit 3: Collecting Data

"Not everything that can be counted counts;
and not everything that counts can be counted" ~George Gallup

UNIT 3 OVERVIEW

The first three units introduced some of the basics of exploratory data analysis. In this unit, we'll learn about the second major topic in statistics – planning and conducting a study. Because it is difficult to perform a descriptive statistical analysis without data, we need to learn appropriate ways to produce data. We will start by learning the difference between a population and a sample. Then, we will study sampling techniques and learn how to identify potential sources of bias. Our goal is to collect data that is representative of the population we wish to study. Therefore, it is important that our data collection techniques do not systematically over- or under-represent any segment of the population. We will focus on two important types of studies: observational studies and experiments. We will learn a number of different ways to design experiments so we can draw conclusions of cause and effect. Finally, we will conclude this unit by reviewing some cautions about using studies wisely. This unit contains a lot of vocabulary and concepts. Be sure to study them, as proper use of terms is important for strong statistical communication!

Sections in Unit 3
Section 3A: Introduction to Data Collection
Section 3B: Sampling and Surveys
Section 3C: Experiments

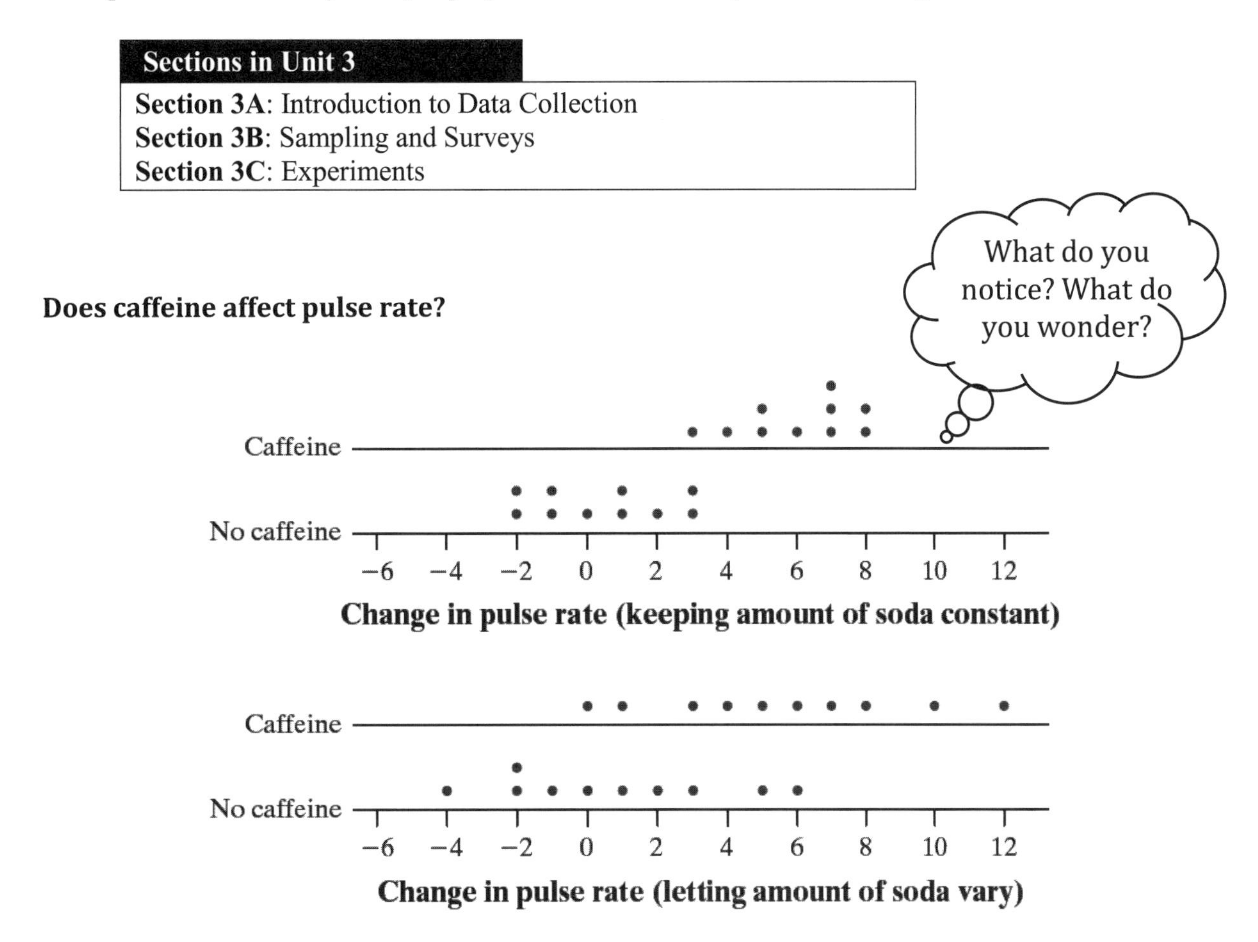

PRACTICE FOR MASTERY

Use the following *suggested* guide to the pages and exercises in your text to practice for mastery!
Note: your teacher may assign different problems. Be sure to follow their instructions!

Day	Topics	Read	Do
1	• Populations and Samples • Observational Studies and Experiments • Inference from Observational Studies	p. 240-245	**3A**: 1, 3, 5, 7, 9, 11, 14, 15
2	• Simple Random Samples	p. 249-252	**3B**: 1, 3, 39, 40
3	• Other Sampling Methods	p. 253-259	**3B**: 5, 7, 9, 11, 13
4	• Sampling Poorly: Convenience and Voluntary Response Samples • Other Sources of Bias in Surveys	p. 259-263	**3B**: 17, 19, 23, 25, 27, 31, 33
5	AP Classroom Topic Questions (Topics 3.1-3.4)		**3B**: Ex. 15, 21, 29, 35 – 38
6	• Confounding • The Language of Experiments	p. 271-274	**3C**: 1, 3, 5, 7, 53, 56, 59
7	• Comparison and Control Groups • Blinding and the Placebo Effect • Random Assignment • Controlling Other Variables	p. 274-282	**3C**: 9, 13, 15, 19, 23, 25, 27
8	• Randomized Block Designs • Matched Pairs Designs	p. 282-287	**3C**: 31, 33, 35, 37, 54, 55, 57
9	• Inference for Experiments • Putting It All Together: The Scope of Inference	p. 288-292	**3C**: 39, 41, 43, 45, 58
10	AP Classroom Topic Questions (Topics 3.5-3.7)		**3C**: 11, 17, 21, 29, 60, 61
11	Unit 3 AP® Statistics Practice Test (p. 309)		Unit 3 Review Exercises (p. 307)
12	Unit 3 Test: Celebration of learning!		Cumulative AP® Practice Test 1 (p. 313)

Section 3A: Introduction to Data Collection

Section Summary

In this section, you will learn four important terms which are foundational to the study of statistics: population, sample, observational study, and experiment. Suppose we want to know the percentage of registered voters who vote in presidential elections. There are simply too many registered voters to survey all of them. Therefore, we select a random sample of registered voters and survey the voters in the sample. In doing so, we hope that the sample percentage accurately reflects the population percentage. However, because we are not surveying all registered voters, there is a little mystery about how accurately the sample percentage estimates the population percentage.

You will also learn about the difference between observational studies and experiments, and the conclusions that can be drawn from these two types of studies. This distinction is just as important as the distinction between categorical and quantitative data, so work hard to become good at recognizing what type of study is being conducted and identifying the type of data that is collected.

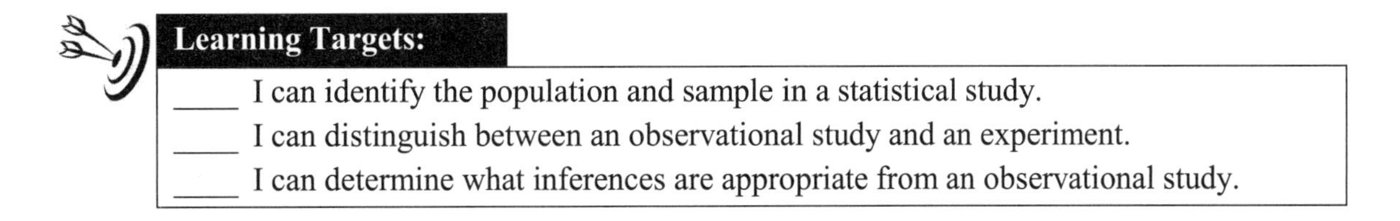

Learning Targets:

____ I can identify the population and sample in a statistical study.
____ I can distinguish between an observational study and an experiment.
____ I can determine what inferences are appropriate from an observational study.

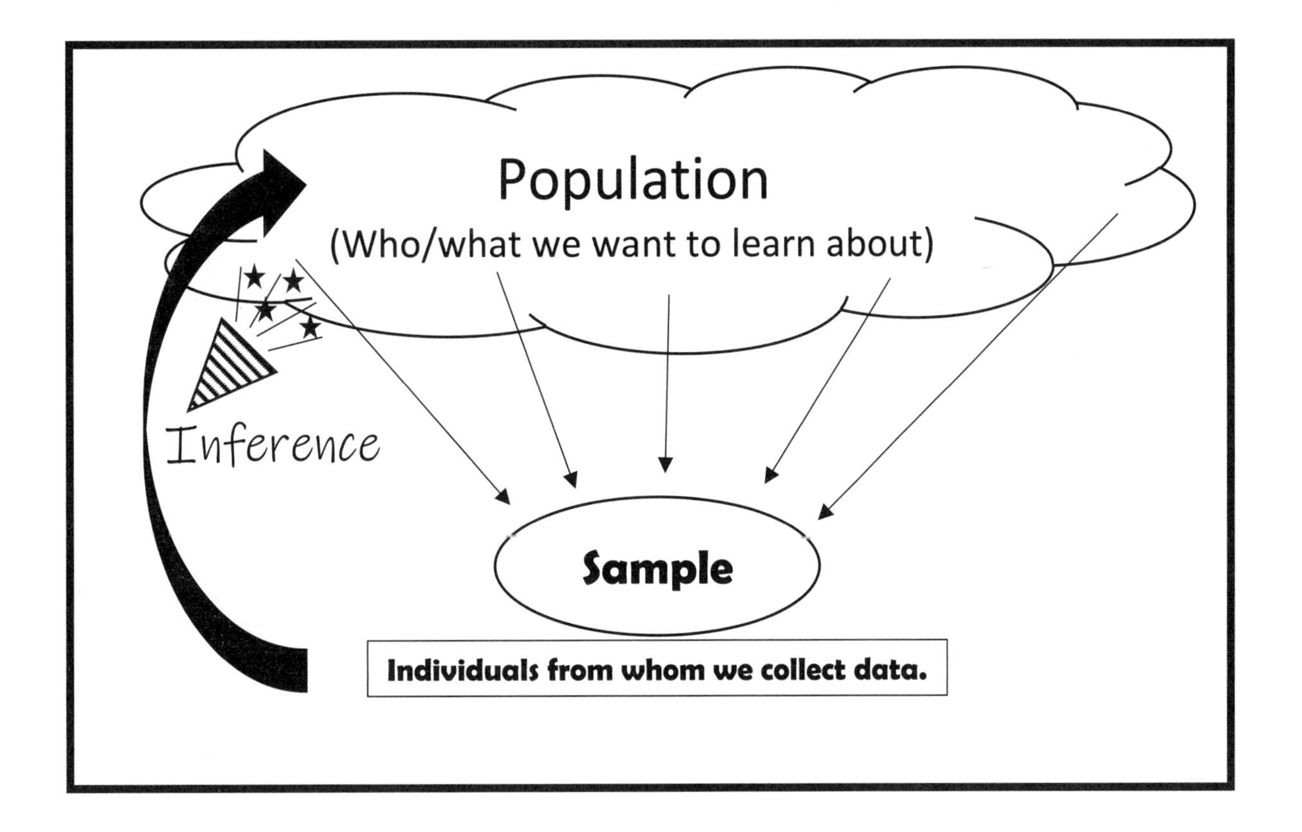

Read: Vocabulary and Key Concepts

- The ___________________ in a statistical study is the entire group of individuals we want information about.
- A _________________ collects data from every individual in the population.
- A ___________________ is a subset of individuals in the population from which we collect data.
- A ___________________________ is a study that collects data from a sample to learn about the population from which the sample was selected.
- The first step in planning a sample survey is to decide what __________________ we want to describe. The second step is to decide what we want to ______________________.
- ______________________________ involves using a chance process to determine which members of a population are chosen for the sample.
- An _____________________________________ observes individuals and measures variables of interest but does not attempt to influence the responses.
- Observational studies that examine existing data for a sample of individuals are called ____________________.
- Observational studies that track individuals into the future are called _____________________.
- An _____________________ deliberately imposes treatments on experimental units to measure their responses.
- The goal of an observational study is to ______________ some group or situation, or to examine ___________________ between variables.
- The purpose of an experiment is to determine whether the treatment ______________ a change in the response.
- When the sample is randomly selected from a larger population, we can ________________ the results of the study to the population. If the study does not use a random sample, we can only apply the results to individuals like _________________________.
- When the goal is to understand cause and effect, __________________ are the only source of fully convincing data.

Watch: Go to bfwpub.com/TPS7eStrive. Select Unit 3 and watch the following Example videos:

Video	Topic	Done
Unit 3A, Sampling monitors and voters (p. 241)	*Populations and samples*	
Unit 3A, Family dinners and background music (p. 244)	*Observational studies and experiments*	
Unit 3A, A little something sweet (p. 245)	*Inference from observational studies*	

Do: Check for Understanding

Concept 1: Population and Sample

The purpose of sampling is to provide information about a population without actually gathering the information from every single member of the population. A sample is a part of the population from which we collect data. This data is used to make an inference about the population.

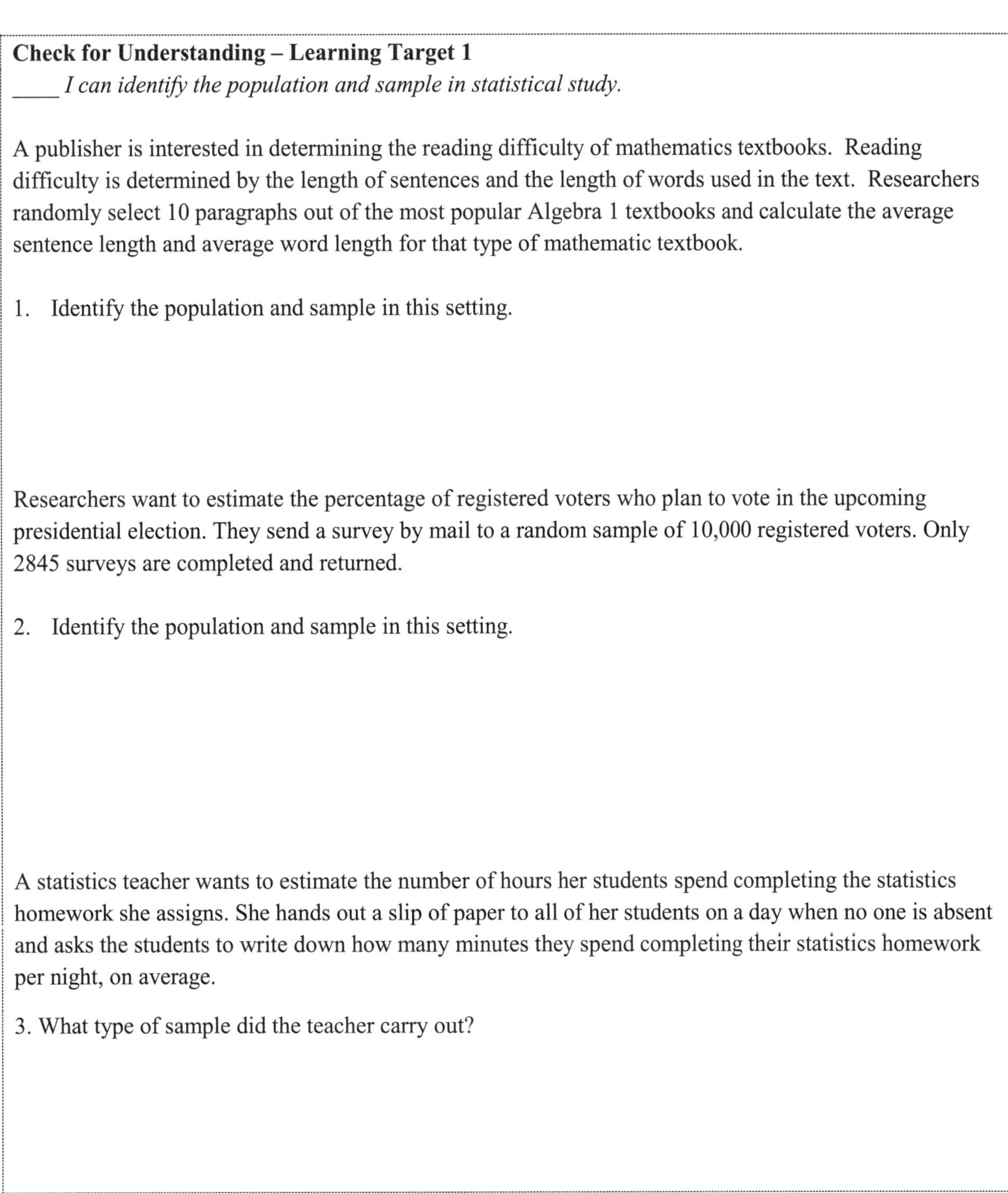

Check for Understanding – Learning Target 1

____ *I can identify the population and sample in statistical study.*

A publisher is interested in determining the reading difficulty of mathematics textbooks. Reading difficulty is determined by the length of sentences and the length of words used in the text. Researchers randomly select 10 paragraphs out of the most popular Algebra 1 textbooks and calculate the average sentence length and average word length for that type of mathematic textbook.

1. Identify the population and sample in this setting.

Researchers want to estimate the percentage of registered voters who plan to vote in the upcoming presidential election. They send a survey by mail to a random sample of 10,000 registered voters. Only 2845 surveys are completed and returned.

2. Identify the population and sample in this setting.

A statistics teacher wants to estimate the number of hours her students spend completing the statistics homework she assigns. She hands out a slip of paper to all of her students on a day when no one is absent and asks the students to write down how many minutes they spend completing their statistics homework per night, on average.

3. What type of sample did the teacher carry out?

Concept 2: Observational Studies vs. Experiments

Sample surveys are examples of observational studies. Their goal is to describe the population and examine relationships between variables. Often, however, we wish to determine whether or not a cause-and-effect relationship exists between an explanatory and a response variable. Observational studies cannot be used to establish this relationship because of confounding between the explanatory variable and one or more other variables.

Concept 3: What Inferences are Appropriate from an Observational Study?

When the sample is randomly selected from the population, we can generalize the results of the study to the population from which the sample was selected. However, when the goal is to understand cause and effect, well-designed experiments are the only source of fully convincing data.

Check for Understanding – Learning Targets 2 and 3

____ *I can distinguish between an observational study and an experiment.*

____ *I can determine what inferences are appropriate from an observational study.*

Does shoe size affect spelling ability? A recent study was conducted in a suburban school district to answer this question. Thirty students from grades 1 through 8 were randomly selected. Each student was administered a spelling test and reported their shoe size. Spelling test scores were plotted against shoe size and a strong, positive relationship was observed.

1. Was this an observational study or an experiment? Explain your answer.

2. What are the explanatory and response variables?

3. What is the largest population to which we can generalize the results of this study?

4. Based on this study, should we conclude that we could increase spelling test scores by stretching the students' feet? Explain your answer.

Section 3B: Sampling and Surveys

Section Summary

Often in statistics, our goal is to draw a conclusion about a population based on information gathered from a sample. In order to make a valid inference about the population, we must feel confident that the sample is representative of the population. There are a number of different ways to select samples from a population—some better than others. In this section, you will explore ways in which you can sample badly, sample well, and review cautions to consider when sampling. You will be introduced to a number of different sampling methods. Be sure to familiarize yourself with how to select samples using each of the methods and how to explain potential advantages and disadvantages of each.

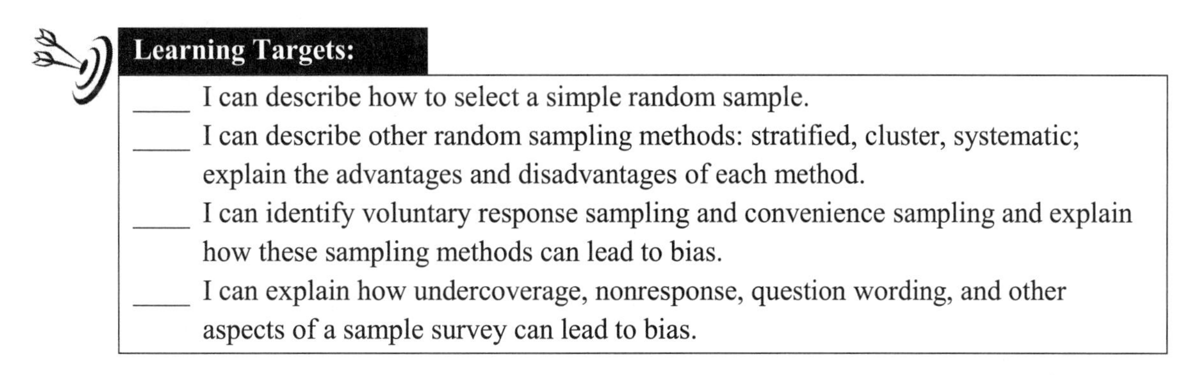

Learning Targets:

____ I can describe how to select a simple random sample.

____ I can describe other random sampling methods: stratified, cluster, systematic; explain the advantages and disadvantages of each method.

____ I can identify voluntary response sampling and convenience sampling and explain how these sampling methods can lead to bias.

____ I can explain how undercoverage, nonresponse, question wording, and other aspects of a sample survey can lead to bias.

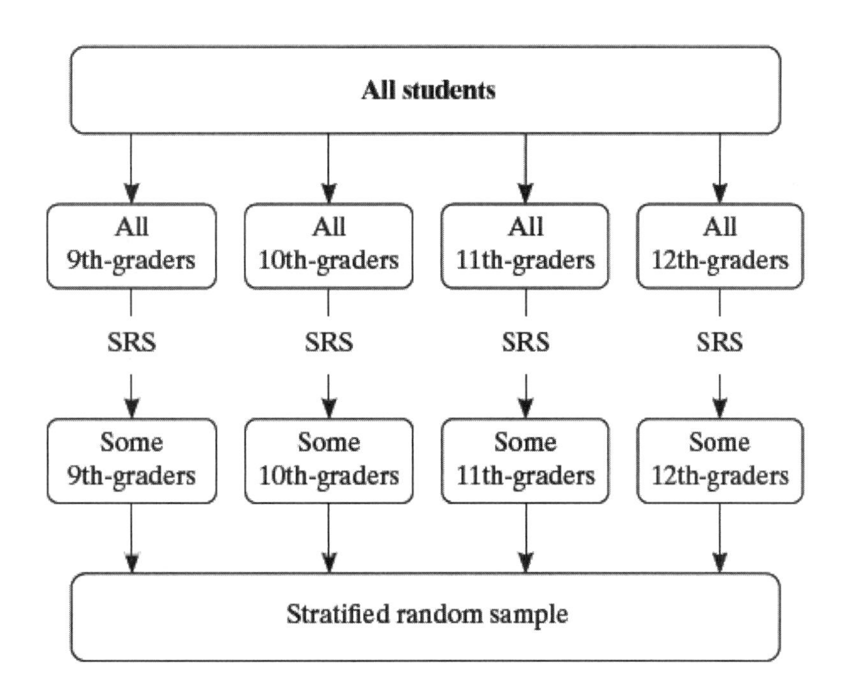

Stratified random sampling works best when the individuals within each stratum are similar (homogeneous) with respect to what is being measured and when there are large differences between strata.

Read: Vocabulary and Key Concepts

- A ____________________________ (SRS) of size *n* is a sample chosen in such a way that every group of *n* individuals in the population has an equal chance to be selected as the sample.
- When sampling ______________________________, an individual from a population can be selected only once.
- When sampling ___________________________, an individual from a population can be selected more than once.
- ______________ are groups of individuals in a population that share characteristics thought to be associated with the variables being measured in a study. The singular form of *strata* is *stratum*.
- A ____________________________________ is a sample selected by choosing an SRS from each stratum and combining the SRSs into one overall sample.
- A ________________ is a group of individuals in the population that are located near each other.
- A ___________________________ is a sample selected by randomly choosing clusters and including each member of the selected clusters in the sample.
- A ____________________________________ is a sample selected from an ordered arrangement of the population by randomly selecting one of the first *k* individuals and choosing every *k*th individual thereafter.
- A ___________________________ consists of individuals from the population who are easy to reach.
- The design of a statistical study shows _________________ if it is very likely to underestimate or very likely to overestimate the value you want to know.
- A _____________________________ consists of people who choose to be in the sample by responding to a general invitation. Voluntary response samples are sometimes called *self-selected* samples.
- _______________________ occurs when some members of the population are less likely to be chosen or cannot be chosen in a sample.
- _______________________ occurs when an individual chosen for the sample can't be contacted or refuses to participate.
- _______________________ occurs when there is a consistent pattern of inaccurate responses to a survey question.

Watch: Go to bfwpub.com/TPS7eStrive. Select Unit 3 and watch the following Example videos:

Video	Topic	Done
Unit 3B, Attendance audit (p. 251)	*Simple random samples*	
Unit 3B, Sampling at a school assembly (p. 257)	*Other random sampling methods*	
Unit 3B, Boaty McBoatface (p. 260)	*Convenience and voluntary response samples*	
Unit 3B, Wash your hands! (p. 263)	*Other sources of bias in surveys*	

Do: Check for Understanding

Concept 1: Simple Random Sampling

The key to sampling is to use a method that helps ensure that the sample is as representative of the population as possible. Because sampling methods like voluntary response and convenience samples can systematically favor certain outcomes in a population, we say they are biased. An unbiased sampling method is one that does not favor any element of the population. We rely on the use of chance to select unbiased samples. The easiest way to do this is to select a simple random sample (SRS) from the population of interest. One way we could select an SRS would be to write the name of each individual in the population on a piece of paper, put the pieces in a hat, mix them well, and draw out the necessary number of slips for our sample. We could also select an SRS by labeling each individual in the population with a number of the same length (ex. one-, two-, three- digit number depending on the size of the population) and then generate random numbers from technology or a random number table until you reach the desired sample size. The labeled individuals that match the generated numbers, ignoring repeats, make up the sample.

Check for Understanding – Learning Target 1

____ *I can describe how to select a simple random sample.*

The Simonetti family is an adventurous bunch. They decide to use an SRS to select their family's next 3 vacation destinations. Here are the options as ranked as the top 15 destination locations in the U.S.:

1. Orlando, FL	6. Cape Cod, MA	11. Lake Tahoe
2. Yellowstone National Park	7. Grand Canyon	12. Gulf Shores, AL
3. Destin, FL	8. Washington, DC	13. Outer Banks, NC
4. Anaheim – Disneyland	9. Yosemite National Park	14. Honolulu, HI
5. Wisconsin Dells	10. San Diego, CA	15. Gatlinburg, TN

1. Describe how to use a table of random digits to select a simple random sample (SRS) of 3 vacation destinations from this list.

2. Use the random digits here to choose the sample.

19220 95034 05756 28709 96149 12531 42544 82853
73676 47150 99400 11327 27754 40548 82421 76290

(*Need Tech Help?* View the video **Tech Corner 10: Choosing an SRS** at bfwpub.com/TPS7eStrive)

Concept 2: Stratified Random Sampling, Cluster Sampling, Systematic Random Sampling

While simple random samples give each group of n individuals in the population an equal chance to be selected, they are not always the easiest to obtain. Several other sampling methods are available that can be used instead of an SRS. If the population consists of several groups of individuals that are likely to produce similar responses within groups, but systematically different responses between groups, consider selecting a stratified random sample. Another method, cluster sampling, divides the population into smaller groups (clusters) that mirror the overall population and selects an SRS of those clusters. A systematic random sample selects every kth person, starting at a randomly selected person from the 1st to the kth, in an ordered arrangement of the population. This method is much easier than simple random sampling because you can survey people who are lined up in some manner, like leaving a store, finishing a race, or entering an amusement park.

Check for Understanding – Learning Target 2

____ *I can describe other random sampling methods: stratified, cluster, systematic; and explain the advantages and disadvantages of each method.*

A pollster would like to select a random sample of U.S. Senators to ask their opinion on a new legislation. There are 100 U.S. Senators. The Republicans are seated in the 3 right-most sections and the Democrats are seated in the 3 left-most sections. In February 2024, the Senate consisted of 49 Republicans and 51 Democrats.

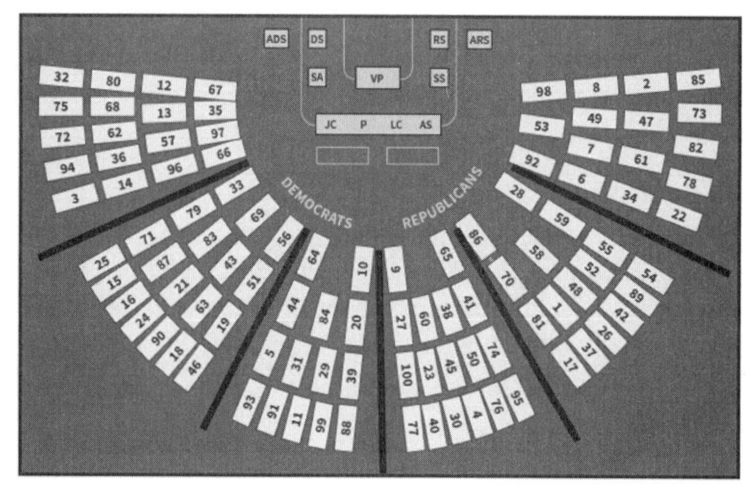

1. Describe how to obtain a random sample of 10 Senators using stratified sampling. Explain your choice of strata.

2. Describe how to obtain a random sample of at least 10 Senators using cluster sampling. Explain your choice of clusters.

3. Describe how to obtain a random sample of 10 Senators using systematic sampling.

Concept 3: Voluntary Response Sampling and Convenience Sampling

When samples are selected poorly, we should expect biased results, no matter how large the sample size may be. Two bad sampling methods are voluntary response sampling and convenience sampling. Convenience sampling tends to produce unreliable conclusions because the members of the sample often differ from the population in ways that affect their responses. Voluntary response sampling leads to voluntary response bias when the members of the sample differ from the population in ways that affect their responses.

Check for Understanding – Learning Target 3

____ *I can identify voluntary response sampling and convenience sampling and explain how these sampling methods can lead to bias.*

A local radio talk-show host asks listeners to call in and vote for or against a proposed plan to raise the prices charged by municipal parking meters in a downtown shopping district. From the final tally, 75% of the respondents were opposed to the increase

Identify the type of bias in this poll and state whether the sample proportion overestimates or underestimates the true proportion of listeners that are opposed to the increase.

Concept 4: Undercoverage, Nonresponse, Question wording

When designing a sample survey, random sampling helps avoid bias. But there are other kinds of mistakes in the sampling process that can lead to inaccurate information about the population. For instance, if the sampling method is designed in a way that leaves out certain segments of the population, the sample survey suffers from undercoverage. When selected individuals cannot be contacted or refuse to participate, the survey suffers from nonresponse. If people give incorrect or misleading responses, a survey suffers from response bias. Finally, if a question is worded in a way that favors certain responses, the survey suffers from question wording bias. Be sure to design sample surveys to avoid these issues and make sure you can identify them in existing studies.

Check for Understanding – Learning Target 4

____ *I can explain how undercoverage, nonresponse, question wording, and other aspects of a sample survey can lead to bias.*

An SRS of 400 recent high school graduates was asked, "Did you ever cheat on an exam during your senior year?" Of the 400 respondents, 64 percent admitted to cheating on at least one test.

What is one practical problem with this survey? What is the likely direction of the bias?

Section 3C: Experiments

Section Summary

Observational studies provide a snapshot of the population but cannot be used to establish any sort of cause-effect relationship. Observational studies only allow us to describe the population, compare groups, or examine basic relationships between variables. In this section, we will move beyond sampling to study the elements of experimental design. Experiments allow us to produce data in a way that can lead to conclusions about causation. There are a lot of vocabulary terms in this section. It is easy to confuse sampling terms and experimental design terms. Make sure you understand each vocabulary term or concept!

Learning Targets:

____ I can explain the concept of confounding and how it limits the ability to make cause- and effect conclusions.
____ I can identify experimental units and treatments in an experiment.
____ I can explain the purpose of a control group in an experiment.
____ I can describe the placebo effect and explain the purpose of blinding in an experiment.
____ I can describe how to randomly assign treatments in an experiment and explain the purpose of random assignment.
____ I can explain the purpose of controlling other variables in an experiment.
____ I can describe a randomized block design for an experiment and explain the benefits of blocking in an experiment.
____ I can describe a matched pairs design for an experiment.
____ I can explain the meaning of statistically significant in the context of an experiment.
____ I can identify when it is appropriate to make an inference about a population and when it is appropriate to make an inference about cause and effect.

Read: Vocabulary and Key Concepts

- A ____________________________ measures an outcome of a study.
- An _________________________________ may help explain or predict changes in a response variable.
- ____________________ occurs when two variables are associated in such a way that their effects on a response variable cannot be distinguished from each other.
- A ______________________ is a specific condition applied to the individuals in an experiment. If an experiment has several explanatory variables, a treatment is a combination of specific values of these variables.
- An ___________________________ is the object to which a treatment is randomly assigned. When the experimental units are human beings, they are often called ________________.
- A _______________ is a treatment that has no active ingredient but is otherwise like other treatments.
- In an ____________________, a factor is an explanatory variable that is manipulated and may cause a change in the response variable.
- The different values of a factor are called _____________.
- In an experiment, a _____________________ is used to provide a baseline for comparing the effects of other treatments. Depending on the purpose of the experiment, a control group may be given an inactive treatment (placebo), an active treatment, or no treatment at all.

- The _____________________ describes the fact that some subjects in an experiment will respond favorably to any treatment, even an inactive treatment.
- In a _______________________ experiment, neither the subjects nor those who interact with them and measure the response variable know which treatment a subject is receiving.
- In a _______________________ experiment, either the subjects or the people who interact with them and measure the response variable don't know which treatment a subject is receiving.
- In an experiment, ___________________________ means that treatments are assigned to experimental units (or experimental units are assigned to treatments) using a chance process.
- In a _______________________________________, the experimental units are assigned to the treatments completely at random.
- In an experiment, __________________ means giving each treatment to enough experimental units so that a difference in the effects of the treatments can be distinguished from chance variation due to the random assignment.
- In an experiment, __________________ means keeping other variables constant for all experimental units.
- A _____________ is a group of experimental units that are known before the experiment to be similar in some way that is expected to affect the response to the treatments.
- In a ____________________________, the random assignment of experimental units to treatments is carried out separately within each block.
- A ____________________________________ is a common experimental design for comparing two treatments that uses blocks of size 2. In some matched pairs designs, two very similar experimental units are paired, and the two treatments are randomly assigned within each pair. In others, each experimental unit receives both treatments in a random order.
- When an observed difference in responses between the groups in an experiment is so large that it is unlikely to be explained by chance variation in the random assignment, the results are called ___________________________________.

Watch: Go to bfwpub.com/TPS7eStrive. Select Unit 3 and watch the following Example videos:

Video	Topic	Done
Unit 3C, Smoking and ADHD (p. 272)	*Confounding*	
Unit 3C, The five-second rule (p. 273)	*The language of experiments*	
Unit 3C, Preventing malaria (p. 275)	*Comparison and control groups*	
Unit 3C, Do magnets repel pain? (p. 277)	*Blinding and the placebo effect*	
Unit 3C, Is the vaccine effective? (p. 279)	*Random assignment*	
Unit 3C, Multitasking (p. 281)	*Controlling other variables*	
Unit 3C, Should I use the popcorn button? (p. 285)	*Randomized block design*	
Unit 3C, Will an additive improve my mileage? (p. 287)	*Matched pairs design*	
Unit 3C, Distracted driving (p. 290)	*Inference for experiments*	
Unit 3C, When will I ever use this stuff? (p. 292)	*The scope of inference*	

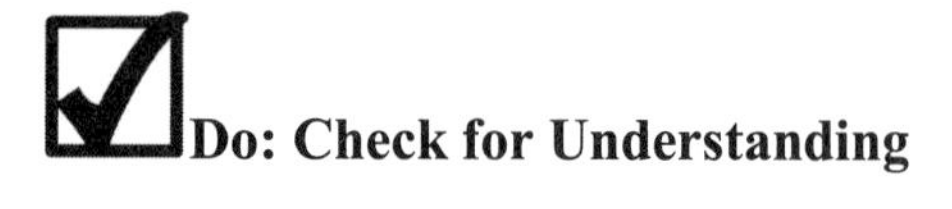

Do: Check for Understanding

Concept 1: Confounding

It is very difficult to show causation from an observational study due to confounding. Prior to administering a test, Mrs. Pavelka asked her students to record how many hours of sleep they had during the previous night. After scoring the test, the teacher found there was a strong positive correlation between the number of hours of sleep and examination score. However, Mrs. Pavelka cannot conclude that getting more sleep caused the increase in exam scores. It is possible that students who got more sleep also have better study habits and that study habits lead to better exam scores. Observational studies are at risk of confounding.

Check for Understanding – Learning Target 1

____ *I can explain the concept of confounding and how it limits the ability to make cause-and-effect conclusions.*

Many utility companies have introduced programs to encourage energy conservation. A water company creates a program that will send customers a daily email to update them on what their water bill would be if their usage that day continued for a month. It sends usage emails to 100 customers for one year and then compares the average water use in these customers' homes this year to the previous year. The data showed that customers' average water use decreased by 10%.

Explain why this is not strong evidence that the app caused customers to decrease their water use.

Concept 2: Experimental Units and Treatments

In an experiment, the object to which the treatments are randomly assigned are called the experimental units. If the experimental units are humans, we give them the more dignified title of "subjects." We apply treatments to the subjects to observe the response.

Concept 3: Control group

A control group is a group in an experiment that is given no treatment or an inactive treatment. The purpose of including a control group is to provide a baseline for comparing the effects of the other treatments. Not all experiments have a control group. A well-designed experiment needs to compare at least 2 treatments.

Concept 4: Placebo Effect and Blinding

Placebo is Latin for "I will please", which describes the fact that people tend to feel better even after receiving a treatment that is designed to have no medical benefit. One way to counteract the placebo effect is to use blinding in an experiment. Experiments can be single- or double-blind. In a single-blind experiment either the subjects or the people who interact with them and measure the response variable don't know which treatment the subjects received. In a double-blind experiment both the subjects and those who interact with them don't know which treatment the subjects received.

Concept 5: Random Assignment, Completely Randomized Design

Random assignment is a necessary component of a well-designed experiment. By randomly assigning the treatments to the subjects we create groups that are roughly equivalent at the beginning of the experiment. Therefore, the only systematic difference between the groups are the treatments. If one of the treatments significantly outperforms the others, we can say that the treatments *caused* the difference in the response. A completely randomized design is an experimental design in which the subjects are randomly assigned to the treatments, just like drawing names from a hat.

Concept 6: Controlling other variables

While random assignment is the "gold standard" in experimental design, controlling other variables is like going the extra mile. When we control other variables, we make an intentional effort to keep as many variables as possible the same for the treatment groups. There are two big benefits: Controlling other variables helps to prevent confounding, making it easier to determine if one treatment is more effective than another. Controlling other variables also reduces the variability in the response variable, making it easier to determine if one treatment is more effective than another. Note that the principle of control isn't the same thing as including a control group, which is an aspect of comparison.

Check for Understanding – Learning Targets 2, 3, 4, 5, and 6

____ *I can identify the experimental units and treatments in an experiment.*

____ *I can explain the purpose of a control group in an experiment.*

____ *I can describe the placebo effect and explain the purpose of blinding in an experiment.*

____ *I can describe how to randomly assign treatments in an experiment and explain the purpose of random assignment.*

____ *I can explain the purpose of controlling other variables in an experiment.*

Mr. Tyson teaches statistics to 150 students. He is interested in knowing whether or not listening to classical music while studying results in higher test scores than listening to no music. He wishes to design an experiment to answer this question. All students who are assigned to listen to classical music will be given the same 45-minute classical music playlist. Students are told to study for no more than 45 minutes. All students were given the same instruction by Mr. Tyson and the same test.

1. What are the experimental units, explanatory variable, treatments, and response variable?

2. Why is it necessary to include a control group of students who do not listen to classical music when studying?

3. Can this study be single-blind? Double-blind?

4. Describe a completely randomized design for Mr. Tyson's experiment.

5. Identify two variables that are kept the same during the experiment. Why is it important to keep these variables the same?

Concepts 7 and 8: Randomized Block Design and Matched Pairs Design

When groups of subjects share a common characteristic that might systematically affect their responses to the treatments, we can use blocking to control the effects of this variable. For example, suppose researchers are conducting an experiment to compare the effectiveness of a new medicine for treating high blood pressure with the most commonly prescribed drug. If they believe that older and younger people may respond differently to such medications, the researchers can separate the subjects into blocks of older and younger people and randomly assign treatments within each block. This randomized block design helps isolate the variation in responses due to age, which makes it easier for researchers to find evidence of a treatment effect. If we are only comparing two treatments, we can sometimes conduct a matched pairs design by creating blocks of two similar individuals and then randomly assigning each subject to a treatment. Another type of matched pairs design involves assigning two treatments to each subject in a random order. In each type of experimental design, our goal is to ensure control, random assignment, and replication. If these principles are addressed in our design, we can establish good evidence of causal relationships.

Check for Understanding – Learning Targets 7 and 8

____ *I can describe a randomized block design for an experiment and explain the benefits of blocking in an experiment.*

____ *I can describe a matched pairs design for an experiment.*

Refer to the previous Check for Understanding.

1. Describe an experimental design involving blocking that will help answer Mr. Tyson's question. Explain your choice of a blocking variable and describe why a randomized block design is preferable to a completely randomized design.

2. Could Mr. Tyson carry out the experiment using a matched pairs design? If so, explain the method of pairing and how he could randomly assign the students to the treatments.

Concept 9 and 10: Statistical Significance and Scope of Inference

When the members of a sample are selected at random from a population, we can use the sample results to infer things about the population. That is, well-designed samples allow us to make inferences about the population from which we sampled. Random sampling is what allows us to generalize our results with confidence. If our goal is to make an inference about cause and effect, we must use a randomized experiment and obtain statistically significant results. If the experimental units were not randomly selected from a larger population of interest, we cannot extend our conclusions beyond individuals like those who took part in the experiment.

Check for Understanding – Learning Targets 9 and 10

_____ *I can explain the meaning of statistically significant in the context of an experiment.*

_____ *I can identify when it is appropriate to make inference about a population and when it is appropriate to make an inference about cause and effect.*

According to Louann Brizendine, author of *The Female Brain*, women say nearly 3 times as many words per day as men. Skeptical researchers devised a study to test this claim. They used electronic devices to record the talking patterns of 396 university students who volunteered to participate. The device was programmed to record 30 seconds of sound every 12.5 minutes without the carrier's knowledge. According to a published report in *Scientific American*, "Men showed a slightly wider variability in words uttered. . . . But in the end, the sexes came out just about even in the daily averages: women at 16,215 words and men at 15,669." This difference was not statistically significant.

1. Is it reasonable to make a cause-and-effect conclusion based on this study? Explain your answer.

2. What is the largest population to which we can generalize the results of this study? Explain your answer.

3. Explain the meaning of "This difference was not statistically significant."

Unit 3 Summary: Collecting Data

This unit is an important one in your study of statistics. After all, we cannot describe or analyze data without collecting it first! Because one of the major goals of statistics is to make inferences that go beyond the data, it is critical that we produce data in a way that will allow for such inferences. Biased data production methods can lead to incorrect inferences. Random sampling allows us to make an inference about the population as a whole. Well-designed experiments in which we randomly assign treatments and control for other variables allow us to make inferences about cause and effect. We will learn how to perform these inferences in later units. Your goal in this unit is to be able to describe good sampling and experimental design techniques and recognize when sampling or experimental design has been done poorly. There is always a question about sampling or experimental design on the free-response portion of the AP exam. Study the vocabulary and concepts in this unit so you can answer that question with confidence!

How well do you understand each of the learning targets?

Learning Target	Got It!	Almost There	Needs Work
I can identify the population and sample in a statistical study.			
I can distinguish between an observational study and an experiment.			
I can determine what inferences are appropriate from an observational study.			
I can describe how to select a simple random sample.			
I can describe other random sampling methods: stratified, cluster, systematic; and explain the advantages and disadvantages of each method.			
I can identify voluntary response sampling and convenience sampling, and explain how these sampling methods can lead to bias.			
I can explain how undercoverage, nonresponse, question wording, and other aspects of a sample survey can lead to bias.			
I can explain the concept of confounding and how it limits the ability to make cause-and-effect conclusions.			
I can identify experimental units and treatments in an experiment.			
I can explain the purpose of a control group in an experiment.			
I can describe the placebo effect and the purpose of blinding in an experiment.			
I can describe how to randomly assign treatments in an experiment and explain the purpose of random assignment.			
I can explain the purpose of controlling other variables in an experiment.			
I can describe a randomized block design for an experiment and explain the benefits of blocking in an experiment.			
I can describe a matched pairs design for an experiment.			
I can explain the meaning of statistical significance in the context of an experiment.			
I can identify when it is appropriate to make an inference about a population and when it is appropriate to make an inference about cause and effect.			

Unit 3 Multiple Choice Practice

Directions. Identify the choice that best completes the statement or answers the question. Check your answers and note your performance when you are finished.

1. A researcher is testing a company's new stain remover. He has contracted with 40 families who have agreed to test the product. He randomly assigns 20 families to the group that will use the new stain remover and 20 to the group that will use the company's current product. The most important reason for this random assignment is that
   (A) randomization makes the analysis easier since the data can be collected and entered into the computer in any order.
   (B) randomization eliminates the impact of any confounding variables.
   (C) randomization is a good way to create two groups of 20 families that are as similar as possible, except for the treatments they receive.
   (D) randomization ensures that the study is double-blind.
   (E) randomization reduces the impact of outliers.

2. A researcher observes that, on average, the number of traffic violations in cities with Major League Baseball teams is larger than in cities without Major League Baseball teams. The most plausible explanation for this observed association is that the
   (A) presence of a Major League Baseball team causes the number of traffic incidents to rise (perhaps due to the large number of people leaving the ballpark).
   (B) high number of traffic incidents is responsible for the presence of Major League Baseball teams (more traffic incidents mean more people have cars, making it easier for them to get to the ballpark).
   (C) association is due to the presence of another variable (Major League teams tend to be in large cities with more people, hence a greater number of traffic incidents).
   (D) association makes no sense since many people take public transit or walk to baseball games.
   (E) observed association is purely coincidental. It is implausible to believe the observed association could be anything other than accidental.

3. A researcher is testing the effect of a new fertilizer on crop growth. He marks 30 plots in a field, splits the plots in half, and randomly assigns the new fertilizer to one half of the plot and the old fertilizer to the other half. After 4 weeks, he measures the crop yield and compares the effects of the two fertilizers. This design is an example of
   (A) matched pairs experiment
   (B) completely randomized comparative experiment
   (C) cluster experiment
   (D) double-blind experiment
   (E) this is not an experiment

4. A large suburban school wants to assess student attitudes towards their mathematics textbook. The administration randomly selects 15 mathematics classes and gives the survey to every student in the class. This is an example of a
   (A) multistage sample
   (B) stratified sample
   (C) cluster sample
   (D) simple random sample
   (E) convenience sample

5. Eighty volunteers who currently use a certain brand of medication to reduce blood pressure are recruited to try a new medication. The volunteers are randomly assigned to one of two groups. One group continues to take their current medication, the other group switches to the new experimental medication. Blood pressure is measured before, during, and after the study. Which of the following best describes a conclusion that can be drawn from this study?
   (A) We can determine whether the new drug reduces blood pressure more than the old drug for anyone who suffers from high blood pressure.
   (B) We can determine whether the new drug reduces blood pressure more than the old drug for individuals like the subjects in the study.
   (C) We can determine whether the blood pressure improved more with the new drug than with the old drug, but we can't establish cause and effect.
   (D) We cannot draw any conclusions, since the all the volunteers were already taking the old drug when the experiment started.
   (E) We cannot draw any conclusions because there was no control group.

6. To determine employee satisfaction at a large company, the management selects an SRS of 200 workers from the marketing department and a separate SRS of 50 workers from the sales department. This kind of sample is called a
   (A) simple random sample.
   (B) simple random sample with blocking.
   (C) multistage random sample.
   (D) stratified random sample.
   (E) random cluster sample.

7. For a certain experiment you have 8 subjects, of which 4 are female and 4 are male. The names of the subjects are listed below:

   Males: Atwater, Bacon, Chu, Diaz. *Females: Johnson, King, Liu, Moore*

   There are two treatments, A and B. If a randomized block design is used with the subjects blocked by their gender, which of the following is *not* a possible group of subjects who receive treatment A?
   (A) Atwater, Chu, King, Liu
   (B) Bacon, Chu, Liu, Moore
   (C) Atwater, Diaz, Liu, King
   (D) Atwater, Bacon, Chu, Johnson
   (E) Atwater, Bacon, Johnson, King

8. An article in the student newspaper of a large university had the headline "A's swapped for evaluations?" Results showed that higher grades directly corresponded to a more positive evaluation. Which of the following would be a valid conclusion to draw from the study?
   (A) A teacher can improve his or her teaching evaluations by giving good grades.
   (B) A good teacher, as measured by teaching evaluations, helps students learn better, resulting in higher grades.
   (C) Teachers of courses in which the mean grade is higher apparently tend to have above-average teaching evaluations.
   (D) Teaching evaluations should be conducted before grades are awarded.
   (E) All of the above.

9. A new cough medicine was given to a group of 25 subjects who had a cough caused by the common cold. 30 minutes after taking the new medicine, 20 of the subjects reported that their coughs had disappeared. From this information you conclude
   (A) that the remedy is effective for the treatment of coughs.
   (B) nothing, because the sample size is too small.
   (C) nothing, because there is no control group for comparison.
   (D) that the new treatment is better than the old medicine.
   (E) that the remedy is not effective for the treatment of coughs.

10. A study is conducted using 100 volunteers who suffer from anxiety. Half of the volunteers are selected at random and assigned to receive a new drug that is thought to be extremely effective in reducing anxiety. The other 50 are given an existing anti-anxiety drug. A doctor evaluates anxiety levels after two months of treatment to determine if there has been a larger reduction in the anxiety levels of those who take the new drug. This would be double blind if
   (A) both drugs looked the same.
   (B) neither the subjects nor the doctor knew which treatment any subject had received.
   (C) the doctor couldn't see the subjects and the subjects couldn't see the doctor.
   (D) there was a third group that received a placebo.
   (E) all of the above

Check your answers below. If you got a question wrong, check to see if you made a simple mistake or if you need to study that concept more. After you check your work, identify the concepts you feel very confident about and note what you will do to learn the concepts in need of more study.

#	Answer	Concept	Right	Wrong	Simple Mistake?	Need to Study More
1	C	Why we randomize				
2	C	Confounding				
3	A	Matched Pairs				
4	C	Cluster Sampling				
5	B	Inference about the population				
6	D	Stratified random sampling				
7	D	Blocking				
8	C	Surveys vs Experiments				
9	C	Control				
10	B	Definition of Experiments				

FRAPPY! Free Response AP® Problem, Yay!

The following problem is modeled after actual Advanced Placement Statistics free response questions. Your task is to generate a complete, concise response in 15 minutes. After you generate your response, read over the solution and scoring guide. Score your response and note what, if anything, you would do differently to increase your own score.

A large school district is interested in determining student attitudes about their co-curricular offerings such as athletics and fine arts. The district consists of students attending 4 elementary schools (2000 students total), 1 middle school (1000 students total), and 2 high schools (2000 students total).

The administration is considering two sampling plans. The first consists of taking a simple random sample of students in the district and surveying them. The second consists of taking a stratified random sample of students and surveying them.

(a) Describe how you would select a simple random sample of 200 students in the district.

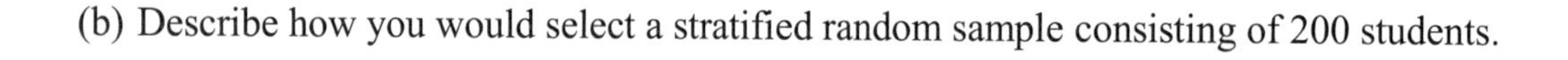

(b) Describe how you would select a stratified random sample consisting of 200 students.

(c) Describe the statistical advantage of using a stratified random sample over the simple random sample in this study.

FRAPPY! Scoring Rubric

Use the following rubric to score your response. Each part receives a score of "Essentially Correct," "Partially Correct," or "Incorrect." When you have scored your response, reflect on your understanding of the concepts addressed in this problem. If necessary, note what you would do differently on future questions like this to increase your score.

Intent of the Question

The goal of this question is to determine your ability to describe sampling methods and explain the advantages of stratifying over simple random sampling

Solution

(a) Write each student's name on a slip of paper. Place the slips of paper in a hat and mix well. Select 200 different slips of paper and the students in the sample are those whose names were selected. OR Label each student with a number from 0001 to 5000. Use a random number table or technology to produce random 4-digit numbers, ignoring repeats, any numbers from 5001 to 9999, and 0000, until 200 different numbers are identified. These 200 numbers correspond to the individuals who will be surveyed.

(b) Because student attitudes may differ by level of school (elementary, middle, or high school), we should stratify by level. The names of each elementary student should be placed on slips of paper in a hat, mixed well and then 80 student names selected without replacement. A similar process would be done for the middle school students with 40 names selected from a different hat without replacement, then the same would be done for the high school students with 80 different names selected. The students from each level whose names were selected will be combined to form the sample of 200 students. This ensures each level is represented in the same proportion as the overall student enrollments.

(c) Stratifying ensures no level is over or underrepresented in the sample. It is possible to select very few (or even no!) students from one level in a simple random sample. The opinions of students at one level may not reflect the opinions of all students in the district. Stratifying ensures each level is fairly represented, while an SRS does not ensure fair representation for each strata.

Scoring:

Parts (a), (b), and (c) are scored as essentially correct (E), partially correct (P), or incorrect (I).

Part (a) is essentially correct if the response describes an appropriate method of selecting a simple random sample. This method should include labeling the individuals and employing a sufficient means of random selection that could be replicated by someone knowledgeable in statistics and avoids duplicate selection.
Part (a) is partially correct if random selection is used correctly, but the description does not provide sufficient detail for implementation. Incorrect, otherwise.

Part (b) is essentially correct if the response describes selecting strata based on a reasonable variable (such as school level) and indicates randomly selecting individuals from each stratum to be a part of

the survey. The method can result in a nearly equal number of students from each level OR proportional representation based on the strata.
Part (b) is partially correct if a reasonable variable is identified, but the random selection method is unclear or does not ensure proportional or nearly equal representation. Incorrect, otherwise.

Part (c) is essentially correct if the response (1) provides a reasonable statistical advantage of a stratified random sample, (2) provides that statistical advantage in context with clear communication, (3) notes that the advantage stated for a stratified random sample is not also true for an SRS or provides a relevant disadvantage of an SRS.
Part (c) is partially correct if the response includes 2 of the 3 components above. Incorrect, otherwise.

4 Complete Response
All three parts essentially correct

3 Substantial Response
Two parts essentially correct and one part partially correct

2 Developing Response
Two parts essentially correct and no parts partially correct
One part essentially correct and one or two parts partially correct
Three parts partially correct

1 Minimal Response
One part essentially correct and no parts partially correct
No parts essentially correct and two parts partially correct

My Score:
What I did well:
What I could improve:
What I should remember if I see a problem like this on the AP Exam:

Unit 3 Formula Sheet

Create a one-page summary of important concepts and formulas found in this unit. This will serve as a valuable resource as you prepare for the AP Exam.

Unit 3 Crossword Puzzle

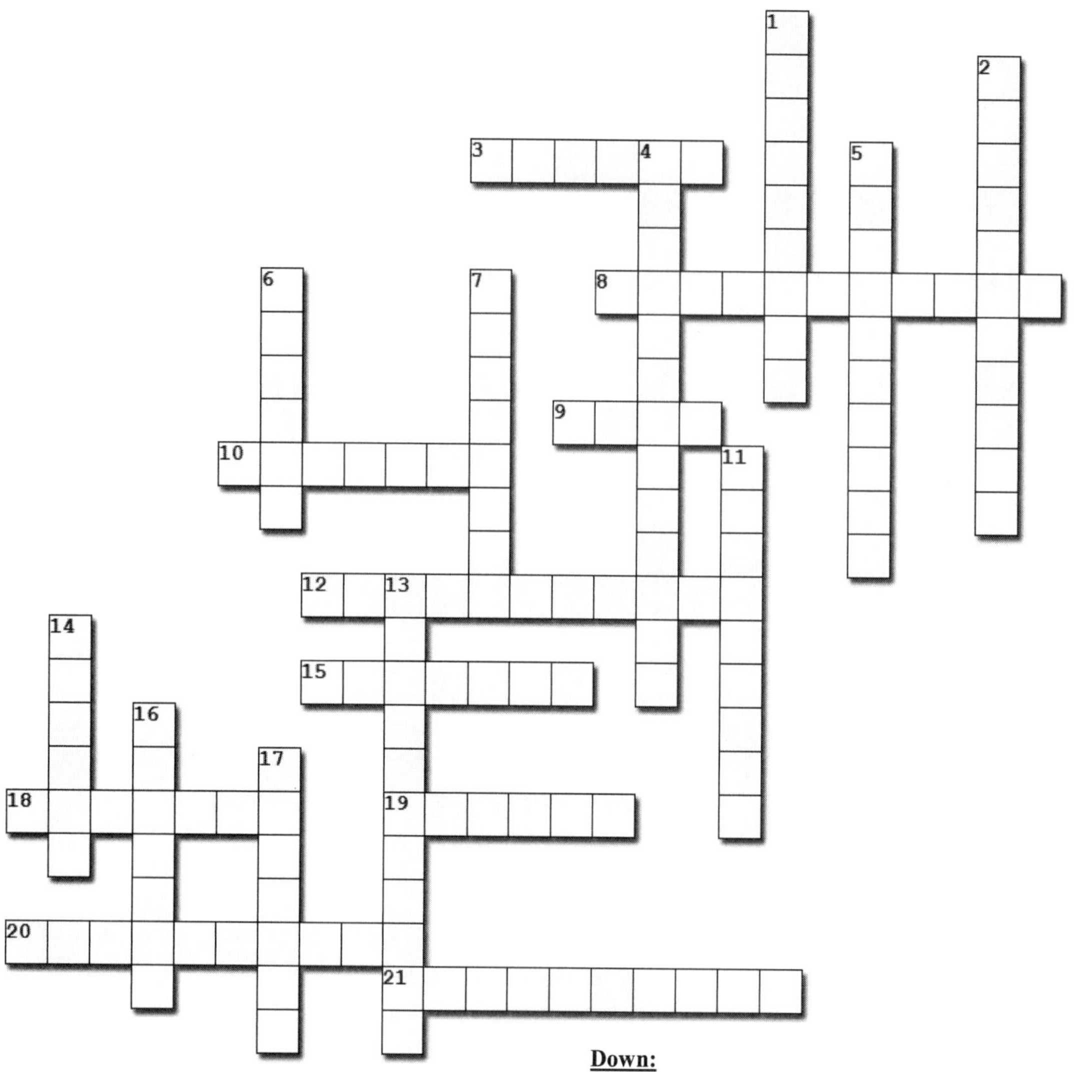

Across:

3. In an experiment, a ____ is an explanatory variable that is manipulated and may cause a change in the response variable.
8. When sampling with ____, an individual from a population can be selected more than once.
9. A statistical study shows ____ if it is likely to underestimate or likely to overestimate the value you want to know.
10. A ____ is a treatment that has no active ingredient but is otherwise like other treatments.
12. A ____ sample consists of individuals from the population who are easy to reach.
15. In an experiment, ____ means keeping other variables constant for all experimental units.
18. In an experiment, a ____ group is used to provide a baseline for comparing the effects of other treatments.
19. A ____ is a subset of individuals in the population from which we collect data.
20. The ____ in a statistical study is the entire group of individuals we want information about.
21. A ____ random sample is selected by choosing an SRS from each stratum and combining the SRSs into one overall sample.

Down:

1. A ____ response sample consists of people who choose to be in the sample by responding to a general invitation.
2. An ___ variable may help explain or predict changes in a response variable.
4. An ___ study observes individuals and measures variables of interest but does not attempt to influence the responses.
5. An ____ deliberately imposes treatments on experimental units to measure their responses.
6. A ____ random sample gives every possible sample of a given size the same chance to be chosen.
7. A ____ variable measures an outcome of a study.
11. Random selection of individuals justifies ____ about the population from which the individuals were selected.
13. ____ occurs when an individual chosen for the sample can't be contacted or refuses to participate.
14. In an experiment, ____ assignment means that treatments are assigned to experimental units using a chance process.
16. When sampling ____ replacement, an individual from a population can be selected only once.
17. A ____ is a group of individuals in the population that are located near each other.

Unit 4, Part I: Probability

"The most important questions of life are, for the most part, really questions of probability." Pierre-Simon LaPlace

UNIT 4 PART I OVERVIEW

Now that we have learned how to collect data and how to analyze it graphically and numerically, we turn our study to probability, the mathematics of chance. Probability is the basis for the fourth and final theme in AP Statistics, inference.

In this unit, you will learn the definition of probability as a long-term relative frequency. You will study how to use simulation to answer probability questions as well as some basic rules to calculate probabilities of events. You will also learn two concepts that will reappear later in our studies: conditional probability and independence. Your goal with probability is to be able to answer the question, "What would happen if we did this many, many times?" so you can make an informed statistical inference.

Sections in this Unit 4, Part I

Section 4A: Randomness, Probability, and Simulation
Section 4B: Probability Rules
Section 4C: Conditional Probability and Independent Events

PRACTICE FOR MASTERY

Use the following *suggested* guide to the pages and exercises in your text to practice for mastery!
Note: your teacher may assign different problems. Be sure to follow their instructions!

Day	Topics	Read	Do
1	◆ The Idea of Probability	p. 319-323	**4A**: 1, 3, 5, 7, 20, 21, 24
2	◆ Estimating Probabilities Using Simulation	p. 323-327	**4A**: 9, 11, 13, 15, 22, 23
3	◆ Probability Models ◆ Basic Probability Rules ◆ Mutually Exclusive Events	p. 333-337	**4B**: 1, 3, 5, 7, 9, 11, 26, 30
4	◆ Two-Way Tables and Probability ◆ General Addition Rule	p. 338-342	**4B**: 13, 15, 17, 19, 21, 31
5	AP Classroom Topic Questions (Topics 4.1-4.4)		**4A**: 17, 25, **4B**: 23, 27-29, 32
6	◆ Conditional Probability ◆ Conditional Probability and Independence	p. 348-354	**4C**: 1, 5, 7, 11, 13, 15, 46, 47, 49
7	◆ The General Multiplication Rule ◆ Tree Diagrams and Probability	p. 355-360	**4C**: 19, 21, 23, 25, 27, 29, 48
8	◆ Multiplication Rule for Independent Events ◆ Use Probability Rules Wisely	p. 361-365	**4C**: 31, 33, 35, 37, 39, 41, 45
9	Unit 4, Part I AP® Statistics Practice Test (p. 376)		Unit 4, Part I Review Exercises (p.374)
10	Unit 4, Part I Test: Celebration of learning!		

Section 4A: Randomness, Probability, and Simulation

Section Summary

This section introduces the basic definition of probability as a long-term relative frequency. That is, probability answers the question, "How often would we expect to see a particular outcome if we repeated a random process many times?" The big idea that emerges when we study random processes is that random behavior is unpredictable in the short run but has a regular and predictable pattern in the long run. This idea seems pretty straightforward, yet there are a lot of common misconceptions about probability. Several are discussed in this section. Be sure to avoid falling for these common myths!

The second topic in this section addresses the use of simulation to estimate probabilities. Simulation is a powerful tool for modeling random behavior that can be used to illustrate many of the inference ideas you'll study later in the course.

Learning Targets:

____ I can interpret probability as a long-run relative frequency.
____ I can estimate probabilities using simulation.

Read: Vocabulary and Key Concepts

- A random process is unpredictable in the ______________________ but has a regular and predictable pattern in the ______________________.
- A ____________________________ generates outcomes that are determined purely by chance.
- The ____________________ of any outcome of a random process is a number between 0 and 1 (inclusive) that describes the proportion of times the outcome would occur in a very long series of trials.
- The ________________________ says that if we observe more and more trials of any random process, the proportion of times that a specific outcome occurs approaches its probability.
- A ______________________ imitates a random process in such a way that simulated outcomes are consistent with real-world outcomes.
- To perform a simulation: (1) Describe how to set up and use a ___________________ to perform one trial of the simulation. Identify what you will ___________ at the end of each trial. (2) Perform _______ trials. (3) Use the results of your simulation to ______________________________.

Watch: Go to bfwpub.com/TPS7eStrive. Select Unit 4, Part I and watch the following Example videos:

Video	Topic	Done
Unit 4A, Who drinks coffee? (p. 322)	*The idea of probability*	
Unit 4A, Superhero comics and cereal boxes (p. 324)	*Estimating probabilities using simulation*	
Unit 4A, Golden ticket parking lottery (p. 326)	*Estimating probabilities using simulation*	

Do: Check for Understanding

Concept 1: The Idea of Probability

When we observe random behavior over a long series of repetitions, a useful fact emerges. While random behavior is unpredictable in the short term, a regular and predictable pattern becomes evident in the long run. The law of large numbers tells us that as we observe more and more repetitions of a random behavior, the proportion of times a specific outcome occurs will "settle down" around a single value. This long-term proportion is the probability of the outcome occurring. The probability of an event is always described as a value between 0 and 1 where 0 represents an event that never occurs and 1 means that the event will occur on every repetition.

Check for Understanding – Learning Target 1

____ *I can interpret probability as a long-run relative frequency.*

At a carnival there is a "duck pond" which is filled with plastic ducks. 10% of the ducks have a red dot on the bottom. If you select a duck with a red dot, you win a prize!

1. Explain what the probability 0.10 means in the context of problem.

2. Does this mean if we repeatedly select a duck, replace it, let the ducks mix up, and select again for a total of 100 times that we will select a duck with a red dot on the bottom exactly 10 times? Why or why not?

Concept 2: Simulation

In this section, we learn how we can use simulation to estimate the probability of an event occurring. A three-step process can be used to perform a simulation of a question of interest by: 1) describing how to use a random device to imitate a repetition of the random behavior, 2) performing many repetitions of the simulation, and 3) using the results of the simulation to answer the original question. While simulations don't provide exact theoretical probabilities, the use of random numbers and other random devices to imitate random behavior can be a useful tool for estimating the likelihood of events.

Check for Understanding – Learning Target 2

____ *I can estimate probabilities using simulation.*

A popular airline knows that, in general, 95% of individuals who purchase a ticket for a 200-seat commuter flight actually show up for the flight. In an effort to ensure a full flight, the airline sells 205 tickets for each flight. What is the probability that the flight will be overbooked? That is, more than 200 passengers show up for the flight. To help answer this question, we want to perform a simulation to estimate the number of passengers who will show up for the flight.

1. Describe how to use a random number generator to perform one trial of the simulation.

A researcher carried out 100 trials of the simulation and noted the number of passengers simulated to show up in each trial. The dotplot shows the results.

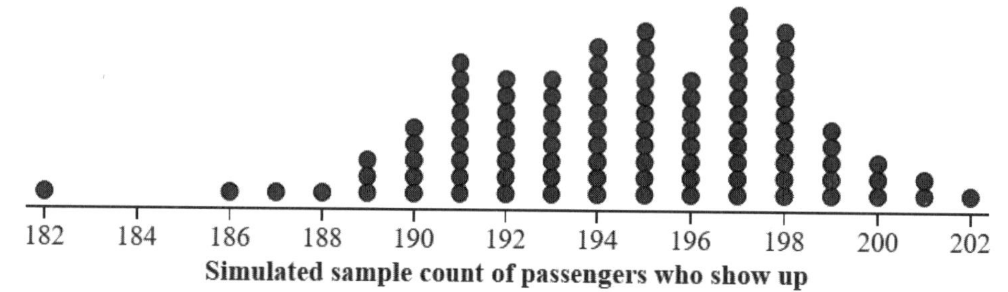

2. Explain what the dot at 202 represents.

3. Use the results of the simulation to estimate the probability that the flight will be overbooked.

Section 4B: Probability Rules

Section Summary

Now that you have the basic idea of probability down, you will learn how to describe probability models and use probability rules to calculate the likelihood of events. You will also learn how to organize information into two-way tables and Venn diagrams to help in determining probabilities. Understanding probability is important for understanding inference. Make sure you are comfortable with the definitions and rules in this section, as it will make your study of probability much easier!

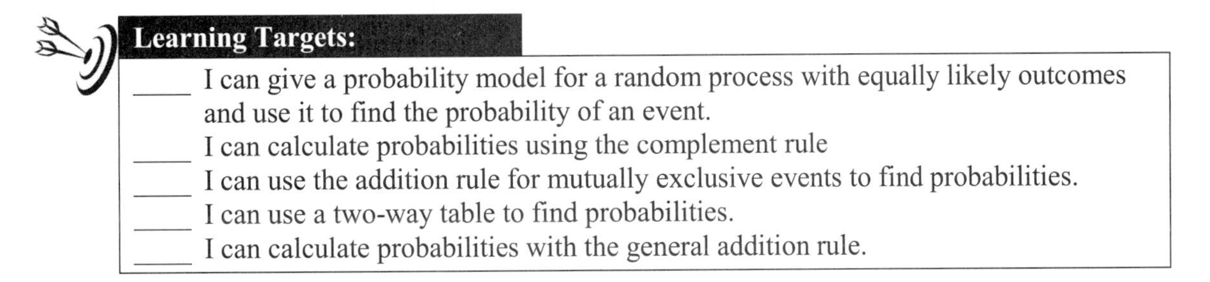

Learning Targets:

_____ I can give a probability model for a random process with equally likely outcomes and use it to find the probability of an event.
_____ I can calculate probabilities using the complement rule
_____ I can use the addition rule for mutually exclusive events to find probabilities.
_____ I can use a two-way table to find probabilities.
_____ I can calculate probabilities with the general addition rule.

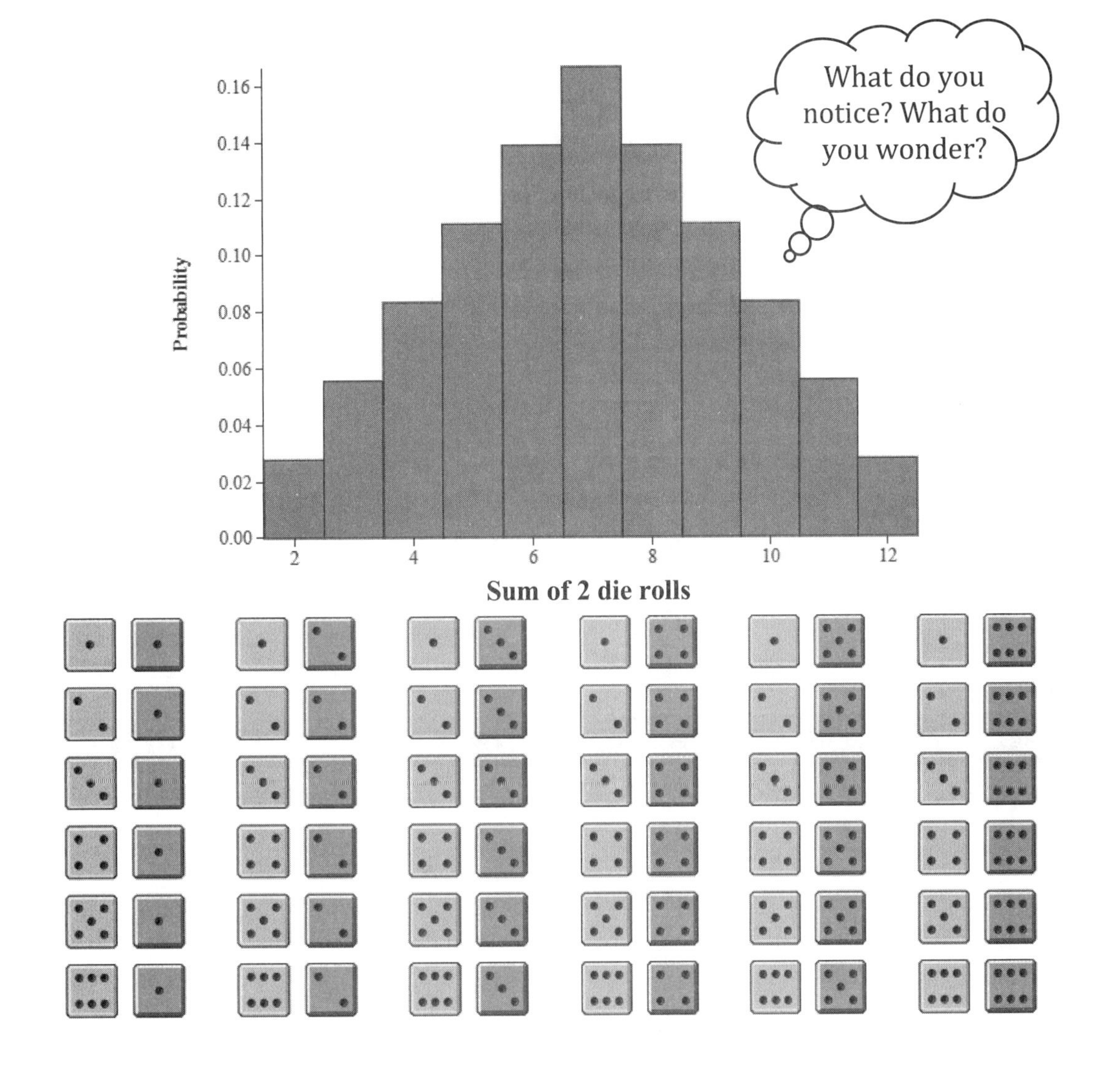

Read: Vocabulary and Key Concepts

- A ______________________________ is a description of a random process that consists of two parts: a list of all possible outcomes and the probability of each outcome.
- The list of all possible outcomes is called the ____________________.
- An ________________________ is a subset of the possible outcomes from the sample space of a random process.
- The probability of any event is a number between ____ and ____ (inclusive).
- All possible outcomes together must have probabilities that add up to ____.
- The ______________________________ says that $P(A^C) = 1 - P(A)$, where A^C is the complement of event A; that is, the event that A does not occur.
- Two events A and B are ____________________________________ if they have no outcomes in common and so can never occur together — that is, if $P(\text{A and B}) = 0$.
- The ____ A and B says that $P(\text{A or B}) = P(A) + P(B)$
- "A or B" means one or the other or ____________.
- If A and B are any two events resulting from the same random process, the ______________________ ______________________ says that $P(\text{A or B}) = P(A) + P(B) - P(\text{A and B})$.
- A ________________________ consists of one or more circles surrounded by a rectangle. Each circle represents an event. The region inside the rectangle represents the sample space of the random process.
- The event "A and B" is called the __________________________ of events A and B. It consists of all outcomes that are common to both events, and is denoted $A \cap B$.
- The event "A or B" is called the ____________________________ of events A and B. It consists of all outcomes that are in event A, event B, or both, and is denoted $A \cup B$.
- Here's a way to keep the symbols straight: ____ for **u**nion; ____ for i**n**tersection.

Watch: Go to bfwpub.com/TPS7eStrive. Select Unit 4, Part I and watch the following Example videos:

Video	Topic	Done
Unit 4B, Spin the spinner (p. 334)	*Probability models*	
Unit 4B, Avoiding blue M&M'S® (p. 336)	*Basic probability rules*	
Unit 4B, More M&M'S® (p. 337)	*Mutually exclusive events*	
Unit 4B, Who can roll their tongue or raise one eyebrow? (p. 338)	*Two-way tables and probability*	
Unit 4B, Facebook versus Instagram (p. 340)	*General addition rule*	

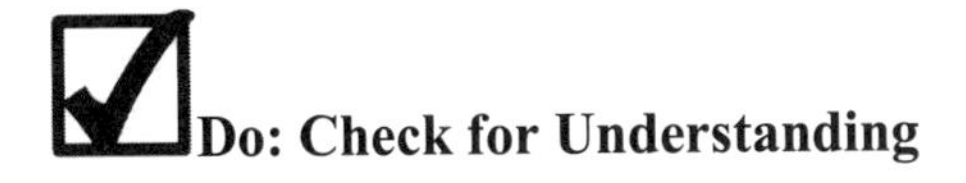

Do: Check for Understanding

Concepts 1, 2, and 3: Probability Models and the Basic Rules of Probability

Random behavior can be described using a probability model. This model provides two pieces of information: a list of possible outcomes (sample space) and the likelihood of each outcome. By describing random behavior with a probability model, we can find the probability of an event—a particular outcome or collection of outcomes. Probability models must obey some basic rules of probability:

- For any event A, $0 \leq P(A) \leq 1$.
- If S is the sample space in a probability model, $P(S) = 1$.
- In the case of equally likely outcomes, $P(A)$ = (# outcomes in event A) / (# outcomes in S).
- Complement Rule: $P(A^C) = 1 - P(A)$.
- If A and B are mutually exclusive events, $P(A \text{ or } B) = P(A) + P(B)$.

Check for Understanding – Learning Targets 1, 2, and 3

____ *I can give a probability model for a random process with equally likely outcomes and use it to find the probability of an event.*

____ *I can calculate probabilities using the complement rule.*

____ *I can use the addition rule for mutually exclusive events to find probabilities.*

Ten equal-sized cards are placed in a hat, and are mixed well. One card is selected at random.

1. How many possible outcomes are in the sample space? What is the probability of each outcome?

Now, each letter in the word STATISTICS is written on one of the 10 cards. The cards are placed in the hat and mixed well.

2. Select one letter card at random. What outcomes are in the sample space? What is the probability of each outcome?

3. Let V = a vowel is selected. Find $P(V)$.

4. Find and interpret $P(V^C)$.

Concepts 4 and 5: Two-Way Tables, General Addition Rule

Sometimes we'll need to find probabilities involving two events. In these cases, it may be helpful to organize and display the sample space using a two-way table or Venn diagram. This can be especially helpful when two events are not mutually exclusive. When dealing with two events A and B, it is important to be able to describe the union (or collection of all outcomes in A, B, or both) and the intersection (the collection of outcomes in both A and B). The general addition rule expands upon the basic rules presented in this section to help us find the probability of two events that are not mutually exclusive.

For any two events A and B, $P(\text{A or B}) = P(\text{A}) + P(\text{B}) - P(\text{A and B})$.

Check for Understanding – Learning Targets 4 and 5

____ *I can use a two-way table to find probabilities.*

____ *I can calculate probabilities with the general addition rule.*

Ben Roethlisberger played quarterback for the Pittsburgh Steelers from 2004 to 2021. During this this period, the Steelers played a total of 289 games, of which they won 188. Ben played in 249 games, of which the Steelers won 165. Define the following events: B: Ben played and W: the Steelers won.

1. Use a two-way table to display the sample space.

2. Use a Venn diagram to display the sample space.

3. One of the 289 games is selected at random. Find $P(\text{B or W})$. Show your work.

4. What is the complement of $P(\text{B or W})$? Write the complement using probability notation, then calculate and interpret the probability.

Section 4C: Conditional Probability and Independent Events

Section Summary

Two important concepts are introduced in this section: conditional probability and independence. These concepts will reappear throughout the remainder of your studies in statistics, so it is important that you understand what they mean! This section will also introduce you to several rules for calculating probabilities: the conditional probability formula, the general multiplication rule, and the multiplication rule for independent events. Not only do you want to know how to use these rules, but also when. As you perform probability calculations, make sure you can justify why you are using a particular rule!

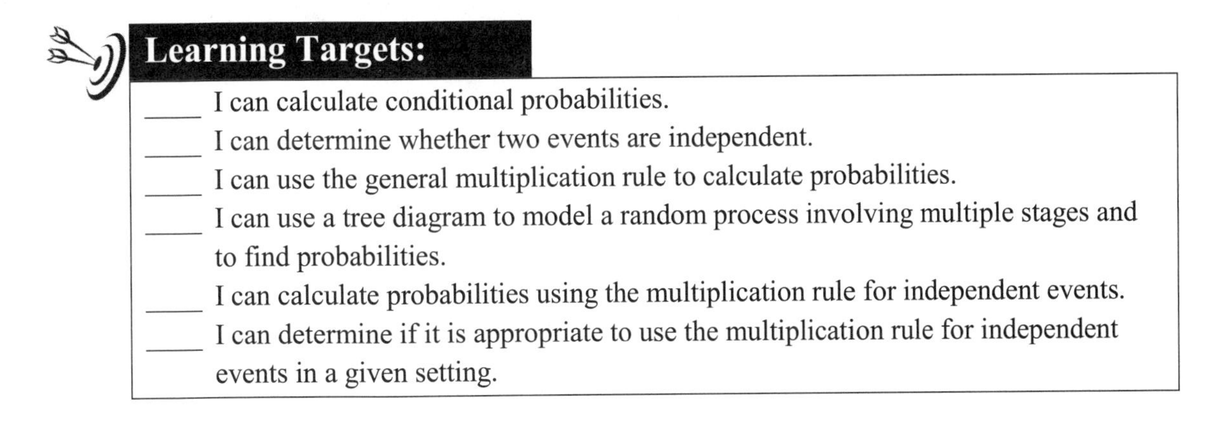

Learning Targets:

____ I can calculate conditional probabilities.
____ I can determine whether two events are independent.
____ I can use the general multiplication rule to calculate probabilities.
____ I can use a tree diagram to model a random process involving multiple stages and to find probabilities.
____ I can calculate probabilities using the multiplication rule for independent events.
____ I can determine if it is appropriate to use the multiplication rule for independent events in a given setting.

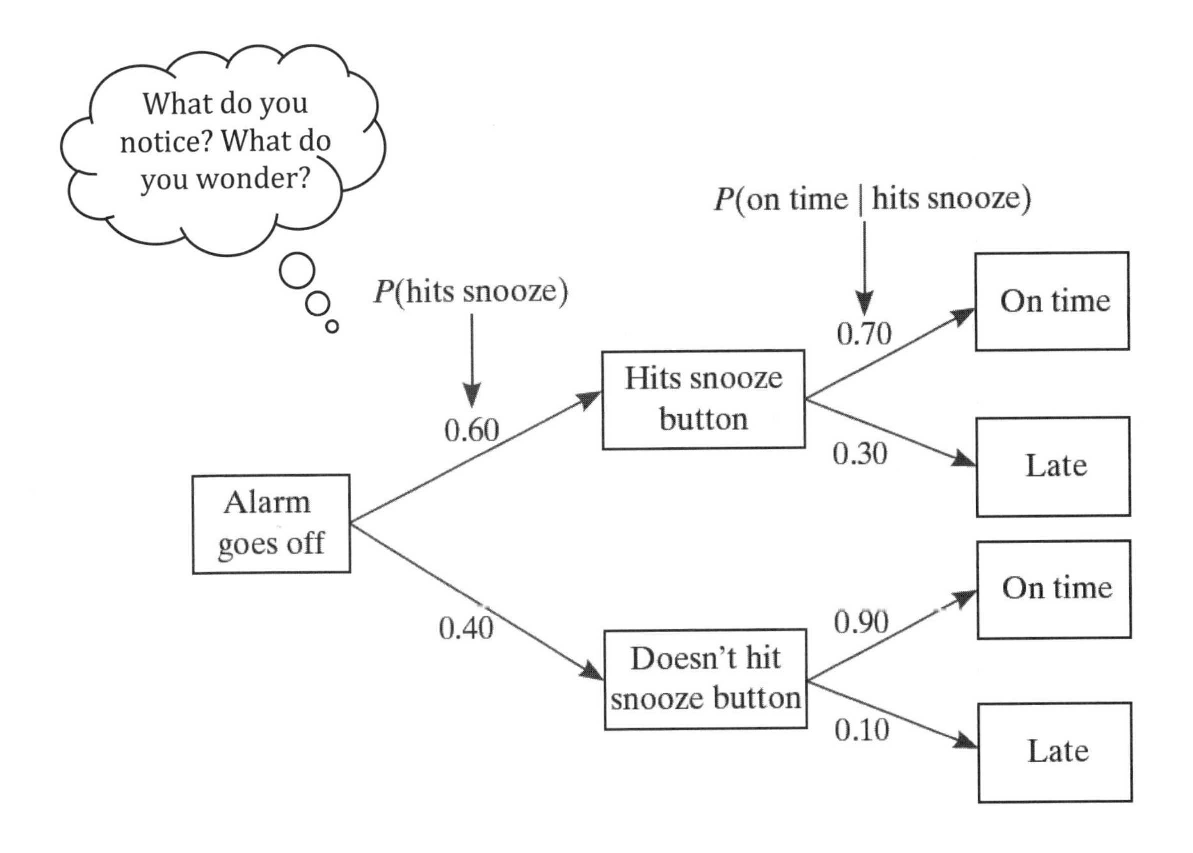

Read: Vocabulary and Key Concepts

- The probability that one event happens given that another event is known to have happened is called a __________________________________. The conditional probability that event A happens given that event B has happened is denoted by $P(\text{A} \mid \text{B})$.
- We often use the phrase "________________" to signal the condition.
- Conditional probability formula: $P(\text{A} \mid \text{B}) =$ ____
- A and B are _____________________________________ if knowing whether or not one event has occurred does not change the probability that the other event will happen. In other words, events A and B are independent if ____. Alternatively, events A and B are independent if ____.
- For any random process, the probability that events A and B both occur can be found using the general multiplication rule: $P(\text{A and B}) =$ ____.
- A __________________________ shows the sample space of a random process involving multiple stages. The probability of each outcome is shown on the corresponding branch of the tree. All probabilities after the first stage are conditional probabilities.
- The multiplication rule for independent events says that if A and B are independent events, the probability that A and B both occur is $P(\text{A and B}) =$ __________________________________.
- Events A and B are ________________________ if and only if $P(\text{A and B}) = P(\text{A}) \cdot P(\text{B})$.
- The multiplication rule for independent events can also be used to help find $P(\text{at least one})$ because the events "at least one" and "none" are ____.
- Two mutually exclusive events with nonzero probabilities can *never* be ______________________, because if one event happens, the other event is guaranteed not to happen.

Watch: Go to bfwpub.com/TPS7eStrive. Select Unit 4, Part I and watch the following Example videos:

Video	Topic	Done
Unit 4C, A Titanic disaster, revisited (p. 349)	*Conditional probability*	
Unit 4C, Facebook versus Instagram (p. 352)	*Conditional probability*	
Unit 4C, Are more math teachers left-handed? (p. 353)	*Conditional probability and independence*	
Unit 4C, Losing your marbles? (p. 356)	*The general multiplication rule*	
Unit 4C, Do people read more ebooks or print books? (p. 358)	*Tree diagrams and probability*	
Unit 4C, How reliable are mammograms? (p. 359)	*Tree diagrams and probability*	
Unit 4C, The Challenger disaster (p. 362)	*Multiplication rule for independent events*	
Unit 4C, Watch the weather! (p. 364)	*Use probability rules wisely*	

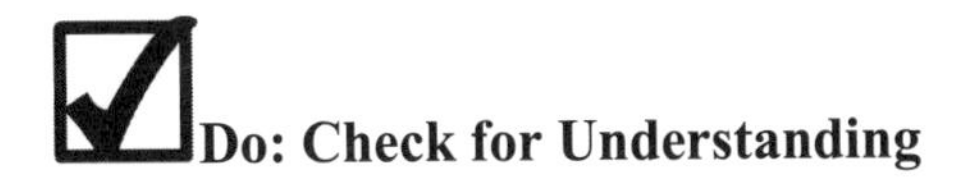

Do: Check for Understanding

Concept 1: Conditional Probability

A conditional probability describes the chance that an event will occur given that another event is already known to have happened. To note that we are dealing with a conditional probability, we use the symbol | to mean "given that." For example, suppose we draw one card from a shuffled deck of 52 playing cards. We could write "the probability that the card is an ace given that it is a red card" as *P*(ace | red).

Concept 2: Independence

We say that two events are independent when knowing that one event has occurred has no effect on the probability of another event occurring. That is, events A and B are independent if the probability that A occurs, *P*(A), is the same whether B occurs, *P*(A | B), or not, $P(A \mid B^C)$. Suppose you are holding a well-shuffled deck of playing cards. You hold out the cards and I select one at random. What is the probability that I selected an ace? Yes, 4/52 because 4 of the 52 cards are aces. Without showing you my card, I give you one clue. My card is red. You know there are 26 red cards in the deck and there are 2 red aces. What is *P*(ace | red)? It's 2/26, which is the same as 4/52. Did the clue "the card is red" affect the probability that the card is an ace? No! That's because "ace" and "red card" are independent events. The probability that the card is an ace is the same whether the card is red or not.

Check for Understanding – Learning Targets 1 and 2

_____ *I can calculate conditional probabilities.*

_____ *I can determine whether two events are independent.*

The sinking of the Titanic is a well-known tragedy. A lesser known tragedy was the sinking of a ship by a German U-boat during World War II which contained 100 children who were being evacuated from England to Canada. The table shows the relationship between type of passenger and outcome.

		Type of Passenger		
		Child	Adult	Total
Outcome	Survived	19	129	148
	Died	81	177	258
	Total	100	306	406

Suppose we select one passenger at random.

1. What is the probability the passenger survived?

2. What is the probability the passenger survived given that the passenger is a child?

3. Are the events "survived" and "child" independent? Justify your answer.

Concept 3: General Multiplication Rule

When events are dependent, we can calculate the probability of "A and B" using the general multiplication rule. This rule considers the fact that $P(\text{B})$ changes given that we know A has occurred. General multiplication rule: $P(\text{A and B}) = P(\text{A}) \cdot P(\text{B} \mid \text{A})$.

Concept 4: Tree Diagrams

When random behavior involves a sequence of events, we can model it using a tree diagram. A tree diagram provides a branch for each outcome of an event along with the associated probabilities of those outcomes. Successive branches represent particular sequences of outcomes. To find the probability of an event, we multiply the probabilities on the branches that make up the event.

Check for Understanding – Learning Targets 3 and 4

____ *I can use the general multiplication rule to calculate probabilities.*

____ *I can use a tree diagram to model a random process involving multiple stages and to find probabilities.*

A statistics teacher decides to raffle off 2 homework passes. Of the 25 students in the class, 10 are athletes who really want to win because they are so busy after school with sports. The other 15 students are not athletes. Two students will be selected at random by drawing names from a hat.

1. What is the probability that two athletes will win the homework passes?

2. Draw a tree diagram to model this random process.

3. What is the probability that an athlete won the second homework pass?

4. What is the probability that the first homework pass went to an athlete given that second homework pass went to an athlete?

Concepts 5 and 6: Multiplication Rule for Independent Events

If events A and B are independent, then $P(B)$ does not change regardless of whether A happened or not. Therefore, if A and B are independent the general multiplication rule – $P(\text{A and B}) = P(\text{A}) \cdot P(\text{B} \mid \text{A})$ – is simplified to, $P(\text{A and B}) = P(\text{A}) \cdot P(\text{B})$. Likewise, if we know that $P(\text{A})$ times $P(\text{B})$ is equal to $P(\text{A and B})$ then we can say that events A and B are independent. Hooray!

Check for Understanding – Learning Targets 5 and 6

____ *I can calculate probabilities using the multiplication rule for independent events.*

____ *I can determine if it is appropriate to use the multiplication rule for independent events in a given setting.*

Have you ever played Yahtzee? In this fun dice-rolling game, you roll 5 dice. If all 5 dice show the same number, then you get "Yahtzee" which is worth 50 points.

1. What is the probability of getting all sixes when you roll 5 dice?

2. You get Yahtzee regardless of the number showing, as long as all dice show the same number. This means there are 6 ways you can get a Yahtzee (all 6's, all 5's, all 4's, all 3's, all 2's, or all 1's). What is the probability of getting Yahtzee when you roll 5 dice?

Note: In the actual game of Yahtee, you get three chances on each turn, where you can choose which outcomes to keep and which dice you want to roll again, so the probability of rolling Yahtzee on a single turn (4.61%) is much greater than the probability of rolling Yahtzee on a single roll.

Checking independence.

3. Suppose events A and B are such that $P(\text{A}) = 0.60$, $P(\text{B}) = 0.75$, and $P(\text{A and B}) = 0.45$. Are events A and B independent? Justify your answer.

4. Since 1997, NBA players have made 73% of the first of two foul shots in a one-and-one situation. If the first shot is missed, the ball goes to the team that rebounds the basketball. If the player makes the first shot, they get to shoot a second time. Is it reasonable to say that the probability of making two foul shots in a row is (0.73)(0.73)? Why or why not?

Unit 4, Part I Summary: Probability

Probability describes the long-term behavior of random processes. Because random occurrences display patterns of regularity after many repetitions, we can use the rules of probability to determine the likelihood of observing particular results. At this point, you should be comfortable with the basic definition and rules of probability. In the next unit, you will study some further concepts in probability so we can build the foundation necessary for statistical inference.

Note that the AP exam may contain several questions about the probability of particular events. Make sure you understand how and when to apply each formula. More importantly, make sure you show your work when calculating probabilities so anyone reading your response understands exactly how you arrived at your answer!

How well do you understand each of the learning targets?

Target	Got It!	Almost There	Needs Work
I can interpret probability as a long-run relative frequency.			
I can estimate probabilities using simulation.			
I can give a probability model for a random process with equally likely outcomes and use it to find the probability of an event.			
I can calculate probabilities using the complement rule.			
I can use the addition rule for mutually exclusive events to find probabilities.			
I can use a two-way table to find probabilities.			
I can calculate probabilities with the general addition rule.			
I can calculate conditional probabilities.			
I can determine whether two events are independent.			
I can use the general multiplication rule to calculate probabilities.			
I can use a tree diagram to model a random process involving multiple stages and to find probabilities.			
I can calculate probabilities using the multiplication rule for independent events.			
I can determine if it is appropriate to use the multiplication rule for independent events in a given setting.			

Unit 4, Part I Multiple Choice Practice

Directions. Identify the choice that best completes the statement or answers the question. Check your answers and note your performance when you are finished.

1. The probability that you will win a prize in a carnival game is about 1/7. During the last nine attempts, you have failed to win. You decide to give it one last shot. Assuming the outcomes are independent from game to game, what is the probability that you will win?
   (A) 1/7
   (B) $(1/7) - (1/7)^9$
   (C) $(1/7) + (1/7)^9$
   (D) 1/10
   (E) 7/10

2. A friend has placed a large number of plastic disks in a hat and invited you to select one at random. He informs you that half are red, and half are blue. If you draw a disk, record the color, replace it, and repeat 100 times, which of the following is true?
   (A) It is unlikely you will choose red more than 50 times.
   (B) If you draw 10 blue disks in a row, it is more likely you will draw a red on the next try.
   (C) The overall proportion of red disks drawn should be close to 0.50.
   (D) The chance that the 100th draw will be red depends on the results of the first 99 draws.
   (E) All of the above are true.

3. The two-way table below shows the relationship between political party and survival status for the 46 presidents of the United States as of January 2024.

	None	Federalist	Democratic Republican	Whig	Democrat	Republican	Total
Dead	1	1	4	4	13	17	40
Alive	0	0	0	0	4	2	6
Total	1	1	4	4	17	19	46

   You select one U.S. President at random. Which of the following statements is true about the events "Alive" and "Whig"?
   (A) The events are mutually exclusive and independent.
   (B) The events are mutually exclusive but not independent.
   (C) The events are not mutually exclusive, but they are independent.
   (D) The events are not mutually exclusive, and they are not independent.
   (E) There is not enough information to determine if the events are independent.

4. People with type O-negative blood are universal donors. That is, any patient can receive a transfusion of O-negative blood. Only 7.2% of the American population has O-negative blood. If 10 people appear at random to give blood, what is the probability that at least 1 of them is a universal donor?
   (A) 0.280
   (B) 0.474
   (C) 0.526
   (D) 0.720
   (E) 1

5. A die is loaded so that the number 6 comes up three times as often as any other number. What is the probability of rolling a 4, 5, or 6?
   (A) 2/3
   (B) 1/2
   (C) 1/3
   (D) 1/4
   (E) 5/8

6. You draw two candies at random from a bag that has 20 red, 10 green, 15 orange, and 5 blue candies without replacement. What is the probability that both candies are red?
   (A) 0.1551
   (B) 0.1600
   (C) 0.2222
   (D) 0.4444
   (E) 0.8000

7. Event A occurs with probability 0.5. Event B occurs with probability 0.6. The probability that both A and B will occur is 0.1. Which of the following statements is true?
   (A) Events A and B are independent.
   (B) Events A and B are mutually exclusive.
   (C) Either A or B always occurs.
   (D) Events A and B are complementary.
   (E) None of the above is correct.

8. Event A occurs with probability 0.8. The conditional probability that event B occurs, given that A occurs, is 0.5. What is the probability that both A and B occur?
   (A) 0.3
   (B) 0.4
   (C) 0.625
   (D) 0.8
   (E) 1.0

9. At Lakeville South High School, 60% of students have high-speed internet access, 30% have a laptop, and 20% have both. What percentage of students at this high school have neither high-speed internet access nor a laptop?
   (A) 0%
   (B) 10%
   (C) 30%
   (D) 80%
   (E) 90%

10. Experience has shown that a certain lie detector will show a positive reading (indicates a lie) 10% of the time when a person is telling the truth and 95% of the time when a person is lying. Suppose that a random sample of 5 suspects is subjected to a lie detector test regarding a recent one-person crime. What is the probability of observing no positive readings if all suspects plead innocent and are telling the truth?
   (A) 0.409
   (B) 0.735
   (C) 0.00001
   (D) 0.591
   (E) 0.99999

Check your answers below. If you got a question wrong, check to see if you made a simple mistake or if you need to study that concept more. After you check your work, identify the concepts you feel very confident about and note what you will do to learn the concepts in need of more study.

#	Answer	Concept	Right	Wrong	Simple Mistake?	Need to Study More
1	A	Probability Basics				
2	C	Definition of Probability				
3	B	Mutually Exclusive/Independent				
4	C	Probability Calculations				
5	E	Probability Calculations				
6	A	Probability Calculations				
7	C	Probability Basics				
8	B	Conditional Probabilities				
9	C	General Addition Rule				
10	D	Conditional Probabilities				

FRAPPY! Free Response AP® Problem, Yay!

The following problem is modeled after actual Advanced Placement Statistics free response questions. Your task is to generate a complete, concise response in 15 minutes. After you generate your response, read over the solution and scoring guide. Score your response and note what, if anything, you would do differently to increase your own score.

A simple random sample of adults in a metropolitan area was selected and a survey was administered to determine political views. The results are recorded below:

		Political Views		
Age	**Conservative**	**Moderate**	**Liberal**	**Total**
18-29	10	15	30	55
30-44	20	30	35	85
45-59	35	15	20	70
Over 60	20	15	10	45
Total	85	75	95	255

(a) What is the probability that a person chosen at random from this sample will have moderate political views?

(b) What is the probability that a person chosen at random from those in the sample who are between the ages of 30 and 44 will have moderate political views? Show your work.

(c) Based on your answers to (a) and (b), are political views and age independent for the population of adults in this metropolitan area? Why or why not?

FRAPPY! Scoring Rubric

Use the following rubric to score your response. Each part receives a score of "Essentially Correct," "Partially Correct," or "Incorrect." When you have scored your response, reflect on your understanding of the concepts addressed in this problem. If necessary, note what you would do differently on future questions like this to increase your score.

Intent of the Question

The goal of this question is to determine your ability to calculate probabilities and determine whether or not two events are independent.

Solution

(a) $P(\text{moderate}) = \frac{75}{255} \approx 0.2941$

(b) $P(\text{moderate} \mid \text{age } 30\text{-}44) = \frac{30}{85} \approx 0.3529$

(c) If moderate political views and age were independent, the probabilities in (a) and (b) would be the same. Since they are not equal, age and political views are not independent for the individuals in this sample.

Scoring:

Parts (a), (b), and (c) are scored as essentially correct (E), partially correct (P), or incorrect (I).

Part (a) is essentially correct if the probability is correct with supporting work shown. Part (a) is partially correct if the correct formula is shown, but minor arithmetic errors are present, or if the correct answer without supporting work is provided. Otherwise, it is incorrect.

Part (b) is essentially correct if the conditional probability is calculated correctly with supporting work shown. Part (b) is partially correct if the conditioning is reversed and $P(\text{age } 30\text{-}44 \mid \text{moderate}) = \frac{30}{75} = 0.40$ is calculated or if the correct answer without supporting work is provided. Otherwise, it is incorrect.

Part (c) is essentially correct if the response indicates the two variables are not independent and clearly justifies the conclusion based on an appropriate probability argument using the results in parts (a) and (b). Part (c) is partially correct if the response indicates the two variables are not independent, but the argument is weak or based on a different, but appropriate probability argument than comparing results of parts (a) and (b). Otherwise, it is incorrect.

4 Complete Response
All three parts essentially correct

3 Substantial Response
Two parts essentially correct and one part partially correct

2 Developing Response
Two parts essentially correct and no parts partially correct
One part essentially correct and one or two parts partially correct
Three parts partially correct

1 Minimal Response
One part essentially correct and no parts partially correct
No parts essentially correct and two parts partially correct

My Score:
What I did well:
What I could improve:
What I should remember if I see a problem like this on the AP Exam:

Unit 4, Part I Formula Sheet

Create a one-page summary of important concepts and formulas found in this unit. This will serve as a valuable resource as you prepare for the AP Exam.

Unit 4, Part I Crossword Puzzle

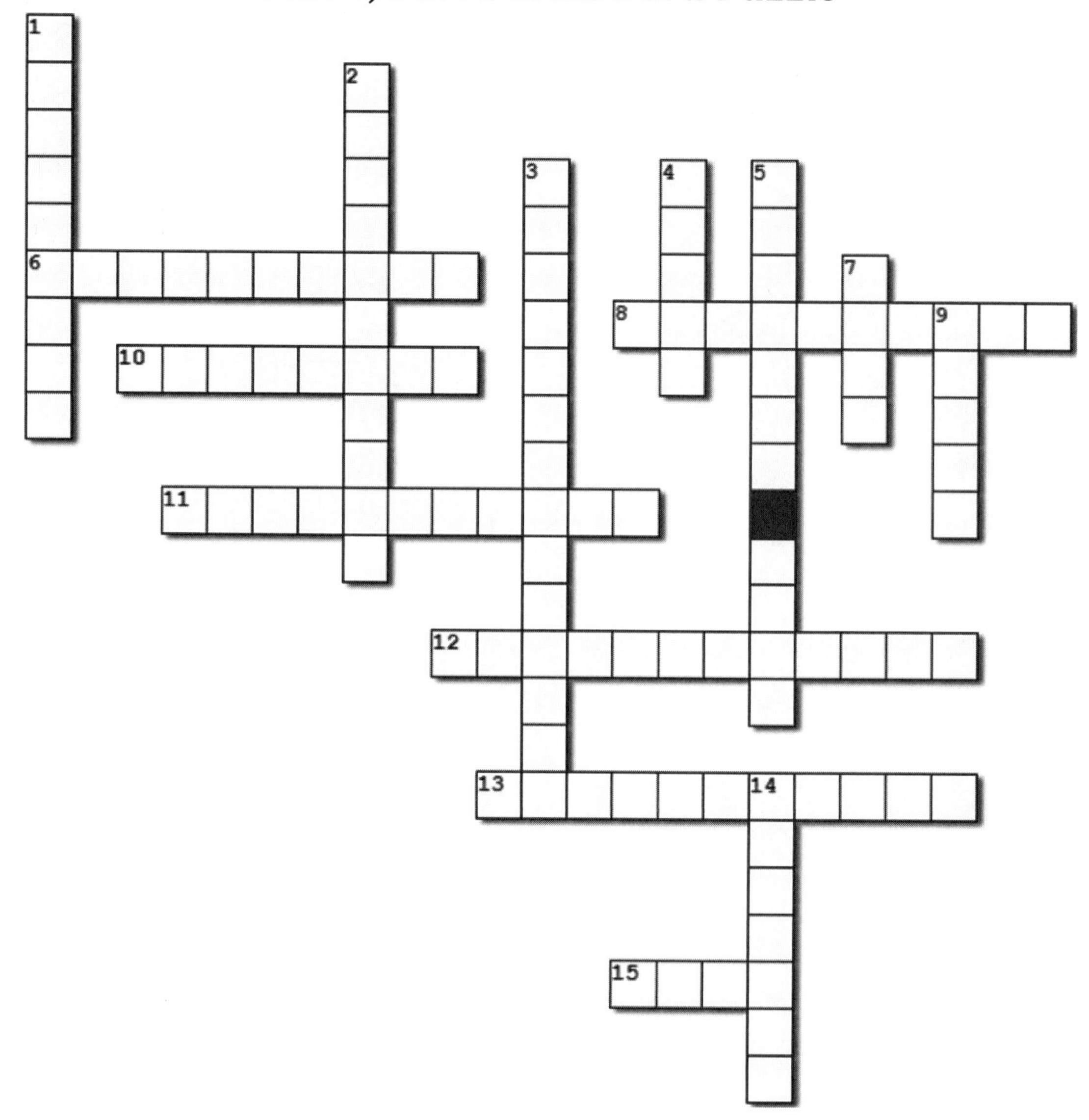

Across:

6. A ____ imitates a random process in such a way that simulated outcomes are consistent with real-world outcomes.
8. The event that event A does not happen is the ____ of event A.
10. *P*(A or B) can be found using the general ____ rule.
11. The proportion of times an outcome would occur in a very long series of repetitions.
12. The event "A and B" is known as the ____ of event A and event B.
13. Events A and B are ____ if $P(A) = P(A \mid B)$.
15. A ____ diagram can help model chance behavior that involves a sequence of outcomes.

Down:

1. Events A and B are mutually ____ if *P*(A and B) = 0.
2. The probability that one event happens given another event is known to have happened.
3. The probability that two events both occur can be found using the general ____ rule.
4. The event "A or B" is known as the ____ of A and B.
5. The set of all possible outcomes for a chance process. (2 words).
7. A ____ diagram can be used to display the sample space and help find probabilities for a random process involving two events.
9. A collection of outcomes from a chance process.
14. The law of large ____ says that the proportion of times that a specific outcome occurs approaches its probability.

Unit 4, Part II: Probability, Random Variables, and Probability Distributions

"We must become more comfortable with probability and uncertainty." Nate Silver

UNIT 4 PART II OVERVIEW

In the last unit we learned the basic rules of probability. We continue our study of probability by exploring situations that involve the assignment of a numerical value to each possible outcome of a random process. In this unit, you will learn how to calculate probabilities of events involving random variables as well as how to describe their probability distributions. Specifically, you will learn formulas to determine the mean and standard deviation of individual random variables as well as the combination of several independent random variables. Finally, you'll explore two special random variables – binomial and geometric – and learn how to calculate probabilities of events in binomial and geometric settings. This unit involves a lot of formulas, so you may want to familiarize yourself with the formula sheet provided on the AP® Statistics exam. As in earlier units, you should focus less on memorizing formulas or calculator keystrokes and more on how to apply the formulas and interpret results.

Sections in Unit 4, Part II
Section 4D: Introduction to Discrete Random Variables
Section 4E: Transforming and Combining Random Variables
Section 4F: Binomial and Geometric Random Variables

Your teacher announces a 10-question multiple choice pop quiz. There are 5 choices for each question. How many are you likely to get right if you guess on all of them?

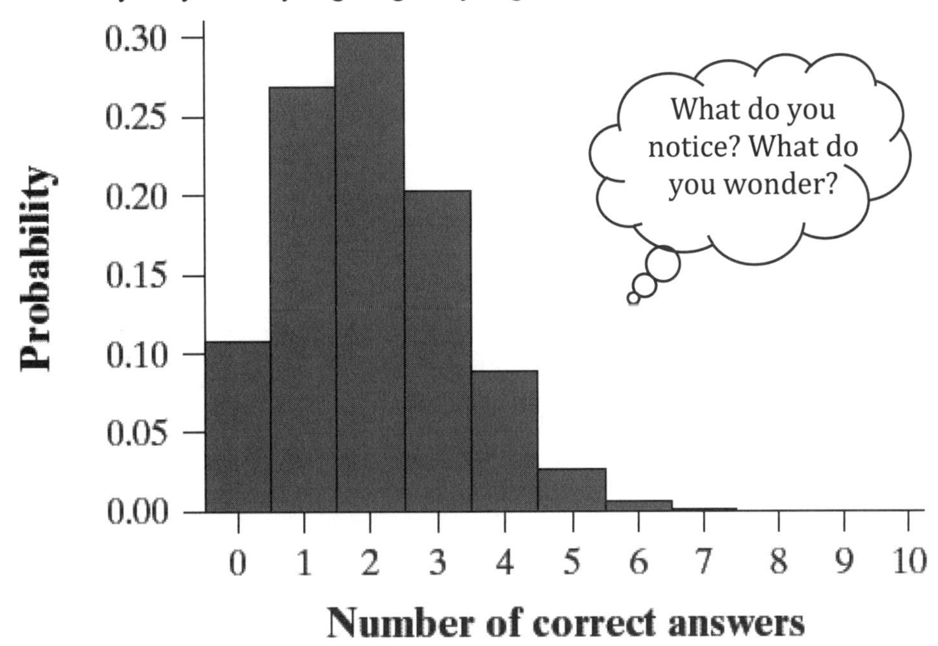

PRACTICE FOR MASTERY

Use the following *suggested* guide to the pages and exercises in your text to practice for mastery!
Note: your teacher may assign different problems. Be sure to follow their instructions!

Day	Topics	Read	Do
1	◆ Discrete Random Variables and Probability ◆ Displaying Discrete Probability Distributions	p. 381-385	**4D**: 1, 3, 5, 7, 9, 27, 28
2	◆ Measuring Center: The Mean, or Expected Value, of a Discrete Random Variable ◆ Measuring Variability: The Standard Deviation of a Discrete Random Variable	p. 386-391	**4D**: 11, 13, 15, 17, 19
3	◆ Transforming Random Variables ◆ Combining Random Variables: The Mean ◆ Combining Random Variables: The Standard Deviation	p. 397-410	**4E**: 1, 3, 7, 9, 11, 13, 15, 19, 24, 25
4	AP Classroom Topic Questions (Topics 4.7-4.9)		**4D**: 23 – 26 **4E**: 5, 17, 21, 26, 27
5	◆ Binomial Random Variables ◆ Calculating Binomial Probabilities	p. 415-424	**4F**: 1, 3, 5, 7, 9, 11, 13, 15
6	◆ Describing a Binomial Distribution: Shape, Center, and Variability	p. 425-430	**4F**: 17, 19, 21, 23
7	◆ Geometric Random Variables and Probability ◆ Describing a Geometric Distribution ◆ Shape, Center, and Variability	p. 430-436	**4F**: 25, 27, 29, 31, 37 – 40
8	Unit 4, Part II AP® Statistics Practice Test (p. 448)		Unit 4, Part II Review Exercises (p.446)
9	Unit 4, Part II Test: Celebration of learning!	Note: Since "Test" can sound daunting, think of this as a way to show off all you have learned. It's your own celebration of learning!	

Section 4D: Introduction to Discrete Random Variables

Section Summary

A random variable takes numerical values that describe the outcomes of a random process. In the last unit, we learned that a probability model describes the possible outcomes for a random process and the probability of each outcome. A random variable does the same thing, describing the possible values that the variable takes and the probability of each. Random variables fall into two categories: discrete and continuous. What differentiates the two is the set of values the random variable can take. If the set is limited to fixed values with gaps between, it is discrete. If the variable can take on any value in an interval, it is continuous. Regardless of the type, we are interested in describing the shape of the random variable's probability distribution, its center, and its variability. Knowing these characteristics will give us a sense of what to expect in repeated observations of the random variable as well as what can be considered likely and unlikely results. This idea forms the basis for inferential thinking in later units, so you want to get used to thinking along these lines in this section!

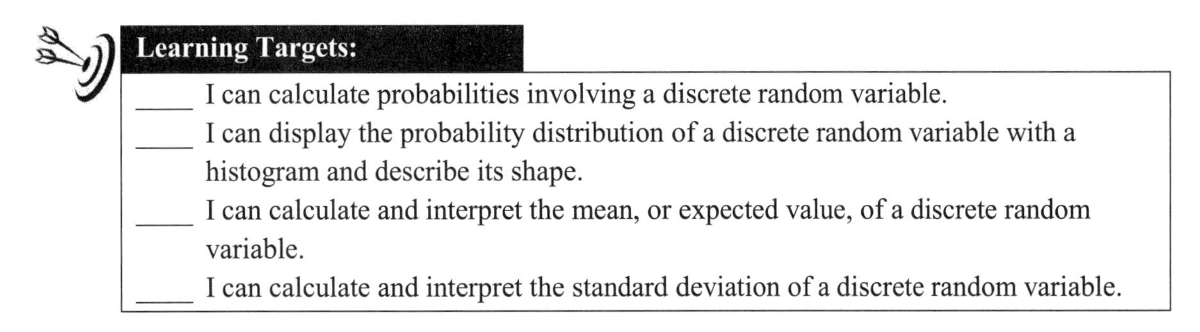

Learning Targets:

____ I can calculate probabilities involving a discrete random variable.

____ I can display the probability distribution of a discrete random variable with a histogram and describe its shape.

____ I can calculate and interpret the mean, or expected value, of a discrete random variable.

____ I can calculate and interpret the standard deviation of a discrete random variable.

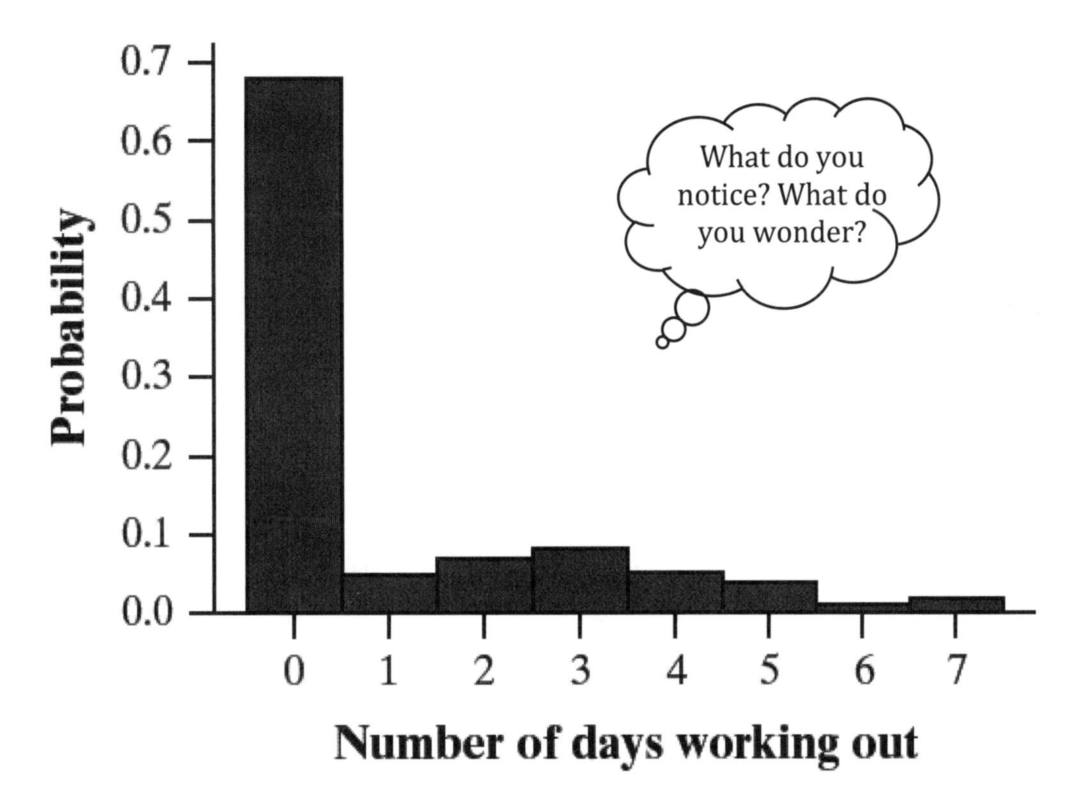

Read: Vocabulary and Key Concepts

- A ______________________ takes numerical values that describe the outcomes of a random process.
- The ________________________________ of a random variable gives its possible values and their probabilities.
- The probability distribution of the random variable X models the ___________________ distribution of the quantitative variable
- A _______________________________________ takes a countable set of possible values with gaps between them on a number line.
- For a probability distribution to be valid, the probabilities must satisfy two requirements:
  1. Every probability is a number between ____ and ____ (inclusive).
  2. The sum of the probabilities is ____.
- The __________, or_________________, of a discrete random variable is its average value over many, many trials of the same random process.
- To find the mean, or expected value, of X, multiply each possible value of X by its probability, then add all the products: $\mu_X = E(X) =$ ____.
- The _____________ of a discrete random variable is the 50th percentile of its probability distribution.
- We can find the median from a ____.
- The ____________________________ of a discrete random variable measures how much the values of the variable typically vary from the mean in many, many trials of the random process.
- Write the formula. $\sigma_X =$ ____

Watch: Go to bfwpub.com/TPS7eStrive. Select Unit 4, Part II and watch the following Example videos:

Video	Topic	Done
Unit 4D, Apgar scores: Babies' health at birth (p. 383)	*Discrete random variables and probability*	
Unit 4D, Pete's Jeep Tours (p. 385)	*Displaying discrete probability distributions*	
Unit 4D, Apgar scores: what's typical? (p. 367)	*The mean, or expected value, of a discrete random variable*	
Unit 4D, How much do Apgar scores vary? (p. 389)	*The standard deviation of a discrete random variable*	

Do: Check for Understanding

Concepts 1 and 2: Discrete Random Variables, Probability Distribution Histogram

A random variable can be classified as either discrete or continuous depending on its possible values. If it takes a fixed (finite or infinite) set of possible values with gaps in between, then we call it a discrete random variable. To describe a random variable, we follow the same process as describing a probability model. First, define the random variable, X, as a numerical outcome of a random process. Next, indicate the possible values of the variable. Finally, give the probability that each value occurs using a table or graph.

Check for Understanding – Learning Targets 1 and 2

____ *I can calculate probabilities involving a discrete random variable.*

____ *I can display the probability distribution of a discrete random variable with a histogram and describe its shape.*

Consider two 4-sided dice, each having sides labeled 1, 2, 3, 4.
Let X = the sum of the numbers that appear after a roll of the dice.

First roll

Second roll	1	2	3	4
1				
2				
3				
4				

1. Complete the probability distribution table, then sketch the probability histogram distribution of X. Describe the shape of the distribution.

X							
$P(X)$							

2. If somebody rolled the dice 10 times and got a sum of 3 or less each time, would you be surprised? Why or why not?

Concepts 3 and 4: Mean and Standard Deviation of a Discrete Random Variable

In Unit 1, we learned that when describing distributions of quantitative data, we should always note the shape, center, and variability. The same holds true for random variables. In order to make inferences, we need to know what a "typical" value of the random variable is as well as how much variation around that value we can expect to see. As with distributions of quantitative data, the center of a random variable's probability distribution can be described by calculating the mean. However, in the case of random variables, the mean (or expected value) is a *weighted* average, taking into account the probability of each outcome occurring. Similarly, the standard deviation of a random variable takes into account the probability of each outcome occurring, giving more weight to those outcomes that are more likely.

Check for Understanding – Learning Targets 3 and 4

____ *I can calculate and interpret the mean, or expected value, of a discrete random variable.*

____ *I can calculate and interpret the standard deviation of a discrete random variable.*

Suppose the random variable Y = number of goals in a randomly selected high school hockey game has the following probability distribution:

Goals	0	1	2	3	4
Probability	0.155	0.295	0.243	0.233	0.074

1. Calculate the mean of Y. Interpret this parameter.

2. Find the median. Show your method clearly.

3. Calculate the standard deviation of S. Interpret this parameter.

(*Need Tech Help?* View the video **Tech Corner 11: Analyzing Discrete Random Variables** at bfwpub.com/TPS7eStrive)

Section 4E: Transforming and Combining Random Variables

Section Summary

This section introduces two distinct topics. First, you will explore transforming a single random variable. That is, you will learn how to describe the shape, center, and variability of the probability distribution of a random variable when a linear transformation (such as adding a constant to each value or multiplying each value by a constant) is applied. Second, you will learn how to combine two or more random variables. This topic is critical as many of the statistical inference problems we will explore involve observing the difference between two random variables. You will learn how to describe the mean and standard deviation of the sum and difference of independent random variables. There are a lot of formulas to keep straight in this section, so you may wish to keep your AP® formula sheet handy!

Learning Targets:

____ I can describe the effect of a linear transformation – adding or subtracting a constant and/or multiplying or dividing by a constant – on the probability distribution of a random variable.

____ I can calculate the mean of a sum, difference, or other linear combination of random variables.

____ I can, if appropriate, calculate the standard deviation of a sum, difference, or other linear combination of random variables.

Read: Vocabulary and Key Concepts

Linear Transformations:

- Adding the same positive number *a* to or subtracting a from each value of a random variable:
  - Adds *a* to or subtracts *a* from the measures of ____________ (mean, median).
  - Does not change measures of _________________ (range, *IQR*, standard deviation).
  - Does not change the _____________ of the probability distribution.
- Multiplying or dividing each value of random variable by the same positive number *b*:
  - ______________________________ the measures of center (mean, median) by *b*.
  - ______________________________ measures of variability (range, *IQR*, standard deviation) by *b*.
  - _________________________________ the shape of the distribution.
- If $Y = a + bX$ is a linear transformation of the random variable *X*,
  - The probability distribution of *Y* has _________________________ as the probability distribution of *X* if $b > 0$.
  - The mean of $Y = a + b$(the mean of *X*). Written in symbols: ____________________
  - The SD of $Y = |b|$ (the SD of *X*). Written in symbols: ____________________

    Note: "*a*" is not part of this formula because adding a constant to each value of *X* does not affect variability.

Combining Random Variables: The Mean

- For any two random variables X and Y,
  - If $S = X + Y$, the mean of the sum of X and Y is $\mu_S = \mu_X + \mu_Y$. In other words, the mean of the sum of two random variables is equal to the ______________________________.
  - If $D = X - Y$, the mean of the difference of X and Y is $\mu_D = \mu_X - \mu_Y$. In other words, the mean of the difference of two random variables is equal to the ______________________________.

    Note: The order of subtraction matters! $\mu_D = \mu_X - \mu_Y$ is not the same as $\mu_D = \mu_Y - \mu_X$.

Combining Random Variables: The Standard Deviation

- If knowing the value of X does not help us predict the value of Y, then X and Y are ______________ ______________________. In other words, two random variables are independent if knowing the value of one variable does not change the probability distribution of the other variable.
- When we add two independent random variables, their standard deviations ________________.
- The variance of the sum of two independent random variables is the sum of their ______________.
- Regardless of whether we are finding the standard deviation of a sum or the standard deviation of a difference of two independent random variables, the formula is the same: $\sigma_S = \sigma_D =$ ____________.
- *Interpretation*: In many, many random randomly selected [context], the total (or difference in) [context] typically varies from the mean of [value] by about [SD value].

Linear Combinations of Random Variables

- If $aX + bY$ is a linear combination of the random variables X and Y,
  - Its ____________ is $a(\mu_X) + b(\mu_Y)$
  - Its ______________________ is $\sqrt{a^2\sigma_X^2 + b^2\sigma_Y^2}$ if X and Y are independent.

Watch: Go to bfwpub.com/TPS7eStrive. Select Unit 4, Part II and watch the following Example videos:

Video	Topic	Done
Unit 4E, How much does college cost? (p. 401)	*Transforming random variables*	
Unit 4E, Comparing college costs (p. 404)	*Combining random variables: The mean*	
Unit 4E, Comparing college costs (p. 407)	*Combining random variables: The standard deviation*	

Do: Check for Understanding

Concept 1: Transforming a Random Variable

We learned how transformations affect the shape, center, and variability of distributions of quantitative data back in Unit 1, Part II. Similar rules apply to random variables. That is, when we add (or subtract) a constant (a) to each value of a random variable, the shape and variability of the probability distribution do not change. However, the measures of center will increase (or decrease) by a. When we multiply (or divide) each value of a random variable by a positive constant (b), the shape of the probability distribution does not change. However, measures of center are multiplied (divided) by b and measures of variability are multiplied (divided) by b.

Check for Understanding – Learning Target 1

____ *I can describe the effect of a linear transformation – adding or subtracting a constant and/or multiplying or dividing by a constant – on the probability distribution of a random variable.*

A carnival game involves tossing a ball into numbered baskets with the goal of having your ball land in a high-numbered basket. The probability distribution of X = value of the basket on a randomly selected toss.

Points earned	0	1	2	3
Probability	0.3	0.4	0.2	0.1

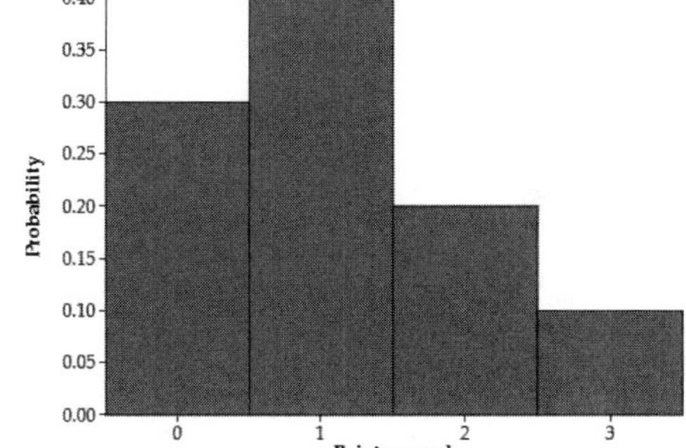

The expected value of X is 1.1 and the standard deviation of X is 0.943.

Suppose it costs \$2 to play and you earn \$1.50 for each point earned on your toss.

Define Y to be the amount of profit you make on a randomly selected toss.

1. What shape does the probability distribution of Y have?

2. Find the mean of Y.

3. Calculate the standard deviation of Y.

Concepts 2 and 3: Combining Random Variables: The Mean and Standard Deviation

Many situations involve two or more random variables. Understanding how to describe the center and variability of the probability distribution for the sum or difference of two random variables is an important skill to have. When given two independent random variables, we can describe the mean and standard deviation of the sum (or difference) of the random variables using the formulas in this section. When adding or subtracting two or more random variables (whether they are independent or not), the mean of the sum or difference of those random variables will be the sum or difference of their means. However, to describe the variability of the sum or difference of *independent* random variables, we must perform two steps. First, we find the *variance* of the sum or difference of two or more independent random variables by *adding* their variances. Then, we take the square root of the variance to find the standard deviation. A common mistake is to simply add standard deviations. Remember to *always add* variances!

Check for Understanding – Learning Targets 2 and 3

____ *I can calculate the mean of the sum, difference, or other linear combination of random variables.*

____ *I can, if appropriate, calculate the standard deviation of a sum, difference, or other linear combination of random variables.*

Students in Mrs. Taylor's class are expected to check their homework in groups of 4 at the beginning of class each day. Students must check it as quickly as possible, one at a time. The means and standard deviations of the time it takes to check homework for the 4 students in one group are noted below. Assume their times are independent.

	Mean	SD
Ann	1.4 min	0.1 min
Barb	1.2 min	0.4 min
Luke	0.9 min	0.8 min
Doug	1.0 min	0.7 min

1. If each student checks one after the other, what are the mean and standard deviation of the total time necessary for these four students to check their homework on a randomly chosen day?

2 Suppose Ann and Doug like to race to see who can check their homework faster. What are the mean and standard deviation for the difference between their times (Doug – Ann)? Interpret these values in the context of the situation.

Section 4F: Binomial and Geometric Random Variables

Section Summary

In the previous two sections, you learned how to describe the probability distributions of discrete random variables as well as how to transform and combine one or more random variables. In this section, you will focus on two special cases of discrete random variables: binomial and geometric. Binomial random variables count the number of successes that occur in a fixed number of independent trials of some random process with a constant probability of success on each trial, while geometric random variables count the number of trials needed to get a success. Binomial random variables appear often on the AP® exam, so you will want to pay particular attention to this topic.

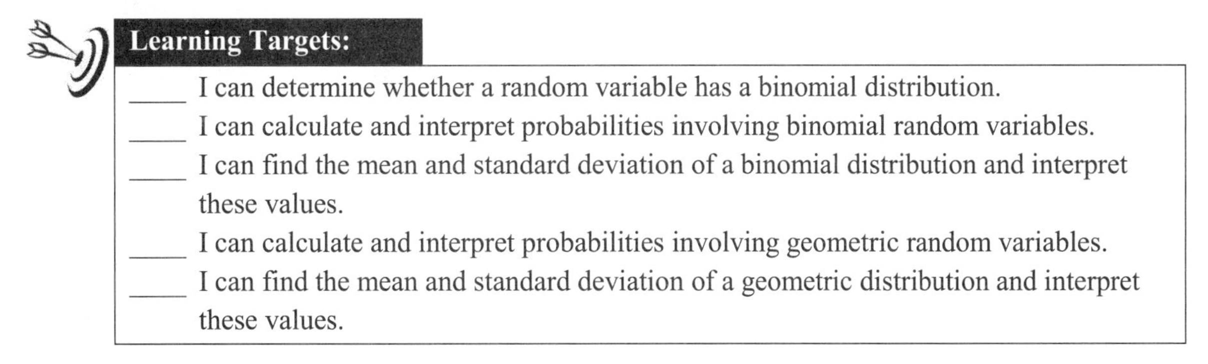

Learning Targets:

____ I can determine whether a random variable has a binomial distribution.
____ I can calculate and interpret probabilities involving binomial random variables.
____ I can find the mean and standard deviation of a binomial distribution and interpret these values.
____ I can calculate and interpret probabilities involving geometric random variables.
____ I can find the mean and standard deviation of a geometric distribution and interpret these values.

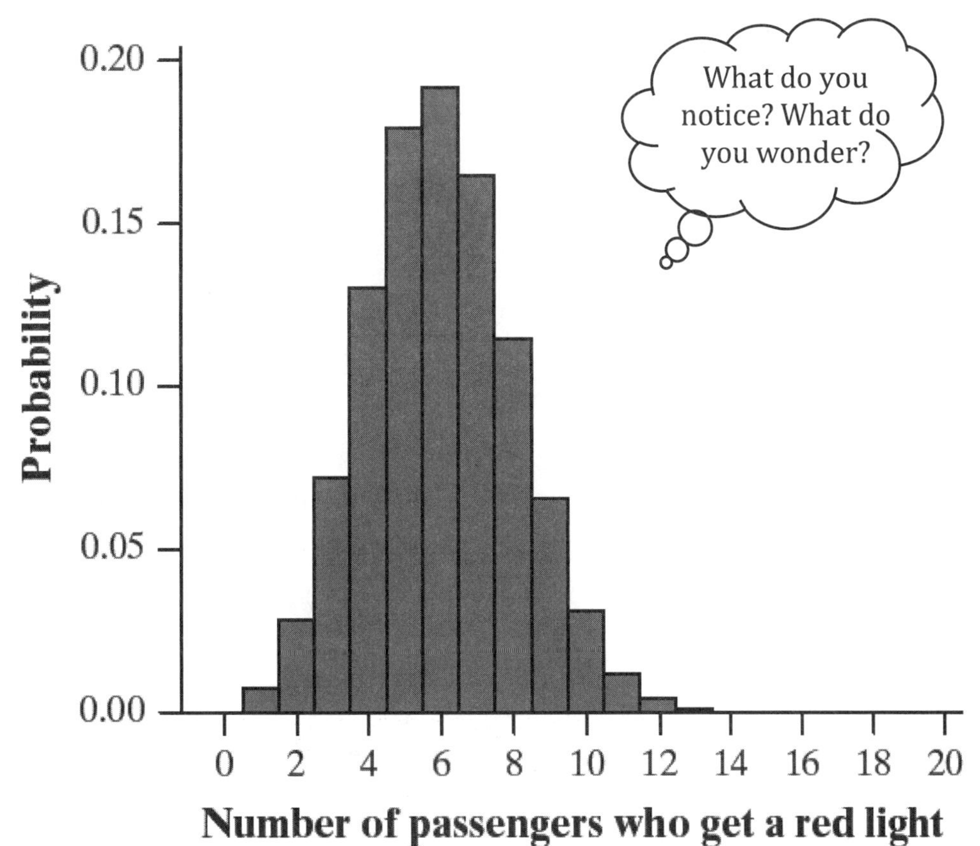

Read: Vocabulary and Key Concepts

Binomial Random Variables and Probability

- A ______________________ arises when we perform n independent trials of the same random process and count the number of times that a particular outcome (called a "success") occurs.
- The four conditions for a binomial setting are:
  - ___________? The possible outcomes of each trial can be classified as "success" or "failure."
  - _______________? Trials must be independent. That is, knowing the outcome of one trial must not tell us anything about the outcome of any other trial.
  - _______________? The number of trials n of the random process must be fixed in advance.
  - _____________________? There is the same probability of success p on each trial.
- The count of __________________ X in a binomial setting is a binomial random variable. The possible values of X are ____________________.
- The probability distribution of X is a __________________________. Any binomial distribution is completely specified by two numbers: the number of trials ____ of the random process and the probability _____ of success on each trial.
- The number of ways to arrange x successes among n trials is given by the ____________________.
- **Binomial Probability Formula:** Suppose that X is a binomial random variable with n trials and probability p of success on each trial. The probability of getting exactly x successes in n trials ($x = 0$, $1, 2, \ldots, n$) is ______________________________ where ______________________________.
- Two steps for calculating probabilities involving binomial, geometric, and normal distributions:

  (1) Define the ______________________ of interest, state how it is ___________________ and identify the ______________________ of interest.

  (2) Perform calculations – show your ___________!

  Be sure to answer the __________________ that was asked.
- How do you use the complement rule to find $P(X \geq 4)$? ____________________________
- When finding the probability of getting less than, more than, or at least a certain number of successes, students sometimes have trouble identifying the third input in the binomcdf command. Here is a tip: Write out the ________________________ of the variable. Circle the ones you ______________ ______________________________ of, and _____________ the rest.
- **Shape of a Binomial Distribution:** When n is small, the probability distribution of a binomial random variable will be roughly symmetric if p is close to _____, right-skewed if p is much less than _____ and left-skewed if p is much greater than _____.
- **Mean and SD of a Binomial Distribution:** If a count X of successes has a binomial distribution with number of trials n and probability of success p, the mean of X is $\mu_X =$ ____ and the standard deviation of X is $\sigma_X =$ ____________________.

- The ____________________ states that when selecting a random sample of size n from a population of size N, we can treat individual observations as independent when performing calculations as long as $n < 0.10N$.

Geometric Random Variables and Probability

- A _______________________ arises when we perform independent trials of the same random process and record the _________________________ it takes to get one success. On each trial, the probability p of success must be the same.
- The number of trials X that it takes to ______________________ in a geometric setting is a geometric random variable.
- The probability distribution of X is a ___________________________ with probability of success ___ on any trial. The possible values of X are ________________.
- **Geometric Probability Formula:** If X has the geometric distribution with probability of success p on each trial, the possible values of X are 1, 2, 3, ... If x is any one of these values, $P(X = x) =$ ___________________________.
- **Geometric Distribution Shape:** Every geometric distribution is ___________________________. This is because the most likely value of any geometric random variable is ____. The probability of each successive value decreases by a factor of ____________.
- If X is a geometric random variable with probability of success p on each trial, then its mean (expected value) is $\mu_X =$ _________ and its standard deviation is $\sigma_X =$ _____________.

Watch: Go to bfwpub.com/TPS7eStrive. Select Unit 4, Part II and watch the following Example videos:

Video	Topic	Done
Unit 4F, From blood types to aces (p. 417)	*Binomial random variables*	
Unit 4F, How many correct on the pop quiz? (p. 421)	*Calculating binomial probabilities*	
Unit 4F, Did Hannah cheat on the pop quiz? (p. 422)	*Calculating binomial probabilities*	
Unit 4F, Free lunch? (p. 424)	*Calculating binomial probabilities*	
Unit 4F, Return of the pop quiz (p. 428)	*Describing a binomial distribution*	
Unit 4F, The Lucky Day game (p. 432)	*Geometric random variables and probability*	
Unit 4F, Waiting for a free lunch (p. 435)	*Describing a geometric distribution*	

Do: Check for Understanding

Concepts 1 and 2: Binomial Random Variables, Probabilities

When we observe a fixed number of repeated trials of the same random process, we are often interested in how many times a particular outcome occurs. This process is the basis for a binomial setting. That is, a binomial setting arises when we perform n independent trials of the same random process and count the number of times a particular outcome occurs.

To be a binomial setting, four conditions must be met. First, we are interested in outcomes that can be classified in one of two ways – success or failure. The particular outcome of interest is considered a success, while anything else is considered a failure. Next, each observed trial of the random process must be independent of the other trials. Third, the number of trials we observe must be fixed in advance. Last, the probability of success on each trial must be the same.

If these conditions are met, we can use the binomial probability formula to determine the likelihood of observing a certain number of successes in a fixed number of trials of the binomial random variable:

$$P(X = x) = \binom{n}{x} p^x (1 - p)^{n-x}$$

Note that the formula uses the multiplication rule for independent events by multiplying the probabilities of successes and failures across the fixed number of trials. However, the formula also considers the number of ways in which we can arrange those successes across our trials, the binomial coefficient. (*Need Tech Help?* View the video **Tech Corner 12: Calculating Binomial Coefficients** at bfwpub.com/TPS7eStrive)

Check for Understanding – Learning Targets 1 and 2

____ *I can determine whether a random variable has a binomial distribution.*

____ *I can calculate and interpret probabilities involving binomial random variables.*

You are one of 10 players at a basketball team practice. The coach places every player's name on a card, put the cards in a hat, and mixed well. Every 15 minutes, the coach will select one name at random. The selected player will do 30 pushups. The card will be returned to the hat and mixed again. This process will continue for a total of 8 selections. Let X = the number of times you are selected to do pushups.

1. Show that X is a binomial random variable. State its probability distribution.

2. Find the probability that you are selected at least once.

(*Need Tech Help?* View the video **Tech Corner 13: Calculating Binomial Probabilities** at bfwpub.com/TPS7eStrive)

Concept 3: Mean and Standard Deviation of a Binomial Distribution

Like other discrete random variables, we can calculate the mean and standard deviation of binomial random variables. This will give us a better sense of what we'd expect to see in the long run as well as how much variability we can expect to observe in the observed number of successes. If a random variable X has a binomial probability distribution based on a random process with n trials each having probability of success p, we can calculate the mean of X by multiplying np. We can find the standard deviation of X by taking the square root of the product $np(1 - p)$. Modified versions of these formulas are useful when trying to make inferences about the proportion of successes in a population, when a sample is selected without replacement. If we select an SRS of size n without replacement from a population, we can treat the observations as independent as long as the sample size is less than 10% of the size of the population.

Check for Understanding – Learning Target 3

____ *I can find the mean and standard deviation of a binomial distribution and interpret these values.*

Suppose 72% of students in the U.S. would give their teachers a positive rating if asked to score their effectiveness. A survey is conducted in which 500 students are randomly selected and asked to rate their teachers. Let X = the number of students in the sample who would give their teachers a positive rating.

1. Explain why the Independent condition for a binomial setting is not met.

2. Explain why it is appropriate to treat the individual observations as independent even though the Independent condition is not met.

3. What probability distribution does X have?

4. Calculate and interpret the mean of X.

5. Calculate and interpret the standard deviation of X.

(*Need Tech Help?* View the video **Tech Corner 14: Graphing Binomial Probability Distributions** at bfwpub.com/TPS7eStrive)

Concepts 4 and 5: Geometric Random Variables – Probabilities, Mean, Standard Deviation

In a binomial setting, we are interested in knowing how many successes will occur in a fixed number of trials. Sometimes we are interested in knowing how long it will take until a success occurs. When we perform independent trials of a random process with the same probability of success on each trial, and record how long it takes to get a success, we have a geometric setting. We can describe the number of trials it takes to get a success using a geometric random variable. As with other random variables, we can describe the probability distribution of a geometric random variable and calculate its mean and standard deviation. Using what we learned in Unit 4, Part I, we can calculate the probability of observing the first success on the xth trial by multiplying the probabilities of $(x - 1)$ consecutive failures by the probability of a success:

$$P(X = x) = (1 - p)^{x-1}p.$$

Check for Understanding – Learning Targets 4 and 5

____ *I can calculate and interpret probabilities involving geometric random variables.*

____ *I can find the mean and standard deviation of a geometric distribution and interpret these values.*

Suppose 20% of Super Crunch cereal boxes contain a secret decoder ring. Let X = the number of boxes of Super Crunch that must be opened until a ring is found.

1. Show that X is a geometric random variable.

2. Find the probability that you will have to open 7 boxes to find a ring.

3. Find the probability that it will take fewer than 4 boxes to find a ring.

4. How many boxes would you expect to have to open to find a ring?

5. What is the standard deviation of the number of boxes you would expect to have to open to find a ring? Interpret this value.

(*Need Tech Help?* View the video **Tech Corner 15: Calculating Geometric Probabilities** at bfwpub.com/TPS7eStrive)

Unit 4, Part II Summary: Discrete Random Variables

In Unit 4, Part I we learned the basic definition and rules of probability. We continued our study of probability in Unit 4, Part II by exploring situations that involve assigning a numerical value to each possible outcome of a random process. You learned how to calculate probabilities of events involving random variables as well as how to describe their probability distributions. You learned formulas to determine the mean and standard deviation of individual random variables as well as the combination of several independent random variables. Finally, you explored two special random variables – binomial and geometric – and learned how to calculate probabilities of events in binomial and geometric settings.

Like earlier units, you should focus less on memorizing formulas or calculator keystrokes and more on how to apply the formulas and interpret results. Make sure you understand how and when to apply each formula. More importantly, make sure you show your work when calculating probabilities so anyone reading your response understands exactly how you arrived at your answer!

How well do you understand each of the learning targets?

Learning Target	Got It!	Almost There	Needs Work
I can calculate probabilities involving a discrete random variable.			
I can display the probability distribution of a discrete random variable with a histogram and describe its shape.			
I can calculate and interpret the mean, or expected value, of a discrete random variable.			
I can calculate and interpret the standard deviation of a discrete random variable.			
I can describe the effect of a linear transformation — adding or subtracting a constant and/or multiplying or dividing by a constant — on the probability distribution of a random variable.			
I can calculate the mean of a sum, difference, or other linear combination of random variables.			
I can, if appropriate, calculate the standard deviation of a sum, difference, or other linear combination of random variables.			
I can determine whether a random variable has a binomial distribution.			
I can calculate and interpret probabilities involving binomial random variables.			
I can find the mean and standard deviation of a binomial distribution and interpret these values.			
I can calculate and interpret probabilities involving geometric random variables.			
I can find the mean and standard deviation of a geometric distribution and interpret these values.			

Unit 4, Part II Multiple Choice Practice

Directions. Identify the choice that best completes the statement or answers the question. Check your answers and note your performance when you are finished.

1. A marketing survey compiled data on the number of cars in households. If X = the number of cars in a randomly selected household, and we omit the rare cases of more than 5 cars, then X has the following probability distribution:

X	0	1	2	3	4	5
$P(X)$	0.24	0.37	0.20	0.11	0.05	0.03

What is the probability that a randomly chosen household has at least two cars?
(A) 0.19
(B) 0.20
(C) 0.29
(D) 0.39
(E) 0.61

2. What is the expected value of the number of cars in a randomly selected household?
(A) 2.5
(B) 0.1667
(C) 1.45
(D) 1
(E) Cannot be determined

3. The famous game, "Rock-Paper-Scissors" can be purchased in dice form. A rock-paper-scissors die is a fair 12-sided die with 4 sides that say "rock," 4 sides that say "paper," and 4 sides that say "scissors." Let Y be the number of "rock" outcomes that occur in 10 rolls. Which of the following best describes this setting?
(A) Y has a geometric distribution with $n = 10$ observations and probability of success $p = 1/3$.
(B) Y has a geometric distribution with $n = 12$ observations and probability of success $p = 1/4$.
(C) Y has a binomial distribution with $n = 12$ observations and probability of success $p = 1/3$.
(D) Y has a binomial distribution with $n = 12$ observations and probability of success $p = 1/4$.
(E) Y has a binomial distribution with $n = 10$ observations and probability of success $p = 1/3$.

4. What is the variance of the sum of two random variables X and Y?
(A) $\sigma_X + \sigma_Y$
(B) $(\sigma_X)^2 + (\sigma_Y)^2$
(C) $\sigma_X + \sigma_Y$, but only if X and Y are independent.
(D) $(\sigma_X)^2 + (\sigma_Y)^2$, but only if X and Y are independent.
(E) None of these

5. In the town of Lakeville, the number of cell phones in a household is a random variable W with the following probability distribution:

Value w_i	0	1	2	3	4	5
Probability p_i	0.1	0.1	0.25	0.3	0.2	0.05

What is the standard deviation of the number of cell phones in a randomly selected house?

(A) 1.32
(B) 1.7475
(C) 2.5
(D) 0.09
(E) 2.9575

6. A random variable Y has the following probability distribution:

Y	-1	0	1	2
$P(Y)$	$4C$	$2C$	0.07	0.03

What is the value of the constant C?

(A) 0.10
(B) 0.15
(C) 0.20
(D) 0.25
(E) 0.75

7. It is known that about 90% of the widgets made by Simonetti Industries meet specifications. Every hour a sample of 18 widgets is selected at random for testing and the number of widgets that meet specifications is recorded. What is the approximate mean and standard deviation of the number of widgets meeting specifications?

(A) $\mu = 1.62$; $\sigma = 1.414$
(B) $\mu = 1.62$; $\sigma = 1.265$
(C) $\mu = 16.2$; $\sigma = 1.62$
(D) $\mu = 16.2$; $\sigma = 4.025$
(E) $\mu = 16.2$; $\sigma = 1.273$

8. A stock investment of $100 offers returns of $0, $500, $1000, or $2000 after one year. Let R be a random variable that represents the return after one year. Here is the probability distribution of R:

Return after 1 year	$0	$500	$1000	$2000
Probability	0.60	0.22	0.13	0.05

What are the expected earnings for a single $100 investment, E, where $E = R - 100$?

(A) $0
(B) $240
(C) $340
(D) $400
(E) $775

9. Let N = the number of successes in a binomial setting. The histogram of the probability distribution of N appears below.

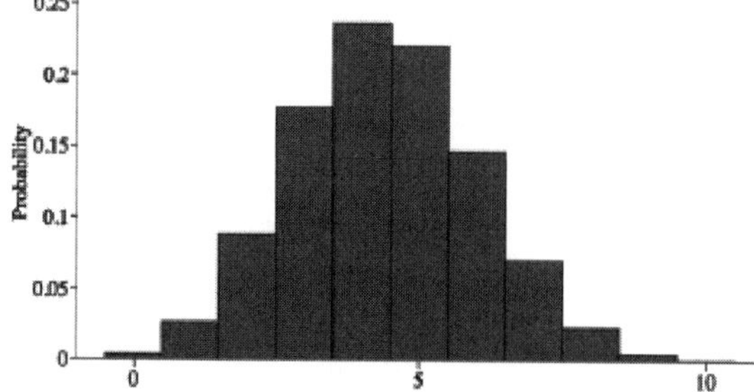

Which of the following provides the *correct* n, p, and description of shape?

(A) $n = 11$. $p = 0.6$. The graph is fairly symmetric with a single peak at $N = 4$.

(B) $n = 11$. $p = 0.4$. The graph is skewed to the right with a single peak at $N = 4$.

(C) $n = 11$. $p = 0.4$. The graph is fairly symmetric with a single peak at $N = 4$.

(D) $n = 11$. $p = 0.6$. The graph is skewed to the right with a single peak at $N = 4$.

(E) None of these provides the correct set of responses.

10. Which of the following random variables is geometric?

(A) The number of digits I will read beginning at a randomly selected starting point in a table of random digits until I find a 7.

(B) The number of times I have to roll a six-sided die to get two 5s.

(C) The number of phone calls received in a one-hour period.

(D) The number of 7s in a row of 40 random digits.

(E) All four of the above are geometric random variables.

Check your answers below. If you got a question wrong, check to see if you made a simple mistake or if you need to study that concept more. After you check your work, identify the concepts you feel very confident about and note what you will do to learn the concepts in need of more study.

#	Answer	Concept	Right	Wrong	Simple Mistake?	Need to Study More
1	D	Discrete Random Variable				
2	C	Expected Value of a Discrete RV				
3	E	Binomial Settings				
4	D	Combining Random Variables				
5	A	Standard Deviation of a Discrete RV				
6	B	Probability Distribution				
7	E	Binomial Mean and SD				
8	B	Transformation of a RV				
9	C	Binomial Histogram				
10	A	Geometric Random Variables				

FRAPPY! Free Response AP® Problem, Yay!

The following problem is modeled after actual Advanced Placement Statistics free response questions. Your task is to generate a complete, concise response in 15 minutes. After you generate your response, read over the solution and scoring guide. Score your response and note what, if anything, you would do differently to increase your own score.

A recent study revealed that electric vehicles may need to be repaired up to 3 times during the car's usable lifespan, typically for battery replacements. Let R represent the number of repairs necessary over the lifetime of a randomly selected electric vehicle. The probability distribution of the number of repairs necessary is given below.

Number of repairs	0	1	2	3
Probability	0.4	0.3	0.2	0.1

(a) Compute and interpret the mean and standard deviation of R.

(b) Gas powered vehicles may also require repairs over their lifetime. The mean and standard deviation of the number of repairs for gas powered vehicles over their lifetime are 2 and 1.2, respectively. Assuming that electric vehicles and gas powered vehicles need repairs independently of each other, compute and interpret the mean and standard deviation of the difference (gas – electric) in the number of repairs necessary for the two types of vehicles.

(c) On average, gas powered vehicle repairs costs \$9,500 and electric vehicle repairs cost \$20,000 on average. Compute the mean and standard deviation of the difference in cost one could expect to pay for repairs over the life of the vehicles.

FRAPPY! Scoring Rubric

Use the following rubric to score your response. Each part receives a score of "Essentially Correct," "Partially Correct," or "Incorrect." When you have scored your response, reflect on your understanding of the concepts addressed in this problem. If necessary, note what you would do differently on future questions like this to increase your score.

Intent of the Question

The goal of this question is to determine your ability to calculate and interpret the mean and standard deviation of a discrete random variable, combine random variables, and describe linear transformations of random variables.

Solution

(a) Mean: $\mu_R = 0(0.4) + 1(0.3) + 2(0.2) + 3(0.1) = 1$
Standard deviation: $\sigma_R = \sqrt{(0-1)^2(0.4) + (1-1)^2(0.3) + (2-1)^2(0.2) + (3-1)^2(0.1)} = 1$
We can expect to have to repair an electric vehicle once over its lifetime, but that can vary on average by about 1 repair.

(b) If D = the difference in the number of repairs across the two types of vehicles, the mean of D will be $\mu_D = 2 - 1 = 1$ repair and the standard deviation is $\sigma_D = \sqrt{1.2^2 + 1^2} = 1.562$ repairs.
The expected difference (gas – electric) in the number of repairs is 1 repair and that difference typically varies by about 1.562 repairs from the mean of 1 repair for randomly selected gas and electric vehicles.

(c) The difference (gas – electric) in cost we can expect to pay in repairs will be \$9500(2) – \$20,000(1) = –\$1000. If we were to repeat the process of randomly selecting one gas powered vehicle and one electric vehicle and finding the difference (gas – electric) in their repair costs many, many times, the average difference would be about –\$1000.
The standard deviation of the difference in repair costs is $\sqrt{20{,}000^2(1^2) + 9500^2(1.2)^2}$ = \$23.020.86.
If we were to repeat the process of randomly selecting one gas powered vehicle and one electric vehicle and finding the difference (gas – electric) in their repair costs many, many times, the costs would typically vary from the mean of –\$1000 by about \$23,020.86.

Scoring:

Parts (a), (b), and (c) are scored as essentially correct (E), partially correct (P), or incorrect (I).

Part (a) is essentially correct if the mean and standard deviation are calculated correctly AND interpreted correctly AND supporting work is shown for at least one of the calculations.
Part (a) is partially correct if the correct values are stated with no interpretation, or an incorrect interpretation is provided OR if only one calculation/interpretation pair is correct OR if both values and interpretations are correct with no supporting work.

Part (b) is essentially correct if the mean and standard deviation are calculated correctly AND interpreted correctly.
Part (b) is partially correct if no interpretation or an incorrect interpretation is provided OR if only one calculation/interpretation pair is correct.

Part (c) is essentially correct if the mean and standard deviation are calculated correctly AND interpreted correctly.
Part (c) is partially correct if no interpretation or an incorrect interpretation is provided OR if only one calculation/interpretation pair is correct.

4 Complete Response

All three parts essentially correct

3 Substantial Response

Two parts essentially correct and one part partially correct

2 Developing Response

Two parts essentially correct and no parts partially correct

One part essentially correct and one or two parts partially correct

Three parts partially correct

1 Minimal Response

One part essentially correct and no parts partially correct

No parts essentially correct and two parts partially correct

My Score:
What I did well:
What I could improve:
What I should remember if I see a problem like this on the AP Exam:

Unit 4, Part II Formula Sheet

Create a one-page summary of important concepts and formulas found in this unit. This will serve as a valuable resource as you prepare for the AP Exam.

Unit 4, Part II Crossword Puzzle

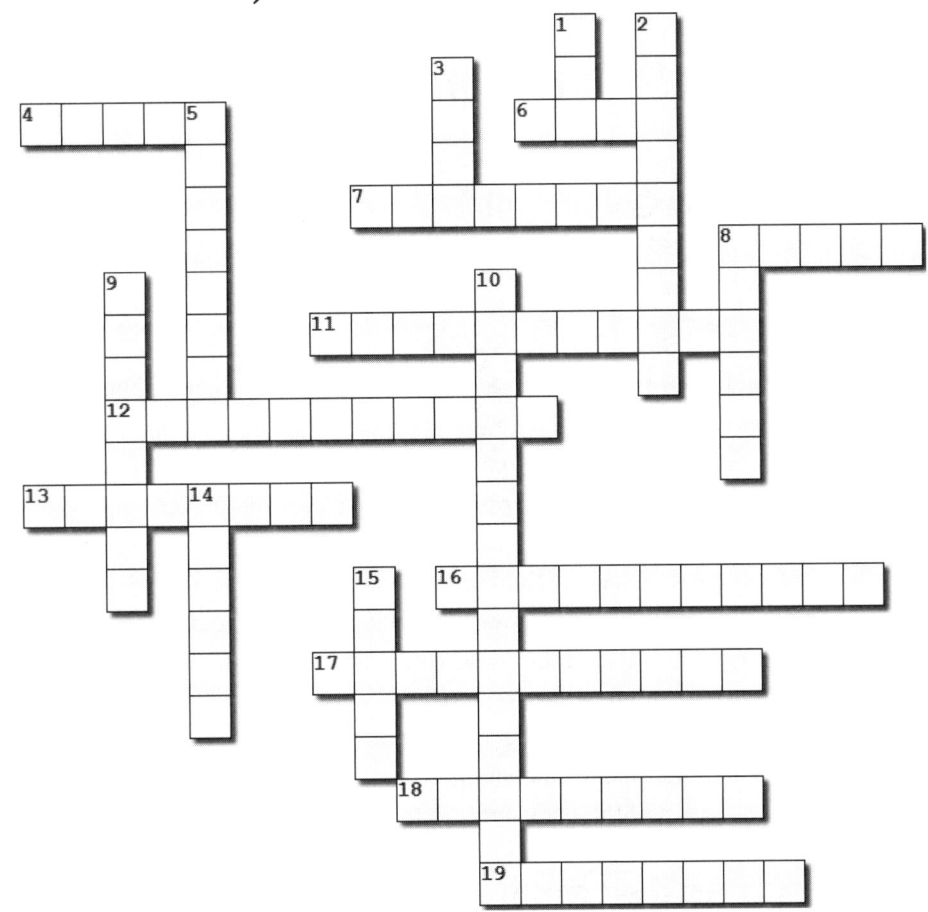

Across:

4. When multiplying or dividing by a constant, the ____ does not change.
6. Every probability must be a value between ___ and 1, inclusive. (spell out the number)
7. A ____ random variable takes a countable set of possible values with gaps between them on a number line.
8. Geometric distributions are always skewed to the ____.
11. When finding the difference in means for 2 random variables, the order of ____ is important.
12. In order to find the SD of a sum or difference of two random variables, the random variables must be ____.
13. In a ____ distribution, the possible outcomes are $X = 0, 1, 2, 3\ldots$
16. The ____ distribution of a random variable gives its possible values and their probabilities.
17. Adding or subtracting a constant does not change the shape or the ____ of the distribution.
18. A ____ setting arises when we perform independent trials of a chance process and record the number of trials to obtain the first success.
19. The ____ deviation of a discrete RV measures how much the value of the variable typically varies from the mean.

Down:

1. The sum of the probabilities must add to ____. (spell out the number)
2. In a ____ distribution the possible outcomes are $X = 1, 2, 3, \ldots$
3. Acronym to help you remember the conditions for a binomial distribution.
5. Another word for the mean of a discrete random variable is its ____ value.
8. A ____ variable takes numerical values that describe the outcomes of a random process.
9. The standard deviation is the square root of the ____.
10. Adding, subtracting, multiplying, or dividing by a constant are examples of linear ____.
14. The ____ of a discrete random variable is the 50th percentile of its probability distribution.
15. The mean of a sum of two random variables is equal to the sum of their ____.

Unit 5: Sampling Distributions

"Statistics may be defined as 'a body of methods for making wise decisions in the face of uncertainty.'" W.A. Wallis

UNIT 5 OVERVIEW

In Units 1-2, you learned how to explore a set of data. Unit 3 introduced you to methods for producing data. Unit 4 focused on the basics of probability and random variables. In this unit, you will study the final piece necessary for statistical inference – sampling distributions. The foundation of statistical inference lies in the concept of the sampling distribution. In order to draw a conclusion about a population based on information from a sample, you need to be able to answer the question, "What results would I expect to see if I sampled repeatedly from the population of interest?" Sampling distributions provide an answer to that question by helping to produce an estimate for the population parameter and also by estimating how much variability there is from sample to sample. In this unit, you will learn how to describe sampling distributions for sample proportions as well as sampling distributions for sample means. You will also learn an important theorem for sample means—the central limit theorem. Units 6 through 9 will present formal methods for statistical inference. The better you understand sampling distributions, the easier your study of inference will be. Be sure to get a good grasp of these concepts before moving on!

Sections in Unit 5

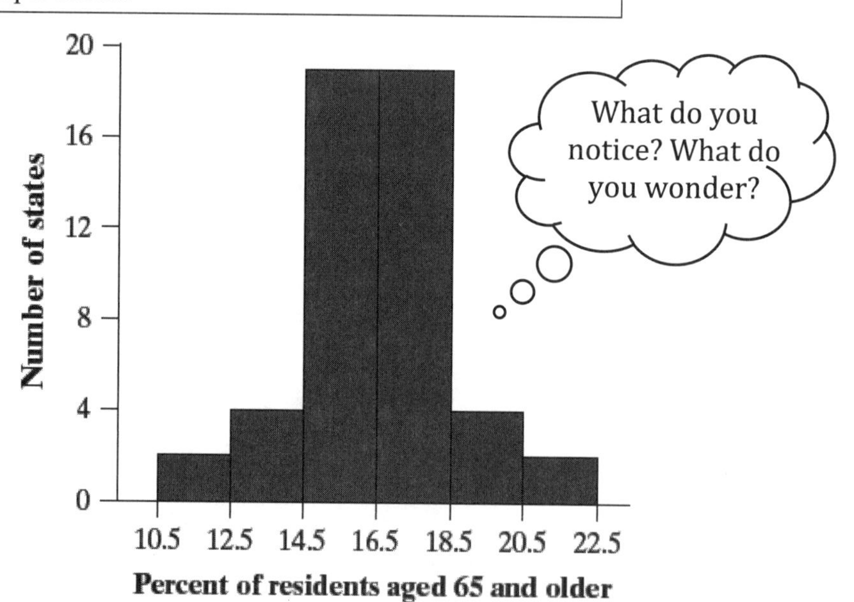

PRACTICE FOR MASTERY

Use the following *suggested* guide to the pages and exercises in your text to practice for mastery!
Note: your teacher may assign different problems. Be sure to follow their instructions!

Day	Topics	Read	Do
1	◆ Normal Random Variables ◆ Combining Normal Random Variables	p. 453-461	**5A**: 3, 5, 7, 9, 11, 13, 17 – 20
2	◆ Parameters and Statistics ◆ The Idea of a Sampling Distribution ◆ Using Sampling Distributions to Evaluate Claims	p. 466-472	**5B**: 1, 3, 5, 7, 9, 11, 13, 15
3	◆ Biased and Unbiased Estimators ◆ Variability of an Estimator	p. 472-479	**5B**: 19, 21, 23, 25, 27
4	AP Classroom Topic Questions (Topics 5.1-5.4)		**5A**: 21, 22 **5B**: 17, 29 – 32, 33
5	◆ Center and Variability ◆ Shape ◆ Using the Normal Approximation for $\hat{p}$	p. 484-491	**5C**: 1, 3, 5, 9, 11, 13, 15
6	◆ The Sampling Distribution of a Difference Between Two Proportions	p. 492-495	**5C**: 19, 21, 23
7	AP Classroom Topic Questions (Topics 5.5-5.6)		**5C**: 7, 17, 27 – 30
8	◆ Center and Variability ◆ Shape ◆ Probabilities Involving $\bar{x}$.	p. 500-508	**5D**: 1, 3, 5, 7, 9, 13, 15, 17
9	◆ The Sampling Distribution of a Difference Between Two Means	p. 509-512	**5D**: 19, 21, 23, 27 – 30
10	Unit 5 AP® Statistics Practice Test (p. 522)		Unit 5 Review Exercises (p.521)
11	Unit 5 Test: Celebration of learning!	Note: Since "Test" can sound daunting, think of this as a way to show off all you have learned. It's your own celebration of learning!	

Section 5A: Normal Distributions, Revisited

Section Summary

In the last unit, we focused on the shape, center, and variability of transformations of probability distributions of discrete random variables. We also learned how to find the mean and standard deviation of a sum or difference of two independent discrete random variables. In Section 5A, we will apply those same rules to continuous random variables, specifically uniform distributions and normal distributions. When finding the mean of a sum or difference of two random variables, just add (or subtract) the mean of each distribution. If the two random variables are independent, you can find the standard deviation of the sum (or difference) of the random variables by applying "the Pythagorean Theorem of Statistics." Specifically, $\sigma_{\text{Sum}} = \sigma_{\text{Difference}} = \sqrt{\sigma_X^2 + \sigma_Y^2}$. The same concepts that you applied to discrete random variables also apply to continuous random variables. Hooray!

Learning Targets:

____ I can calculate probabilities and percentiles involving normal random variables.

____ I can find probabilities involving a sum, a difference, or another linear combination of independent, normal random variables.

Read: Vocabulary and Key Concepts

- A ______________________________ can take any value in a specified interval on the number line.
- A __________________ models the probability distribution of a continuous random variable and has two characteristics:
  1. Is always ____________________ the horizontal axis.
  2. Has an area of ____________________ underneath it.
- The _________ under the density curve and above any specified interval of values on the horizontal axis gives the __________________ that the random variable falls within that interval.
- All continuous probability distributions assign probability _____ to every individual outcome. Therefore, for continuous probability distributions, $P(X < k)$ is the same as $P(X \leq k)$.
- Unlike for discrete random variables, the probability distribution of a continuous random variable assigns probabilities to _________________________ rather than to _________________________.
- A ______________________________ is a continuous random variable whose probability distribution is described by a normal curve.

Watch: Go to bfwpub.com/TPS7eStrive. Select Unit 5 and watch the following Example videos:

Video	Topic	Done
Unit 5A, The baby and the bathwater (p. 457)	*Normal random variables*	
Unit 5A, Will the lid fit? (p. 459)	*Combining normal random variables*	

Do: Check for Understanding

Concepts 1 and 2: Combining Normal Random Variables, Finding Probabilities

If the random variables of interest are normally distributed, we can calculate the probability of observing particular outcomes using the skills we learned in Unit 1, Part II. To do so, we rely on one important fact. When combining independent normal random variables, the resulting distribution is also normal! We can find the mean and standard deviation of the resulting distribution using the formulas we learned in the previous unit. Then we can calculate and interpret probabilities.

Check for Understanding – Learning Targets 1 and 2

____ *I can calculate probabilities and percentiles involving normal random variables.*

____ *I can find probabilities involving a sum, a difference, or another linear combination of independent, normal random variables.*

Mr. Molesky is an avid video game golfer. Mr. Molesky's time to complete a particular course on his favorite game X is normally distributed with a mean of 110 minutes and standard deviation of 10 minutes.

1. Calculate and interpret the probability that Mr. Molesky will complete the course in less than 95 minutes.

2. Find the 90th percentile of X. Describe this value in words.

Ms. Liberty plays the same video golf game. Ms. Liberty's course completion time Y is normally distributed with mean 100 minutes and standard deviation of 8 minutes. Mr. Molesky's and Ms. Liberty's completion times are independent.

3. Find the probability that Mr. Molesky will finish his game before Ms. Liberty on a randomly selected day.

Section 5B: What Is a Sampling Distribution?

Section Summary

This section will introduce you to the big ideas behind sampling distributions. First, you will learn how to distinguish between population parameters and statistics derived from samples. Next, you will explore the fact that statistics vary from sample to sample. This fact is the reason we study sampling distributions. By describing the shape, center, and variability of the sampling distribution of a statistic, we can determine the critical information necessary to perform statistical inference.

Learning Targets:

- ____ I can distinguish between a parameter and a statistic.
- ____ I can create a sampling distribution using all possible samples from a small population.
- ____ I can use the sampling distribution of a statistic to evaluate a claim about a parameter.
- ____ I can determine if a statistic is an unbiased estimator of a population parameter.
- ____ I can describe the relationship between sample size and the variability of an estimator.

Read: Vocabulary and Key Concepts

- A ________________ is a number that describes some characteristic of a sample.
- A ________________ is a number that describes some characteristic of a population.
- Complete the table by filling in the appropriate symbols.

Sample statistic		Population parameter
_____ (the sample mean)	estimates	_____ (the population mean)
_____ (the sample proportion)	estimates	_____ (the population proportion)
_____ (the sample SD)	estimates	_____ (the population SD)

- ________________________ refers to the fact that different random samples of the same size from the same population produce different values for a statistic.
- The __________________________ of a statistic is the distribution of values taken by the statistic in all possible samples of the same size from the same population.
- There are three different types of "distributions."
    1. The distribution of the ________________. (one word)
    2. The distribution of a __________________________. (two words)
    3. The _______________ distribution, which is the distribution of all possible samples for a particular statistic.

- A statistic used to estimate a parameter is an ______________________ if the mean of its sampling distribution is equal to the value of the parameter being estimated. The mean of the sampling distribution is also known as the ___________________________ of the estimator.
- Because the sample range is consistently smaller than the population range, the sample range is a ______________ estimator of the population range.
- The sampling distribution of any estimator (statistic) will have less variability when the sample size is ________________.

Choosing an Estimator:

- If we want to estimate the population mean (μ), use the ____________________ ($\bar{x}$).
- If we want to estimate a population proportion (p), use the ________________________ ($\hat{p}$).

Label each dart board diagram as showing high or low bias and high or low variability.

________ bias	________ bias	________ bias	________ bias
________ variability	________ variability	________ variability	________ variability

- Ideally, we'd like our estimates to be ______________ (unbiased) and ______________ (have low variability).

Watch: Go to bfwpub.com/TPS7eStrive. Select Unit 5 and watch the following Example videos:

Video	Topic	Done
Unit 5B, From ghosts to cold cabins (p. 467)	*Parameters and statistics*	
Unit 5B, Sampling heights (p. 469)	*The idea of a sampling distribution*	
Unit 5B, Reaching for chips (p. 470)	*Using sampling distributions to evaluate claims*	
Unit 5B, Estimating the range (p. 475)	*Biased and unbiased estimators*	
Unit 5B, Battery life (p. 477)	*Variability of an estimator*	

Do: Check for Understanding

Concept 1: Parameters and Statistics

One of the most powerful skills we learn from statistics is how to answer a question about a population based on information gathered from a random sample. That is, we can use a statistic calculated from a sample to draw a conclusion about the corresponding parameter in the population. However, we must note that the statistics we calculate from a sample may differ somewhat from the population characteristic we are trying to measure. Further, the value of the statistic would likely differ from sample to sample. This sample-to-sample variability poses a problem when we try to generalize our findings to the population. However, based on what we learned in the last unit we can view a sample statistic as a random variable. That is, while we have no way of predicting exactly what statistic value we will get from a single sample, we know how those values will behave in repeated random sampling. With the probability distribution of this random variable in mind, we can use a sample statistic to estimate the population parameter.

Check for Understanding – Learning Target 1

_____ *I can distinguish between a parameter and a statistic.*

For each of the following situations, identify the population, the parameter, the sample, and the statistic.

1. A medical researcher is interested in exploring the effects of a new medicine on blood pressure. The researcher recruits 500 adults with high blood pressure and gives them the new drug. After two months, their blood pressure is measured again and the average reduction in blood pressure is calculated.

2. A study is conducted to determine whether or not the dangerous activity of texting while driving is a common practice. The researcher randomly selects 1500 drivers who are 16 to 24 years old and asks whether or not they text while driving. Of the 1500 drivers, 12% indicate they text while driving. Bonus: Is the result of this study trustworthy? If not, what type of bias may be present?

Concept 2: Describing Sampling Distributions

You need to be familiar with three types of "distributions". First, is the population distribution. The population distribution shows the value of every member of the population. If the question of interest produces quantitative data (Ex: How tall are you, in inches?) then we could make a dotplot, stemplot, histogram, or boxplot to display the population distribution. If the question of interest produces categorical data (Ex: What is your favorite subject? Statistics, of course!) then we could make a bar graph to display the population distribution.

The distribution of a sample is a graph that displays the values for the members of a single random sample from the population.

The sampling distribution is a graph that displays the distribution of the mean, proportion, range, median (or whatever statistic we calculate) for every possible random sample of a certain size. If the population is small, we can create a sampling distribution, otherwise, we may use a simulation to create an appropriate sampling distribution.

Check for Understanding – Learning Target 2

____ *I can create a sampling distribution using all possible samples from a small population.*

Nick wants to buy a power bank from an office supply store. The store has 5 power banks in stock. Here are the costs: $26, $20, $30, $29, and $50.

1. Make a dotplot to display the population distribution of power bank cost.

2. Suppose you randomly selected 2 power banks from the 5 possible power bank options. Make a dotplot to display the distribution of one possible random sample. (Answers will vary).

3. Now, list all 10 possible samples of size $n = 2$ from the population of 5 power banks. Calculate the mean cost for each sample. Display the sampling distribution of the sample mean on a dotplot.

Concept 3: Using a Sampling Distribution to Evaluate a Claim

To draw a conclusion about a population proportion p, we select a random sample and calculate the sample proportion $\hat{p}$. Likewise, to reach a conclusion about a population mean μ, we select a random sample and calculate the sample mean $\bar{x}$. Because of chance variation in random sampling, the values of our sample statistic will vary from sample to sample. The sampling distribution describes sampling variability and provides a foundation for performing inference. We can use simulated sampling distributions or approximate sampling distributions to evaluate a claim about the population from which the samples were selected.

Check for Understanding – Learning Target 3

____ *I can use the sampling distribution of a statistic to evaluate a claim about a parameter.*

Inexpensive bathroom scales are not consistently accurate. A manufacturer of an inexpensive bathroom scale claims that when a 150-pound weight is placed on any scale produced in the factory, the weight indicated is normally distributed with a mean of 150 pounds and a standard deviation of 2 pounds.

A consumer advocacy group randomly selects 12 scales from this manufacturer and places a 150-pound weight on each one. They get a mean weight of 151.5 pounds, which makes them suspicious about the company's claim.

To determine if it is unusual to get a mean weight of 151 pounds from a random sample of 12 scales that function properly, they used a computer to simulate 200 samples of $n = 12$ scales from a normal population with a mean of 150 pounds and a standard deviation of 2 pounds. The dotplot shows this simulated sampling distribution.

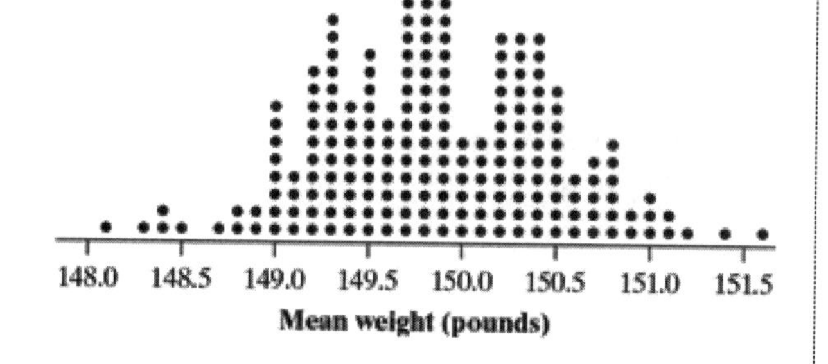

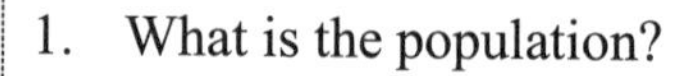

1. What is the population?

2. What parameter is being evaluated?

3. What does the simulated sampling distribution tell us about the population parameter?

4. We have information about one random sample of $n = 12$ scales produced by this manufacturer. The mean weight of the 12 randomly selected scales was 151.5 pounds. Is that sample mean surprising? Do you think the manufacturer is correct about the accuracy of its bathroom scales?

Concepts 4 and 5: Bias and Variability

When trying to estimate a parameter, we want a statistic that produces minimum sampling variability and no bias. What a challenge!

But have no fear…random sampling helps avoid bias and larger samples help minimize sampling variability. The good news doesn't stop there. When selecting a sample at random from a population, we can rest easy knowing that the sample proportion ($\hat{p}$) is an unbiased estimator of the population proportion. But wait, there's more! When selecting a sample at random from a population, we can also breathe easy knowing that the sample mean ($\bar{x}$) is an unbiased estimator of the population mean. But don't try to use the sample median to estimate the population median, or the sample range to estimate the population range. That will lead to trouble!

Check for Understanding – Learning Targets 4 and 5

____ *I can determine if a statistic is an unbiased estimator of a population parameter.*

____ *I can describe the relationship between sample size and the variability of an estimator.*

In a random sample of 50 U.S. high school students, the average amount of change found in their lockers is $0.42.

1. Do you think the average amount of change per locker for all U.S. high school students is $0.42? Explain your answer.

2. Which would be more likely to give an estimate close to the population mean amount of change in the lockers of all U.S. high school students: a random sample of 50 students or a random sample of 200 students? Explain your answer.

Section 5C: Sample Proportions

Section Summary

The objective of some statistical questions is to reach a conclusion about a population proportion p. For example, we may try to estimate the percentage of U.S. residents who approve of the current president, or we may test a claim about the proportion of defective light bulbs in a shipment based on a random sample. Since p is unknown to us, we must base our conclusion on a sample proportion, $\hat{p}$. However, as we have noted, we know that the value of $\hat{p}$ will vary from sample to sample. The amount of variability will depend on the size of the sample. In this section, you will learn how to describe the shape, center, and variability of the sampling distribution of $\hat{p}$ and $\hat{p}_1 - \hat{p}_2$.

Learning Targets:

____ I can calculate and interpret the mean and standard deviation of the sampling distribution of a sample proportion $\hat{p}$.

____ I can determine if the sampling distribution of $\hat{p}$ is approximately normal.

____ I can, if appropriate, use a normal distribution to calculate probabilities involving sample proportions.

____ I can describe the shape, center, and variability of the sampling distribution of a difference in sample proportions $\hat{p}_1 - \hat{p}_2$.

Read: Vocabulary and Key Concepts

- The ____ $\hat{p}$ describes the distribution of values taken by the sample proportion $\hat{p}$ in all possible samples of the same size from the same population.

Sampling Distribution of $\hat{p}$: Center and Variability

- Let $\hat{p}$ be the sample proportion of successes in an SRS of size n from a population of size N with proportion p of successes. Then:
- The mean of the sampling distribution of $\hat{p}$ is ____________.
- The standard deviation of the sampling distribution of $\hat{p}$ is approximately ____________________ as long as the ______________________ is met: $n < 0.10N$.
  - The value $\mu_{\hat{p}}$ gives the ______________ value of $\hat{p}$ in all possible samples of a given size from a population.
  - The value $\sigma_{\hat{p}}$ measures the ______________________ between a ______________________ $\hat{p}$ and the ______________________________ p in all possible samples of a given size from a population.
- The sample proportion $\hat{p}$ is ____________________ when the sample size is ____________. Specifically, multiplying the sample size by _____ cuts the standard deviation in _________.

- When we sample ____________ replacement, the standard deviation of the sampling distribution of $\hat{p}$ is exactly $\sigma_{\hat{p}} = \sqrt{\frac{p(1-p)}{n}}$. When we sample _______________ replacement, the observations are not independent, and the actual standard deviation of the sampling distribution of $\hat{p}$ is ______________ than the value given by the formula.

Sampling Distribution of $\hat{p}$: Shape

- Let $\hat{p}$ be the proportion of successes in a random sample of size n from a population with proportion of successes p. The ___________________________ says that the sampling distribution of $\hat{p}$ will be approximately normal when _____ ≥ 10 and __________ ≥ 10 .
- We call it the "Large Counts" condition because np is the _________________________ of successes in the sample and $n(1-p)$ is the _________________________ of failures in the sample.

Sampling Distribution of a Difference Between Two Proportions

- Let $\hat{p}_1$ be the sample proportion of successes in an SRS of size n_1 from Population 1 of size N_1 with proportion of successes p_1 and let $\hat{p}_2$ be the sample proportion of successes in an independent SRS of size n_2 from Population 2 of size N_2 with proportion of successes p_2. Then:
  - The mean of the sampling distribution of $\hat{p}_1 - \hat{p}_2 =$ ___________________
  - The standard deviation of the sampling distribution of $\hat{p}_1 - \hat{p}_2$ is approximately $\sigma_{\hat{p}_1-\hat{p}_2} =$ ______________________________ as long as the samples are ________________ and the ____________________ is met for both samples: $n_1 < 0.10N_1$ and $n_2 < 0.10N_2$.
- The sampling distribution of $\hat{p}_1 - \hat{p}_2$ is approximately normal if the __________________________ is met for both samples: n_1p_1 , $n_1(1-p_1)$, n_2p_2 ,and $n_2(1-p_2)$ are all at least 10.
- The formula for the standard deviation is exactly correct only when we have 2 types of independence:
  - _____________________ samples, so that we can add the variances of $\hat{p}_1$ and $\hat{p}_2$.
  - Independent ___________________________________. When sampling _______________ replacement, the actual value of the standard deviation is ___________ than the formula suggests. However, if the 10% condition is met for both samples, the difference is negligible.

Watch: Go to bfwpub.com/TPS7eStrive. Select Unit 5 and watch the following Example videos:

Video	Topic	Done
Unit 5C, Backing the pack (p. 487)	*Center and variability*	
Unit 5C, Backing the pack (p. 489)	*Shape*	
Unit 5C, Going to college (p. 490)	*Using the normal approximation for $\hat{p}$*	
Unit 5C, Yummy goldfish! (p. 493)	*The sampling distribution of $\hat{p}_1 - \hat{p}_2$*	

Do: Check for Understanding

Concepts 1 and 2: The Sampling Distribution of $\hat{p}$: Shape, Center, Variability

If we take repeated samples of the same size n from a population with a proportion of interest p, the sampling distribution of $\hat{p}$ will have the following characteristics:

1) The shape of the sampling distribution of $\hat{p}$ will be approximately normal as long as $np \geq 10$ and $n(1-p) \geq 10$.
2) The mean of the sampling distribution of $\hat{p}$ is $\mu_{\hat{p}} = p$.
3) The standard deviation of the sampling distribution of $\hat{p}$ is $\sigma_{\hat{p}} = \sqrt{\frac{p(1-p)}{n}}$.

Note: The formula for the standard deviation is exactly correct only if we are sampling from an infinite population or sampling *with replacement* from a finite population. When we sample without replacement from a finite population, the formula provides a value that is approximately correct as long as the 10% condition is satisfied. That is, the sample size must be less than 10% of the size of the population.

Check for Understanding – Learning Targets 1 and 2

____ *I can calculate and interpret the mean and standard deviation of the sampling distribution of a sample proportion $\hat{p}$.*

____ *I can determine if the sampling distribution of $\hat{p}$ is approximately normal.*

Your job at a potato chip factory is to check each shipment of potatoes for quality assurance. Suppose that a truckload of potatoes contains 5% that are unacceptable for processing. You randomly select and inspect 250 potatoes. Let $\hat{p}$ be the sample proportion of unacceptable potatoes.

1. Calculate and interpret the mean of the sampling distribution of $\hat{p}$?

2. Verify that the 10% condition is met. Then calculate and interpret the standard deviation of the sampling distribution of $\hat{p}$.

3. Describe the shape of the sampling distribution of $\hat{p}$. Justify your answer.

Concept 3: Using the Normal Approximation for $\hat{p}$

When the sample size n is large enough for np and $n(1-p)$ to both be at least 10, the shape of the sampling distribution of $\hat{p}$ will be approximately normal. In that case, we can use normal calculations to determine the probability that an SRS will generate a value of $\hat{p}$ in a particular interval.

Check for Understanding – Learning Target 3

____ *I can, if appropriate, use a normal distribution to calculate probabilities involving sample proportions.*

A tech company is interested in creating a new social media app for teenagers. They ask an SRS of 1000 U.S. high school students whether they use social media. Suppose 70% of all U.S. high school students use social media. What is the probability that the random sample selected by the company will result in a value of $\hat{p}$ that is greater than 75%?

Concept 4: The Sampling Distribution of $\hat{p}_1 - \hat{p}_2$

If we take a simple random sample of size n_1 from Population 1 with a proportion of successes p_1 and an independent simple random sample of size n_2 from Population 2 with a proportion of successes p_2 the sampling distribution of $\hat{p}_1 - \hat{p}_2$ will have the following characteristics:

1) The shape of the sampling distribution of $\hat{p}_1 - \hat{p}_2$ will be approximately normal as long as the Large Counts Condition is met for both samples: $n_1p_1 \geq 10$, $n_1(1 - p_1) \geq 10$, $n_2p_2 \geq 10$, and $n_2(1 - p_2) \geq 10$.
2) The mean of the sampling distribution of $\hat{p}_1 - \hat{p}_2$ is $\mu_{\hat{p}_1-\hat{p}_2} = p_1 - p_2$.
3) The standard deviation of the sampling distribution of $\hat{p}_1 - \hat{p}_2$ is approximately

   $\sigma_{\hat{p}_1-\hat{p}_2} = \sqrt{\dfrac{\hat{p}_1(1-\hat{p}_1)}{n_1} + \dfrac{\hat{p}_2(1-\hat{p}_2)}{n_2}}$ as long as the 10% condition is met for both samples: $n_1 < 0.10N_1$ and $n_2 < 0.10N_2$.

Check for Understanding – Learning Target 4

____ *I can describe the shape, center, and variability of the sampling distribution of a difference in sample proportions* $\hat{p}_1 - \hat{p}_2$.

At West High, 45% of the students play sports while at East High 40% of the students play sports. Random samples of 50 students from each large high school are selected. Let $\hat{p}_W - \hat{p}_E$ be the difference in the sample proportion of students that play sports.

1. What is the shape of the sampling distribution of $\hat{p}_W - \hat{p}_E$? Why?

2. Calculate and interpret the mean of the sampling distribution of $\hat{p}_W - \hat{p}_E$.

3. Verify both types of independence in this setting. Then calculate and interpret the standard deviation of the sampling distribution of $\hat{p}_W - \hat{p}_E$.

4. What is the probability that the sample proportion of students from East High who play sports is greater than the sample proportion of students from West High who play sports?

Section 5D: Sample Means

Section Summary

When the goal of a statistical application is to reach a conclusion about a population mean μ we will rely on a sample mean $\bar{x}$. However, the value of $\bar{x}$ will vary from sample to sample. As we observed with sample proportions, the amount of variability will depend on the size n of the sample. In this section, you will learn how to describe the shape, center, and variability of the sampling distribution of $\bar{x}$ and $\bar{x}_1 - \bar{x}_2$.

Learning Targets:

____ I can calculate and interpret the mean and standard deviation of the sampling distribution of a sample mean $\bar{x}$.

____ I can determine if the shape of the sampling distribution of $\bar{x}$ is approximately normal.

____ I can, if appropriate, use a normal distribution to calculate probabilities involving sample means.

____ I can describe the shape, center, and variability of the sampling distribution of a difference in sample means $\bar{x}_1 - \bar{x}_2$.

Read: Vocabulary and Key Concepts

- The ______________________________ of the ______________ $\bar{x}$ describes the distribution of values taken by the sample mean $\bar{x}$ in all possible samples of the same size from the same population.

The Sampling Distribution of $\bar{x}$: Center and Variability

- Let $\bar{x}$ be the sample mean in an SRS of size n from a population of size N with mean μ and standard deviation σ. Then:
  - The ________________ of the sampling distribution of $\bar{x}$ is ___________________.
  - The standard deviation of the sampling distribution of $\bar{x}$ is approximately $\sigma_{\bar{x}} =$ ___________ as long as the ____________________ is met: $n < 0.10N$.
- The value of $\mu_{\bar{x}}$ measures the __________________ distance between a _________________ mean ($\bar{x}$) and the ______________________ mean (μ) in all possible samples of a given size from a population.
- The sample mean ($\bar{x}$) is ____________ variable when the sample size is ___________________.
- When we sample ______________ replacement, the standard deviation of the sampling distribution of $\bar{x}$ is exactly $\sigma_{\bar{x}} = \frac{\sigma}{\sqrt{n}}$. When we sample ____________________ replacement, the observations are not independent and the actual standard deviation of the sampling distribution of $\bar{x}$ is ___________ than the value given by the formula.

The Sampling Distribution of $\bar{x}$: Shape

- If the shape of the _______________________ distribution is approximately normal, the shape of the __________________________ of $\bar{x}$ will also be approximately normal for ______ samples of size n.

- It is a remarkable fact that when the ________________ distribution is ________________, the ________ distribution of $\overline{x}$ looks more like a ________ distribution as the sample size increases.
- Suppose we select an SRS of size n from any population with mean μ and standard deviation σ. The ________________________ (CLT) says that when n is sufficiently __________, the sampling distribution of the sample mean $\overline{x}$ is ________________________.
- The sample size needed for the sampling distribution of $\overline{x}$ to be approximately normal, as guaranteed by the central limit theorem, depends on the __________ of the ________________ distribution. To be safe, we'll require that n be at least _____ to invoke the CLT.
- If the ______________ distribution is ________________, the ______________ distribution of $\overline{x}$ will also be ______________________, no matter what the sample size n is.
- If the ______________ distribution is ______________, the ____________ distribution of $\overline{x}$ will be approximately normal when the sample size is ____________ ($n \geq 30$ in most cases).
- When the population is non-normal and the sample size is __________, the sampling distribution of $\overline{x}$ will be ________ skewed than the population distribution, but still not ________________.

The Sampling Distribution of a Difference Between Two Means: Center, Variability, Shape

- Let x_1 be the sample mean in an SRS of size n_1 from Population 1 of size N_1 with mean μ_1 and standard deviation σ_1 and let x_2 be the sample mean in an independent SRS of size n_2 from Population 2 of size N_2 with mean μ_2 and standard deviation σ_2. Then:
  - The ________ of the sampling distribution of $\overline{x}_1 - \overline{x}_2$ is $\mu_{\bar{x}_1 - \bar{x}_2} =$ ______________.
  - The ________________________ of the sampling distribution of $\overline{x}_1 - \overline{x}_2$ is approximately $\sigma_{\bar{x}_1 - \bar{x}_2} =$ ________________ as long as the samples are ________________ and the ______________ is met for both samples: $n_1 < 0.10N_1$ and $n_2 < 0.10N_2$.
  - The sampling distribution of $\overline{x}_1 - \overline{x}_2$ is approximately ____________ if both ____________ distributions are approximately normal or if both sample sizes are ______ ($n_1 \geq 30$ and $n_2 \geq 30$).

Watch: Go to bfwpub.com/TPS7eStrive. Select Unit 5 and watch the following Example videos:

Video	Topic	Done
Unit 5D, Been to the movies recently? (p. 502)	*Center and variability*	
Unit 5D, Going back to the movies (p. 506)	*Shape*	
Unit 5D, Free oil changes (p. 507)	*Probabilities involving $\overline{x}$*	
Unit 5D, Medium or large drink? (p. 510)	*The sampling distribution of $\overline{x}_1 - \overline{x}_2$*	

Do: Check for Understanding

Concepts 1 and 2: Sampling Distribution of $\overline{x}$

If we take repeated random samples of the same size n from a population with mean μ, the sampling distribution of $\bar{x}$ will have the following characteristics:

1) The shape of the sampling distribution depends upon the shape of the population distribution. If the population is normally distributed, the sampling distribution of $\overline{x}$ will be normally distributed. If the population distribution is non-normal, the sampling distribution of $\overline{x}$ will become more and more normal as n increases.
2) The mean of the sampling distribution is $\mu_{\overline{x}} = \mu$.
3) The standard deviation of the sampling distribution is $\sigma_{\bar{x}} = \dfrac{\sigma}{\sqrt{n}}$

Note: The formula for the standard deviation is exactly correct only if we are sampling from an infinite population or *with replacement* from a finite population. When sampling without replacement from a finite population, the formula provides a value that is approximately correct as long as the 10% condition is satisfied. That is, the sample size must be less than 10% of the size of the population.

Check for Understanding – Learning Targets 1 and 2

____ *I can calculate and interpret the mean and standard deviation of the sampling distribution of a sample mean $\overline{x}$.*

____ *I can determine if the sampling distribution of $\overline{x}$ is approximately normal.*

At the end of a school year, the guidance counselor reports that the number of absences for all students at their large high school are approximately normal with a mean of 6.8 absences and a standard deviation of 2.1 absences. Suppose we select an SRS of 40 students from this high school and calculate $\overline{x}$ = the mean number of absences for the members of the sample.

1. Calculate and interpret the mean of the sampling distribution of $\overline{x}$.

2. Verify that the 10% condition is met. Then calculate and interpret the standard deviation of the sampling distribution of $\overline{x}$.

3. What is the shape of the sampling distribution of $\overline{x}$? Justify your answer.

Concept 3: The Central Limit Theorem, Calculating Probabilities

When the population is normally distributed, we know that the sampling distribution of $\bar{x}$ will be normally distributed, so we can use normal calculations. However, most population distributions are not normally distributed. Thankfully, a pretty remarkable fact about sample means helps us out: when the sample size n is large, the shape of the sampling distribution of $\bar{x}$ will be approximately normal no matter what the shape of the population distribution may be. For our purposes, we'll define "large" to be a sample size of at least 30. So, if $n \geq 30$, we can be safe in assuming that the sampling distribution of $\bar{x}$ will be approximately normal and we can proceed to perform normal calculations. If $n < 30$, we must consider the shape of the population distribution because the sampling distribution of $\bar{x}$ will retain some characteristics of the population distribution.

Check for Understanding – Learning Target 3

____ *I can, if appropriate, use a normal distribution to calculate probabilities involving sample means.*

The blood cholesterol level of adult men has mean 188 mg/dl and standard deviation 41 mg/dl. An SRS of 250 men is selected and the mean blood cholesterol level in the sample is calculated.

Describe the sampling distribution of $\bar{x}$ and calculate the probability that the sample mean will be greater than 193 mg/dl.

Concept 4: Sampling Distribution of $\bar{x}_1 - \bar{x}_2$

If we select an SRS of size n_1 from Population 1 with mean μ_1 and standard deviation σ_1 and an independent SRS of size n_2 from Population 2 with mean μ_2 and standard deviation σ_2, the sampling distribution of $\bar{x}_1 - \bar{x}_2$ will have the following characteristics:

1) The sampling distribution of $\bar{x}_1 - \bar{x}_2$ is normal if both population distributions are normal. It is approximately normal if both sample sizes are large ($n_1 \geq 30$ and $n_2 \geq 30$) or if one population is normally distributed and the other sample size is large.
2) The mean of the sampling distribution of $\bar{x}_1 - \bar{x}_2$ is $\mu_{\bar{x}_1 - \bar{x}_2} = \mu_1 - \mu_2$.
3) The standard deviation of the sampling distribution of $\bar{x}_1 - \bar{x}_2$ is approximately $\sigma_{\bar{x}_1 - \bar{x}_2} = \sqrt{\frac{\sigma_1^2}{n_1} + \frac{\sigma_2^2}{n_2}}$ as long as the 10% condition is met for both samples: $n_1 < 0.10N_1$ and $n_2 < 0.10N_2$.

Check for Understanding – Learning Target 4

____ *I can describe the shape, center, and variability of the sampling distribution of a difference in sample means $\bar{x}_1 - \bar{x}_2$.*

At West High, the amount of time that students spend on homework has a mean of 95 minutes and a standard deviation of 25 minutes. At East High, the amount of time that students spend on homework has a mean of 55 minutes and a standard deviation of 35 minutes. Random samples of 50 students from each large high school are selected. Let $\bar{x}_W - \bar{x}_E$ be the difference in the sample mean amount of time the students spend on homework.

1. Is the shape of the sampling distribution of $\bar{x}_W - \bar{x}_E$ approximately normal? Justify your answer.

2. Calculate and interpret the mean of the sampling distribution of $\bar{x}_W - \bar{x}_E$.

3. Verify both types of independence in this setting. Then calculate and interpret the standard deviation of the sampling distribution of $\bar{x}_W - \bar{x}_E$.

4. What is the probability that the mean time spent on homework for the random sample of students from East High is greater than the mean time spent on homework for the random sample of students from West High?

Unit 5 Summary: Sampling Distributions

This unit introduced you to a key concept for inferential thinking – sampling distributions. Because we are interested in drawing conclusions about population proportions and means, it is important to know how $\overline{x}$, $\hat{p}$, $\overline{x}_1 - \overline{x}_2$, and $\hat{p}_1 - \hat{p}_2$ will behave in repeated random sampling. Being able to describe the sampling variability for sample statistics allows us to estimate and test claims about population parameters. This unit provided us with some key facts about the shape, center, and variability of sample statistics that will help as we begin our formal study of inference. We also learned that statistics will vary from sample to sample. The larger the sample size(s), the smaller the variability of the sampling distribution. The sampling distributions of $\hat{p}$ and $\overline{x}$ will be centered at p and μ, respectively. Finally, the variability of these sampling distributions can be computed (as long as the 10% condition is met).

One important fact about sample means was revealed in this unit. When sampling from a normal population, the sampling distribution of $\overline{x}$ will be normal. However, as long as our sample size is at least 30, the shape of the sampling distribution of $\overline{x}$ will be approximately normal--no matter what the population distribution looks like. The central limit theorem is a powerful fact that will be revisited in the coming units. Now that we have a grasp of the basic concept of sampling distributions, we are ready to begin the formal study of statistical inference!

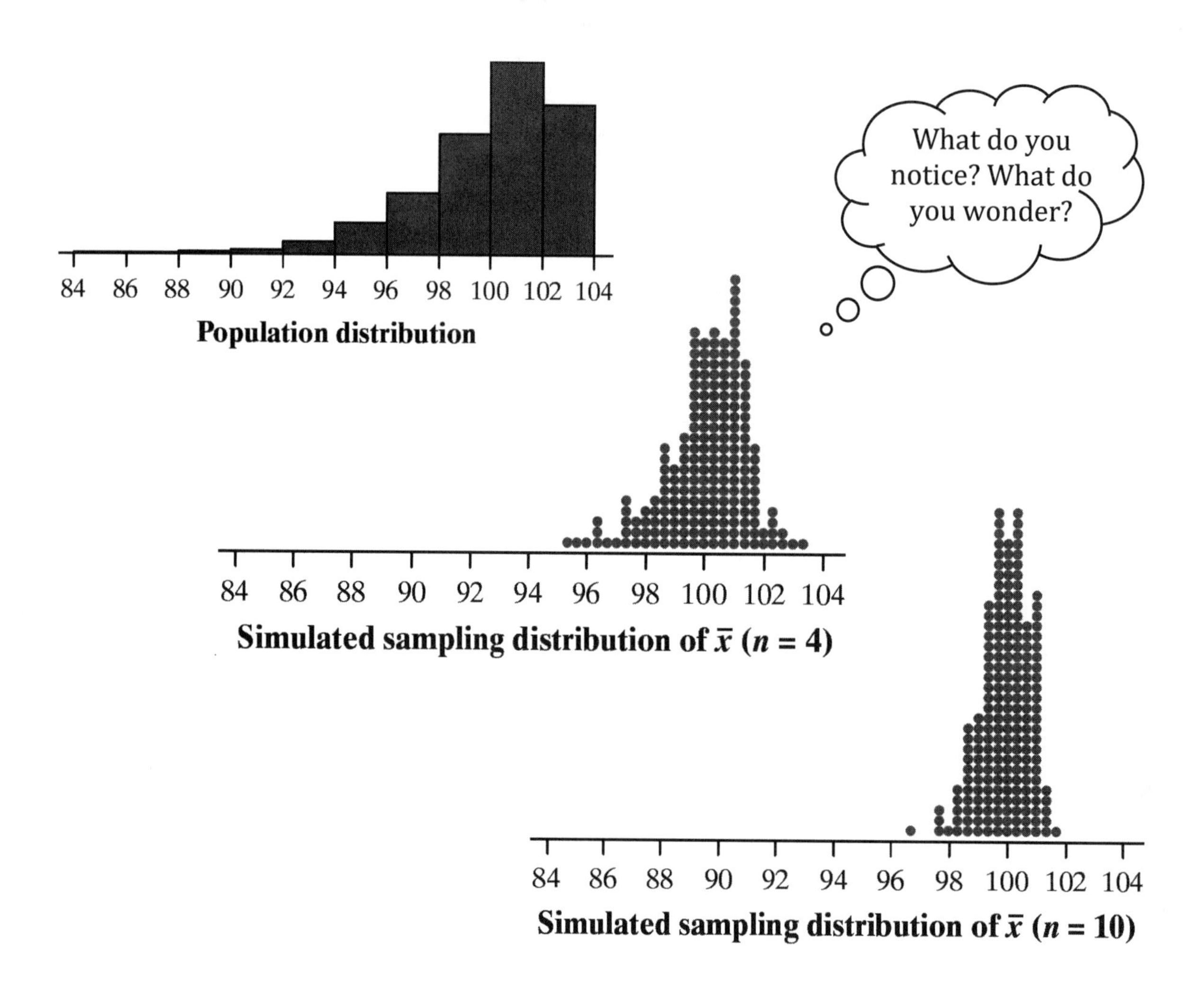

How well do you understand each of the learning targets?

Learning Target	Got It!	Almost There	Needs Work
I can calculate probabilities and percentiles involving normal random variables.			
I can find probabilities involving a sum, a difference, or another linear combination of independent, normal random variables.			
I can distinguish between a parameter and a statistic.			
I can create a sampling distribution using all possible samples from a small population.			
I can use the sampling distribution of a statistic to evaluate a claim about a parameter.			
I can determine if a statistic is an unbiased estimator of a population parameter.			
I can describe the relationship between sample size and the variability of an estimator.			
I can calculate and interpret the mean and standard deviation of the sampling distribution of a sample proportion $\hat{p}$.			
I can determine if the sampling distribution of $\hat{p}$ is approximately normal.			
I can, if appropriate, use a normal distribution to calculate probabilities involving sample proportions.			
I can describe the shape, center, and variability of the sampling distribution of a difference in sample proportions $\hat{p}_1 - \hat{p}_2$.			
I can calculate and interpret the mean and standard deviation of the sampling distribution of a sample mean $\bar{x}$.			
I can determine if the sampling distribution of $\bar{x}$ is approximately normal.			
I can, if appropriate, use a normal distribution to calculate probabilities involving sample means.			
I can describe the shape, center, and variability of the sampling distribution of a difference in sample means $\bar{x}_1 - \bar{x}_2$.			

Unit 5 Multiple Choice Practice

Directions. Identify the choice that best completes the statement or answers the question. Check your answers and note your performance when you are finished.

1. The variability of a statistic is described by
   (A) the spread of its sampling distribution.
   (B) the amount of bias present.
   (C) the vagueness in the wording of the question used to collect the sample data.
   (D) probability calculations.
   (E) the stability of the population it describes.

2. Here are dot plots of the values taken by three different statistics in 30 samples from the same population. The true value of the population parameter is marked with an arrow.

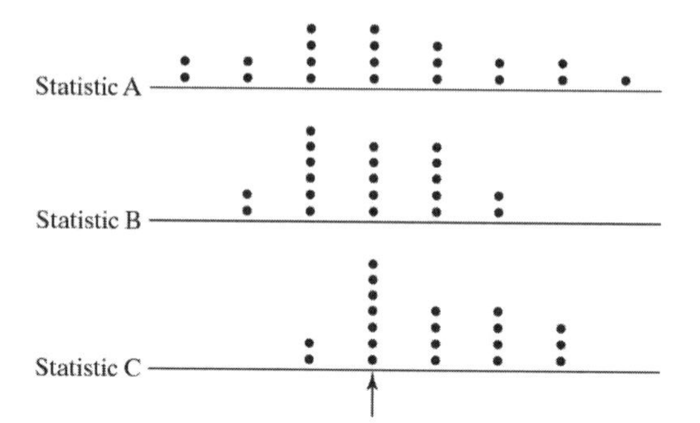

   Which statistic has the largest *bias* among these three?
   (A) Statistic A
   (B) Statistic B
   (C) Statistic C
   (D) A and B have similar bias, and it is larger than the bias of C.
   (E) B and C have similar bias, and it is larger than the bias of A.

3. According to a recent poll, 27% of Americans prefer to read their news in a physical newspaper instead of online. Let's assume this is the parameter value for the population. If you take a simple random sample of 25 Americans and let $\hat{p}$ = the proportion in the sample who prefer a newspaper, is the shape of the sampling distribution of $\hat{p}$ approximately normal?
   (A) No, because $p < 0.50$
   (B) No, because $np = 6.75$
   (C) Yes, because we can reasonably assume there are more than 250 individuals in the population.
   (D) Yes, because we took a simple random sample.
   (E) Yes, because $n(1 - p) = 18.25$

4. The time it takes students to complete a statistics quiz has a mean of 20.5 minutes and a standard deviation of 15.4 minutes. What is the probability that a random sample of 40 students will have a mean completion time greater than 25 minutes?
   (A) 0.0323
   (B) 0.0344
   (C) 0.3851
   (D) 0.6149
   (E) 0.9678

5. A fair coin (one for which both the probability of heads and the probability of tails are 0.5) is tossed 60 times. What is the probability that more than 1/3 of the tosses are heads?
   (A) 0.0049
   (B) 0.0357
   (C) 0.8090
   (D) 0.9643
   (E) 0.9951

6. The histogram to the right was obtained from data on 750 high school basketball games in a regional athletic conference. It represents the number of three-point baskets made in each game.

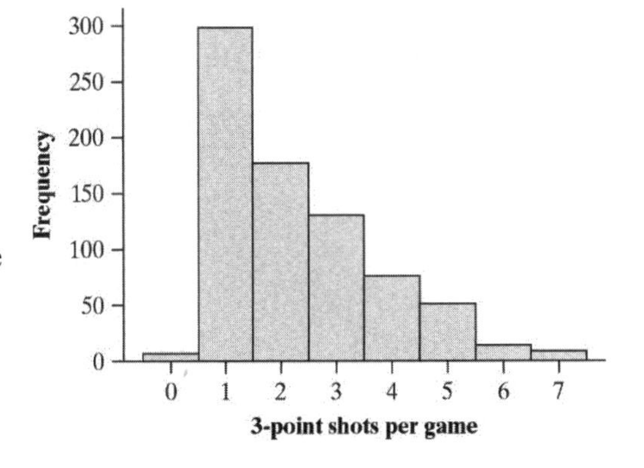

   What is the range of sample sizes a researcher could take from this population without violating conditions required for performing normal calculations with the sampling distribution of $\overline{x}$?
   (A) $0 \leq n \leq 30$
   (B) $30 \leq n \leq 50$
   (C) $30 \leq n \leq 75$
   (D) $30 \leq n \leq 750$
   (E) $75 \leq n \leq 750$

7. The incomes in a certain large population of college teachers have a normal distribution with mean $60,000 and standard deviation $5000. Four teachers are selected at random from this population to serve on a salary review committee. What is the probability that their average salary exceeds $65,000?
   (A) 0.0228
   (B) 0.1587
   (C) 0.8413
   (D) 0.9772
   (E) essentially 0

8. A random sample of size 25 is to be taken from a population that is normally distributed with mean 60 and standard deviation 10. The mean $\bar{x}$ of the observations in our sample is to be computed. The sampling distribution of $\bar{x}$
   (A) is normal with mean 60 and standard deviation 10.
   (B) is normal with mean 60 and standard deviation 2.
   (C) is approximately normal with mean 60 and standard deviation 2.
   (D) has an unknown shape with mean 60 and standard deviation 10.
   (E) has an unknown shape with mean 60 and standard deviation 2.

9. The scores of individual students on a college entrance examination have a left-skewed distribution with mean 18.6 and standard deviation 6.0. At Millard North High School, 36 seniors take the test. The sampling distribution of mean scores for random samples of 36 students is
   (A) approximately normal.
   (B) symmetric and mound-shaped, but non-normal.
   (C) skewed left, just like the population distribution.
   (D) neither normal nor non-normal. It depends on the particular 36 students selected.
   (E) exactly normal.

10. The distribution of prices for home sales in Minnesota is skewed to the right with a mean of $290,000 and a standard deviation of $145,000. Suppose you take a simple random sample of 100 home sales from this (very large) population. What is the probability that the mean of the sample is above $325,000?
    (A) 0.0015
    (B) 0.0027
    (C) 0.0079
    (D) 0.4046
    (E) 0.4921

Check your answers below. If you got a question wrong, check to see if you made a simple mistake or if you need to study that concept more. After you check your work, identify the concepts you feel very confident about and note what you will do to learn the concepts in need of more study.

#	Answer	Concept	Right	Wrong	Simple Mistake?	Need to Study More
1	A	Sampling Variability				
2	C	Bias and Variability				
3	B	Normality Condition				
4	A	Normal Probability Calculation				
5	E	Normal Probability Calculation				
6	C	10% Condition and CLT				
7	A	Normal Probability Calculation				
8	B	Sampling Distribution for Means				
9	A	Sampling Distribution for Means				
10	C	Normal Probability Calculation				

FRAPPY! Free Response AP® Problem, Yay!

The following problem is modeled after actual Advanced Placement Statistics free response questions. Your task is to generate a complete, concise response in 15 minutes. After you generate your response, read over the solution and scoring guide. Score your response and note what, if anything, you would do differently to increase your own score.

A television producer must schedule a selection of paid advertisements during each hour of programming. The lengths of the advertisements are slightly skewed to the right with a mean of 28 seconds and standard deviation of 5 seconds. During each hour of programming, 45 minutes are devoted to the program and 15 minutes are set aside for advertisements. To fill in the 15 minutes, the producer randomly selects 30 advertisements from the population of 450 advertisements on file.

(a) Describe the shape of the sampling distribution of sample mean length for random samples of 30 advertisements. Justify your answer.

(b) Find the mean and standard deviation of the sampling distribution of the sample mean length for random samples of 30 advertisements.

(c) If 30 advertisements are randomly selected, what is the probability that the total time needed to air them will exceed the 15 minutes available? Show your work.

FRAPPY! Scoring Rubric

Use the following rubric to score your response. Each part receives a score of "Essentially Correct," "Partially Correct," or "Incorrect." When you have scored your response, reflect on your understanding of the concepts addressed in this problem. If necessary, note what you would do differently on future questions like this to increase your score.

Intent of the Question

The goal of this question is to determine your ability to describe the sampling distribution of a sample mean and use it to perform a probability calculation.

Solution:

(a) The producer selects an SRS of size 30 from a population that is slightly skewed to the right with mean 28 seconds and standard deviation 5 seconds. Because $n = 30 \geq 30$, the sampling distribution of $\bar{x}$ will be approximately normal.

(b) The sampling distribution of the sample mean length for random samples of 30 advertisements has a mean of 28 seconds. Because $n = 30$ is less than 10% of the population of 450 advertisements on file, the standard deviation is approximately $\sigma_{\bar{x}} = \frac{5}{\sqrt{30}} = 0.913$ second.

(c) The probability that a random sample of 30 advertisements will exceed the allotted time is equivalent to the probability that the sample mean length of the 30 advertisements is greater than 900/30 = 30 seconds.

In part (a), we determined that the sampling distribution is normal with mean = 28 seconds and standard deviation = 0.913 second.

$$z = \frac{30-28}{0.913} = 2.191$$

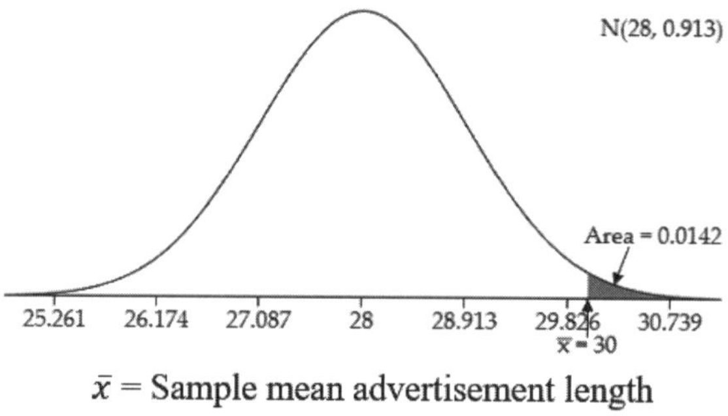

N(28, 0.913)

Area = 0.0142

25.261 26.174 27.087 28 28.913 29.826 30.739

$\bar{x} = 30$

$\bar{x}$ = Sample mean advertisement length

Using Table A: 1 – 0.9857 = 0.0143

Using technology: normalcdf(lower: 30, upper: 1000, mean: 28, SD: 0.913) = 0.0142.

There is a 1.42% chance the randomly selected advertisements will exceed the allotted time.

Scoring:

Parts (a), (b), and (c) are scored as essentially correct (E), partially correct (P), or incorrect (I).

Part (a) is essentially correct if the response correctly identifies the shape of the sampling distribution (approximately normal) and bases the justification on the fact that the sample size is at least 30. Part (a) is partially correct if the shape is correctly identified, but the justification is incorrect or incomplete. Incorrect otherwise.

Part (b) is essentially correct if the response (1) correctly identifies the center of the sampling distribution (mean = 28 seconds), (2) correctly identifies the variability of the sampling distribution (standard deviation = 0.913 seconds, and (3) the calculation of the standard deviation is shown. Part (b) is partially correct if the solution correctly identifies the mean and standard deviation but fails to show the calculation for the standard deviation. Incorrect otherwise.

Part (c) is essentially correct if the response sets up and performs a correct probability calculation with all parameters clearly identified. Part (b) is partially correct if the response includes a correctly set up calculation but fails to calculate the correct value OR if it sets up an incorrect, but plausible, calculation but carries it through correctly OR sets up an appropriate calculation that does not clearly identify the parameters.

4 Complete Response
All three parts essentially correct

3 Substantial Response
Two parts essentially correct and one part partially correct

2 Developing Response
Two parts essentially correct and no parts partially correct
One part essentially correct and one or two parts partially correct
Three parts partially correct

1 Minimal Response
One part essentially correct and no parts partially correct
No parts essentially correct and two parts partially correct

My Score:
What I did well:
What I could improve:
What I should remember if I see a problem like this on the AP Exam:

Unit 5 Formula Sheet

Create a one-page summary of important concepts and formulas found in this unit. This will serve as a valuable resource as you prepare for the AP Exam.

Unit 5 Crossword Puzzle

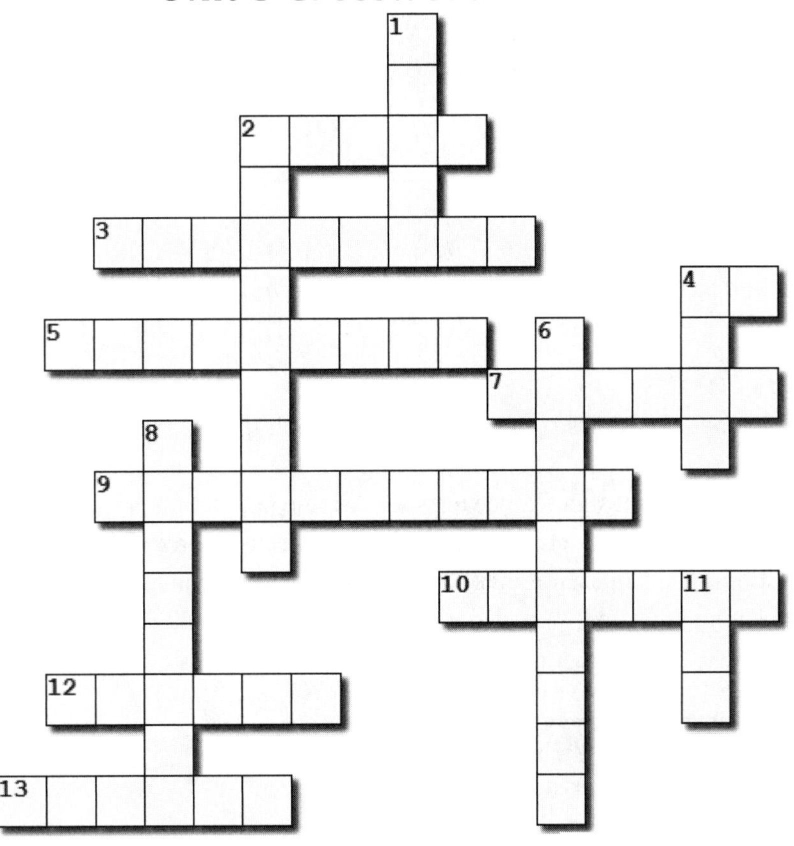

Across:

2. Greek letter used for the population standard deviation.
3. A ____ is a number that describes some characteristic of a population.
4. Greek letter used for the population mean.
5. When estimating a parameter, choose a ____ with low or no bias and minimum variability.
7. When the sample size is large, the sampling distribution of the sample mean is approximately ____.
9. The of a statistic is described by the spread of the sampling distribution.
10. A ____ curve always has an area of exactly 1 underneath it.
12. Rule of thumb for using the central limit theorem: The sample size should be at least ____.
13. The sampling distribution of any estimator will have less variability when the sample size is ____.

Down:

1. Central ____theorem: When n is large, the sampling distribution of the sample mean is approximately normal.
2. A ____ is a number that describes some characteristic of a sample.
4. Center of a sampling distribution.
6. A ____ random variable can take any value on a specified interval on the number line.
8. ____ distribution: Distribution of values taken by the statistic in all possible samples of the same size from the same population.
11. The sampling distribution of $\hat{p}$ will be approximately normal when np and $n(1-p)$ are both at least ___. (spell the number)

Unit 6, Part I: Inference for Categorical Data: Proportions

"Do not put your faith in what statistics say until you have carefully considered what they do not say." William W. Watt

UNIT 6, PART I OVERVIEW

Now that you have learned the basics of probability and sampling distributions, you are ready to begin a formal study of inference. In this unit, you will use your knowledge of sampling distributions to construct confidence intervals and carry out significance tests for a single population proportion. In later units, you'll learn how to carry out these same inference procedures for means, paired data, differences in means and proportions, distributions, and slopes of regression lines. Each of these procedures uses the same approach as the ones you will learn in this unit, so you'll want to get a solid foundation in the basics here!

Sections in Unit 6, Part I
Section 6A: Confidence Intervals: The Basics
Section 6B: Confidence Intervals for a Population Proportion
Section 6C: Significance Tests: The Basics
Section 6D: Significance Tests for a Population Proportion

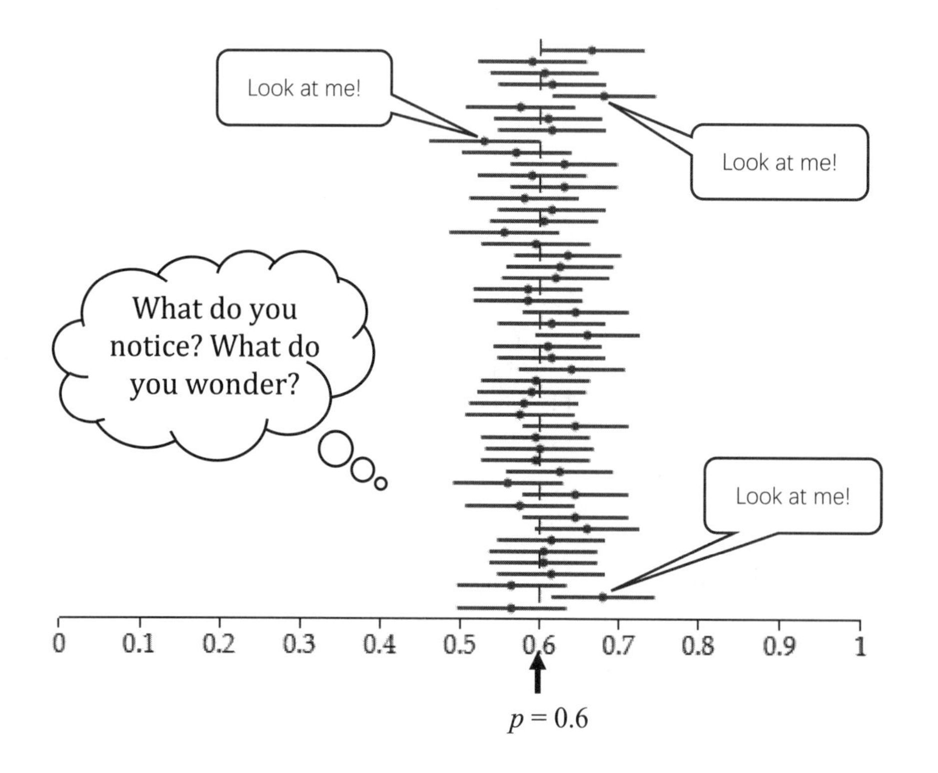

PRACTICE FOR MASTERY

Use the following *suggested* guide to the pages and exercises in your text to practice for mastery!
Note: your teacher may assign different problems. Be sure to follow their instructions!

Day	Topics	Read	Do
1	◆ Interpreting a Confidence Interval ◆ Using Confidence Intervals to Justify a Claim ◆ Interpreting Confidence Level	p. 530-537	**6A**: 1, 3, 5, 7, 9, 11, 13, 16, 17
2	◆ Checking Conditions for a Confidence Interval for *p* ◆ Calculating a Confidence Interval for *p* ◆ One-Sample *z* Interval for *p*	p. 541-549	**6B**: 1, 3, 5, 7, 11, 13, 15, 17
3	◆ Factors that Affect the Margin of Error ◆ Determining the Sample Size	p. 549-555	**6B**: 21, 23, 25, 27, 31 – 34
4	AP Classroom Topic Questions (Topics 6.1-6.3)		**6B**: 9, 19, 29, 35, 36, 37, 38
5	◆ Stating Hypotheses ◆ Interpreting *P*-values ◆ Making Conclusions	p. 561-569	**6C**: 1, 3 5, 7, 9, 11, 13, 18 – 20
6	◆ Checking Conditions for a Test about *p* ◆ Calculating the Standardized Test Statistic and *P*-value for a Test about *p*	p. 573-581	**6D**: 1, 3, 5, 7, 9, 36, 37, 38
7	◆ One-Sample *z* Test for *p* ◆ Type I and Type II Errors	p. 581-586	**6D**: 11, 13, 15, 17, 19, 21
8	◆ The Power of a Test	p. 586-591	**6D**: 25, 27, 29, 31
9	AP Classroom Topic Questions (Topics 6.4-6.7)		**6D**: 23, 39, 40, 41, 42, 43
10	Unit 6, Part I AP® Statistics Practice Test (p. 604)		Unit 6, Part I Review Exercises (p.602)
11	Unit 6, Part I Test: Celebration of learning!	Note: Since "Test" can sound daunting, think of this as a way to show off all you have learned. It's your own celebration of learning!	

Section 6A: Confidence Intervals: The Basics

Section Summary

In this section you will be introduced to the basic ideas behind constructing and interpreting a confidence interval for a population parameter. You will learn how we can obtain a point estimate for a population parameter and use what we know about sampling variability to construct an interval of plausible values for the parameter. You will focus on the big ideas in this section. Make sure you understand the different components of a confidence interval as well as the correct interpretation of both the interval and the confidence level. The next two sections will build upon the concepts presented here and focus on the details for estimating a population proportion.

Learning Targets:

____ I can interpret a confidence interval in context.
____ I can use a confidence interval to make a decision about the value of a parameter.
____ I can interpret a confidence level in context.

Read: Vocabulary and Key Concepts

- A ____________________________ is a statistic that provides an estimate of a population parameter.
- The value of a point estimator from a sample is called a ________________________.
- A ______________________________, also known as an interval estimate, gives a set of plausible values for a parameter based on sample data.
- To interpret a *C*% confidence interval for an unknown parameter, say, "____________________ ____"
- The ____________________________ *C* gives the approximate percentage of confidence intervals that will capture the population parameter in repeated random sampling with the same sample size.
- To interpret a confidence level *C*, say, "If we were to select ____________________________ of the same ________ from the same _______________ and construct a *C*% ______________________ using each sample, about ____% of the intervals would _____________ the [parameter in context]."
- The confidence level does *not* tell us the ________________ that a ________________ confidence interval captures the ______________________________.
- A particular confidence interval either includes the parameter (probability = ____) or doesn't include the parameter (probability = ___).

Watch: Go to bfwpub.com/TPS7eStrive. Select Unit 6, Part I and watch the following Example videos:

Video	Topic	Done
Unit 6A, What's your blood pressure? (p. 533)	*Interpreting a confidence interval*	
Unit 6A, Is your blood pressure too high? (p. 534)	*Using confidence intervals to justify a claim*	
Unit 6A, More blood pressure (p. 536)	*Interpreting confidence level*	

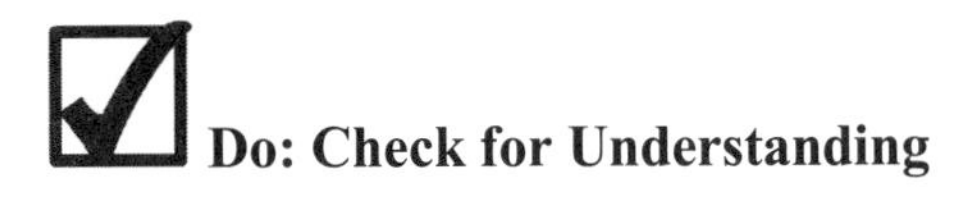

Do: Check for Understanding

Concepts 1 and 2: Confidence Interval, Confidence Level

When our goal is to estimate a population parameter, we often must rely on a sample statistic to provide a "point estimate." However, as we learned in the last unit, that estimate will vary from sample to sample. A confidence interval takes that variation into account to provide an interval of plausible values, based on the statistic, for the true parameter. All confidence intervals have a confidence level C that reports the success rate of the method used to construct the interval in capturing the parameter in repeated constructions. For example, "C% confident" means C% of all samples of the same size from the population of interest would yield an interval that captures the true parameter. We can then interpret the interval itself to say, "We are C% confident that the interval from a to b captures the true value of the population parameter."

Concept 3: Using confidence intervals to justify a claim

A confidence interval gives an interval of plausible values for the population parameter we wish to estimate. Therefore, we can use a confidence interval to justify a claim about a parameter. Suppose Kathy thinks that she takes at least 10,000 steps per day, on average. She has been wearing a watch that counts her steps for the past few months. Using a random sample, she calculates a 95% confidence interval and for the mean number of steps she takes per day to be 7850 to 8890 steps. Is it reasonable for her to continue to think that she takes at least 10,000 steps per day? No!

Check for Understanding – Learning Targets 1, 2, and 3

____ *I can interpret a confidence interval in context.*

____ *I can use a confidence interval to make a decision about the value of a parameter.*

____ *I can interpret a confidence level in context.*

How much do the volumes of bottles of water vary? A random sample of fifty "20 oz." water bottles is selected, and the contents are measured. A 90% confidence interval for the population mean μ is 19.10 ounces to 20.74 ounces.

1. Interpret the confidence interval in context.

2. Interpret the confidence level in context.

3. Based on this interval, is there convincing evidence that the mean contents of all "20 oz." water bottles are less than 20 ounces?

Section 6B: Confidence Intervals for a Population Proportion

Section Summary

In the last section, you learned the basic ideas behind confidence intervals. In this section, you will learn how to construct and interpret a confidence interval for a population proportion. You will start by learning how to make sure the conditions for inference are met, then you will focus on the mechanics of constructing a confidence interval for a single population proportion. After you have the basics down, you will learn a four-step process that will set you up for success on the AP® Statistics Exam.

Learning Targets:

____ I can check the conditions for constructing a confidence interval for a population proportion.
____ I can calculate a confidence interval for a population proportion.
____ I can construct and interpret a one-sample z interval for a proportion.
____ I can describe how the sample size and confidence level affect the margin of error.
____ I can determine the sample size required to obtain a confidence interval for a population proportion with a specified margin of error.

Read: Vocabulary and Key Concepts

Conditions for Calculating a Confidence Interval for a Population Proportion

- Random: The data come from a ________________________ from the population of interest.
    - 10%: When sampling _____________ replacement, $n < 0.10N$.
- Large Counts: Both _______ and ____________ are at least 10.

- If the data come from a __________________ sample or if other sources of _______ are present in the data collection process, there's no reason to calculate a confidence interval for p. You should have *no* confidence in an interval when the ________________ condition is violated.
- The actual capture rate is almost always ________________ than the reported confidence level when the 10% condition is violated. (This is a good thing!)
- The actual capture rate will almost always be ______________ than the stated confidence level when the Large Counts condition is violated.
- The structure of a confidence interval is _____________________ ± ______________________.
- The __________________________ of an estimate describes how far, at most, we expect the point estimate to vary from the population parameter.
- The margin of error is calculated using two factors: the __________________ and the __________________ of the statistic used to estimate the parameter.
- The critical value is a multiplier that makes the interval _____ enough to have the stated capture rate.
- The standard error describes how much the sample proportion $\hat{p}$ ___________________ from the population proportion p in repeated random samples of size n.

Calculating a Confidence Interval for a Population Proportion

- When the conditions are met, a $C\%$ confidence interval for the unknown population proportion p is

Formula: ______________________________

where z^* is the critical value for the standard normal curve with $C\%$ of its area between $-z^*$ and z^*.

Decreasing the Margin of Error

- We prefer an estimate with a _________ margin of error. The margin of error gets smaller when:
    - *The confidence level* ____________. To obtain a smaller margin of error from the same data, you must be willing to accept a smaller capture rate.
    - *The sample size n* _____________. In general, increasing the sample size n reduces the margin of error for any fixed confidence level.
- The width of a confidence interval for a proportion is proportional to $\frac{1}{\sqrt{n}}$, so quadrupling the sample size cuts the margin of error in _______.
- The margin of error accounts for *only* the variability we expect from _______________ sampling.
- The margin of error does not account for any sources of bias in the _________________________.

Calculating the Sample Size for a Desired Margin of Error When Estimating *p*

- To determine the sample size n that will yield a $C\%$ confidence interval for a population proportion p with a maximum margin of error ME, solve the following inequality for n:

$$z^*\sqrt{\frac{\hat{p}(1-\hat{p})}{n}} \leq ME$$

where $\hat{p}$ is a __________ value for the sample proportion. The margin of error will always be less than or equal to ME if you use $\hat{p}$ = ____.

Watch: Go to bfwpub.com/TPS7eStrive. Select Unit 6, Part I and watch the following Example videos:

Video	Topic	Done
Unit 6B, The beads (p. 541)	*Checking conditions for a confidence interval for p*	
Unit 6B, Read any good books lately? (p. 546)	*Calculating a confidence interval for p*	
Unit 6B, Distracted walking (p. 548)	*One-sample z interval for p*	
Unit 6B, Distracted walking (p. 552)	*Factors that affect the margin of error*	
Unit 6B, Customer satisfaction (p. 554)	*Determining the sample size*	

Do: Check for Understanding

Concept 1: Conditions for Estimating p

When constructing a confidence interval for p, it is critical that you begin by checking that the conditions are met. First, check to make sure that the sample was randomly selected. If sampling was done without replacement be sure to check the 10% condition, which states that the size of the sample must be less than 10% of the size of the population. Because the construction of the interval is based on the sampling distribution of $\hat{p}$, next you must ensure that the Large Counts Condition is met. That is, check to see that $n\hat{p}$ and $n(1-\hat{p})$ are both at least 10. If all conditions are met, you can construct and interpret a confidence interval for a population proportion p.

Concepts 2 and 3: Constructing a Confidence Interval for p, The Four Step Process

To construct a confidence interval for a population proportion p, you should follow a four-step process.

- **State:** State the parameter you want to estimate and the confidence level.
- **Plan:** Identify the appropriate inference method and check the conditions.
- **Do:** If the conditions are met, perform calculations. Formula: $\hat{p} \pm z^* \sqrt{\frac{\hat{p}(1-\hat{p})}{n}}$
- **Conclude:** Interpret your interval in the context of the problem.

Check for Understanding – Learning Targets 1, 2, and 3

____ *I can check the conditions for calculating a confidence interval for a population proportion.*
____ *I can calculate a confidence interval for a population proportion.*
____ *I can construct and interpret a one-sample z interval for a proportion.*

Can you name the 5 freedoms that are protected by the First Amendment? A researcher quizzed a random sample of 300 adults and found that 126 could not name a single freedom that is protected by the First Amendment. Use the four-step process to construct and interpret a 90% confidence interval for the proportion of adults who cannot name a single freedom that is protected by the First Amendment.

(*Need Tech Help?* View the video **Tech Corner 16: Confidence Intervals for a Proportion** at bfwpub.com/TPS7eStrive)

Concept 4: Margin of Error

Our goal with confidence intervals is to provide as precise an estimate as possible. That is, we wish to construct a narrow interval that we are confident captures the parameter of interest. We can achieve this in two ways: by decreasing our confidence or by increasing our sample size. Decreasing your confidence is not an ideal solution. However, if you are willing and able to pay the costs and do the work necessary to increase the sample size you can have a high confidence level and a small margin of error.

Concept 5: Choosing the Sample Size

As noted previously, our goal is to estimate the parameter as precisely as possible. We want high confidence and a low margin of error. To achieve that, we can determine how large a sample size is necessary before proceeding with the data collection. To calculate the sample size necessary to achieve a set margin of error at a confidence level *C*, we solve the following inequality for *n*:

$$z^* \sqrt{\frac{\hat{p}(1-\hat{p})}{n}} \leq ME$$

where the sample proportion $\hat{p}$ is estimated based on a previous study or set to 0.5 to account for the maximum possible margin of error. Be sure to *round up* to the next integer in order for the margin of error to be less than or equal to the desired value!

Check for Understanding – Learning Targets 4 and 5

____ *I can describe how the sample size and confidence level affect the margin of error.*

____ *I can determine the sample size required to obtain a confidence interval for a population proportion with a specified margin of error.*

In the previous Check for Understanding, we were 90% confident that the interval from 0.373 to 0.467 captures the proportion of all adults who cannot name a single freedom that is protected by the First Amendment. This interval was based on a random sample of 300 adults.

1. Explain what would happen to the width of the interval if the confidence level were decreased to 80%.

2. How would the width of a 90% confidence interval based on a sample of size 100 adults compare to the original 90% confidence interval, assuming the sample proportion remained the same?

3. Suppose we wanted to repeat this study but want the margin of error to be no more than 0.02 at a 95% confidence level. How large a sample is needed?

Section 6C: Significance Tests: The Basics

Section Summary

In this section, you will learn the basic ideas and logic behind a significance test. You will be introduced to stating hypotheses, checking conditions, calculating a standardized test statistic, and drawing a conclusion. *P*-values will be introduced as a means of weighing the strength of evidence against a claim. As you work through this section, remember the similarity between significance testing and a court case. Someone makes a claim, and you get to be the judge and jury! Their claim (provided in the null hypothesis) is innocent until you have convincing evidence that the claim is guilty. The evidence that you are presented with is the sample data. Based upon the evidence, you must make a decision. Is there convincing evidence that the defendant (the null hypothesis) is guilty (reject H_0) or do "you the jury" find the defendant not guilty (do you fail to reject the null hypothesis)?

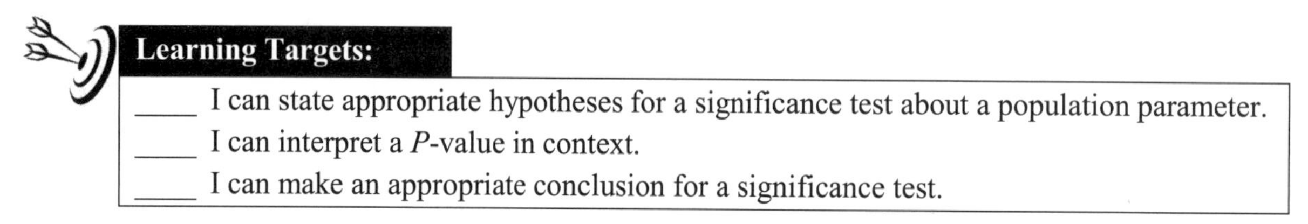

Learning Targets:

____ I can state appropriate hypotheses for a significance test about a population parameter.
____ I can interpret a *P*-value in context.
____ I can make an appropriate conclusion for a significance test.

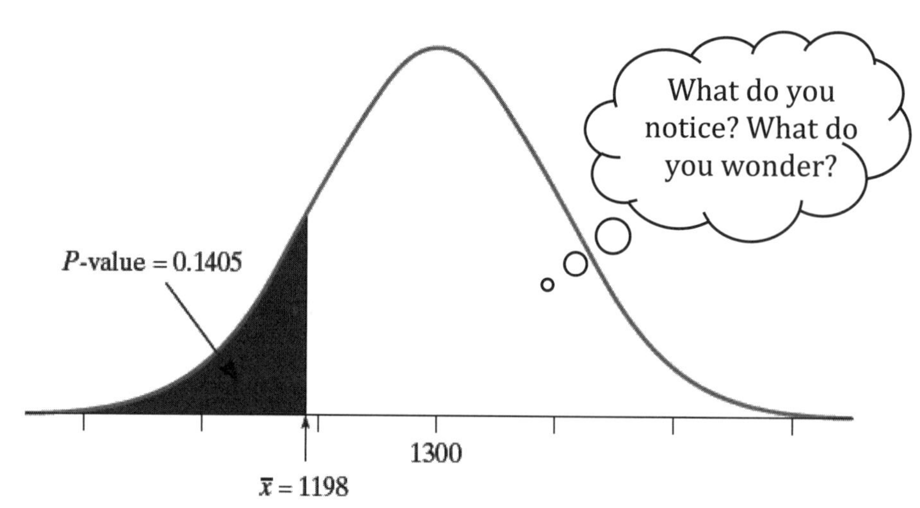

Read: Vocabulary and Key Concepts

- A ____________________________ is a formal procedure for using observed data to decide between two competing claims (called hypotheses). The claims are usually statements about ______________ parameters.

Stating Hypotheses

- The claim that we weigh evidence against in a significance test is called the ________ hypothesis (H_0).
- The claim that we are trying to find evidence for is the _______________ hypothesis (H_a).
- The alternative hypothesis is _____________ if it states that a parameter is *greater than* the null value or if it states that the parameter is *less than* the null value.

- The alternative hypothesis is ____________ if it states that the parameter is *different from* the null value (it could be either greater than or less than).
- The null hypothesis has the form H_0: parameter ___ null value.
- A one-sided alternative hypothesis has one of the forms H_a: parameter ___ null value or H_a: parameter ___ null value.
- A two-sided alternative hypothesis has the form H_a: parameter ___ null value.
- The hypotheses should express the belief or suspicion we have _________ we see the data.

Interpreting *P*-values

- The ____________ of a test is the probability of getting evidence for the alternative hypothesis H_a as strong or stronger than the ____________ evidence when the _______ hypothesis H_0 is true.
- Example phrasing for interpreting a *P*-value: __________ that the player makes 80% of their free throws in the long run, there is about a 0.0075 probability of getting a _________ proportion of 0.64 or less just by _________ in a random sample of 50 shots.
- Small *P*-values give ________________ evidence for H_a because they say that the observed result is _____________ to occur when H_0 is true. Large *P*-values ___________ give convincing evidence for H_a because they say that the observed result is __________ to occur by chance alone when H_0 is true.

Making Conclusions

- If the *P*-value is _______, we _______ H_0 because the observed result is unlikely to occur when H_0 is true. In this case, there is ________________ evidence for H_a.
- If the *P*-value is ___________, we _______________ H_0 because the observed result is at least somewhat likely to occur when H_0 is true. In this case, there is ______ convincing evidence for H_a.
- The __________________________ α is the value that we use as a boundary to decide if an observed result is unlikely to happen by chance alone when the null hypothesis is true.
- The *P*-value of 0.0003 gives much ____________ evidence against H_0 and in favor of H_a than the *P*-value of 0.03. This is why we always include the *P*-value in our conclusions, and not just a statement about significance.
- Never "_________ H_0" or conclude that H_0 is _____!
- The significance level should be stated ____________ the data are produced.

Watch: Go to bfwpub.com/TPS7eStrive. Select Unit 6, Part I and watch the following Example videos:

Video	Topic	Done
Unit 6C, Juicy pineapples (p. 564)	*Stating hypotheses*	
Unit 6C, Healthy bones (p. 565)	*Interpreting P-values*	
Unit 6C, More healthy bones (p. 568)	*Making conclusions*	

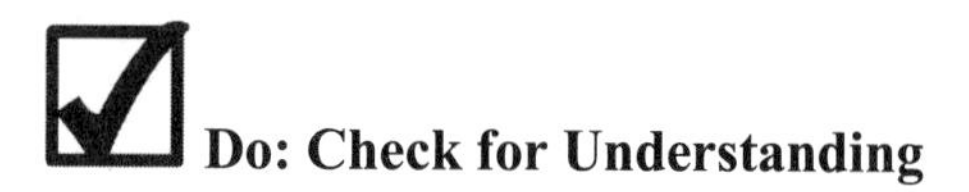

Do: Check for Understanding

Concept 1: Stating Hypotheses

A significance test is a procedure that allows us to test a claim about a population parameter by studying a statistic from a random sample. If an observed statistic is "far" away from a hypothesized claim about a parameter, we have some evidence that the claim is wrong. Whether or not the statistic is "far enough" away depends on the sampling distribution of the statistic. That is, statistics will vary from sample to sample. A significance test answers the question, "What are the chances we would observe a sample statistic at least this extreme, assuming the claim about the parameter is true?" If the chances are low, there is evidence that the claim may be wrong. If the chances are pretty good, then there is little evidence to suggest the claim is wrong. Of course, to even begin making that argument, you must start by setting up a null hypothesis (or claim about the parameter you are trying to find evidence against) and an alternative hypothesis (the claim about the parameter you are trying to find evidence for). Note: the null hypothesis will always be a statement of equality. That is, it says, "Let's assume the parameter is equal to _____." The alternative hypothesis is set up to test whether or not the actual value of the parameter is greater than, less than, or simply not equal to the value in the null hypothesis.

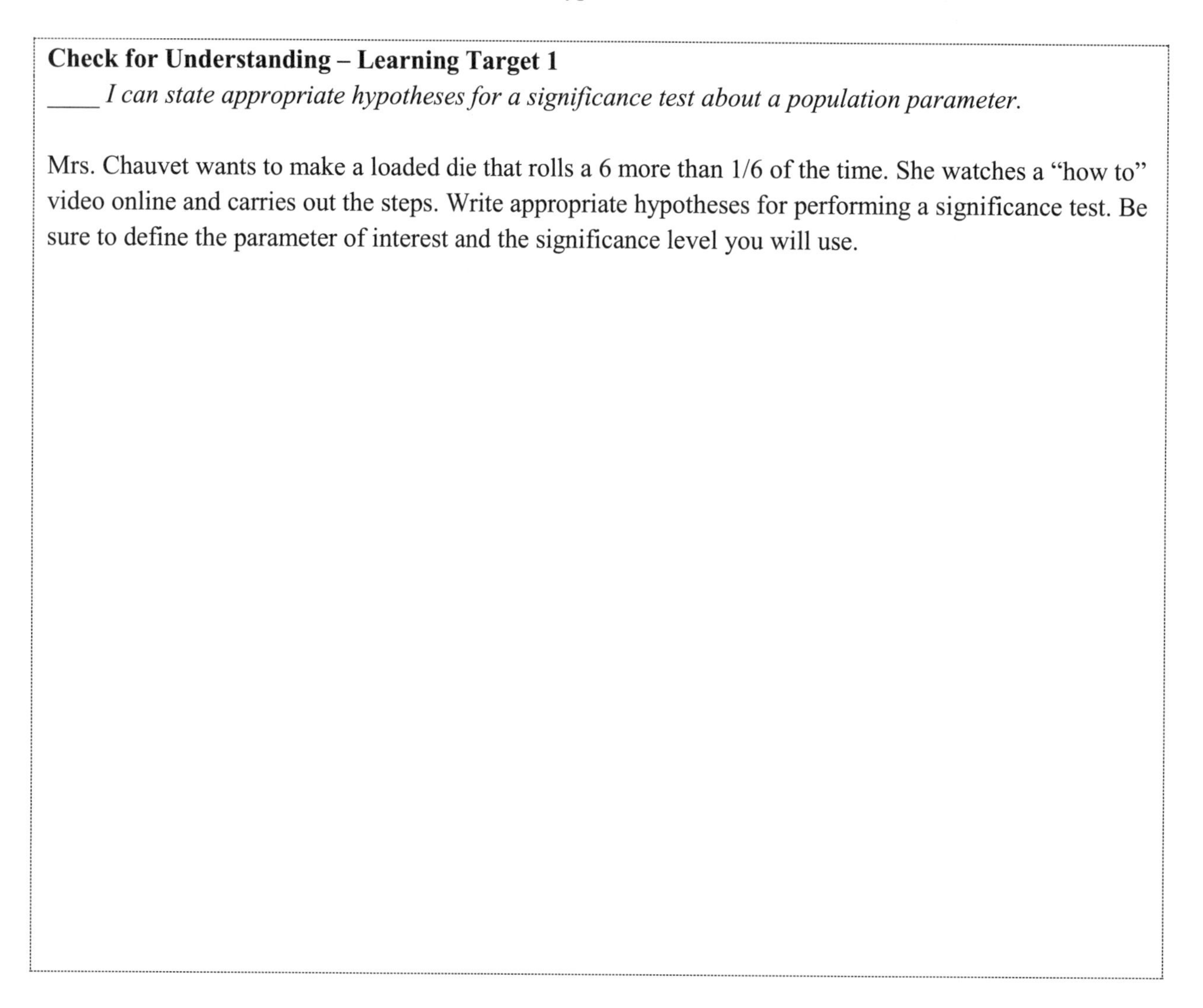

Check for Understanding – Learning Target 1

____ *I can state appropriate hypotheses for a significance test about a population parameter.*

Mrs. Chauvet wants to make a loaded die that rolls a 6 more than 1/6 of the time. She watches a "how to" video online and carries out the steps. Write appropriate hypotheses for performing a significance test. Be sure to define the parameter of interest and the significance level you will use.

Concepts 2 and 3: P-values, Making Conclusions

To conclude that we have convincing evidence against a null hypothesis, the sample statistic must have a small likelihood of occurring under the assumption that the null hypothesis is true. To determine this, we must consider the sampling distribution of the statistic under the assumption that the null hypothesis is true and calculate the probability of observing a statistic at least as extreme as the one observed. If the conditions for inference are met, we can describe the sampling distribution and calculate a standardized test statistic. This test statistic can then be used to determine a *P*-value. If this *P*-value is small enough, we can say the data are statistically significant and we have evidence to reject the null hypothesis. The general rule of thumb is to use a significance level of 5%, although some situations may specify levels of 1% or even 10%. If the *P*-value is smaller than the chosen significance level, we have convincing evidence to reject the null hypothesis and conclude that the alternative is true. The reasoning behind this is that if we observe an outcome that has an extremely low chance of occurring under a given assumption, then there is evidence that the assumption may be false. If we observe an outcome that has a fairly likely chance of occurring under a given assumption, then we have little reason to doubt the assumption!

Check for Understanding – Learning Targets 2 and 3

____ *I can interpret a P-value in context.*

____ *I can make an appropriate conclusion for a significance test.*

Refer to the hypotheses you wrote in the previous check for understanding. Mrs. Chauvet rolled the "loaded" die 100 times and got 38 sixes. The *P*-value of the significance test is less than 0.001.

1. Explain what it means for the null hypothesis to be true in this setting.

2. Interpret the *P*-value in context.

3. Do these data provide convincing evidence against the null hypothesis at a 1% significance level? Explain.

Section 6D: Significance Tests for a Population Proportion

Section Summary

In this section, you will apply the logic of significance tests and the four-step process to test a claim about a population proportion. You will learn how to check the conditions for a significance test as well as how to calculate a standardized test statistic and *P*-value. Finally, you will learn how to write an appropriate conclusion in context. Your study will also include an introduction to two-sided tests to find evidence that a population proportion is different than a hypothesized value.

Because we base our conclusion on a sample statistic that varies from sample to sample, we must be aware of the fact that our conclusion could be wrong. This section ends with a discussion of the types of errors that could occur when performing a significance test and how to deal with them. Be sure to get a good understanding of the logic and format for a significance test. You will be using them throughout the rest of the course!

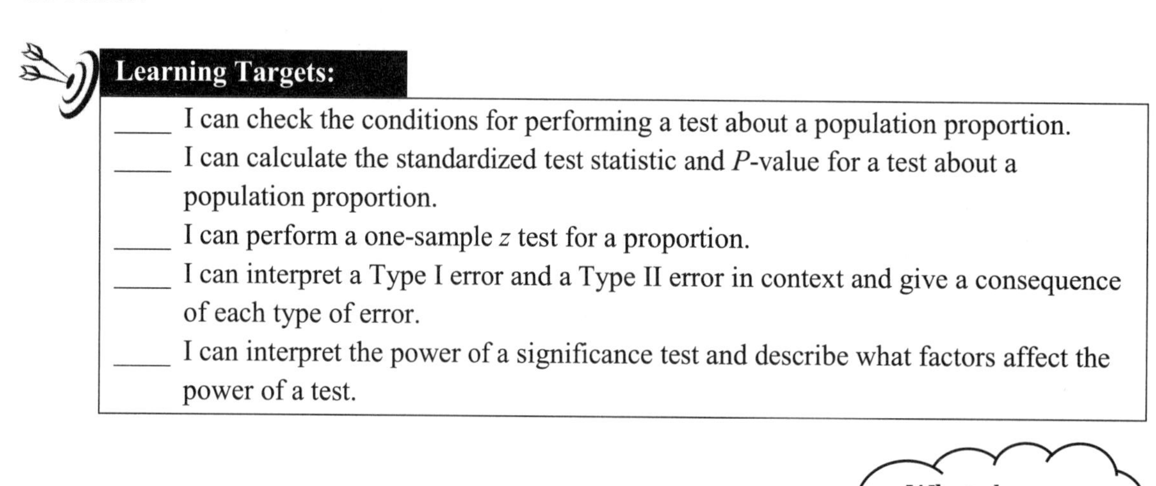

Learning Targets:

____ I can check the conditions for performing a test about a population proportion.

____ I can calculate the standardized test statistic and *P*-value for a test about a population proportion.

____ I can perform a one-sample z test for a proportion.

____ I can interpret a Type I error and a Type II error in context and give a consequence of each type of error.

____ I can interpret the power of a significance test and describe what factors affect the power of a test.

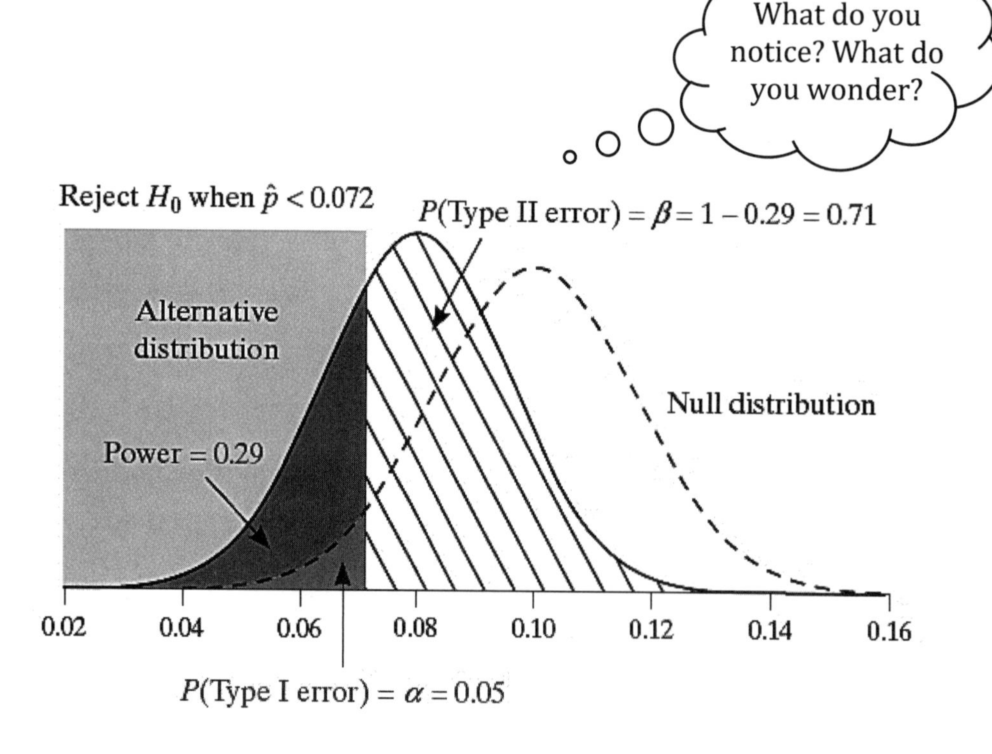

Read: Vocabulary and Key Concepts

Conditions for Performing a Significance Test About a Population Proportion

- Random: The data come from a ______________ sample from the population of interest.
  - 10%: When sampling _____________ replacement, $n < 0.10N$.
- Large Counts: Both np_0 and $n(1-p_0)$ are at least ____.
- Note: In this section we use the ________________ value specified by the null hypothesis (denoted p_0) when checking the Large Counts condition. We use p_0 instead of $\hat{p}$ because we are assuming the null hypothesis is _______ when performing the test.
- The values of np_0 and $n(1-p_0)$ represent the ______________ numbers of successes and failures, assuming the null hypothesis is true.
- A __________________________________ measures how far a sample statistic is from what we would expect if the null hypothesis H_0 were true, in standard deviation units. That is, standardized test statistic = ______________________________

Calculating the Standardized Test Statistic and *P*-value In a Test about a Population Proportion

- To perform a test of H_0: $p = p_0$, calculate the ________________ test statistic

$$z = \frac{\hat{p} - p_0}{\sqrt{\frac{p_0(1-p_0)}{n}}}$$

Find the *P*-value by calculating the _______________ of getting a z statistic this large or _________ in the direction specified by the _______________ hypothesis H_a in the standard normal distribution.

Two-Sided Tests

- In a two-sided test, the alternative hypothesis has the form H_a: p ____ p_0.
- The *P*-value in such a test is the probability of getting a sample proportion as far as or farther from p_0 in _____________ direction than the observed value of $\hat{p}$. As a result, you have to find the area in ________ tails of a standard normal distribution to get the *P*-value. For this reason, a two-sided test is sometimes called a *two-tailed* test.

Confidence Intervals Give More Information

- There is a link between confidence intervals and ________________ tests. For example, a 95% confidence interval gives an approximate set of p_0's that should __________________ by a two-sided test at the $\alpha = 0.05$ significance level. Any value of p _____________ the interval should be ______________ as implausible.

Four Step Process

- **State:** State the _______________, ________________, and ____________________.
- **Plan:** Identify the appropriate ________________ method and check the ________________.
- **Do:** If the conditions are met, perform ____________________.

- Calculate the ________________________.
- Find the ___________.

◆ **Conclude:** Make a __________________ about the hypotheses in the context of the problem.

Type I and Type II Errors

◆ A Type I error occurs if we __________ H_0 when H_0 is ________. That is, the data give convincing evidence that H_a is true when it really isn't.

◆ A Type II error occurs if we _________________ H_0 when H_a is ________. That is, the data do not give convincing evidence that H_a is true when it really is.

◆ The probability of making a Type I error in a significance test is equal to the significance level _____.

◆ There is a trade-off between *P*(Type I error) and *P*(Type II error): as one increases, the other _______________, assuming everything else remains the same.

Power

◆ The ___________ of a test is the probability that the test will find convincing evidence for *Ha* when a specific alternative value of the parameter is true.

◆ The power of a test is a __________________ probability.

◆ Power = *P*(reject H_0 | parameter = some specific alternative value).

◆ To interpret the power of a test in a given setting, just interpret the relevant conditional probability.

◆ The power of a test to detect a specific alternative parameter value is related to the probability of a Type ____ error for that alternative: power = 1 – *P*(Type II error) and *P*(Type II error) = 1 – power

◆ When H_0 is false and H_a is true, a significance test based on a random sample of size n and significance level α will have greater power and a smaller probability of making a Type II error when:
  - The sample size n is __________.
  - The significance level α is __________.
  - The null and alternative parameter values are ___________ apart.
  - The standard error of the statistic is ____________.

Watch: Go to bfwpub.com/TPS7eStrive. Select Unit 6, Part I and watch the following Example videos:

Video	Topic	Done
Unit 6D, Get a job! (p. 574)	*Checking conditions for a test about p*	
Unit 6D, Part-time jobs (p. 577)	*Calculating the standardized test statistic and P-value for a test about p*	
Unit 6D, One potato, two potato (p. 582)	*One-sample z test for p*	
Unit 6D, Perfect potatoes (p. 584)	*Type I and Type II errors*	
Unit 6D, Powerful potatoes (p. 590)	*The power of a test*	

Do: Check for Understanding

Concepts 1, 2, and 3: One-Sample z Test for a Population Proportion

Both significance tests and confidence intervals are based on the same concept—sampling distributions. Therefore, when performing inference with either of these methods, it is important to follow a process that ensures the calculations are justified. To test a claim about a population parameter, the following four-step process should be used:

1) **STATE: State the hypotheses, parameter(s), and significance level.**
   When testing a claim about a population proportion, we start by defining hypotheses about the parameter. The null hypothesis assumes that the population proportion is equal to a particular value while the alternative hypothesis is that the population proportion is greater than, less than, or not equal to that value:
   H_0: $p = p_0$
   H_a: $p > p_0$ OR $p < p_0$ OR $p \neq p_0$
   Define the parameter, in context. State a significance level.

2) **PLAN: Identify the appropriate inference method and check the conditions.**
   We must check the Random, 10%, and Large Counts conditions before proceeding with inference. Make sure a random sample was selected. If sampling is done without replacement, verify that the sample size is less than 10% of the size of the population. To ensure that the sampling distribution of $\hat{p}$ is approximately normal, make sure np_0 and $n(1 - p_0)$ are both at least 10.

3) **DO: If conditions are met, perform calculations.**
   Calculate the z test statistic

$$z = \frac{\hat{p} - p_0}{\sqrt{\frac{p_0(1 - p_0)}{n}}}$$

   and find the P-value by calculating the probability of observing a z statistic at least this extreme in the direction of the alternative hypothesis.

4) **CONCLUDE: Make a conclusion about the hypotheses in the context of the problem.**
   If the P-value is smaller than the stated significance level, we have convincing evidence to reject the null hypothesis. If the P-value is greater than the significance level, then we fail to reject the null hypothesis and we do not have convincing evidence for H_a.

This four-step procedure can be used with *any* significance test. The details change slightly depending on the inference procedure, but the overall process is the same for all significance tests. Be sure to get a good grasp of this process here as you will be using it a lot in the coming units!

Check for Understanding – Learning Targets 1, 2, and 3

____ *I can check conditions for performing a significance test about a population proportion.*

____ *I can calculate the standardized test statistic and P-value for a test about a population proportion.*

____ *I can perform a one-sample z test for a proportion.*

It is interesting to read about old laws that have never been removed from law books. For example, in Alaska, it is illegal to whisper in someone's ear while they are moose hunting. Alaska legislators agree that they will remove the law from the books if there is convincing evidence at the $\alpha = 0.01$ level that this behavior has happened to less than 5% of moose hunters.

A random sample of 250 moose hunters are selected from the large population of Alaskan moose hunters and the results show that 4 of them have had someone whisper in their ear while they were moose hunting. Do the data provide convincing evidence to allow for the removal of the law from the Alaskan law books?

(*Need Tech Help?* View the video **Tech Corner 17: Significance Tests for a Proportion** at bfwpub.com/TPS7eStrive)

Concept 4: Type I and Type II Errors

Because the conclusion we make in a significance test is based on what we would expect to see in a sampling distribution, there is a chance the sample statistic does not accurately reflect the true value of the parameter. That is, due to sampling variability, it is possible that the random sample that is collected will lead to an incorrect conclusion about the parameter. If we obtain a very small *P*-value and reject the null when, in fact, the null hypothesis is true, we commit a Type I error. The probability of committing a Type I error is equal to the significance level, α. If we obtain a relatively large *P*-value and fail to reject the null when, in fact, the null hypothesis is false, we commit a Type II error. Make sure you know how to interpret both Type I and Type II errors in the context of the problem and explain the consequences of each.

Check for Understanding – Learning Target 4

____ *I can interpret a Type I error and a Type II error in context and give a consequence of each type of error.*

Refer to the previous Check for Understanding.

1. Describe a Type I error in this setting. Give a possible consequence of making a Type I error.

2. Describe a Type II error in this setting. Give a possible consequence of making a Type II error.

Concept 5: The Power of a Test

The power of a test is the probability that the test will find convincing evidence for H_a when a specific alternative value of the parameter is true. You can think of this as the probability of avoiding a Type II error. Therefore, for a specific alternative value, Power = 1 – P(Type II error).

There are a few ways we can increase the power of a significance test. We can increase the sample size, increase the significance level, or increase the difference we wish to detect between the null and a specific alternative parameter value. The design of our study can also increase power. Specifically, controlling for other variables and blocking in experiments or stratified random sampling can increase power.

Check for Understanding – Learning Target 5

____ *I can interpret the power of a significance test and describe what factors affect the power of a test.*

A computer manufacturer requires that the processors they install meet high quality standards. If the manufacturer finds convincing evidence that greater than 3% of the processors in any given shipment use more than 70W of power, the shipment will be sent for expensive additional testing. A quality control engineer inspects a random sample of processors from each shipment and performs a test of H_0: $p = 0.03$ versus H_a: $p > 0.03$, where p = the proportion of all processors in the shipment that use more than 70 watts of power.

1. The power of the test to detect that $p = 0.04$, based on a random sample of 30 processors and a significance level $\alpha = 0.05$, is 0.12. Interpret this value.

Determine if each of the following changes would increase or decrease the power of the test. Explain your answers.

2. Change the significance level to $\alpha = 0.01$.

3. Select a random sample of 50 processors, instead of 30.

4. The true proportion is $p = 0.05$ instead of $p = 0.04$.

Unit 6, Part I Summary: Inference for Categorical Data: Proportions

Statistical inference is the practice of drawing a conclusion about a population based on information gathered from a sample. Part I of this unit introduced us to the practice of estimating a parameter based on a statistic. By using what we learned about sampling distributions, we constructed an interval around a point estimate that we are confident captures the parameter of interest. The confidence level itself tells what would happen if we constructed many intervals for samples of the same size. In essence, the confidence level is the capture rate for all of the constructed intervals. So, when we build a level C interval, we can interpret it by saying "We are C% confident the interval from a to b captures the true parameter of interest."

A significance test tells us whether or not a certain sample provides convincing evidence against a claim about a population parameter. The test answers the question, "How likely is it to observe this particular sample statistic (or one more extreme) if the claim about the parameter is true?" This probability, the P-value, gives us an idea of just how surprising an observed statistic is under the assumption that the null hypothesis is true. If the P-value is small (less than 5% or some other chosen significance level), we have enough evidence to reject the null hypothesis, suggesting the actual parameter value may be greater than, less than, or different from the claim. If the P-value is greater than or equal to a specified significance level, we do not have convincing evidence against the claim about the parameter. Keep in mind, however, that because we are basing our decision on a likelihood calculated from a sampling distribution, it is possible that we may be wrong. Type I and Type II errors occur when we mistakenly reject a null hypothesis that is true or fail to reject a null hypothesis that is false, respectively. You should be able to define each of these errors in the context of the situation and explain the consequences of each.

Unit 6, Part I not only introduced you to the logic behind confidence intervals and significance tests, but also introduced a four-step procedure for carrying them out. This procedure will be used throughout the rest of the course, so you need to be comfortable with it. If not, be sure to practice a few more tests. If you understand and follow those steps, you will have very few problems performing significance tests in the upcoming units!

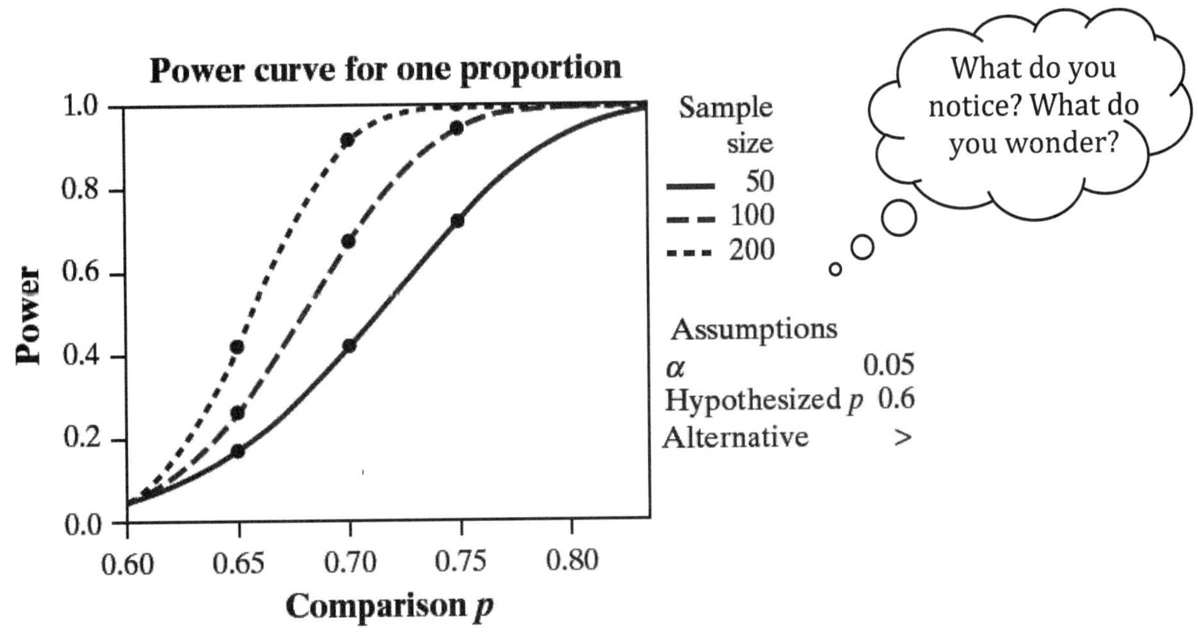

How well do you understand each of the learning targets?

Learning Target	Got It!	Almost There	Needs Work
I can interpret a confidence interval in context.			
I can use a confidence interval to make a decision about the value of a parameter.			
I can interpret a confidence level in context.			
I can check conditions for calculating a confidence interval for a population proportion.			
I can calculate a confidence interval for a population proportion.			
I can construct and interpret a one-sample z interval for a proportion.			
I can describe how the sample size and confidence level affect the margin of error.			
I can determine the sample size required to obtain a confidence interval for a population proportion with a specified margin of error.			
I can state appropriate hypotheses for a significance test about a population parameter.			
I can interpret a *P*-value in context.			
I can make an appropriate conclusion for a significance test.			
I can check the conditions for performing a test about a population proportion.			
I can calculate the standardized test statistic and *P*-value for a test about a population proportion.			
I can perform a one-sample z test for a proportion.			
I can interpret a Type I error and a Type II error in context and give a consequence of each type of error.			
I can interpret the power of a significance test and describe which factors affect the power of a test.			

Unit 6, Part I Multiple Choice Practice

Directions. Identify the choice that best completes the statement or answers the question. Check your answers and note your performance when you are finished.

1. Gallup Poll interviews 1600 people. Of these, 8% say that they jog regularly. A news report gives the 95% confidence interval for the true proportion individuals who jog regularly. They also add: "The poll had a margin of error of plus or minus three percentage points." You can safely conclude
   (A) 95% of all Gallup Poll samples like this one give estimates within ±3% of the population parameter.
   (B) The percent of the population who jog is certain to be between 5% and 11%.
   (C) 95% of the population jog between 5% and 11% of the time.
   (D) We can be 3% confident that the sample result is true.
   (E) If Gallup selected many random samples of this same size, 95% of them would find that 8% of the people in the sample jog.

2. An agricultural researcher plants 25 plots with a new variety of corn. A 90% confidence interval for the mean yield for these plots is found to be 162.72 ± 4.47 bushels per acre. Which of the following is the correct interpretation of the interval?
   (A) There is a 90% chance the interval from 158.25 to 167.19 captures the population mean yield.
   (B) 90% of sample mean yields will be between 158.25 and 167.19 bushels per acre.
   (C) We are 90% confident the population mean yield is 162.72 bushels per acre.
   (D) 90% of the time, the population mean yield will fall between 158.25 and 167.19 bushels per acre.
   (E) We are 90% confident the interval from 158.25 to 167.19 bushels per acre captures the population mean yield.

3. You are told that the proportion of those who answered "yes" to a poll about internet use is 0.70, and that the standard error is 0.0458. What is the sample size?
   (A) 50
   (B) 99
   (C) 100
   (D) 200
   (E) cannot be determined from the information given.

4. A newspaper conducted a statewide survey concerning the 2024 race for president. The newspaper took a random sample (assume it is an SRS) of 1200 registered voters and found that 620 would vote for the Republican candidate. Let p represent the proportion of registered voters in the state that would vote for the Republican candidate. What is a 90% confidence interval for p?
   (A) 0.517 ± 0.014
   (B) 0.517 ± 0.022
   (C) 0.517 ± 0.024
   (D) 0.517 ± 0.028
   (E) 0.517 ± 0.249

5. After a college's football team once again lost a football game to the college's archrival, the alumni association decided to conduct a survey to see if alumni were in favor of firing the coach. Let p represent the proportion of all living alumni who favor firing the coach. Which of the following is the smallest sample size needed to provide an estimate that's within $\pm$ 0.025 of p at a 95% confidence level?
   (A) 269
   (B) 385
   (C) 538
   (D) 768
   (E) 1537

6. An opinion poll asks a random sample of 200 adults how they feel about voting for an amendment in an upcoming election. In all, 150 say they are in favor of the amendment. We want to know if the poll provides convincing evidence that the proportion p of adults who are in favor of the amendment is greater than 60%? What are the null and alternative hypotheses?
   (A) H_0: $p = 0.6$ against H_a: $p > 0.6$
   (B) H_0: $p = 0.6$ against H_a: $p \neq 0.6$
   (C) H_0: $p = 0.6$ against H_a: $p < 0.6$
   (D) H_0: $p = 0.6$ against H_a: $p = 0.75$
   (E) H_0: $p = 0.75$ against H_a: $p < 0.6$

7. A test of significance produces a P-value of 0.035. Which conclusion is appropriate?
   (A) Accept H_a at the $\alpha = 0.05$ level
   (B) Reject H_a at the $\alpha = 0.01$ level
   (C) Fail to reject H_0 at the $\alpha = 0.05$ level
   (D) Reject H_0 at the $\alpha = 0.05$ level
   (E) Accept H_0 at the $\alpha = 0.01$ level

8. A Type II error is
   (A) rejecting the null hypothesis when it is true.
   (B) failing to reject the null hypothesis when it is false.
   (C) rejecting the null hypothesis when it is false.
   (D) failing to reject the null hypothesis when it is true.
   (E) more serious than a Type I error.

9. A researcher plans to conduct a significance test at the $\alpha = 0.05$ significance level. She designs her study to have a power of 0.85 for a particular alternative value of the parameter. What is the probability the researcher will commit a Type II error?
   (A) 0.05
   (B) 0.15
   (C) 0.80
   (D) 0.95
   (E) equal to "1 – P-value" and cannot be determined until the data have been collected.

10. A claimed psychic was presented with 200 cards face down and asked to determine if the card was one of five symbols: a star, cross, circle, square, or three wavy lines. The "psychic" was correct in 50 cases. To determine if he has ESP, we test the hypotheses H_0: $p = 0.20$, H_a: $p > 0.20$, where p = the true proportion of cards for which the psychic would correctly identify the symbol in the long run. The conditions for inference are met. What is the *P*-value of this test?
    (A) between 0.05 and 0.10
    (B) between 0.025 and 0.05
    (C) between 0.01 and 0.025
    (D) between 0.001 and 0.01
    (E) below 0.001

Check your answers below. If you got a question wrong, check to see if you made a simple mistake or if you need to study that concept more. After you check your work, identify the concepts you feel very confident about and note what you will do to learn the concepts in need of more study.

#	Answer	Concept	Right	Wrong	Simple Mistake?	Need to Study More
1	A	Interpreting a Confidence Level				
2	E	Interpret a Confidence Interval				
3	C	Standard Error of $\hat{p}$				
4	C	Confidence Interval for p				
5	E	Choosing Sample Size				
6	A	Stating Hypotheses				
7	D	Making a Conclusion				
8	B	Type II Error				
9	B	Power and Type II Error				
10	B	*P*-value				

FRAPPY! Free Response AP® Problem, Yay!

The following problem is modeled after actual Advanced Placement Statistics free response questions. Your task is to generate a complete, concise response in 15 minutes. After you generate your response, read over the solution and scoring guide. Score your response and note what, if anything, you would do differently to increase your own score.

A pollster conducts a survey in the Eastern U.S. to determine the percentage of U.S. adults from this area who approve of our current president. The results of the survey show that 429 of 1000 randomly selected participants approve.

(a) Construct a 95% confidence interval for the proportion of U.S. adults in the Eastern U.S. who approve of our current president.

(b) Shortly thereafter, a similar study is conducted in the western U.S. with the intention of constructing a 95% confidence interval for the proportion of adults from the Western U.S. who approve of our current president. What is the smallest sample size that can be used to guarantee that the margin of error will be less than or equal to 0.02?

FRAPPY! Scoring Rubric

Use the following rubric to score your response. Each part receives a score of "Essentially Correct," "Partially Correct," or "Incorrect." When you have scored your response, reflect on your understanding of the concepts addressed in this problem. If necessary, note what you would do differently on future questions like this to increase your score.

Intent of the Question

The goal of this question is to determine your ability to construct and interpret a confidence interval and correctly determine the necessary sample size.

Solution

(a) Name: 95% one–sample *z* interval for $p =$ the proportion of all adults from the Eastern U.S. who approve of the current president.

Conditions:

- Random: A random sample of 1000 adults from the Eastern U.S. was selected.
- 10%: 1000 < 10% of the population of U.S. adults in the Eastern U.S.
- Large Counts: $n\hat{p} \geq 10$: $429 \geq 10$ and $n(1-\hat{p}) \geq 10$: $571 \geq 10$

Calculations: $0.429 \pm 1.960\sqrt{\frac{0.429(1-0.429)}{1000}} = (0.398, 0.460)$

Interpretation: Based on the sample, we are 95% confident that the interval from 0.398 to 0.460 captures the proportion of all adults from the Eastern U.S. who approve of the president.

(b) Because we need to find the sample size required to provide that the margin of error will be no larger than 0.02, we will use the value $\hat{p} = 0.5$ in this calculation.

$1.96\sqrt{\frac{0.5(1-0.5)}{n}} \leq 0.02 \rightarrow n = 2401$

The margin of error will be 0.01 or less when *n* is at least 2401.

Scoring

Parts (a) is split into two parts of which each part, along with part (b) are scored as essentially correct (E), partially correct (P), or incorrect (I).

Part (a) – Name and Conditions: This is essentially correct if the response correctly checks the conditions AND correctly names the confidence interval (one-sample *z* confidence interval for a proportion) by name or formula. Part (a) is partially correct if the conditions are not properly checked but the interval is correctly named OR if the conditions are properly checked and at most one part of naming the confidence interval is missing or incorrect.

Part (a) – Calculations and Conclusion: This is essentially correct if the response correctly calculates the interval AND writes a correct conclusion in context that references the parameter ("population proportion of all adults from the Eastern U.S. who approve…". Part (a) is partially correct if one of these requirements is not met.

Part (b) is essentially correct if the response correctly calculates the necessary integer sample size AND shows their work (formula with correct values) AND doesn't round down. Part (b) is partially correct if a value other than 0.5 is used OR the answer is not a whole number or is rounded down.

4 Complete Response
All three parts essentially correct

3 Substantial Response
Two parts essentially correct and one part partially correct

2 Developing Response
Two parts essentially correct and no parts partially correct
One part essentially correct and one or two parts partially correct
Three parts partially correct

1 Minimal Response
One part essentially correct and no parts partially correct
No parts essentially correct and two parts partially correct

My Score:
What I did well:
What I could improve:
What I should remember if I see a problem like this on the AP Exam:

FRAPPY! Free Response AP® Problem, Yay!

The following problem is modeled after actual Advanced Placement Statistics free response questions. Your task is to generate a complete, concise response in 15 minutes. After you generate your response, read over the solution and scoring guide. Score your response and note what, if anything, you would do differently to increase your own score.

During a recent movie promotion, Fruity O's cereal placed mini action figures in some of its boxes. The advertisement on the box states 1 out of every 4 boxes contains an action figure. A group of promotional-toy collectors suspects the proportion of boxes containing the action figure may be less than 0.25. The group randomly selected and purchased 70 boxes of cereal and found 12 action figures. Do the data provide convincing evidence to support the toy collector's belief that the proportion of boxes containing the figure is less than 0.25? Provide statistical evidence to support your answer. Use $\alpha = 0.05$.

FRAPPY! Scoring Rubric

Use the following rubric to score your response. Each part receives a score of "Essentially Correct," "Partially Correct," or "Incorrect." When you have scored your response, reflect on your understanding of the concepts addressed in this problem. If necessary, note what you would do differently on future questions like this to increase your score.

Intent of the Question

The goal of this question is to determine your ability to conduct a significance test for a single proportion.

Solution

The solution should contain:

- **Hypotheses** must be stated appropriately. H_0: $p = 0.25$ and H_a: $p < 0.25$ where p = proportion of all boxes which contain an action figure.
- **Name of test and conditions**: The test must be identified by name or formula as a one-sample z test for a population proportion. A random sample of cereal boxes is selected. 10%: 70 < 10% of the population of all cereal boxes. Large Counts: $70(0.25) = 17.5$ and $70(0.75) = 52.5$ are both at least 10.
- **Calculations**: $z = -1.52$ and P-value = 0.064
- **Conclusion**: Because the P-value of 0.064 is greater than $\alpha = 0.05$, we fail to reject the null hypothesis. There is not convincing evidence to suggest that the proportion of all boxes with an action figure is less than 0.25.

Scoring

Each element scored as essentially correct (E), partially correct (P), or incorrect (I).

"Hypotheses and Name" is essentially correct if the hypotheses are written correctly, and the response correctly identifies the test by name or formula. This part is partially correct if the test isn't identified or if the hypotheses are written incorrectly. Incorrect otherwise.

"Conditions and Calculations" is essentially correct if the response correctly does 4 of the 5 following things: (1) checks the random condition, (2) checks the 10% condition, (3) checks the large counts condition, (4) correct test statistic, (5) correct P-value. This part is partially correct if 3 or more of those 3 things are done correctly. Incorrect otherwise.

"Conclusion" is essentially correct if the response correctly fails to reject the null hypothesis because the P-value is greater than a significance level of 5% OR if the response is to reject the null hypothesis because the P-value is less than a significance level of 10% and provides an interpretation in context. This part is partially correct if the response fails to justify the decision by comparing the P-value to a significance level OR if the conclusion lacks an interpretation in context. Incorrect otherwise. Note: A correct conclusion consistent with an incorrect P-value can earn credit for this part.

4 Complete Response
All three parts essentially correct

3 Substantial Response
Two parts essentially correct and one part partially correct

2 Developing Response
Two parts essentially correct and no parts partially correct
One part essentially correct and one or two parts partially correct
Three parts partially correct

1 Minimal Response
One part essentially correct and no parts partially correct
No parts essentially correct and two parts partially correct

My Score:
What I did well:
What I could improve:
What I should remember if I see a problem like this on the AP Exam:

Unit 6, Part I Formula Sheet

Create a one-page summary of important concepts and formulas found in this unit. This will serve as a valuable resource as you prepare for the AP Exam.

Unit 6, Part I Crossword Puzzle

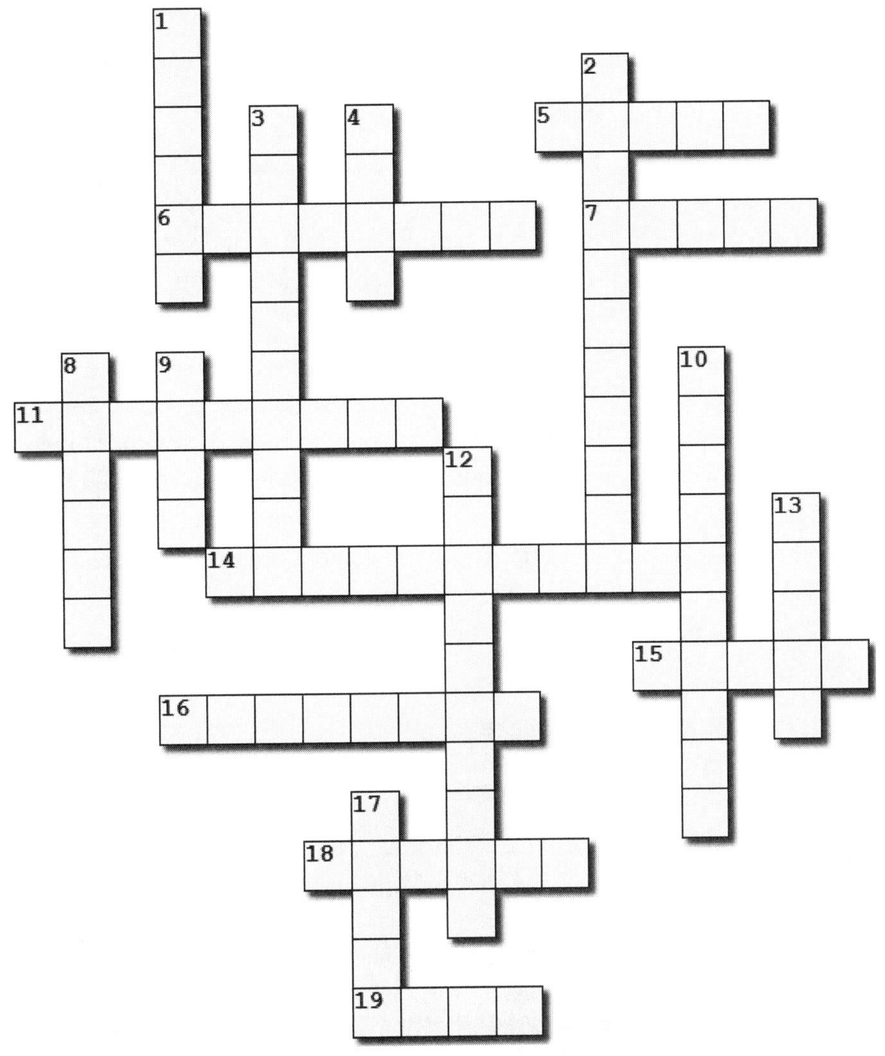

Across:

5. To make a conclusion in a significance test, compare the *P*-value to ____.
6. The four-step process is: State, Plan, Do, ____.
7. The margin of ____ of an estimate describes how far, at most, we expect the point estimate to vary from the population parameter.
11. We use a sample statistic to estimate a population ____.
14. Check the 10% condition when sampling without ____.
15. The confidence ____ is the long-run capture rate (success rate) of the method that produces the interval.
16. The ____ value is a multiplier that makes the confidence interval wide enough to have the stated capture rate.
18. When calculating a one-sample z confidence interval, the large ____ condition says that the number of successes and failures must each be at least 10.
19. A Type I error occurs if we reject H_0 when H_0 is ____.

Down:

1. If the *P*-value is less than α, we should ____ H_0.
2. In a significance test, the claim we are trying to find evidence for is called the ____ hypothesis.
3. A ____ interval gives a set of plausible values for a parameter based on sample data.
4. Quadrupling the sample size cuts the margin of error in ____.
8. The first condition in any inference procedure is that the data must be a ____ sample from the population of interest.
9. A Type II error occurs when we ____ to reject H_0 when H_a is true.
10. In a significance test, the claim we weigh evidence against is called the null ____.
12. We can increase the power of a test by ____ the sample size.
13. ____ = 1 – *P*(Type II error)
17. A ____ estimator is a statistic that provides an estimate of a population parameter.

Unit 6, Part II: Inference for Categorical Data: Proportions

"A statistical analysis, properly conducted, is a delicate dissection of uncertainties, a surgery of suppositions." M.J. Moroney

UNIT 6, PART II OVERVIEW

In Unit 6, Part I, you learned how to estimate a parameter based on a statistic obtained from one random sample. Now you will learn how to use data from two independent random samples to estimate a difference in population proportions and to test a claim about a difference in population proportions. Unit 6, Part II revisits and reviews the skills that you acquired in Unit 6, Part I, but in the context of two samples. In Unit 7 you will revisit and review these skills again, but in the context of means. In each case, you will use the same logic and the same four-step procedure. The first part of this unit contained a lot of new information. Use the time spent on Unit 6, Part II to solidify your understanding as you revisit and deepen your understanding of statistical inference.

Sections in Unit 6, Part II
Section 6E: Confidence Intervals for a Difference in Population Proportions
Section 6F: Significance Tests for a Difference in Population Proportions

PRACTICE FOR MASTERY

Use the following *suggested* guide to the pages and exercises in your text to practice for mastery!
Note: your teacher may assign different problems. Be sure to follow their instructions!

Day	Topics	Read	Do
1	◆ Checking Conditions for a Confidence Interval for p_1-p_2 ◆ Calculating a Confidence Interval for p_1-p_2 ◆ Two-Sample z Interval for p_1-p_2	p. 608-617	**6E**: 1, 3, 5, 9, 11, 17 – 19
2	◆ Stating Hypotheses for a Test about p_1-p_2 ◆ Checking Conditions for a Test about p_1-p_2 ◆ Calculating the Standardized Test Statistic and *P*-value for a Test about p_1-p_2	p. 622-628	**6F**: 1, 3, 5, 7, 9, 11, 22 – 25
3	◆ Two-Sample z Test for p_1-p_2	p. 629-632	**6F**: 13, 15, 17, 19
4	Unit 6, Part II AP® Statistics Practice Test (p.641)		Unit 6, Part II Review Exercises (p.640)
5	Unit 6, Part II Test: Celebration of learning!	Note: Since "Test" can sound daunting, think of this as a way to show off all you have learned. It's your own celebration of learning!	

Section 6E: Estimating a Difference in Proportions

Section Summary

What is the difference in the percentage of high school and college students who work in addition to attending classes? How much might a particular medication reduce the rate of having a heart attack in the population of middle-aged males compared to not taking the medication? In this section, you will learn how to construct and interpret a confidence interval for a difference in two population proportions. Two-sample settings can arise from selecting independent random samples from two different populations (high school students vs. college students) or by generating data from two groups in a randomized experiment (those who do and do not take a particular medication for heart attack prevention). As with a single proportion, you will learn the conditions necessary for performing inference and then how to construct the confidence interval. By the end of this section, you should be able to provide an estimate for the difference in two population proportions.

Learning Targets:

____ I can check the conditions for calculating a confidence inference about a difference between two proportions.

____ I can calculate a confidence interval for a difference between two population proportions.

____ I can construct and interpret a two-sample z interval for a difference in proportions.

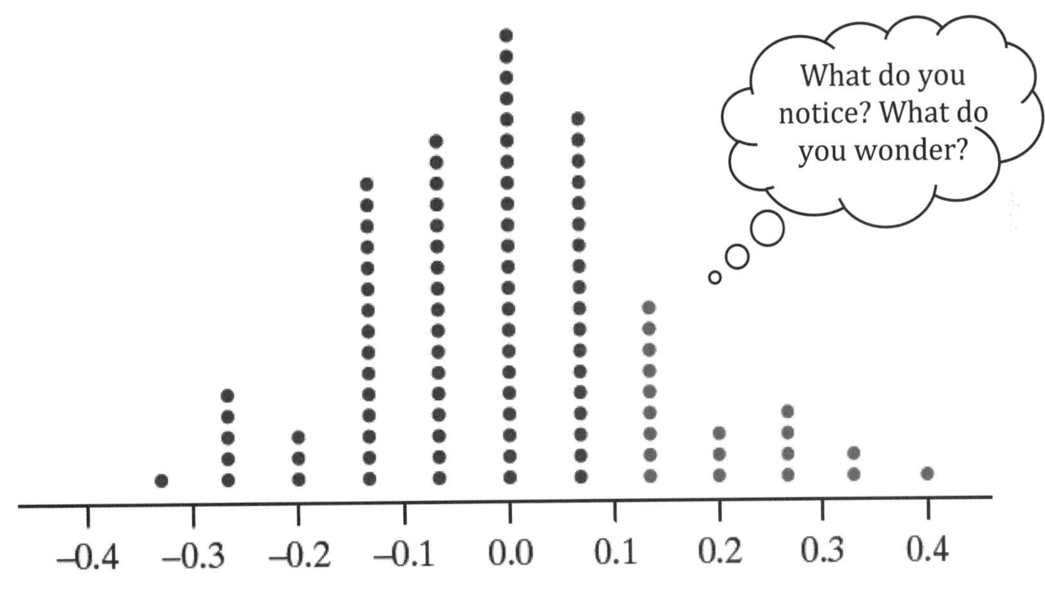

Read: Vocabulary and Key Concepts

Conditions - Confidence Interval for a Difference Between Two Population Proportions

- **Random:** The data come from two ________________ random samples or from two __________ in a randomized experiment.
  - **10%:** When sampling without ________________, $n_1 < 0.10N_1$ and $n_2 < 0.10N_2$.
- **Large Counts:** The counts of "____________" and "____________" in each sample or group — $n_1\hat{p}_1$, $n_1(1-\hat{p}_1)$, $n_2\hat{p}_2$, and $n_2(1-\hat{p}_2)$ — are all at least 10.

- There are now two ways to satisfy the Random condition. First, using ______________ random samples allows us to ____________ our results to the populations of interest. Second, using ________________________ in an experiment permits us to draw cause-and-effect conclusions.

- When sampling without replacement from two populations, the ____________ allows us to view the observations within each sample as ________________. The formula also requires that the __________ themselves are independent, which is why we require *independent* random samples in the ________________ condition when the goal is to generalize our results to the two populations.

- The method we use to calculate a confidence interval for $p_1 - p_2$ requires that the sampling distribution of $\hat{p}_1 - \hat{p}_2$ be approximately __________. This will be true whenever $n_1\hat{p}_1$, $n_1(1-\hat{p}_1)$, $n_2\hat{p}_2$, and $n_2(1-\hat{p}_2)$ are all ________________. Because we don't know the value of p_1 or p_2 when we are estimating $p_1 - p_2$, we use ____ and ____ when checking the Large Counts condition.

Calculating a Confidence Interval for a Difference Between Two Population Proportions

- When the conditions are met, a *C*% confidence interval for $p_1 - p_2$ is

  Formula: ____

  where z^* is the critical value for the standard normal curve with *C*% of the area between $-z^*$ and z^*.
- When a confidence interval does not include _____ (no difference) as a plausible value for $p_1 - p_2$, we have ________________ evidence of a ________________ between the population proportions.
- Never suggest that you believe the difference between the true proportions *is* ______ just because ______ is in the interval! There are other values in the interval as well and $p_1 - p_2$ may be one of them!

Watch: Go to bfwpub.com/TPS7eStrive. Select Unit 6, Part II and watch the following Example videos:

Video	Topic	Done
Unit 6E, Watching at home? (p. 611)	*Checking conditions for a confidence interval for* $p_1 - p_2$	
Unit 6E, Avoiding the theater (p. 612)	*Calculating a confidence interval for* $p_1 - p_2$	
Unit 6E, Treating lower back pain (p. 615)	*Two-sample z interval for* $p_1 - p_2$	

 **Do: Check for Understanding**

Concepts 1, 2, and 3: Conference Intervals for $p_1 - p_2$

To construct a confidence interval for the difference between two proportions $p_1 - p_2$, follow the four-step process. The logic behind its construction is the same as the logic behind the construction of a confidence interval for a single proportion. That is, we will estimate the difference by comparing the sample proportions from two random samples and build an interval around that point estimate by using the standard error of the statistic and a critical z value, z^*. To construct a two-sample z interval for a difference between two proportions:

- **State:** State the parameters of interest and the confidence level you will use to estimate the difference.
- **Plan:** Name the confidence interval you are constructing and verify that the Random, 10%, and Large Counts conditions are satisfied for both samples.
- **Do:** Carry out the calculations using the following formula:

$$(\hat{p}_1 - \hat{p}_2) \pm z^* \sqrt{\frac{\hat{p}_1(1-\hat{p}_1)}{n_1} + \frac{\hat{p}_2(1-\hat{p}_2)}{n_2}}$$

  where z^* is the critical value for the standard normal curve with area C between $-z^*$ and z^*.
- **Conclude:** Interpret the interval in the context of the problem.

Check for Understanding - Learning Targets 1, 2, and 3

____ *I can check conditions for calculating a confidence interval about a difference between two proportions.*

____ *I can calculate a confidence interval for a difference between two population proportions.*

____ *I can construct and interpret a two-sample z interval for a difference in proportions.*

In 2000, 551 of 1500 randomly sampled U.S. adults indicated they smoked cigarettes. In 2020, 352 of 2000 randomly sampled U.S. adults indicated they were cigarette smokers. Use this information to construct and interpret a 95% confidence interval for the difference in the proportion of U.S. adults who smoked cigarettes in 2000 and 2020.

(*Need Tech Help?* View the video **Tech Corner 18: Confidence Intervals for a Difference in Proportions** at bfwpub.com/TPS7eStrive)

Section 6F: Significance Tests For a Difference in Population Proportions

Section Summary

In this section, you will learn how to carry out a significance test about a difference in two population proportions. The logic behind this process is rooted in an understanding of the sampling distribution of a difference between two proportions. As with a single proportion, you will learn the conditions necessary for performing inference and then will learn how to perform a significance test. By the end of this section, you should be able to draw a conclusion about a claim that is made about the difference between two proportions.

Learning Targets:

- ____ I can state appropriate hypotheses for performing a test about a difference between two population proportions.
- ____ I can check conditions for performing a test about a difference between two population proportions.
- ____ I can calculate the standardized test statistic and *P*-value for a test about a difference between two proportions.
- ____ I can perform a two-sample *z* test for a difference in proportions.

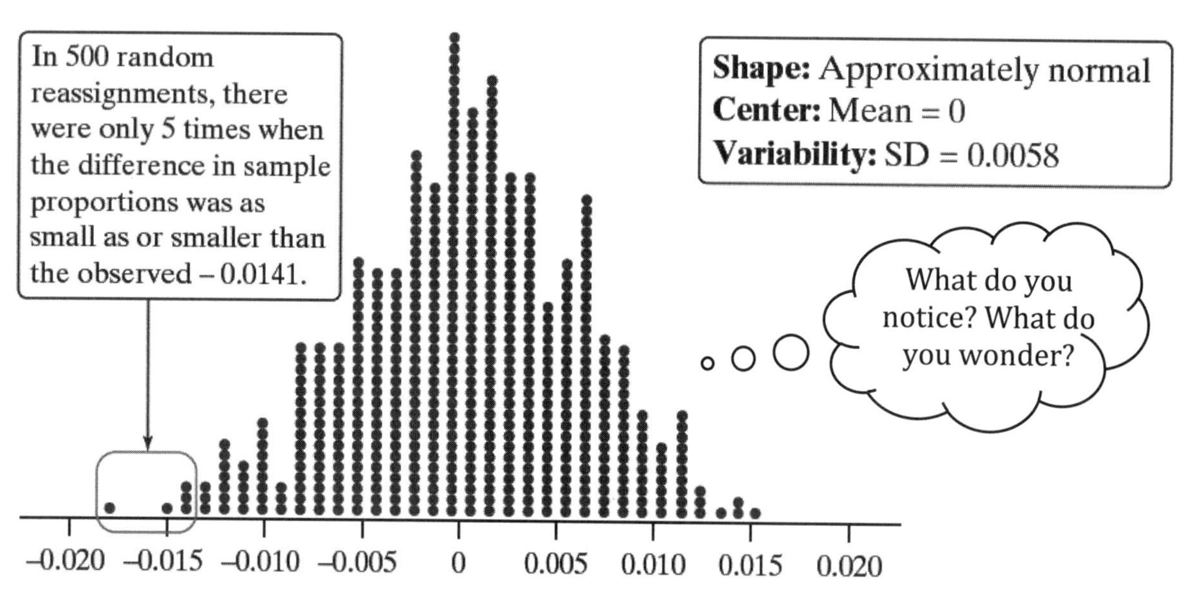

Read: Vocabulary and Key Concepts

- The null hypothesis for a significance test for a difference in population proportions says that there is no difference between the two parameters: H_0: _______________. You will sometimes see this null hypothesis written in the equivalent form H_0: _______________.

- A significance test begins by assuming that the null hypothesis is _______. In that case, $p_1 = p_2$. We call the common value of these two parameters p. To estimate p, we _____________ (or "pool") the data from the two samples as if they came from one larger sample.

$$\hat{p}_C = \frac{\text{number of successes in both samples combined}}{\text{number of individuals in both samples combined}} = \_\_\_\_\_\_\_\_\_\_\_\_$$

- In other words, $\hat{p}_C$ gives the overall proportion of _____________ in the combined samples.

Conditions for Performing a Significance Test About a Difference Between Two Population Proportions

- **Random:** The data come from two independent _____________________ or from two groups in a _________________________________.
  - **10%:** When sampling _______________________________, $n_1 < 0.10N_1$ and $n_2 < 0.10N_2$.
- **Large Counts:** The ____________ numbers of successes and failures in each sample or group — $n_1\hat{p}_C$, $n_1(1-\hat{p}_C)$, $n_2\hat{p}_C$, and $n_2(1-\hat{p}_C)$ — are all _____________.

Calculating the Standardized Test Statistic and *P*-value in a Test About the Difference Between Two Population Proportions

- Suppose the conditions are met. To perform a test of $H_0: p_1 - p_2 = 0$, calculate the _____________ test statistic:

$$z = \frac{(\hat{p}_1 - \hat{p}_2) - 0}{\sqrt{\hat{p}_C(1-\hat{p}_C)\left(\frac{1}{n_1} + \frac{1}{n_2}\right)}} \quad \text{where} \quad \hat{p}_C = \frac{X_1 + X_2}{n_1 + n_2}$$

Find the ______________ by calculating the probability of getting a z statistic this large or larger in the direction specified by the _______________ hypothesis in the standard normal distribution.

Watch: Go to bfwpub.com/TPS7eStrive. Select Unit 6, Part II and watch the following Example videos:

Video	Topic	Done
Unit 6F, Accurate orders (p. 622)	*Stating hypotheses for a test about $p_1 - p_2$*	
Unit 6F, Accurate orders (p. 624)	*Checking conditions about $p_1 - p_2$*	
Unit 6F, Accurate orders (p. 626)	*Calculating the standardized test statistic and P-value for a test about $p_1 - p_2$*	
Unit 6F, Cholesterol and heart attacks (p. 629)	*Two-sample z test for $p_1 - p_2$*	

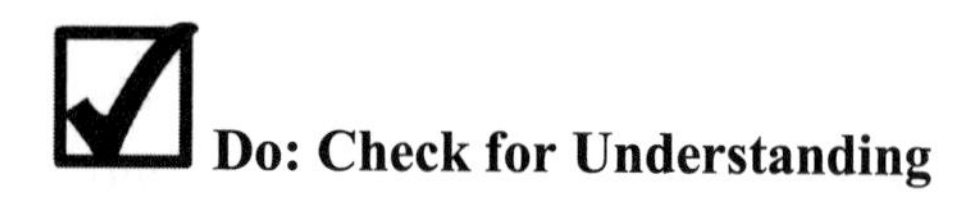

Do: Check for Understanding

Concepts 1, 2, 3, and 4: Significance test for $p_1 - p_2$

To test a claim about the difference between two proportions, we will follow the same four-step process we used in the previous sections. The key difference when performing a test about $p_1 - p_2$ is that we begin by assuming these two parameters are equal (in the null hypotheses), so we need to estimate the value they are equal to. To calculate this "pooled" proportion, simply divide the sum of the *successes* in the two samples by the sum of the *sample sizes*. This pooled proportion is used in the formula for the standard deviation of the sampling distribution. To conduct a two-sample z test for the difference between two proportions:

1) **STATE: State the hypotheses, parameters, and significance level.**
   The null hypothesis assumes the population proportions are equal (no difference) and the alternative hypothesis is that one population proportion is greater than, less than, or not equal to the other:

   H_0: $p_1 = p_2$
   H_a: $p_1 > p_2$ or $p_1 < p_2$ or $p_1 \neq p_2$

   OR

   H_0: $p_1 - p_2 = 0$
   H_a: $p_1 - p_2 > 0$ OR $p_1 - p_2 < 0$ OR $p_1 - p_2 \neq 0$

   Define the parameters in the context of the problem. Make sure you do not refer to the sample while defining the parameters. Then state a significance level.

2) **PLAN: Identify the appropriate inference method and check the conditions.**
   First, verify that the data come from two independent random samples or two groups in a randomized experiment. Next, if sampling without replacement occurred, check that each sample size is less than 10% of its respective population size. Finally, to ensure that the sampling distribution of $\hat{p}_1 - \hat{p}_2$ is approximately normal, check that $n_1\hat{p}_C$, $n_1(1-\hat{p}_C)$, $n_2\hat{p}_C$, and $n_2(1-\hat{p}_C)$ are all at least 10. The value of the pooled proportion is $\hat{p}_C = \dfrac{X_1 + X_2}{n_1 + n_2}$.

3) **DO: If conditions are met, perform calculations.**
   Calculate the test statistic: $z = \dfrac{(\hat{p}_1 - \hat{p}_2) - 0}{\sqrt{\dfrac{\hat{p}_C(1-\hat{p}_C)}{n_1} + \dfrac{\hat{p}_C(1-\hat{p}_C)}{n_2}}}$ where $\hat{p}_C = \dfrac{X_1 + X_2}{n_1 + n_2}$

   and find the P-value by calculating the probability of observing a z statistic at least as extreme in the direction of the alternative hypothesis.

4) **CONCLUDE: Make a conclusion about the hypotheses in the context of the problem.**
   If the P-value is smaller than the stated significance level, you should reject the null hypothesis. The data provide convincing evidence for H_a. If the P-value is greater than or equal to the significance level, then you fail to reject the null hypothesis. The data do not provide convincing evidence for H_a.

   Be sure to explicitly compare the P-value to α, state your decision, then state the conclusion in context.

Check for Understanding – Learning Targets 1, 2, 3, and 4

____ *I can state appropriate hypotheses for performing a test about a difference between two population proportions.*

____ *I can check the conditions for performing a test about a difference between two population proportions.*

____ *I can calculate the standardized test statistic and P-value for a test about a difference between two population proportions.*

____ *I can perform a two-sample z test for a difference in proportions.*

Do healthcare providers practice what they preach? An article claims that healthcare providers are just as likely to smoke as the general population. To investigate, a researcher selects a random sample of 150 healthcare providers and a random sample of 120 adults from the general population. The results show that 30 of the 150 healthcare providers smoke cigarettes while 28 out of 120 adults from the general population smoke. Do these data provide convincing evidence that the proportion of all healthcare providers who smoke cigarettes is different than that of the general population? Use a 5% significance level.

(*Need Tech Help?* View the video **Tech Corner 19: Significance Tests for a Difference in Proportions** at bfwpub.com/TPS7eStrive)

Unit 6, Part II Summary: Inference for Categorical Data: Proportions

When you began your study of statistics, one of the first things you learned was how to distinguish between categorical and quantitative data. Then, you learned how to display a distribution of categorical data (bar graph) and quantitative data (dotplot, stemplot, histogram). You learned how to create numerical summaries of categorical data (percentages) and quantitative data (mean, median, standard deviation, *IQR*, range). Then, you moved on to study the relationships between two categorical variables (two-way tables) and two quantitative variables (scatterplots). Now, your toolbox also contains the skills necessary to perform inference using categorical data. In Unit 6, Part I you learned how to calculate a confidence interval for a population proportion and carry out a significance test for a proportion. In Unit 6, Part II you went a step further and learned how to perform inference when the categorical data come from two independent random samples or two groups in a randomized experiment. In the next unit, you will round out your understanding of statistical inference by learning how to perform inference on quantitative data. Keep up the great work!

How well do you understand each of the learning targets?

Learning Target	Got It!	Almost There	Needs Work
I can check the conditions for calculating a confidence interval about a difference between two population proportions.			
I can calculate a confidence interval for a difference between two population proportions.			
I can construct and interpret a two-sample z interval for a difference in proportions.			
I can state appropriate hypotheses for performing a test about a difference between two population proportions.			
I can check the conditions for performing a test about a difference between two population proportions.			
I can calculate the standardized test statistic and P-value for a test about a difference between two population proportions.			
I can perform a two-sample z test for a difference in proportions.			

Unit 6, Part II Multiple Choice Practice

Directions. Identify the choice that best completes the statement or answers the question. Check your answers and note your performance when you are finished.

1. A teacher would like to know which of two homework incentive plans is more effective. To do so, she randomly assigns the 60 students in her class to one of two incentive plans by having each student flip a coin. Of the 34 students that were assigned to Plan A, 23 successfully completed all homework assignments. The remaining students were assigned to Plan B, of which 20 successfully completed all homework assignments. The teacher would like to construct a 90% confidence interval for $p_A - p_B =$ the difference in the proportion of all students like these that would successfully complete all homework assignments under these two plans. Which of the following conditions for inference, if any, are not met?
   (A) The Random condition is the only condition that is not met.
   (B) The 10% condition is the only condition that is not met.
   (C) The Large Counts condition is the only condition that is not met.
   (D) All conditions for inference are met.
   (E) Both the 10% condition and the Large Counts condition are not met.

2. An SRS of 100 teachers showed that 64 work (another job) outside of school. An SRS of 100 students showed that 80 work a job outside of school. Let p_T be the proportion of all teachers who work outside of school and p_S be the proportion of all students who work outside of school. A 95% confidence interval for $p_T - p_S$ is
   (A) (0.264, 0.056)
   (B) (0.098, 0.222)
   (C) (–0.222, –0.098)
   (D) (–0.264, –0.056)
   (E) (–0.283, –0.038)

3. A school receives textbooks independently from two suppliers. An SRS of 400 textbooks from supplier 1 finds 20 that are defective. An SRS of 100 textbooks from supplier 2 finds 10 that are defective. Let p_1 and p_2 be the proportions of all textbooks from suppliers 1 and 2, respectively that are defective. Which of the following represents a 95% confidence interval for $p_1 - p_2$?

   (A) $-0.05 \pm 1.960\sqrt{\frac{(0.05)(0.95)}{400} - \frac{(0.1)(0.9)}{100}}$

   (B) $-0.05 \pm 1.960\sqrt{\frac{(0.05)(0.95)}{400} + \frac{(0.1)(0.9)}{100}}$

   (C) $0.05 \pm 1.960\sqrt{\frac{(0.05)(0.95)}{400} - \frac{(0.1)(0.9)}{100}}$

   (D) $0.05 \pm 1.960\sqrt{\frac{(0.05)(0.95)}{400} + \frac{(0.1)(0.9)}{100}}$

   (E) $-0.05 \pm 1.960\sqrt{\frac{(0.06)(0.94)}{500}}$

4. I selected two independent random samples of the same size from two large populations and used the data collected to compute a 95% confidence interval for the difference in population proportions. Which of the following would produce a wider confidence interval, assuming the sample proportions remain the same?
   (A) Use a larger confidence level.
   (B) Use a smaller confidence level.
   (C) Increase the population sizes.
   (D) Increase the sample sizes.
   (E) Nothing can ensure that you will get a wider interval.

5. A random sample of n_1 high school students in the U.S. showed that x_1 are on Facebook. A separate random sample of n_2 adults in the U.S. showed that x_2 are on Facebook. Let p_1 and p_2 be the proportion of all high school students and adults in the U.S., respectively, who are on Facebook. Suppose the 95% confidence interval for the true difference in the proportion of high school students and adults who are on Facebook is constructed. Which of the following is the best interpretation of the confidence interval?
   (A) 95% of the time, the true difference in the proportion of high school students and adults in the U.S. who are on Facebook is captured in the interval.
   (B) We are 95% confident that the difference in the proportion of high school students and adults in these samples who are on Facebook is captured in the interval.
   (C) The probability that the true difference in the proportion of high school students and adults in the U.S. who are on Facebook is in the constructed interval is 95%.
   (D) 5% of the differences calculated from samples this size will be in the constructed interval.
   (E) We are 95% confident that the constructed interval captures the difference in the proportion of all high school students and all adults in the U.S. who are on Facebook.

6. A student decides to carry out a statistical study as a graduation project. The student would like to study the effect of music on perception. Specifically, he recruited 50 students for study. He randomly assigned the 50 students to two groups of 25. The members of the first group watched a short video of someone being dishonest while happy music played in the background. The members of the second group watched the same short video while sad music played in the background. Of the 25 members of the first group, 12 of the students acknowledged that the person in the video was dishonest, while 22 members of the second group made the same acknowledgement. The student would like to conduct a z test for a difference in two proportions. Which of the following statements is true?
   (A) The Random condition is not met because the students were not randomly selected.
   (B) The 10% condition is not needed in this case because random sampling did not occur.
   (C) The Large Counts condition is met because the expected number of successes and failures in each group is at least 10.
   (D) The Random condition and the Large Counts condition are both violated.
   (E) All conditions for inference are met.

7. Sixty-eight people from a random sample of 128 residents of Uppsala, Sweden had blue eyes. Forty-five people from a random sample of 110 people from Preston, England, had blue eyes. Let p_1 represent the proportion of people in Uppsala with blue eyes and let p_2 represent the proportion of people in Preston with blue eyes. If researchers suspected that the proportion of people that have blue eyes is different in these two countries, which of the following pairs of hypotheses would be appropriate?
   (A) $H_0: p = 0.53, p_2 = 0.41$; $H_a: p_1 \neq 0.53, p_2 \neq 0.41$
   (B) $H_0: p_1 = p_2 = 0.47$; $H_a: p_1 \neq p_2 \neq 0.47$
   (C) $H_0: p_1 = p_2$; $H_a: p_1 > p_2$
   (D) $H_0: p_1 \neq p_2$; $H_a: p_1 = p_2$
   (E) $H_0: p_1 = p_2$; $H_a: p_1 \neq p_2$

8. In a random sample of 80 male students at a college, 28 said they frequently send at least one text message before getting out of bed in the morning. In a random sample of 105 female students at the same college, 57 said they frequently texted before getting out of bed. Let p_1 = the proportion of all male students at this college who frequently text before getting up and let p_2 = the corresponding proportion for female students. For a test of the hypothesis $H_0: p_1 - p_2 = 0$ versus $H_0: p_1 - p_2 \neq 0$, which of the following is a correct expression for the test statistic?

   (A) $z = \dfrac{0.350 - 0.543}{\sqrt{\dfrac{0.350(0.650)}{80}} + \sqrt{\dfrac{0.543(0.457)}{105}}}$

   (B) $z = \dfrac{0.350 - 0.543}{\sqrt{\dfrac{0.350(0.650)}{80} - \dfrac{0.543(0.457)}{105}}}$

   (C) $z = \dfrac{0.350 - 0.543}{\sqrt{\dfrac{(0.530)(0.543)}{185}}}$

   (D) $z = \dfrac{0.350 - 0.543}{\sqrt{(0.459)(0.541)\left(\dfrac{1}{80} + \dfrac{1}{105}\right)}}$

   (E) $z = \dfrac{0.350 - 0.543}{\sqrt{(0.350)(0.543)\left(\dfrac{1}{80} + \dfrac{1}{105}\right)}}$

9. Return to the previous scenario. What is the approximate *P*-value of this test?
   (A) 0.001
   (B) 0.003
   (C) 0.045
   (D) 0.009
   (E) 0.018

10. Was the proportion of marshmallows in Mr. Miller's Lucky Charms cereal greater in the past than it is now? To determine this, you test the hypotheses H_0: $p_{old} = p_{now}$, H_a: $p_{old} > p_{now}$ at the $\alpha = 0.05$ level with p_{old} = the proportion of cereal that was marshmallows in the past and p_{now} = the proportion of cereal that is marshmallows now. You calculate a test statistic of 1.980. Which of the following is the appropriate *P*-value and conclusion for your test?
    (A) *P*-value = 0.024; reject H_0; We have convincing evidence that the proportion of marshmallows has been reduced.
    (B) *P*-value = 0.024; fail to reject H_0; We have convincing evidence that the proportion of marshmallows has not changed.
    (C) *P*-value = 0.024; fail to reject H_0; We do not have convincing evidence that the proportion of marshmallows has been reduced.
    (D) *P*-value = 0.047; reject H_0; We do not have convincing evidence that the proportion of marshmallows has been reduced.
    (E) *P*-value = 0.047; accept H_a; There is convincing evidence that the proportion of marshmallows has been reduced.

Check your answers below. If you got a question wrong, check to see if you made a simple mistake or if you need to study that concept more. After you check your work, identify the concepts you feel very confident about and note what you will do to learn the concepts in need of more study.

#	Answer	Concept	Right	Wrong	Simple Mistake?	Need to Study More
1	C	Conditions for Inference (CI)				
2	E	Confidence Interval for Difference in Proportions				
3	B	Confidence Interval for Difference in Proportions				
4	A	Width of a Confidence Interval				
5	E	Interpretation of a Confidence Interval				
6	B	Conditions for Inference (Test)				
7	E	Stating Hypotheses				
8	D	Test Statistic				
9	D	*P*-value				
10	A	Significance Test, *P*-value, Conclusion				

FRAPPY! Free Response AP® Problem, Yay!

The following problem is modeled after actual Advanced Placement Statistics free response questions. Your task is to generate a complete, concise response in 15 minutes. After you generate your response, read over the solution and scoring guide. Score your response and note what, if anything, you would do differently to increase your own score.

The manager of a quick service convenience store wants to estimate the difference in the proportion of customers who would purchase a specialty drink if a pop-up advertisement appeared on the self-service kiosk screen before the customer completed the order. To investigate, the kiosk was programmed to flip a virtual coin for each customer. If the virtual coin toss came up heads, the kiosk would display an advertisement for a specialty drink. If the virtual coin toss came up tails, no advertisement was displayed. For each customer, the kiosk recorded whether or not an advertisement was displayed and whether or not a specialty drink was ordered.

(a) The manger found that 48 of the 82 customers who received an advertisement purchased a specialty drink and 31 of the 86 customers who did not receive an advertisement purchased a specialty drink. Construct and interpret a 99% confidence interval for the difference (Ad – No Ad) in the true proportion of customers like these who would purchase a specialty drink when shown the advertisement and the proportion who would purchase a specialty drink without being shown the advertisement.

(b) Does this interval provide convincing evidence that there is a difference in the proportion of customers like these who would purchase a specialty drink depending on whether or not they were shown the advertisement? Explain your reasoning.

FRAPPY! Scoring Rubric

Use the following rubric to score your response. Each part receives a score of "Essentially Correct," "Partially Correct," or "Incorrect." When you have scored your response, reflect on your understanding of the concepts addressed in this problem. If necessary, note what you would do differently on future questions like this to increase your score.

Intent of the Question

The goal of this question is to determine your ability to calculate and interpret a confidence interval for a difference in population proportions and use the interval to draw a conclusion.

Solution

(a) **STATE:** 99% CI for $p_1 - p_2$, where p_1 = the true proportion of customers like these who would purchase a specialty drink after being shown the advertisement and p_2 = the true proportion of customers like these who would purchase a specialty drink without being shown the advertisement.

PLAN: Two-sample z interval for $p_1 - p_2$

Random? The data come from two groups of a randomized experiment. ✓

Large Counts? 48, 82 – 48 = 34, 31, 86 – 31 = 55 are all at least 10. ✓

DO: $\hat{p}_1 = \frac{48}{82} = 0.585$ and $\hat{p}_2 = \frac{31}{86} = 0.360$

$$(0.585 - 0.360) \pm 2.576\sqrt{\frac{0.585(1-0.585)}{82} + \frac{0.36(1-0.36)}{86}}$$

0.225 ± 0.193

(0.032, 0.418)

Using technology: 0.031 to 0.418

CONCLUDE: We are 99% confident that the interval from 0.031 to 0.418 captures the true difference (Ad – No Ad) in the proportion of customers like these who would purchase a specialty drink when shown the advertisement and the proportion who would purchase a specialty drink without being shown the advertisement.

(b) Yes. Because the plausible values for the difference in proportions are all positive, there is convincing evidence that a greater proportion of customers like these will purchase a specialty drink when they are shown the advertisement.

Scoring

Each element scored as essentially correct (E), partially correct (P), or incorrect (I).

Part (a) "Name and Conditions" is essentially correct if the response does all 5 of the following correctly: (1) names the inference procedure correctly in words or by formula (2) defines the parameter, (3) checks the Random condition, (4) checks the Large Counts condition, (5) includes context. This part is partially correct if 3 or 4 of those 5 components are done correctly. Incorrect otherwise.

Part (a) "Mechanics and Conclusion" is essentially correct if the response correctly has the 3 following components: (1) correct confidence interval, (2) correct interpretation based upon the calculated interval, (3) includes context. This part is partially correct if 2 of those 3 components are done correctly. Incorrect otherwise.

Part (b) is essentially correct if the response is "yes" and justifies the answer based upon the fact that all of the values in the confidence interval are positive. This part is partially correct if the response is "yes" but has an incorrect justification. Incorrect otherwise. Note: A correct conclusion for an incorrect interval above can earn credit for this part.

Scoring

4 Complete Response
All three parts essentially correct

3 Substantial Response
Two parts essentially correct and one part partially correct

2 Developing Response
Two parts essentially correct and no parts partially correct
One part essentially correct and one or two parts partially correct
Three parts partially correct

1 Minimal Response
One part essentially correct and no parts partially correct
No parts essentially correct and two parts partially correct

My Score:
What I did well:
What I could improve:
What I should remember if I see a problem like this on the AP Exam:

Unit 6, Part II Formula Sheet

Create a one-page summary of important concepts and formulas found in this unit. This will serve as a valuable resource as you prepare for the AP Exam.

Unit 6, Part II Crossword Puzzle

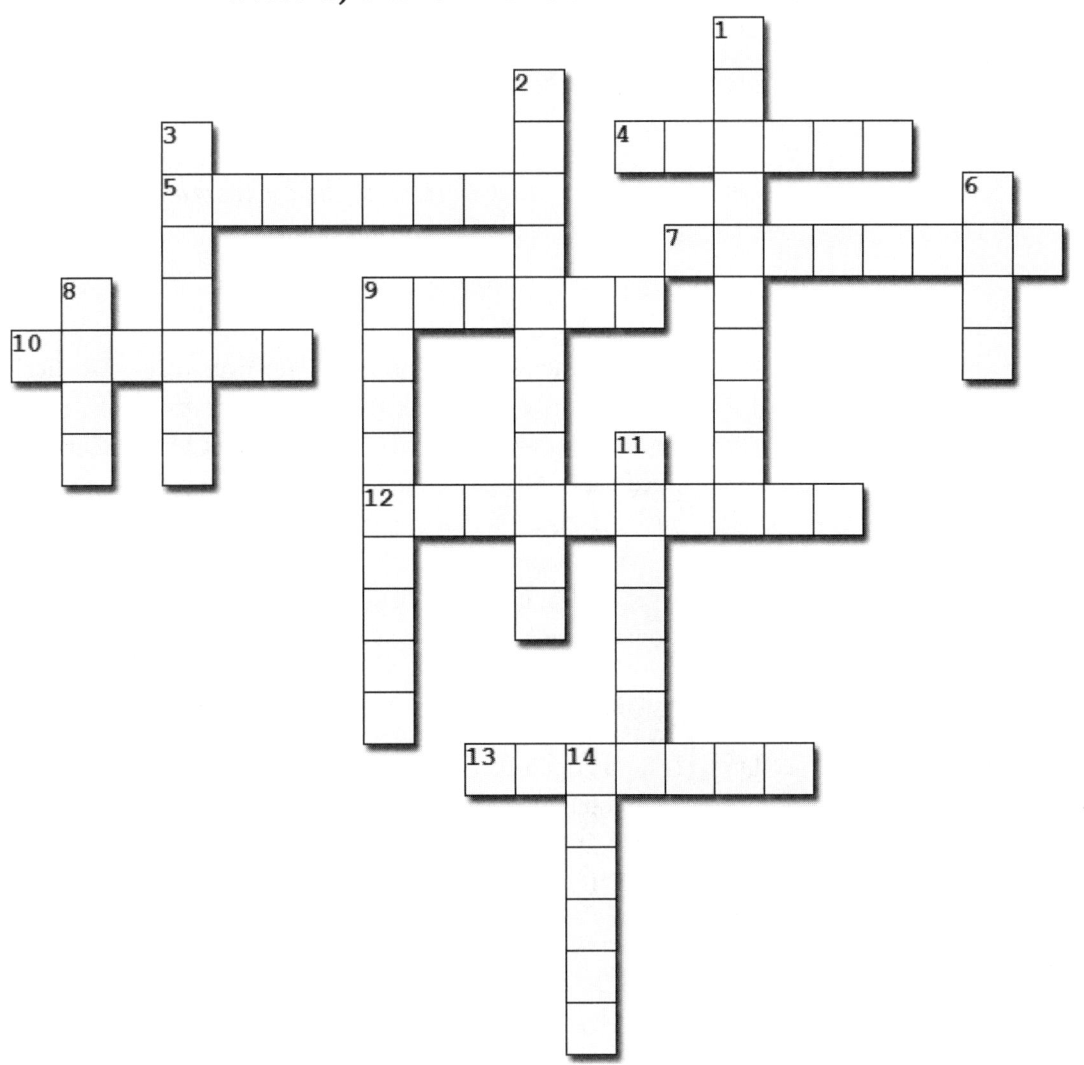

Across:

4. Using ____ assignment in an experiment permits us to draw cause-and-effect conclusions.
5. In a two-sample *z* test, the Large Counts condition finds the ____ numbers of successes and failures in each sample.
7. The value z^* in a confidence interval is also known as a ____ value.
9. Construct a two-____ *z* interval to estimate a difference in two population proportions.
10. If the *P*-value is very small, we should ____ the null hypothesis.
12. The Random condition is met if the data come from two groups in a randomized ____.
13. Conclusions should always be written in ____.

Down:

1. Using independent random samples allows us to ____ our results to the populations of interest.
2. The Random condition is met if the data come from two ____ random samples.
3. Be careful to not say "____" when you mean "percentage points."
6. If the *P*-value is greater than the significance level, we ____ to reject the null hypothesis.
8. When a confidence interval does not include ___, we have convincing evidence of a difference in the population proportions.
9. In a two-sample *z* test, the combined sample proportion gives the overall proportion of ____ in the combined samples.
11. Check the 10% condition when sampling ____ replacement.
14. The Large Counts condition ensures that the sampling distribution of $\hat{p}_1 - \hat{p}_2$ is approximately ____.

Unit 7: Inference for Quantitative Data: Means

"We must be careful not to confuse data with the abstractions we use to analyze them." William James

UNIT 7 OVERVIEW

Up to this point, our studies have focused on inference about proportions. However, many statistical studies involve estimating or carrying out a test about a single mean or a difference in means. In this unit, we will expand our collection of inference procedures to include confidence intervals about a mean and a difference in means as well as a mean difference (paired data). The procedures for means follow the same structure as those we learned for proportions. The only difference is that now you will need to rely on the sampling distribution of the sample mean or difference in sample means to perform your calculations. As in the previous units, you will follow a four-step process for carrying out the inference procedures. However, now your job will increasingly include a decision about *which* inference procedure you should use!

Sections in Unit 7

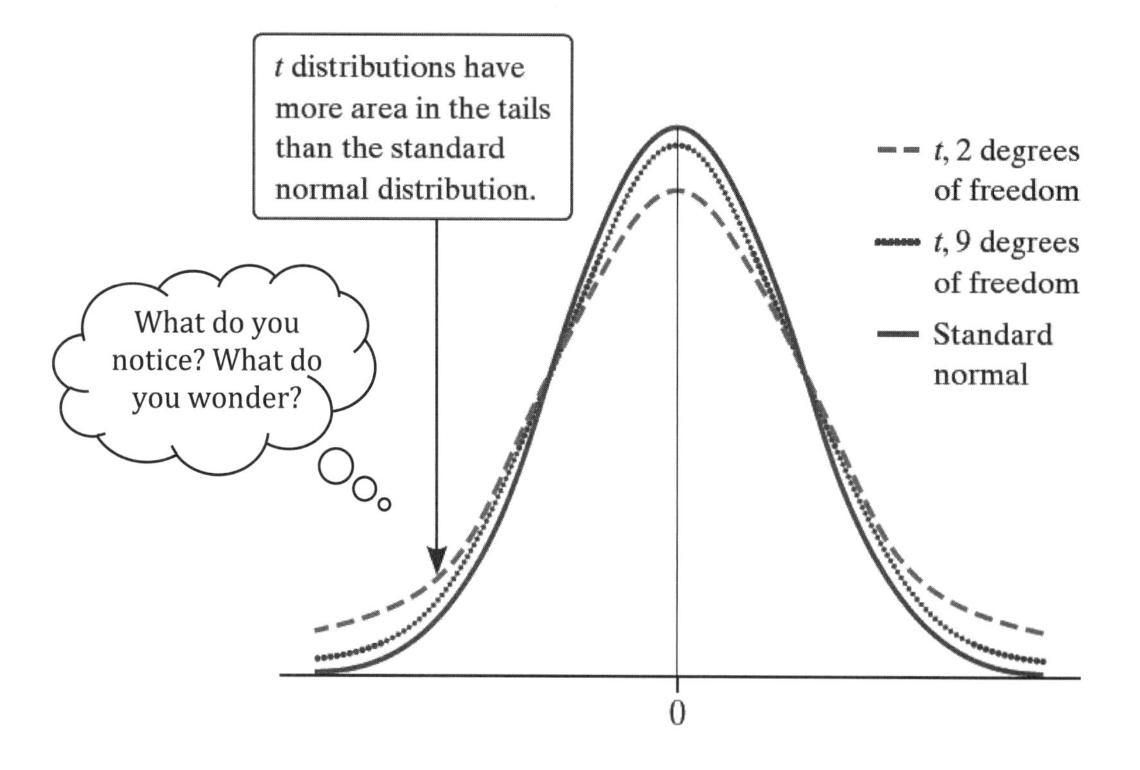

PRACTICE FOR MASTERY

Use the following *suggested* guide to the pages and exercises in your text to practice for mastery!

Note: your teacher may assign different problems. Be sure to follow their instructions!

Day	Topics	Read	Do
1	◆ Determining t^* Critical Values ◆ Checking Conditions for a Confidence Interval for μ	p. 644-653	**7A**: 1, 3, 5, 7, 27, 28
2	◆ Calculating a Confidence Interval for μ ◆ One-Sample t Interval for μ	p. 653-658	**7A**: 9, 11, 13, 15, 29, 30
3	◆ Paired Data: Confidence Intervals for a Population Mean Difference	p. 658-662	**7A**: 17, 19, 21, 23
4	◆ Stating Hypotheses and Checking Conditions for a Test about μ ◆ Calculating the Standardized Test Statistic and *P*-Value for a Test about μ	p. 669-677	**7B**: 1, 3, 5, 7, 9, 11, 13
5	◆ One-Sample t Test for μ	p. 677-679	**7B**: 15, 17, 19
6	◆ Paired Data: Significance Tests about a Population Mean Difference	p. 680-684	**7B**: 21, 23, 25
7	AP Classroom Topic Questions (Topics 7.1-7.5)		**7A**: 33, 34 **7B**: 29 – 33, 34, 35
8	◆ Checking Conditions for a Confidence Interval for $\mu_1 - \mu_2$ ◆ Calculating a Confidence Interval for $\mu_1 - \mu_2$ ◆ Two-Sample t Interval for $\mu_1 - \mu_2$	p. 692-701	**7C**: 1, 5, 7, 9, 13
9	AP Classroom Topic Questions (Topics 7.6-7.7)		**7C**: 3, 11, 16 - 18,
10	◆ Stating Hypotheses and Checking Conditions for a Test about $\mu_1 - \mu_2$ ◆ Calculating the Standardized Test Statistic and *P*-Value for a Test about $\mu_1 - \mu_2$	p. 707-713	**7D**: 1, 3, 5, 7, 9, 24
11	◆ Two-Sample t Test for $\mu_1 - \mu_2$ ◆ Paired Data or Two Samples?	p. 713-721	**7D**: 11, 13, 17, 19, 25 – 27
12	Unit 7 AP® Statistics Practice Test (p.735)		Unit 7 Review Exercises (p.733)
13	Unit 7 Test: Celebration of learning!		Cumulative AP® Practice Test 3 (p.738)

Section 7A: Confidence Intervals for a Population Mean or Mean Difference

Section Summary

In this section, you will continue your study of confidence intervals by learning how to construct and interpret a confidence interval for a mean. While the overall procedure follows the same four-step process as that for a proportion, there is one major difference. When dealing with unknown population means and unknown population standard deviations, we must use a new distribution to determine critical values. You will be introduced to *t* distributions and learn how to use them to construct a confidence interval for a population mean.

Learning Targets:

____ I can determine the critical value t^* for calculating a confidence interval for a population mean.

____ I can check the conditions for calculating a confidence interval for a population mean.

____ I can calculate a confidence interval for a population mean.

____ I can construct and interpret a one-sample *t* interval for a mean.

____ I can, in the special case of paired data, construct and interpret a confidence interval for a population mean difference.

Read: Vocabulary and Key Concepts

- The general form of a confidence interval is
  ______________ ± (________________)(__________________________)
- Use the sample mean ____ as the point estimate for the population mean ____ and use _______ as the standard deviation of the sampling distribution of $\bar{x}$.
- When calculating a confidence interval for a population mean, we use a ___ critical value rather than a ____ critical value whenever we use ____ to estimate ____. We specify a particular *t* distribution by giving its _____________________ (df).
- When we perform inference about a population mean μ using a *t* distribution, the appropriate degrees of freedom are found by subtracting ____ from the sample size ____, making df = _________.
- The *t* distributions have more ______________ — that is, more ______ in the tails — than the standard normal distribution.

How to Find t^* Using Table B

1. Using Table B in the back of the book, find the correct ______________________ at the bottom of the table.
2. On the left side of the table, find the correct number of ______________________. For this type of confidence interval, df = $n - 1$.
3. If the correct df isn't listed, use the greatest df available that is _________________ the correct df.
4. In the body of the table, find the value of ______ that corresponds to the confidence level and df.

Inference for _______________ uses *z*; inference for ______________ uses *t*. That's one reason why distinguishing categorical data from quantitative data is so important.

Checking conditions for a Confidence Interval for μ

- Before constructing a confidence interval for a population mean μ, you must check that the observations in the ___________ can be viewed as ________________ and that the sampling distribution of $\overline{x}$ is _____________________________.
- We check for ___________________ using the Random condition and the 10% condition. If both of these conditions are met, our formula for the ____________________ will be approximately correct.
- If the _____________________ is violated, our formula will overestimate the standard error, resulting in confidence intervals that have a capture rate __________ than the stated confidence level.
- When calculating a confidence interval for a population mean, we check the ___________________ condition to ensure that the sampling distribution of $\overline{x}$ is _________________________ and that we are able to use a t distribution with df $= n - 1$ to calculate the t^* critical value. Violating this condition usually results in confidence intervals that have a capture rate ___________ than the stated confidence level.
- To meet the Normal/Large Sample condition, the ____________ distribution must be approximately normal, or the sample size must be large ($n \geq$ ____).
- If the population distribution has ___________ shape and the sample size is ________ ($n < 30$), we should graph the ______________ and ask, "Is it plausible (believable) that these data came from an approximately normally distributed population?" If we do not see any ________________ or ______________ in the data, then the answer is "Yes."

Conditions for Calculating a Confidence Interval for a Population Mean

- **Random:** The data come from a __________________ from the ________________ of interest.
  - **10%:** When sampling without replacement, ___________.
- **Normal/Large Sample:** The population distribution is ________________________ or the sample size is _______ ($n \geq 30$). If the population distribution has unknown shape and $n < 30$, a graph of the _________________ shows no strong _____________ or ______________.

Calculating a Confidence Interval for μ

- The standard error of the sample mean $\overline{x}$ is an estimate of the standard deviation of the sampling distribution of $\overline{x}$.

  Write the formula here: ____________________

  The standard error of $\overline{x}$ estimates how much the sample mean $\overline{x}$ typically varies from μ in repeated random samples of size n.

- When the conditions are met, a C% confidence interval for the unknown population mean μ is

  Write the formula here: ______________________________
  where t^* is the critical value for a t distribution with df $= n - 1$ and C% of its area between $-t^*$ and t^*.

More About the Margin of Error

- The margin of error $t^* \frac{s_x}{\sqrt{n}}$ in a confidence interval for a population mean _______________ as the sample size _______________, assuming the confidence level and sample standard deviation s_x remain the same.
- The margin of error is proportional to _____ so quadrupling the sample size will cut the margin of error in _______, assuming everything else remains the same.
- The margin of error will be __________ for __________ confidence levels. This makes sense, as wider intervals will have a ___________ capture rate than narrower intervals.
- The margin of error doesn't account for _______ in the data collection process, only __________ variability.

Paired Data: Confidence Intervals for a Population Mean Difference

- _______________ result from recording two values of the same quantitative variable for each individual or for each ________ of similar individuals.

Conditions for Calculating a Confidence Interval for a Population Mean Difference

- **Random:** __________ data come from a random sample from the population of interest or from a randomized _______________.
  - 10%: When sampling without _______________, $n_{\text{Diff}} < 0.10N_{\text{Diff}}$.
- **Normal/Large Sample:** The population distribution of _________________ is approximately normal or the sample size is large ($n_{\text{Diff}} \geq 30$). If the population distribution of ______________ has unknown shape and $n_{\text{Diff}} < 30$, a graph of the sample ______________ shows no strong skewness or outliers.

Calculating a Confidence Interval for a Population Mean Difference

When the conditions are met, a C% confidence interval for the population mean difference μ_{Diff} is

Write the formula here: ________________________________

where t^* is the critical value for a t distribution with df = $n_{\text{Diff}} - 1$ and C% of its area between $-t^*$ and t^*.

Watch: Go to bfwpub.com/TPS7eStrive. Select Unit 7 and watch the following Example videos:

Video	Topic	Done
Unit 7A, How do you find t^*? (p. 650)	*Determining t^* critical values*	
Unit 7A, Engagement rings and pulling wood apart (p. 652)	*Checking conditions for a confidence interval for μ*	
Unit 7A, Put a ring on it! (p. 654)	*Calculating a confidence interval for μ*	
Unit 7A, Video screen tension (p. 656)	*One-sample t interval for μ*	
Unit 7A, Does zinc sink? (p. 661)	*Paired data: Confidence intervals for a population mean difference*	

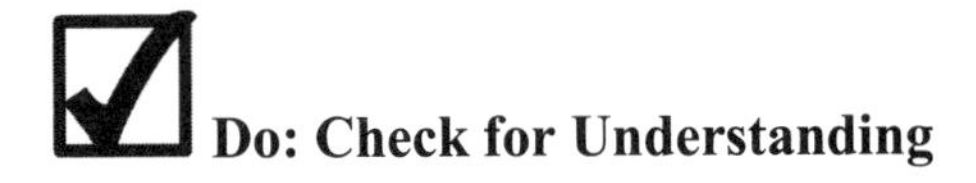

Concept 1: t Distributions

When the population standard deviation is unknown, we can no longer model the distribution of the test statistic with the normal distribution. Therefore, we can't use z^* critical values to determine the margin of error in the confidence interval. Fortunately, it turns out that when the Normal/Large Sample condition is met, the test statistic calculated using the sample standard deviation s_x has a distribution similar in appearance to the normal distribution, but with more area in the tails. That is, the statistic has the t distribution with $(n - 1)$ degrees of freedom. As the sample size increases, the t distribution approaches the standard normal distribution. We calculate standardized t values the same way we calculate z values. However, we must refer to a t table and consider the degrees of freedom when determining critical values. (*Need Tech Help*? View the video **Tech Corner 20: Determining t^* Critical Values** at bfwpub.com/TPS7eStrive)

Concept 2: Conditions for Estimating μ

Like proportions, when constructing a confidence interval for μ, it is critical that you begin by checking that the conditions are met. First, determine if the sample was randomly selected. If the sample was selected without replacement, check the 10% condition. Because the construction of the interval is based on the sampling distribution of $\bar{x}$, ensure that the Normal/Large Sample condition is met. That is, check to see that the population distribution is approximately normal, the sample size is at least 30, or the shape of the distribution of the sample has no strong skewness and no outliers. If all three of these conditions are met, you can safely proceed to construct and interpret a confidence interval about a population mean μ.

Check for Understanding – Learning Targets 1 and 2

____ *I can determine the critical value t^* for calculating a confidence interval for a population mean.*

____ *I can check the conditions for calculating a confidence interval for a population mean.*

1. Find t^* for an 80% confidence interval based upon a random sample of 19 observations.

2. Find t^* for a 95% confidence interval based upon 248 degrees of freedom.

3. Find t^* for a 99% confidence interval based on an SRS of size $n = 30$

4. How much money have college students saved? A random sample of 28 college students was selected. The mean of this sample was \$3482.25 with a standard deviation of \$2775.43. The distribution of the sample data is shown here:

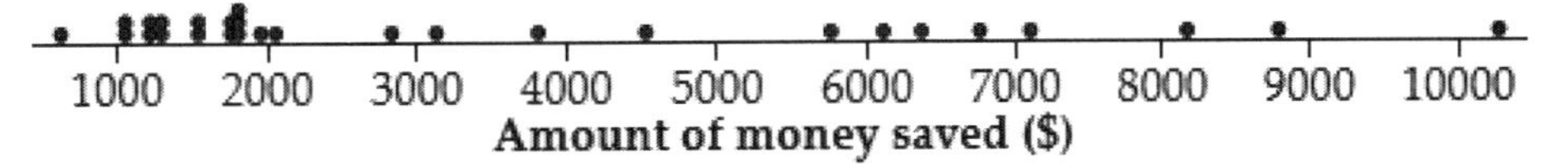

A 95% confidence interval for the mean amount of money in savings for all college students is to be constructed. Are the conditions for constructing this confidence interval satisfied? Explain.

Concepts 3 and 4: Constructing a Confidence Interval for μ

To construct a confidence interval for a population mean μ when the population standard deviation is unknown, you should follow the four-step process.

- **State:** State the parameter you want to estimate and the confidence level.
- **Plan:** Identify the appropriate inference method and check conditions.
- **Do:** If the conditions are met, perform calculations.

  $$\bar{x} \pm t^* \frac{s_x}{\sqrt{n}}$$

  where t^* is the critical value for the t distribution with df $= n - 1$ and area C between $-t^*$ and t^*.
- **Conclude:** Interpret your interval in the context of the problem.

Check for Understanding – Learning Targets 3 and 4

____ *I can calculate a confidence interval for a population mean.*

____ *I can construct and interpret a one-sample t interval for a mean.*

The amount of sugar in soft drinks is increasingly becoming a concern. To test sugar content, Teddy randomly selected 8 soft drinks from a particular manufacturer and measured the sugar content in grams/serving. The following data were produced:

26 31 23 22 11 22 14 31

Use these data to construct and interpret a 95% confidence interval for the mean amount of sugar in this manufacturer's soft drinks.

(*Need Tech Help?* View the video **Tech Corner 21: Confidence Intervals for a Mean** at bfwpub.com/TPS7eStrive)

Concept 5: Confidence Interval for Paired Data

Studies that involve making two observations on the same individual or making an observation for each of two very similar individuals result in paired data. In these types of studies, we are often interested in analyzing the differences in responses within each pair. If the conditions for inference are met, we can use one-sample *t* procedures to estimate or test a claim about the mean difference, μ_{Diff}.

Check for Understanding – Learning Target 5

____ *I can, in the special case of paired data, construct and interpret a confidence interval for a population mean difference.*

A study measured how fast subjects could repeatedly push a button when under the effects of caffeine. Subjects were asked to push a button as many times as possible in two minutes after consuming a typical amount of caffeine. During another test session, they were asked to push the button as many times as possible in two minutes after taking a placebo. The subjects did not know which treatment they were administered each time, and the order of the treatments was randomly assigned. The data, given in presses per two minutes for each treatment, follows. Use a paired *t* procedure to estimate, with 99% confidence, the mean difference in the number of presses for subjects like these.

Subject		1	2	3	4	5	6	7	8	9	10	11
# of Presses	Caffeine	251	284	300	321	240	294	377	345	303	340	408
	Placebo	201	262	283	290	259	291	354	346	283	361	411

Section 7B: Significance Tests for a Population Mean or Mean Difference

Section Summary

In this section, you'll learn how to perform a significance test about a population mean μ. Just like you did for proportions, you'll learn how to set up hypotheses, check the appropriate conditions, calculate a standardized test statistic and P-value, and make a conclusion in context. You will also learn how tests involving "paired data" can be performed using one-sample t procedures.

Learning Targets:

____ I can state appropriate hypotheses and check the conditions for performing a test about a population mean.

____ I can calculate the standardized test statistic and P-value for a test about a population mean.

____ I can perform a one-sample t test for a mean.

____ I can, in the special case of paired data, perform a significance test about a population mean difference.

Read: Vocabulary and Key Concepts

Conditions for Performing a Significance Test About a Population Mean

- **Random:** The data come from a random sample from the ______________________________.
  - **10%:** When ______________ without replacement, $n < 0.10N$.
- **Normal/Large Sample:** The __________________ distribution is approximately __________ or the sample size is large ($n \geq$ ___). If the ______________ distribution has __________ shape and n <30, a graph of the ______________ shows no strong _____________ or _____________.

Calculating the Standardized Test Statistic and *P*-value for a Test about μ

- Write the general formula for a test statistic. Standardized test statistic = ______________________

- When we use the _____________ standard deviation s_x to estimate the unknown _________________ standard deviation σ, the standardized test statistic is denoted by ____.

- Write the specific formula for the test statistic: ______________________________

- Recall that the standardized test statistic tells us __________ the sample result is from the ______ value, and in which ________________, on a standardized scale.

- When the Normal/Large Sample condition is met and the null hypothesis H_0 is true, the standardized test statistic can be modeled by a _______________ with degrees of freedom df = ________.

- Given the limitations of Table B, our advice is to use ________________ to find P-values when carrying out a significance test about a population mean.

Two-sided Tests and Confidence Intervals

- The link between two-sided tests and confidence intervals for a population mean allows us to make a conclusion about H_0 ___________ from a ________________________.
- If a 95% confidence interval for μ ______________________ the null value μ_0, we can ____________ H_0: $\mu = \mu_0$ in a _______________________ at the 5% significance level ($\alpha = 0.05$).
- If a 95% confidence interval for μ _____________ the null value μ_0, then we should ________________________ H_0: $\mu = \mu_0$ in a _______________________ at the 5% significance level.

Paired Data: Significance Tests about a Population Mean Difference

- The null hypothesis says that the population mean difference is 0: H_0: _______ = 0.
- The alternative hypothesis says what kind of difference we expect.

Conditions for Performing a Significance Test About a Population Mean Difference

- **Random:** _________________ come from a _____________ sample from the ________________ of interest or from a ____________________ experiment.
  - **10%:** When sampling without ________________, $n_{\text{Diff}} < 0.10N_{\text{Diff}}$.
- **Normal/Large Sample:** The population distribution of ________________ is approximately normal or the ________________ is large ($n_{\text{Diff}} \geq 30$). If the population distribution of _________________ has unknown shape and $n_{\text{Diff}} < 30$, a graph of the _____________________ shows no strong skewness or outliers.

Calculating the Standardized Test Statistic and *P*-value in a Test about a Population Mean Difference

- Suppose the conditions are met. To test the hypothesis H_0: $\mu_{\text{Diff}} = 0$, compute the standardized test statistic:
- Write the formula here: ______________________
- Find the _____________ by calculating the probability of getting a t statistic this _______________ in the direction specified by the alternative hypothesis H_a in a t distribution with df = $n_{\text{Diff}} - 1$.

- When researchers randomly assign the treatments, they can make an inference about _____________ and ________________.

Watch: Go to bfwpub.com/TPS7eStrive. Select Unit 7 and watch the following Example videos:

Video	Topic	Done
Unit 7B, Golden hamsters (p. 671)	*Stating hypotheses and checking conditions for a test about μ*	
Unit 7B, How heavy are golden hamsters? (p. 675)	*Calculating the standardized test statistic and P-value for a test about μ*	
Unit 7B, Healthy streams (p. 678)	*One-sample t test for μ*	
Unit 7B, Is caffeine dependence real? (p. 681)	*Paired data: Significance tests about a population mean difference*	

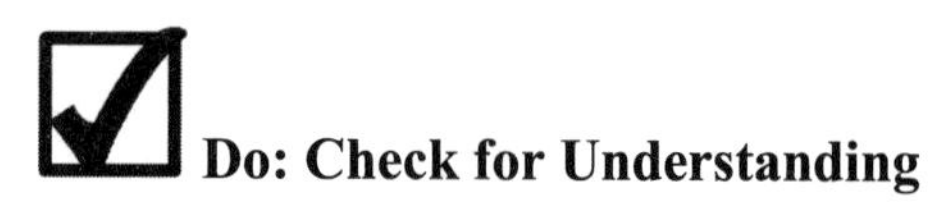

Do: Check for Understanding

Concepts 1, 2, and 3: One-Sample t Test for μ

The process for testing a claim about a population mean follows the same format as the process used for a population proportion. However, like confidence intervals for means, you will base your calculations on the t distributions.

1) **STATE: State the hypotheses, parameter(s), and significance level.**
   The null hypothesis assumes that the population mean is equal to a particular value while the alternative hypothesis is that the population mean is greater than, less than, or not equal to that value:
   H_0: $\mu = \mu_0$
   H_a: $\mu > \mu_0$ OR $\mu < \mu_0$ OR $\mu \neq \mu_0$
   Define the parameter, μ. State a significance level.

2) **PLAN: Identify the appropriate inference method and check the conditions.**
   Check the Random, 10%, and Normal/Large Sample conditions before proceeding with inference. The sample must be selected at random. If sampling is done without replacement, verify that the sample size is less than 10% of the size of the population. To ensure that the sampling distribution of $\bar{x}$ is approximately normal, check that the population distribution is normal or that the sample size is large, $n \geq 30$. If $n < 30$, the condition can only be met if the distribution of the sample data shows no strong skewness or outliers.

3) **DO: If the conditions are met, perform calculations.**
   Compute the standardized test statistic: $t = \dfrac{\bar{x} - \mu_0}{\frac{s_x}{\sqrt{n}}}$
   and find the P-value by calculating the probability of observing a t statistic at least this extreme in the direction of the alternative hypothesis in a t distribution with $n - 1$ degrees of freedom.

   (*Need Tech Help?* View the video **Tech Corner 22: Calculating *P*-values from *t* Distributions** at bfwpub.com/TPS7eStrive)

4) **CONCLUDE: Make a conclusion about the hypotheses in the context of the problem.**
   If the P-value is smaller than the stated significance level, reject H_0. You have convincing evidence for H_a. If the P-value is larger than the significance level, fail to reject H_0. You do not have convincing evidence for H_a.

A confidence interval for a population mean can be used to test a two-sided claim. Further, constructing a confidence interval gives a range of plausible values for μ, while a significance test only allows you to conclude that μ may be different from a particular value.

Check for Understanding – Learning Targets 1, 2, and 3

____ *I can state appropriate hypotheses and check the conditions for performing a test about a population mean*

____ *I can calculate the standardized test statistic and P-value for a test about a population mean.*

____ *I can perform a one-sample t test for a mean.*

Humerus bones from the same species of animal tend to have approximately the same length-to-width ratios. When fossils of humerus bones are discovered, archeologists can often determine the species of animal by examining these ratios. It is known that the species Molekius Primatium exhibits a mean ratio of $\mu = 8.9$. Suppose 41 fossils of humerus bones are unearthed at a site on Minnesota's Iron Range, where a large population of this species was known to have lived. Researchers are willing to view these as a random sample of all such humerus bones. The length-to-width ratios were calculated and are listed below. Is there convincing evidence that the population mean for the species that left these bones differs from 8.9 at $\alpha = 0.05$?

9.73	10.89	9.07	9.20	9.33	9.98	9.84	9.59	9.48	8.71	9.57	10.02	11.67	8.07	8.37
6.85	8.52	8.87	6.23	9.41	6.66	9.35	8.86	9.93	8.91	11.77	10.48	10.39	9.39	9.17
9.89	8.17	8.39	8.80	8.38	8.30	9.17	9.38	9.29	9.94	12.00				

(*Need Tech Help?* View the video **Tech Corner 23: Significance Tests for a Mean** at bfwpub.com/TPS7eStrive)

Concept 4: Inference for Paired Data

Studies that involve making two observations on the same individual or making an observation for each of two very similar individuals result in paired data. In these types of studies, we are often interested in analyzing the difference within each pair. If the conditions for inference are met, we can use one-sample t procedures to estimate or test a claim about the mean difference μ_{Diff}. In this case, we refer to the inference method as a paired t procedure.

Check for Understanding – Learning Target 4

_____ *I can, in the special case of paired data, perform a significance test about a population mean difference.*

A study measured how fast subjects could repeatedly push a button when under the effects of caffeine. Subjects were asked to push a button as many times as possible in two minutes after consuming a typical amount of caffeine. During another test session, they were asked to push the button after taking a placebo. The subjects did not know which treatment they were administered each day, and the order of the treatments was randomly assigned. The data, given in presses per two minutes for each treatment follows. Use a paired t procedure to determine whether or not caffeine results in a higher rate of presses, on average, per two-minute period. Use $\alpha = 0.05$.

Subject		**1**	**2**	**3**	**4**	**5**	**6**	**7**	**8**	**9**	**10**	**11**
# of Presses	**Caffeine**	251	284	300	321	240	294	377	345	303	340	408
	Placebo	201	262	283	290	259	291	354	346	283	361	411
Difference	**Caffeine – Placebo**	50	22	17	31	−19	3	23	−1	20	−21	-3

Section 7C: Confidence Intervals for a Difference in Population Means

Section Summary

In this section, you will learn to estimate the difference in two population means. You will start by learning the conditions necessary to perform inference. Then you will learn how to estimate a difference between two means. As you did with a single mean, you will rely on *t* distributions when performing calculations. Because some of the calculations are complex, you may want to rely on your calculator to do most of the work. As always, though, make sure you can interpret the results your calculator gives you!

Learning Targets:

____ I can check the conditions for calculating a confidence interval for a difference between two population means.
____ I can calculate a confidence interval for a difference between two population means.
____ I can construct and interpret a two-sample *t* interval for a difference in means.

Read: Vocabulary and Key Concepts

- Before constructing a confidence interval for a difference in population means, we must check for ____________________ in the data collection process and confirm that the sampling distribution of $\bar{x}_1 - \bar{x}_2$ is ________________________________.
- To ensure independence, the data should come from ____________________________ from the populations of interest or from ______________ in a randomized experiment (the Random condition). Also, when sampling without replacement, the 10% condition should be met for _______ sample. Remember that it is _____ appropriate to check the 10% condition in experiments with volunteer subjects!
- For the Normal/Large Sample condition to be met, _____ population distributions must be approximately ___________, both sample sizes must be __________ ($n_1 \geq 30$ and $n_2 \geq 30$), or one _____________ distribution must be approximately normal, and the other sample size must be large.

Conditions for Constructing a Confidence Interval for a Difference Between Two Population Means

- **Random:** The data come from two ________________ random samples or from two groups in a randomized ________________.
    - **10%:** When sampling without replacement, $n_1 < 0.10N_1$ and $n_2 < 0.10N_2$.
- **Normal/Large Sample:** For each sample, the data come from an approximately normally distributed _____________ or the sample size is large ($n \geq$ ___). For each sample, if the __________ distribution has unknown shape and $n <$ ____, a graph of the ________ data shows no strong skewness or outliers.

Calculating a Confidence Interval for A Difference Between Two Population Means

- When the conditions are met, a $C\%$ confidence interval for the difference $\mu_1 - \mu_2$ between two population means is

  Write the formula here: ____

  where t^* is the critical value with $C\%$ of its area between $-t^*$ and t^* in the t distribution with degrees of freedom given by technology.

- When a confidence interval does not include 0, we have ________________ evidence that there is a __________________ in the population means.

- The confidence interval provides more information than a significance test because it gives a set of ______________________ values for the difference in population means.

Other Options for a Two-Sample *t* Interval

- If using the conservative approach instead of using technology to find the appropriate degrees of freedom for two-sample t procedures, use df = the _____________ of $n_1 - 1$ and $n_2 - 1$.
- Calculating the degrees of freedom using the smaller of $n_1 - 1$ and $n_2 - 1$ is called the conservative approach because it will *always* result in a __________ df and a __________ t^* value than when using the df given by technology, *making the interval* __________ *than needed* for the given level of confidence.
- The other option is a special version of the two-sample t procedures that assumes the two population distributions have equal ____________. In practice, population variances are rarely equal, so our two-sample t procedures are almost always ________ accurate than the pooled procedures — and they work almost as well when the population variances *are* equal.

Watch: Go to bfwpub.com/TPS7eStrive. Select Unit 7 and watch the following Example videos:

Video	Topic	Done
Unit 7C, Windy City or Big Apple? (p. 693)	*Checking conditions for a confidence interval for* $\mu_1 - \mu_2$	
Unit 7C, Happy customers? (p. 698)	*Calculating a confidence interval for* $\mu_1 - \mu_2$	
Unit 7C, Do portion sizes affect food consumption? (p. 699)	*Two-sample t interval for* $\mu_1 - \mu_2$	

Do: Check for Understanding

Concepts 1, 2, and 3: Conference Intervals for $\mu_1 - \mu_2$

To construct a confidence interval for the difference between two population or treatment means $\mu_1 - \mu_2$, you should follow the familiar four-step process. We will estimate the difference by comparing the sample means and build an interval around that point estimate by using the standard error of the statistic and a critical t value. To construct a two-sample t interval for a difference between two means:

- **State:** State the parameters of interest and the confidence level.
- **Plan:** Identify the appropriate inference method and check the conditions: Two independent random samples or two groups in a randomized experiment, if sampling without replacement: 10% condition for both samples, and the Normal/Large Sample condition for both samples or groups.
- **Do:** If the conditions are met, perform calculations: $(\bar{x}_1 - \bar{x}_2) \pm t^* \sqrt{\frac{s_1^2}{n_1} + \frac{s_2^2}{n_2}}$

  where t^* is the critical value for the t distribution curve having df = smaller of $n_1 - 1$ and $n_2 - 1$ OR given by technology with area C between $-t^*$ and t^*.
- **Conclude:** Interpret the interval in the context of the problem.

Check for Understanding – Learning Targets 1, 2, and 3

_____ *I can check conditions for calculating a confidence interval for a difference between two population means.*

_____ *I can calculate a confidence interval for a difference between two population means.*

_____ *I can construct and interpret a two-sample t interval for a difference in means.*

Researchers are interested in determining the effectiveness of a new diet for individuals with heart disease. 200 heart disease patients are randomly assigned to the new diet, or the current diet used in the treatment of heart disease. The 100 patients on the new diet lost an average of 9.3 pounds with standard deviation 4.7 pounds. The 100 patients continuing with their current diet lost an average of 7.4 pounds with standard deviation of 4 pounds. Construct and interpret a 95% confidence interval for the difference in mean weight loss for the two diets.

(*Need Tech Help?* View the video **Tech Corner 24: Confidence Intervals for a Difference in Means** at bfwpub.com/TPS7eStrive)

Section 7D: Significance Tests for a Difference in Population Means

Section Summary

In this section, you will learn how to carry out a significance test about a difference in two means. Like you did with a single mean, you will rely on t distributions when performing calculations. Because some of the calculations are complex, you may wish to rely on your calculator to do most of the work. As always, though, you will want to make sure you can interpret the results that your calculator gives you!

Learning Targets:

____ I can state appropriate hypotheses and check the conditions for performing a test about a difference between two population means.
____ I can calculate the standardized test statistic and P-value for a test about a difference between two population means.
____ I can perform a two-sample t test for a difference in means.
____ I can determine when it is appropriate to used paired t procedures versus two-sample t procedures.

Read: Vocabulary and Key Concepts

Stating Hypotheses for a Test about $\mu_1 - \mu_2$

- The null hypothesis says that there is no difference between the two parameters: H_0: ___________ = 0.
- The null hypothesis can also be written in the equivalent form H_0: ____ = ____.
- The ________________ hypothesis says what kind of difference we expect.

Checking Conditions for a Test about $\mu_1 - \mu_2$

- **Random:** The data come from two ________________ random samples or from two groups in a randomized ________________.
  - **10%:** When sampling without replacement, ____________ and ____________.
- **Normal/Large Sample:** For each sample, the data come from an approximately normally distributed ________________ or the sample size is _______ ($n \geq 30$). For each sample, if the ________________ distribution has unknown shape and $n < 30$, a graph of the ___________ shows no strong ___________ or _____________.

Calculating the Standardized Test Statistic and P-Value for a Test about $\mu_1 - \mu_2$

- Suppose the conditions are met. To test the hypothesis H_0: $\mu_1 - \mu_2 = 0$, calculate the standardized test statistic:

  Write the formula here: __________________________

- Find the P-value by calculating the probability of getting a t statistic this ______ or ___________ in the direction specified by the alternative hypothesis H_a in a t distribution with df given by technology.

- Sample size strongly affects the _________ of a test. It is easier to detect a _______________ in the effectiveness of two treatments if both are applied to large numbers of subjects.

Other Options for a Two-Sample t Test

- Instead of using technology to find the appropriate degrees of freedom for two-sample t procedures, the ______________________________ uses df = the *smaller* of $n_1 - 1$ and $n_2 - 1$. Technology gives __________, more __________ *P*-values for two-sample t tests than the conservative approach does.

- Pooled Two-Sample t Procedures: This option assumes that the two population distributions have equal _______________ and combines (*pools*) the two sample variances to estimate the common population variance. However, it may not be safe to assume that the population variances are ________. *Don't use pooled two-sample t procedures unless a statistician says it is OK to do so.*

Paired Data or Two Samples?

- We used _____________ t procedures to perform inference about the difference $\mu_1 - \mu_2$ between two _______________ means. These methods require data that come from _______________ random samples from the populations of interest or from _______________ in a completely randomized experiment.

- We use _________ t procedures to perform inference about the population mean difference μ_{Diff}. Recall that paired data come from recording the same ________________ variable twice for each individual or from recording the same quantitative variable for each of ___________________ individuals.

Using Tests Wisely

1. Statistical significance is not the same thing as ____________ importance. When large samples are used, even ________ deviations from the null hypothesis will be _______________.

2. Beware of multiple analyses! Searching data for patterns is a legitimate pursuit. Performing ________ conceivable significance test on a data set with many variables until you obtain a statistically significant result is not.

Watch: Go to bfwpub.com/TPS7eStrive. Select Unit 7 and watch the following Example videos:

Video	Topic	Done
Unit 7D, Big trees, small trees, short trees, tall trees (p. 708)	*Stating hypotheses and checking conditions for a test about* $\mu_1 - \mu_2$	
Unit 7D, Where do the big trees grow? (p. 711)	*Calculating the standardized test statistic and P-value for a test about* $\mu_1 - \mu_2$	
Unit 7D, Calcium and blood pressure (p. 713)	*Two-sample t test for* $\mu_1 - \mu_2$	
Unit 7D, Are you all wet? (p. 717)	*Paired data or two samples?*	

 **Do: Check for Understanding**

Concepts 1, 2, and 3: Significance test for $\mu_1 - \mu_2$

To test a claim about the difference between two population means or the means of two groups in a randomized experiment, we will follow the same four-step process learned previously.

To conduct a two-sample t test for the difference between two means:

1) **STATE: State the hypotheses, parameters, and significance level.**
The null hypothesis usually states that the parameters are equal (no difference) while the alternative hypothesis is that one mean is greater than, less than, or not equal to the other:
H_0: $\mu_1 - \mu_2 = 0$
H_a: $\mu_1 - \mu_2 < 0$ or $\mu_1 - \mu_2 > 0$ or $\mu_1 - \mu_2 \neq 0$
OR
H_0: $\mu_1 = \mu_2$
H_a: $\mu_1 > \mu_2$ or $\mu_1 < \mu_2$ or $\mu_1 \neq \mu_2$
Define the parameters, in context. State the significance level.

2) **PLAN: Identify the appropriate inference method and check the conditions.**
Name the test: Two-sample t test for a difference in means, or 2-sample t test for $\mu_1 - \mu_2$. Next, check the conditions. Verify that the data come from two independent random samples or two groups in a randomized experiment. If the sampling took place without replacement, check that each sample size is less than 10% of its respective population size. To ensure that the sampling distribution of $\bar{x}_1 - \bar{x}_2$ is approximately normal, check that the population distributions are normal OR n_1 and n_2 are both at least 30. If one (or both) of the sample sizes are less than 30 and the population distribution(s) is(are) unknown, then look at a distribution of the sample(s). If there is no strong skewness or outliers, then it is safe to proceed.

3) **DO: If conditions are met, perform calculations.**
Standardized test statistic: $t = \dfrac{(\bar{x}_1 - \bar{x}_2) - 0}{\sqrt{\dfrac{s_1^2}{n_1} + \dfrac{s_2^2}{n_2}}}$
Find the P-value by calculating the probability of observing a t statistic at least this extreme in the direction of the alternative hypothesis.
Use the t distribution with df = smaller of $n_1 - 1$ and $n_2 - 1$ OR use the df that is given by technology.

4) **Conclude by interpreting the results of your calculations in the context of the problem.**
If the P-value is smaller than the stated significance level, reject H_0. You have convincing evidence in favor of the alternative hypothesis. If the P-value is greater than or equal to the significance level, then you fail to reject the null hypothesis. There is not convincing evidence for the alternative hypothesis.

Check for Understanding – Learning Targets 1, 2, and 3

____ *I can state appropriate hypotheses and check the conditions for performing a test about a difference between two population means.*

____ *I can calculate the standardized test statistic and P-value for a test about a difference between two population means.*

____ *I can perform a two-sample t test for a difference in means.*

A student thinks that "country" songs tend be longer than "pop" songs. She has hundreds of each kind in her music collection. She takes a random sample of nine "country" songs and eight "pop" songs. The length of each song is given below (in seconds).

Country:	283	290	305	263	280	289	308	255	265
Pop:	272	283	310	220	237	296	271	250	

Do the data provide convincing evidence that the mean length of all country songs is longer than the mean length of all pop songs?

(*Need Tech Help?* View the video **Tech Corner 25: Significance Tests for a Difference in Means** at bfwpub.com/TPS7eStrive)

Concept 4: Paired t procedures versus a Two-Sample t procedures

To determine whether paired *t* procedures or two-sample *t* procedures are necessary, consider how the data were produced. If the data came from measuring the same quantitative variable twice for each individual or measuring the same quantitative variable for pairs of similar individuals, then the data are paired. If the data come from two independent random samples or two groups in a randomized comparative experiment, then use two-sample *t* procedures. Paired data can never consist of 2 samples or groups of different sizes.

Check for Understanding – Learning Target 4

____ *I can determine when it is appropriate to use paired t procedures versus two-sample t procedures.*

In each of the following scenarios, decide whether you should use paired *t* procedures to perform inference about a mean difference or two-sample *t* procedures to perform inference about a difference in means. Explain your choice.

1. Does listening to Mozart for 10 minutes a day increase intelligence? A student recruited 80 volunteers and randomly assigned 40 of them to listen to Mozart for 10 minutes a day and the other 40 to no Mozart. At the end of the week, the volunteers took an intelligence test and the results between the two groups were compared.

2. Is going through the drive-thru faster than going inside to get your morning coffee? To investigate, Katie takes random samples of 100 drive-thru customers and 100 customers who order inside at a local coffee shop and records how long it takes for them to receive their coffee.

3. Do dogs get more exercise when given access to a fenced-in yard or when walked daily? To investigate, a random sample of 30 dogs who regularly attend a doggie daycare were selected. Each dog was fitted with a pup pedometer, which counted their steps. For each dog, a researcher flipped a coin to determine if they would have access to the fenced-in yard for two weeks or would be walked daily for two weeks. In the following two weeks the treatments were reversed.

Unit 7 Summary: Inference for Quantitative Data: Means

In this unit, you learned the inference procedures that help us estimate a population mean or a difference in population means. Whether you are estimating one mean or a difference in two means, the processes for constructing a confidence interval are the same. Both procedures use a *t* critical value, which is based upon the degrees of freedom of the distribution. Like we learned when we estimated proportions, it is important to identify which procedure you are using, verify that the appropriate conditions are met, perform the necessary calculations, and interpret your results in the context of the problem.

You also learned the inference procedures for carrying out a significance test for a mean as well as a difference in means and paired data. The process for carrying out a significance test for paired data is the same as a significance test for a single mean, except that the single mean comes from a list of differences. When trying to decide between a significance test for two means versus a paired test, determine if the data arose from two independent random samples or two groups in a randomized experiment. If so, then the correct procedure is a *t* test for the difference in two means. If the data are paired (two measurements on each individual) or if physical pairing occurs, then the correct procedure is a paired *t* test.

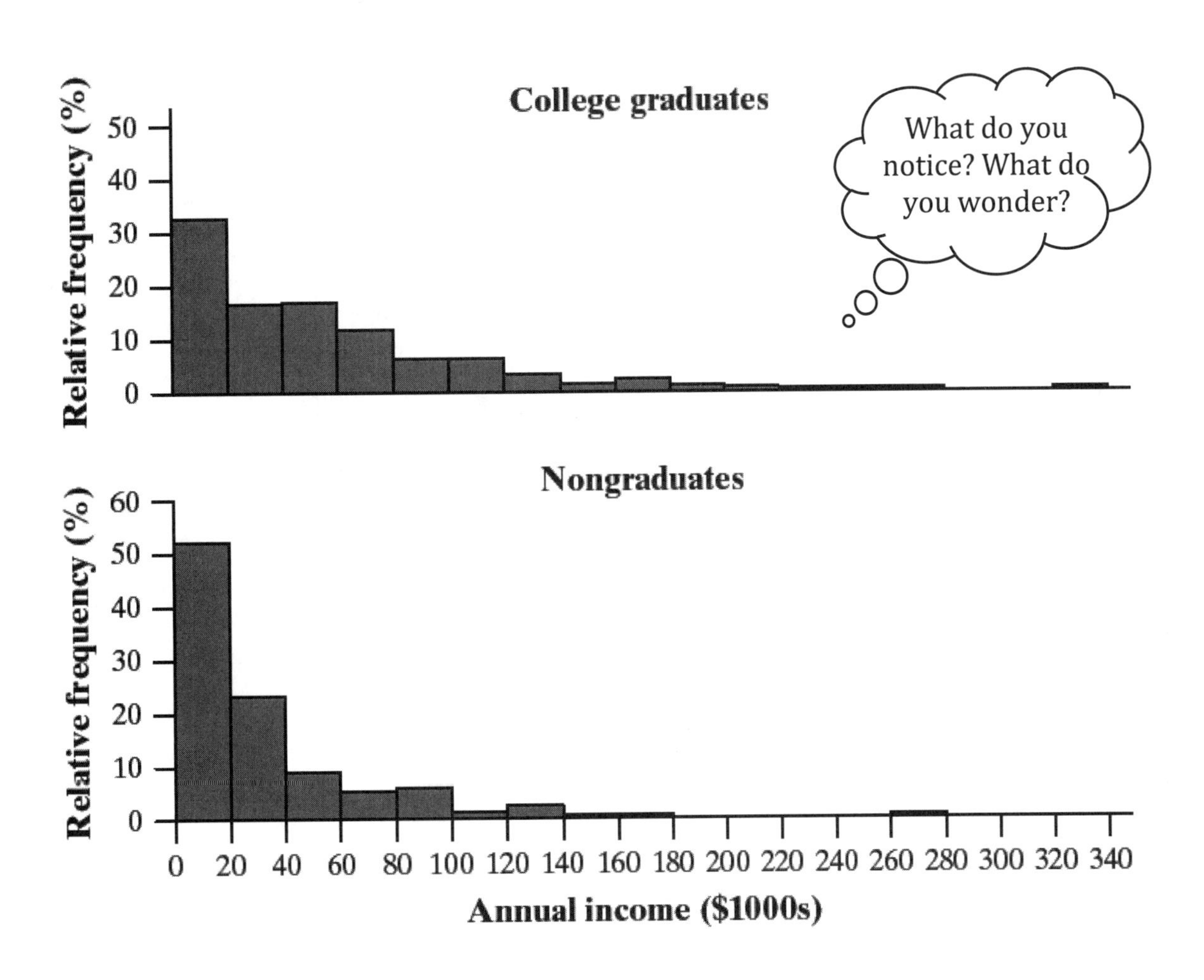

How well do you understand each of the learning targets?

Learning Target	Got It!	Almost There	Needs Work
I can determine the critical value t^* for calculating a confidence interval for a population mean.			
I can check the conditions for calculating a confidence interval for a population mean.			
I can calculate a confidence interval for a population mean.			
I can construct and interpret a one-sample t interval for a mean.			
I can, in the special case of paired data, construct and interpret a confidence interval for a population mean difference.			
I can state appropriate hypotheses and check the conditions for performing a test about a population mean.			
I can calculate the standardized test statistic and P-value for a test about a population mean.			
I can perform a one-sample t test for a mean.			
I can, In the special case of paired data, perform a significance test about a population mean difference.			
I can check the conditions for calculating a confidence interval for a difference between two population means.			
I can calculate a confidence interval for a difference between two population means.			
I can construct and interpret a two-sample t interval for a difference in means.			
I can state appropriate hypotheses and check the conditions for performing a test about a difference between two population means.			
I can calculate the standardized test statistic and P-value for a test about a difference between two population means.			
I can perform a two-sample t test for a difference in means.			
I can determine when it is appropriate to use paired t procedures versus two-sample t procedures.			

Unit 7 Multiple Choice Practice

Directions. Identify the choice that best completes the statement or answers the question. Check your answers and note your performance when you are finished.

1. Pablo would like to estimate the mean amount of time that students in his statistics class studied for the most recent test. To do so, he randomly selected 10 of the 30 students in his class and found that the mean amount of time they spent studying was 2.5 hours with a standard deviation of 0.75 hours. A dotplot of the distribution of study times showed no strong skewness and no outliers. Which of the conditions for inference has been violated, if any?
   (A) The random condition is the only condition that has been violated.
   (B) The 10% condition is the only condition that has been violated.
   (C) The Normal/Large Sample condition is the only condition that has been violated.
   (D) None of the conditions for inference have been violated.
   (E) More than one of the conditions for inference have been violated.

2. An SRS of 100 postal employees found that the average time these employees had worked for the postal service was $\overline{x} = 7$ years with standard deviation $s_x = 0.8$ years. Assume the distribution of the time the population of employees has worked for the postal service is approximately normal. What is a 95% confidence interval for the mean time μ the population of postal service employees has spent with the postal service?
   (A) 7 ± 0.016
   (B) 7 ± 0.019
   (C) 7 ± 0.157
   (D) 7 ± 0.159
   (E) 7 ± 1.587

3. Refer to Question 2. If the sample size were increased to 400, what effect would this have on the width of the confidence interval?
   (A) The width of the interval would double.
   (B) The width of the interval would be cut in half.
   (C) The width of the interval would remain the same.
   (D) The width of the interval would be 4 times as large.
   (E) The width of the interval would be divided by 4.

4. A researcher randomly assigned 50 adult volunteers to two groups of 25 subjects each. Group 1 did a standard step-aerobics workout at a low step height. The mean heart rate at the end of the workout for the subjects in group 1 was 90 beats per minute with a standard deviation of 9 beats per minute. Group 2 did the same workout but at a high step height. The mean heart rate at the end of the workout for the subjects in group 2 was 95.2 beats per minute with a standard deviation of 12.3 beats per minute. Assuming the conditions are met, which of the following could be the 98% confidence interval for the difference in mean heart rates based on these results?
   (A) (2.15, 8.25)
   (B) (–0.77, 11.17)
   (C) (–2.13, 12.54)
   (D) (–2.16, 12.56)
   (E) (–4.09, 14.49)

5. Do students tend to improve their SAT Mathematics (SAT-M) score the second time they take the test? A random sample of four students who took the test twice earned the following scores.

Student	1	2	3	4
First Score	450	520	720	600
Second Score	440	600	720	630

Assume that the change in SAT-M score (second score − first score) for the population of all students taking the test twice is approximately normally distributed with mean μ. What is a 90% confidence interval for μ?

(A) 25.0 ± 118.03
(B) 25.0 ± 64.29
(C) 25.0 ± 47.55
(D) 25.0 ± 43.08
(E) 25.0 ± 33.24

6. Based upon a random sample of 50 adults, a researcher estimates that the true mean amount of time that an individual can focus on a single task is 20 minutes with a standard deviation of 8 minutes. A 95% confidence interval for the mean amount of time that adults can focus on a single task is 17.73 minutes to 22.27 minutes. Which of the following hypotheses would be rejected at the significance level $\alpha = 0.05$?

(A) H_0: $\mu = 19$ versus H_a: $\mu \neq 19$
(B) H_0: $\mu = 20$ versus H_a: $\mu \neq 20$
(C) H_0: $\mu = 21$ versus H_a: $\mu \neq 21$
(D) H_0: $\mu = 22$ versus H_a: $\mu \neq 22$
(E) H_0: $\mu = 23$ versus H_a: $\mu \neq 23$

7. The manager of an assembly line manufacturing plant wants to determine if there is convincing evidence that the mean number of units that are assembled per worker in a given workday is greater than 50 units. To do so, a random sample of 20 employees from this large company is selected. The distribution of the number of units assembled is skewed right, because some employees are far more productive than others. The manager wants to carry out a *t* test for one mean. Are the conditions for inference met?

(A) No, the random condition is the only condition not met.
(B) No, the 10% condition is the only condition not met.
(C) No, the Normal/Large Sample condition is the only condition not met.
(D) No, more than one condition for inference is not met.
(E) Yes, the conditions for inference are met.

8. At a toy store, a package containing a rubber lizard that is 12 inches long claims that when left to soak in water the lizard will grow 600%. This means, when fully grown, the lizard will be 6 feet long! To investigate this claim, a random sample of 30 such lizards are selected. The mean length of the fully-grown lizards, however, is only 40 inches with a standard deviation of 6 inches. We want to know if these data provide convincing evidence that the true mean length for all fully-grown lizards is less than 72 inches (6 feet). What is the test statistic of this significance test?

(A) $t = \dfrac{40-72}{\dfrac{6}{\sqrt{30}}}$ (B) $t = \dfrac{72-40}{\dfrac{6}{\sqrt{30}}}$ (C) $t = \dfrac{40-6}{\dfrac{6(1-6)}{\sqrt{30}}}$

(D) $t = \dfrac{40-6}{\dfrac{6(6-1)}{\sqrt{30}}}$ (E) $t = \dfrac{72-40}{\dfrac{6(1-6)}{\sqrt{30}}}$

9. Some researchers have conjectured that stem-pitting disease in peach tree seedlings might be controlled with weed and soil treatment. An experiment was conducted to compare peach tree seedling growth with soil and weeds treated with one of two herbicides. In a field containing 20 seedlings, 10 were randomly selected from throughout the field and assigned to receive Herbicide A. The remaining 10 seedlings were to receive Herbicide B. Soil and weeds for each seedling were treated with the appropriate herbicide, and at the end of the study period, the height (in centimeters) was recorded for each seedling. A stemplot of each data set showed no indication of non-normality. The following results were obtained:

	$\bar{x}$ (cm)	s_x (cm)
Herbicide A	94.5	10
Herbicide B	109.1	9

Suppose we wished to determine if there is a significant difference in mean height for the seedlings treated with the different herbicides. Based on our data, which of the following is the value of test statistic?
(A) 14.60
(B) 7.80
(C) 3.43
(D) 2.54
(E) 1.14

10. A medical researcher wishes to investigate the effectiveness of exercise versus diet in losing weight. Two groups of 25 overweight adult subjects are used, with a subject in each group matched to a similar subject in the other group. One member of each pair is placed on a regular program of vigorous exercise but with no restriction on diet, and the other is placed on a strict diet but with no requirement to exercise. The weight losses after 20 weeks are determined for each subject, and the difference between each pair of subjects (exercise group – diet group) is computed. The mean of the differences in weight loss is 2 lb. with standard deviation $s_x = 4$ lb. Is this convincing evidence of a difference in mean weight loss for the two methods? To answer this question, you should use
(A) Two-sample t test for a difference in means.
(B) Paired z interval for μ_{Diff}.
(C) Two-sample t interval for a difference in means.
(D) Paired t test for μ_{Diff}.
(E) One-sample z interval for p.

Check your answers below. If you got a question wrong, check to see if you made a simple mistake or if you need to study that concept more. After you check your work, identify the concepts you feel very confident about and note what you will do to learn the concepts in need of more study.

#	Answer	Concept	Right	Wrong	Simple Mistake?	Study More
1	B	Conditions for inference				
2	D	Confidence Interval for μ				
3	B	Sample Size and Interval Width				
4	D	Confidence Interval for Difference in Means				
5	C	Confidence Interval for μ (paired)				
6	E	Two-sided Test and Confidence Intervals				
7	C	Checking Conditions				
8	A	Determine the Correct Test Statistic				
9	C	Significance Test for Difference in Means				
10	D	Choosing the Appropriate Test				

FRAPPY! Free Response AP® Problem, Yay!

The following problem is modeled after actual Advanced Placement Statistics free response questions. Your task is to generate a complete, concise response in 15 minutes. After you generate your response, read over the solution and scoring guide. Score your response and note what, if anything, you would do differently to increase your own score.

An athletic trainer wanted to study the effect of being cheered for on athletic performance. To do so, he used 10 student volunteers. Half of the volunteers were randomly assigned to show up at the track to run one mile in front of a group of their peers who cheered for their classmates as they ran. The other half of the volunteers were assigned to show up at the track at a different time when no one was there other than the trainer, who did not cheer for them. They also each ran one mile. On a different day, the treatments were reversed for each student. Their one-mile times were recorded again. Here are the one-mile times, in minutes, for each student:

Student	1	2	3	4	5	6	7	8	9	10	**Mean**	**SD**
Not Cheered	10.45	7.23	8.85	7.15	8.23	8.78	7.80	8.43	10.00	6.98	**8.39**	**1.176**
Cheered	10.35	6.65	8.40	6.15	7.95	8.38	6.92	8.20	9.38	7.03	**7.94**	**1.295**
Difference	0.10	0.58	0.45	1	0.28	0.40	0.88	0.23	0.62	–0.05	**0.45**	**0.331**

The trainer would like to estimate, with 95% confidence, the true mean difference in one-mile times for students like these using the order "No Cheering – Cheering".

(a) What is the appropriate inference procedure? Explain.

(b) Compute the 95% confidence interval identified in part (a).

(c) Does the confidence interval give reason to believe that students like these run faster when they are being cheered for than when they are not being cheered for?

FRAPPY! Scoring Rubric

Use the following rubric to score your response. Each part receives a score of "Essentially Correct," "Partially Correct," or "Incorrect." When you have scored your response, reflect on your understanding of the concepts addressed in this problem. If necessary, note what you would do differently on future questions like this to increase your score.

Intent of the Question

The goal of this question is to determine your ability to construct and interpret a confidence interval for paired data.

Solution

(a) Paired t confidence interval. We have two data values recorded for each student.

(b) **State**: 95% confidence interval for μ_{Diff} = the true mean difference (no cheering – cheering) in one-mile times.

Plan: one-sample t interval for a mean difference

Random: The volunteers were randomly assigned a treatment order. ✓

Normal/Large Sample: The distribution of differences shows no strong skewness or outliers. ✓

0 0.1 0.2 0.3 0.4 0.5 0.6 0.7 0.8 0.9 1

Difference (No cheering - Cheering) in one-mile time (min)

Do: df = 9, $0.45 \pm 2.262\left(\dfrac{0.331}{\sqrt{10}}\right) = (0.213, 0.687)$.

Conclude: I am 95% confident that the interval from 0.213 minutes to 0.687 minutes contains the true difference (not cheered – cheered) in one-mile run times.

(c) Yes, because all of the values in the confidence interval are positive, we have reason to believe that students like these will run faster when being cheered for than when not being cheered for.

Scoring

Each element scored as essentially correct (E), partially correct (P), or incorrect (I).

"Part (a), name, and conditions" are essentially correct if the response correctly does the following 5 things: (1) identifies the correct inference procedure by name or formula with correct justification, (2) defines the parameter, (3) checks the Random condition, (4) checks the Normal/Large Sample condition, (5) includes context. This part is partially correct if the response correctly does 4 of the 5 items and is incorrect if 3 or fewer of the 5 items are correct.

"Mechanics" is essentially correct if the response does all 3: (1)correctly calculates the confidence interval (2) with supporting work and (3) correctly states the degrees of freedom. This part is partially correct if only 2 of those 3 components are done correctly. Incorrect otherwise.

"Conclusion and part (c)" is essentially correct if the response (1) correctly interprets the interval in context, (2) responds "yes" to part (c), (3) provides correct justification in part (c), (4) includes context. This part is partially correct if 3 of the 4 components are correct. Incorrect otherwise. Note: A correct conclusion for an incorrect confidence interval in (b) can earn credit for this part.

Scoring

4 Complete Response

All three parts essentially correct

3 Substantial Response

Two parts essentially correct and one part partially correct

2 Developing Response

Two parts essentially correct and no parts partially correct
One part essentially correct and one or two parts partially correct
Three parts partially correct

1 Minimal Response

One part essentially correct and no parts partially correct
No parts essentially correct and two parts partially correct

My Score:
What I did well:
What I could improve:
What I should remember if I see a problem like this on the AP Exam:

FRAPPY! Free Response AP® Problem, Yay!

The following problem is modeled after actual Advanced Placement Statistics free response questions. Your task is to generate a complete, concise response in 15 minutes. After you generate your response, read over the solution and scoring guide. Score your response and note what, if anything, you would do differently to increase your own score.

Researchers are interested in whether or not women who are part of a prenatal care program give birth to babies with a higher average birth weight than those who do not take part in the program. A random sample of hospital records indicates that the average birth weight for 75 babies born to mothers enrolled in a prenatal care program was 3100 g with standard deviation 420 g. A separate random sample of hospital records indicates that the average birth weight for 75 babies born to women who did not take part in a prenatal care program was 2750 g with standard deviation 425 g. Do these data provide convincing evidence that mothers who participate in a prenatal care program have babies with a higher average birth weight than those who do not participate? Use $\alpha = 0.05$.

FRAPPY! Scoring Rubric

Use the following rubric to score your response. Each part receives a score of "Essentially Correct," "Partially Correct," or "Incorrect." When you have scored your response, reflect on your understanding of the concepts addressed in this problem. If necessary, note what you would do differently on future questions like this to increase your score.

Intent of the Question

The goal of this question is to determine your ability to conduct a significance test for the difference between two means.

Solution

The solution should contain 4 parts:

- **Hypotheses**: Must be stated correctly: H_0: $\mu_Y - \mu_N = 0$ and H_a: $\mu_Y - \mu_N > 0$ where μ_Y = mean birth weight of babies born to all mothers enrolled in a prenatal care program and μ_N = mean birth weight of babies born to all mothers who are not enrolled in a prenatal care program.
- **Name of Test and Conditions**: The test must be identified by name or formula as a two-sample t test for a difference in population means. Two independent random samples were obtained. ✓ Both sample sizes are greater than 30. ✓ Both sample sizes are less than 10% of their respective population sizes because this is a large hospital. ✓
- **Mechanics**: $t = 5.07$ and P-value ≈ 0 with 147.979 degrees of freedom.
- **Conclusion**: Because the P-value of approximately $0 < \alpha = 0.05$, we reject H_0. There is convincing evidence that the mean birth weight of babies born to all mothers who participate in a prenatal care program is greater than the mean birth weight of babies born to all mothers who do not participate in the prenatal care program.

Scoring

Each element scored as essentially correct (E), partially correct (P), or incorrect (I).

"Hypotheses and Name" is essentially correct if the hypotheses are written correctly, and the response correctly identifies the test by name or formula. This part is partially correct if the test isn't identified correctly or if the hypotheses are written incorrectly. Incorrect otherwise.

"Conditions and Mechanics" is essentially correct if the response correctly does all of the 5 following components: (1) checks the Random condition, (2) checks the 10% condition, (3) checks the Normal/Large Samples condition, (4) correct test statistic, (5) correct P-value. This part is partially correct if 4 of those 5 things are done correctly. Incorrect otherwise. Note: Components 1 – 3 must each clearly be verified for 2 separate/independent samples to be correct.

Conclusion is essentially correct if the response correctly rejects the null hypothesis because the P-value is less than the significance level of 0.05 and provides an interpretation in context. This part is partially correct if the response fails to justify the decision by comparing the P-value to the significance level OR if the conclusion lacks an interpretation in context. Incorrect otherwise.
Note: A conclusion that is correct for an incorrect P-value can earn credit for this part.

Scoring

4 Complete Response

All three parts essentially correct

3 Substantial Response

Two parts essentially correct and one part partially correct

2 Developing Response

Two parts essentially correct and no parts partially correct

One part essentially correct and one or two parts partially correct

Three parts partially correct

1 Minimal Response

One part essentially correct and no parts partially correct

No parts essentially correct and two parts partially correct

My Score:
What I did well:
What I could improve:
What I should remember if I see a problem like this on the AP Exam:

Unit 7 Formula Sheet

Create a one-page summary of important concepts and formulas found in this unit. This will serve as a valuable resource as you prepare for the AP Exam.

Unit 7 Crossword Puzzle

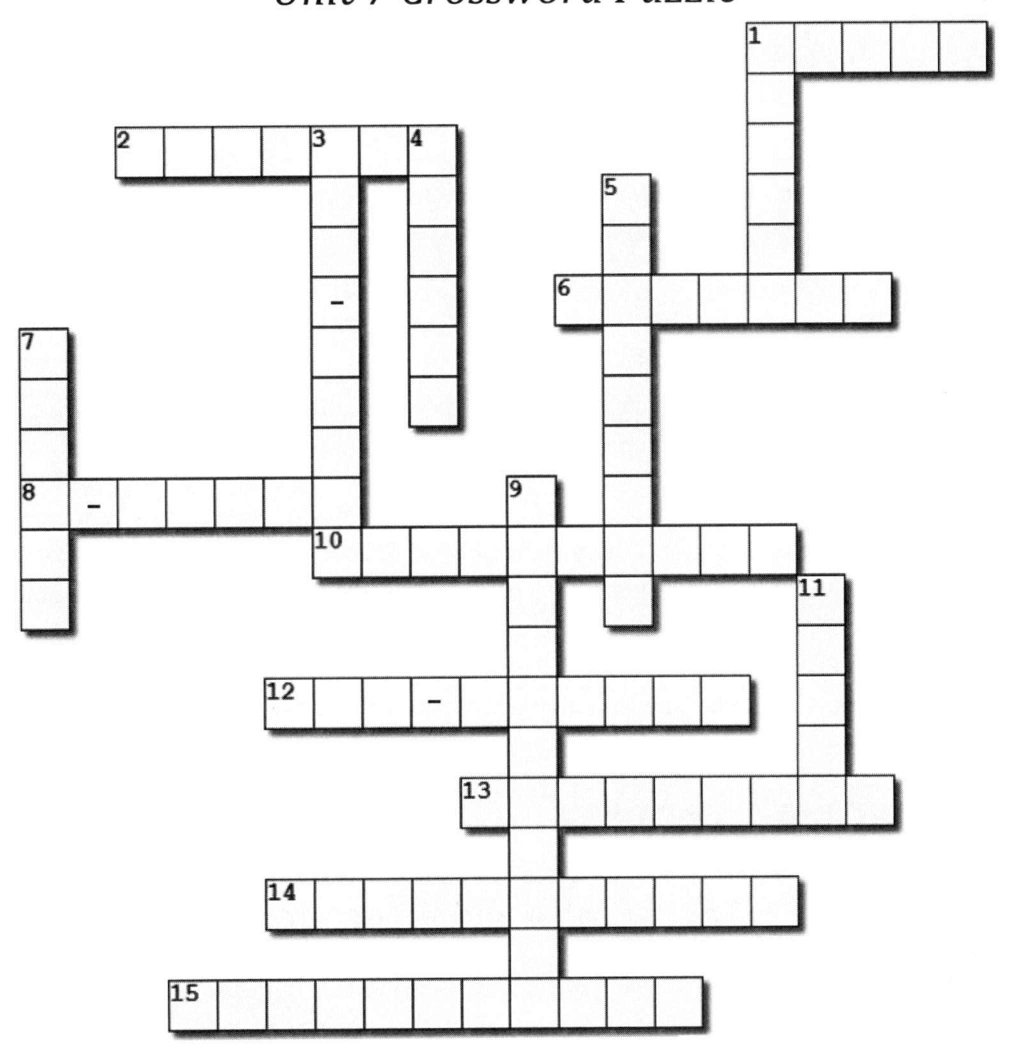

Across:

1. We can increase the ____ of a test by increasing the sample size.
2. If the *P*-value is ____ than the significance level, we fail to reject the null hypothesis.
6. Any *t* distribution is completely specified by its degrees of ____ (df).
8. To draw a conclusion about a test, compare the _-___ to the significance level.
10. In a significance test for a ____ in means, the data may come from two groups in a randomized experiment or two independent random samples.
12. If the data come from 2 groups in a randomized experiment, use __-__ *t* procedures.
13. Increasing the sample size ____ the margin of error.
14. When performing a two-sample *t* test, if the samples are randomly selected, they must be ____.
15. Be sure to specify the order of ____ when defining the parameter in a paired *t* test or two-sample *t* test.

Down:

1. ____ data result from recording 2 values of the same quantitative variable for each individual or pair of similar individuals.
3. A confidence interval for a mean will give consistent results with a __-__ hypothesis test.
4. The data must come from a ____ sample from the population of interest.
5. When stating the hypotheses, always define the ____ of interest.
7. If $n < 30$ and the population distribution has an unknown shape, make a graph of the ___ data.
9. Check the 10% condition when sampling without ____.
11. If the sample size is at least 30, the Normal/____ Sample condition is met for a one-sample *t* test for a mean.

Unit 8: Inference for Categorical Data: Chi-Square

*"A judicious man looks on statistics not to get knowledge,
but to save himself from having ignorance foisted on him." Thomas Carlyle*

UNIT 8 OVERVIEW

So far, our study of inference has focused on how to estimate and test claims about means and proportions from a single random sample as well as about differences between means and proportions for independent random samples or two groups in a randomized experiment.

In this unit, we will shift our focus to inference about distributions of and relationships between categorical variables. You will learn how to perform three different significance tests for distributions of categorical data, allowing you to determine (1) whether or not a sample distribution differs significantly from a hypothesized distribution, (2) whether or not the distribution of a categorical variable differs between multiple populations, and (3) whether or not an association exists between two categorical variables. It is easy to confuse the three tests, so be sure to study the differences between them so you know when to use each one!

Sections in Unit 8
Section 8A: Chi-Square Tests for Goodness of Fit
Section 8B: Chi-Square Tests for Independence or Homogeneity

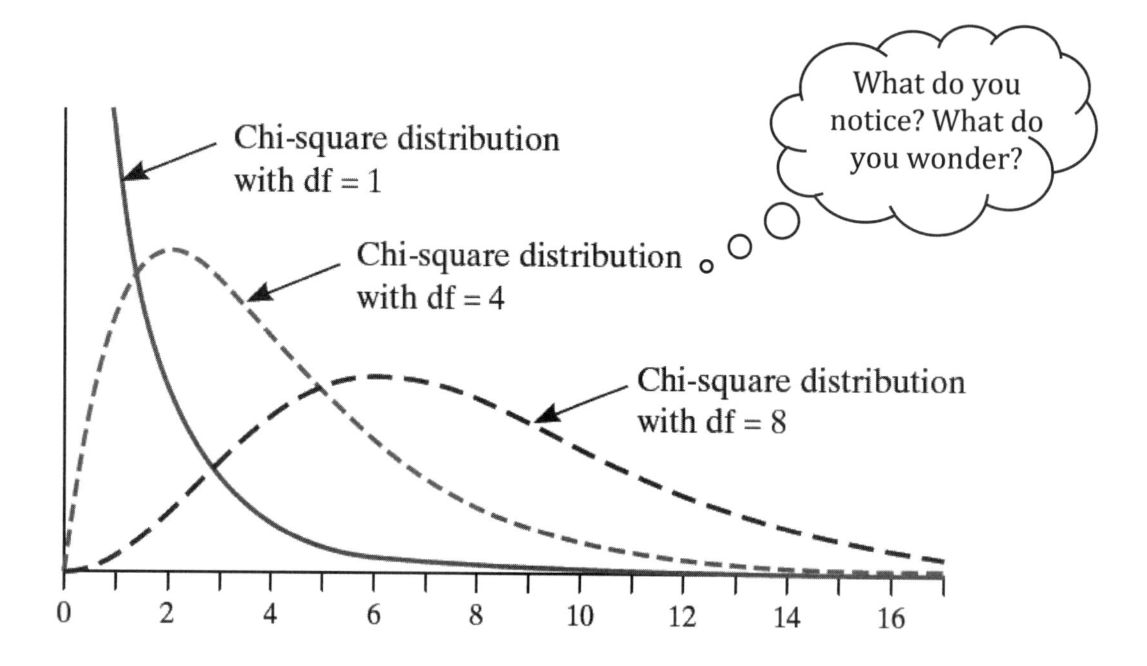

PRACTICE FOR MASTERY

Use the following *suggested* guide to the pages and exercises in your text to practice for mastery!
Note: your teacher may assign different problems. Be sure to follow their instructions!

Day	Topics	Read	Do
1	◆ Stating Hypotheses for Tests about the Distribution of a Categorical Variable ◆ Calculating Expected Counts for Tests about the Distribution of a Categorical Variable ◆ Checking Conditions for Tests about the Distribution of a Categorical Variable	p. 744-751	**8A**: 1, 3, 5, 7, 9, 11, 28
2	◆ Calculating the Test Statistic and P-Value for Tests about the Distribution of a Categorical Variable ◆ The Chi-Square Test for Goodness of Fit	p. 751-760	**8A**: 13, 17, 19, 21, 29-33
3	◆ Stating Hypotheses for Tests about the Relationship Between Two Categorical Variables ◆ Calculating Expected Counts for Tests about the Relationship Between Two Categorical Variables ◆ Checking Conditions for Tests about the Relationship Between Two Categorical Variables ◆ Calculating the Test Statistic and P-Value for Tests about the Relationship Between Two Categorical Variables ◆ The Chi-Square Test for Independence	p. 766-778	**8B**: 1, 3, 5, 7, 9, 11, 13, 15
4	◆ The Chi-Square Test for Homogeneity ◆ Independence or Homogeneity?	p. 778-785	**8B**: 17, 19, 21, 23, 25
5	Unit 8 AP® Statistics Practice Test (p. 799)		Unit 8 Review Exercises (p.798)
6	Unit 8 Test: Celebration of learning!	Note: Since "Test" can sound daunting, think of this as a way to show off all you have learned. It's your own celebration of learning!	

Section 8A: Chi-Square Tests for Goodness of Fit

Section Summary

In this section, you will be introduced to the chi-square distributions and the chi-square test statistic. The chi-square statistic provides us with a way to measure the difference between an observed and a hypothesized distribution of categorical data. When certain conditions are satisfied, we can model the sampling distribution of this statistic with a chi-square distribution and calculate *P*-values. We can use the chi-square goodness of fit test to determine whether or not a significant difference between an observed and hypothesized distribution of a categorical variable exists. Finally, we can use a follow-up analysis to determine which categories contributed the most to the difference.

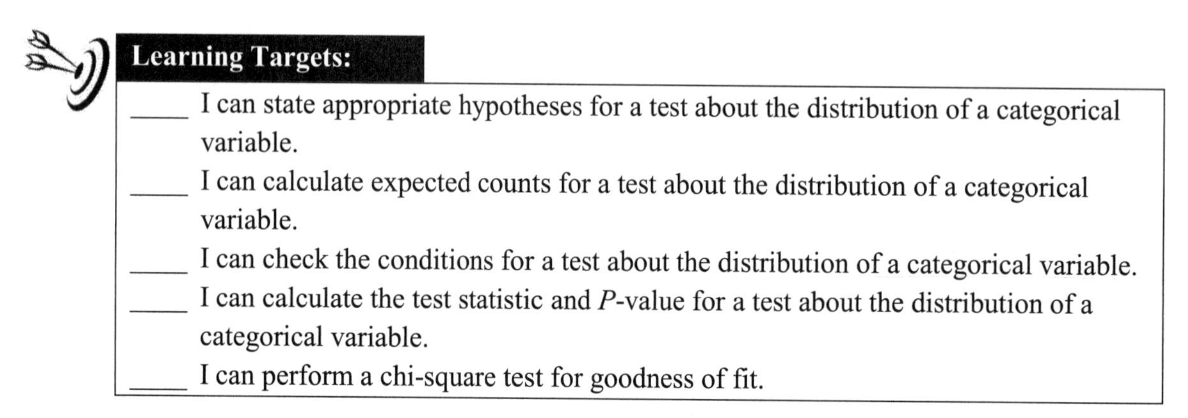

Learning Targets:

- ____ I can state appropriate hypotheses for a test about the distribution of a categorical variable.
- ____ I can calculate expected counts for a test about the distribution of a categorical variable.
- ____ I can check the conditions for a test about the distribution of a categorical variable.
- ____ I can calculate the test statistic and *P*-value for a test about the distribution of a categorical variable.
- ____ I can perform a chi-square test for goodness of fit.

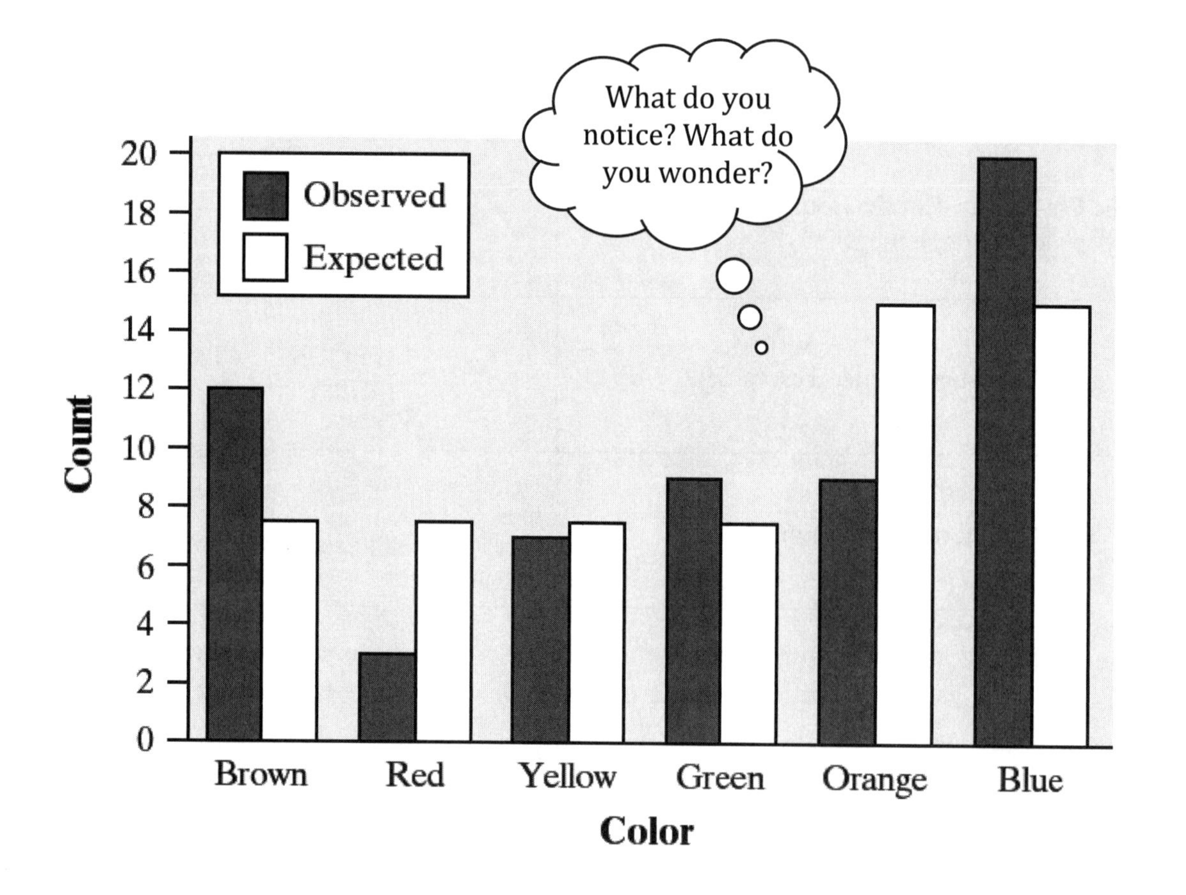

Read: Vocabulary and Key Concepts

Stating Hypotheses for Tests about the Distribution of a Categorical Variable

- The null hypothesis in a test about the distribution of a _______________ variable should state a claim about the ________________ of that variable in the __________________ of interest.
- The alternative hypothesis in a test about the distribution of a categorical variable is that the categorical variable ____________________ the specified distribution.
- H_a: At least _____ of these proportions __________ from the values stated by the null hypothesis.
- Why don't we write the alternative hypothesis as "H_a: At least one of these proportions differs from the values stated by the null hypothesis" instead? If the stated proportion in one category is _______, then the stated proportion in at least one other category must be _________ because the sum of the proportions is always equal to ____.

Calculating Expected Counts for Tests about the Distribution of a Categorical Variable

- The ____________ count for category *i* in the distribution of a categorical variable is ______ where *n* is the overall _______________ and p_i is the ______________ for category *i* specified by the null hypothesis.
- Expected counts are often not ________ numbers and shouldn't be rounded to a _________ number.

Checking Conditions for Tests about the Distribution of a Categorical Variable

- **Random:** The data come from a _________ sample from the _____________ of interest.
  - **10%:** When sampling without _______________, $n < 0.10N$.
- **Large Counts:** All _______________ counts are at least ____.

Calculating the Test Statistic and *P*-Value for Tests about the Distribution of a Categorical Variable

- The symbol χ is the lowercase Greek letter _____, pronounced "kye" like "_____."
- The chi-square test statistic is a measure of how different the _____________ counts are from the _______________ counts, relative to the _____________ counts. The formula for the statistic is:

  Write the formula here: ______________________________

  where the sum is over all possible values of the categorical variable.
- Make sure to use the observed and expected __________, not the observed and expected _____________, when calculating the chi-square test statistic.
- ________ values of χ^2 arise when the differences between the observed counts and the expected counts are _________. Thus, the ___________ the value of χ^2, the ___________ the evidence is for the alternative hypothesis (and against the null hypothesis).

- The sampling distribution of the chi-square test statistic is *not* a ___________ distribution. It is a ________-skewed distribution that allows only ___________ values because χ^2 can never be negative.
- For a chi-square distribution, degrees of freedom (df) = ____________________________.
- The _________ of a particular chi-square distribution is equal to its _________________________.
- For df > 2, the _______ (peak) of the chi-square density curve is at (df – 2).
- Find the __________ by calculating the probability of getting a χ^2 statistic this large or larger in a chi-square distribution with df = number of categories – 1.

Putting It All Together: The Chi-Square Test for Goodness of Fit

- A ___________________________________ is a significance test of the null hypothesis that a categorical variable has a specified distribution in the population of interest.

- If the sample data lead to a ______________________ result, we can conclude that our variable has a distribution ______________ from the one stated. To investigate *how* the distribution is different, start by identifying the categories that _________________________ to the chi-square statistic. Then describe how the observed and expected counts __________ in those categories, noting the _____________ of the difference.

Watch: Go to bfwpub.com/TPS7eStrive. Select Unit 8 and watch the following Example videos:

Video	Topic	Done
Unit 8A, Are more NHL players born earlier in the year? (p. 747)	*Stating hypotheses for tests about the distribution of a categorical variable*	
Unit 8A, Are more NHL players born earlier in the year? (p. 749)	*Calculating expected counts for tests about the distribution of a categorical variable*	
Unit 8A, Are more NHL players born earlier in the year? (p. 750)	*Checking conditions for test about the distribution of a categorical variable*	
Unit 8A, Are more NHL players born earlier in the year? (p. 755)	*Calculating the test statistic and P-value for tests about the distribution of a categorical variable*	
Unit 8A, Trees of New York (p. 758)	*The chi-square test for goodness of fit*	

Do: Check for Understanding

Concepts 1, 2, 3, 4, and 5: Chi-Square Test for Goodness of Fit

To perform a chi-square test for goodness of fit for a claim about a population distribution of categorical data, we will follow the same basic process as we did for means and proportions. That is, to test a claim about the population distribution of a categorical variable:

1) **STATE: State the hypotheses, parameters, and significance level.**
   When testing a claim about a distribution of a categorical variable, we start by defining hypotheses. The null hypothesis assumes that the claimed population distribution is correct while the alternative hypothesis is that the population distribution is different than claimed.

   H_0: The population distribution of the categorical variable is as claimed.
   H_a: The population distribution of the categorical variable is not as claimed.
   OR
   H_0: $p_{[\text{category1}]} =$ _____, $p_{[\text{category2}]} =$ _____, $p_{[\text{category3}]} =$ ______, etc.
   H_a: At least two of these proportions differ from the values stated by the null hypothesis.
   where $p_{[\text{context}]}$ = the population proportion of [context].

   Define the parameters in context. State a significance level.

2) **PLAN: Identify the appropriate inference method and check the conditions.**
   Name: Chi-Square Test for Goodness of Fit
   Conditions: The data must come from a random sample from the population. When sampling without replacement, check that the sample size is less than 10% of the size of the population. All expected counts must be at least 5.

3) **DO: If conditions are met, perform calculations.**
   Compute the chi-square test statistic $\chi^2 = \sum \frac{(\text{observed count-expected count})^2}{\text{expected count}}$
   Find the *P*-value by using a chi-square distribution with df = number of categories – 1.

   (*Need Tech Help?* View the video **Tech Corner 26: Calculating *P*-values from χ^2 Distributions** at bfwpub.com/TPS7eStrive)

4) **CONCLUDE: Make a conclusion about the hypotheses in the context of the problem.**
   If the *P*-value is smaller than the stated significance level, reject H_0. You have convincing evidence in favor of H_a. If the *P*-value is greater than or equal to the significance level, then you fail to reject the null hypothesis. There is not convincing evidence for H_a.

If you have convincing evidence to reject the null hypothesis, you should perform a follow up analysis to determine the components that contributed the most to the chi-square statistic by determining which observed categories differed the most from the expected values.

Check for Understanding – Learning Targets 1, 2, 3, 4, and 5

____ *I can state appropriate hypotheses for a test about the distribution of a categorical variable.*

____ *I can calculate expected counts for a test about the distribution of a categorical variable.*

____ *I can check conditions for a test about the distribution of a categorical variable.*

____ *I can calculate the test statistic and P-value for a test about the distribution of a categorical variable.*

____ *I can perform a chi-square test for goodness of fit.*

Mr. Rothery attempted to make a 6-sided die unfair by modifying the die according to a tutorial he found online. He attempted to make 5 and 6 appear 2 times as often as each of the other outcomes. After modifying the die, he rolled it 400 times. The table shows the results. Do these data provide convincing evidence that the predicted 1:1:1:1:2:2 ratio is incorrect? Use $\alpha = 0.05$.

Outcome	1	2	3	4	5	6
Frequency	42	55	48	57	94	104

(*Need Tech Help?* View the video **Tech Corner 27: Significance Tests for the Distribution of a Categorical Variable** at bfwpub.com/TPS7eStrive)

Section 8B: Chi-Square Tests for Independence or Homogeneity

Section Summary

Chi-square goodness of fit tests allow us to compare the distribution of one categorical variable to a hypothesized distribution. However, sometimes we may be interested in comparing the distribution of a categorical variable across several populations or treatments. Chi-square tests for homogeneity allow us to do this. In this section you will learn how to conduct the chi-square test for homogeneity and how to perform a follow-up analysis, just like you did for the goodness of fit test. Also, you will learn how to conduct a chi-square test for association/independence to determine whether or not there is convincing evidence that two categorical variables are related. While this test is the same as the test for homogeneity in its mechanics, the hypotheses are different. Be sure to note the distinction!

Learning Targets:

____ I can state appropriate hypotheses for a test about the relationship between two categorical variables.
____ I can calculate expected counts for a test about the relationship between two categorical variables.
____ I can check the conditions for a test about the relationship between two categorical variables.
____ I can calculate the test statistic and *P*-value for a test about the relationship between two categorical variables.
____ I can perform a chi-square test for independence.
____ I can perform a chi-square test for homogeneity.
____ I can distinguish between a chi-square test for independence and a chi-square test for homogeneity.

Read: Vocabulary and Key Concepts

- To begin the analysis, make a ________________ bar graph, a ________________ bar graph, or a ____________ plot. Then look for an ___________________. Recall that two variables have an ____________ if knowing the value of one variable helps to predict the value of the other variable.
- An observed association between two categorical variables in a sample may be due to an association between those two categorical variables in the ______________. Or there may be no association between these variables in the population and the association in the sample happened by ___________________.

Stating Hypotheses for Tests about the Relationship Between Two Categorical Variables

- In a test about the relationship between two ______________ variables, the null hypothesis is that there is _________________ between the two variables in the population of interest. The alternative says that there _________________________.
- No association between two variables means that knowing the value of one variable does not help us _____________ the value of the other. That is, the variables are ________________.

Calculating Expected Counts for Tests about the Relationship Between Two Categorical Variables

- As with chi-square tests for goodness of fit, we will compare the ____________ counts with the counts that we would _____________ when the null hypothesis is _______.
- When H_0 is true, the expected count for any cell in a two-way table is ________________________.
- We can complete the table of expected counts by using the formula in ______________ or by using the formula in some cells and ________________ to find the remaining cells.

Checking Conditions for Tests about the Relationship Between Two Categorical Variables

- Random: The data come from a random sample from the _______________ of interest.
    - 10%: When sampling without replacement, _________.
- Large Counts: All ______________ counts are at least 5.

Calculating the Test Statistic and *P*-value for Tests about the Relationship Between Two Categorical Variables

- Suppose the conditions are met. To perform a test of the null hypothesis:

  H_0: There is no association between two categorical variables in the population of interest compute the chi-square test statistic:

  Write the formula here: ____

  where the sum is over all cells in the two-way table (not including the totals).
- Find the *P*-value by calculating the probability of getting a χ^2 statistic this large or larger in a chi-square distribution with df = (__________________ −1)(___________________ −1).

Putting It All Together: The Chi-Square Test for Independence

- A chi-square test for ___________________ is a significance test of the null hypothesis that there is ____ association between _____ categorical variables in the _______________ of interest.
- Whenever you perform a chi-square test for independence, be sure to follow the __________ process.

The Chi-Square Test for Homogeneity

- Another common situation that leads to a two-way table is when we compare the distribution of a ____________ categorical variable for ____________ populations or treatments.
- In general, the null hypothesis in a test comparing the distribution of a categorical variable for two or more populations or treatments says that there is ___________________ in the distributions. As with the chi-square test for independence, the alternative hypothesis says that the null hypothesis _______________.

Conditions:

- Random: The data come from ______________________________ or from __________ in a randomized ______________.
    - 10%: When sampling without ______________, $n < 0.10N$ for each sample.
- Large Counts: All ____________ counts are at least ___.

Test Statistic and *P*-value:

- Suppose the conditions are met. To perform a test of the null hypothesis H_0: There is no difference in the distributions of a categorical variable in the populations of interest or for the treatments in an experiment compute the chi-square test statistic:

  Write the formula here: ________________________________

  where the sum is over all cells in the two-way table (not including the totals).

- Find the __________ by calculating the probability of getting a χ^2 statistic this _______ or ________ in a chi-square distribution with df = (number of rows −1)(number of columns −1) .
- A chi-square test for __________________ is a significance test of the null hypothesis that there is no difference in the distributions of a categorical variable in the __________ of interest or for the ______________ in an experiment.

Independence or Homogeneity?

- The chi-square test for independence tests whether two categorical variables are ________________ in some population of interest.
- A chi-square test for homogeneity tests whether the ________________ of a categorical variable is the same for each of several _____________ or _____________.
- If the data come from a ________________________, with the individuals classified according to two categorical variables, use a chi-square test for independence.
- In contrast, if the data come from ____ or _____________________ in a randomized experiment, use a chi-square test for homogeneity.

Watch: Go to bfwpub.com/TPS7eStrive. Select Unit 8 and watch the following Example videos:

Video	Topic	Done
Unit 8B, Living close to family (p. 768)	*Stating hypotheses for tests about the relationship between two categorical variables*	
Unit 8B, Living close to family (p. 770)	*Calculating expected counts for tests about the relationship between two categorical variable*	
Unit 8B, Living close to family (p. 772)	*Checking conditions for tests about the relationship between two categorical variables*	
Unit 8B, Living close to family (p. 774)	*Calculating the test statistic and P-value for tests about the relationship between two categorical variables.*	
Unit 8B, Anger and heart disease (p. 776)	*The chi-square test for independence*	
Unit 8B, Speaking English (p. 781)	*The chi-square test for homogeneity*	
Unit 8B, A smash or a hit? (p. 784)	*Independence or homogeneity?*	

Do: Check for Understanding

Concepts 1, 2, 3, 4, and 5: Chi-Square Test for Independence

To determine whether or not two categorical variables are related in a population, we can use a chi-square test for association/independence. Use this test when the data come from a single random sample from the population of interest and then is classified according to two categorical variables.

1) **STATE: State the hypotheses and significance level.**
   The hypotheses are defined in terms of an association between the two categorical variables.
   H_0: There is no association between two categorical variables in the population of interest.
   H_a: There is an association between two categorical variables in the population of interest.
   State a significance level.

2) **PLAN: Identify the appropriate inference method and check the conditions.**
   Name: Chi-Square Test for Association/Independence.
   Conditions: The data must come from a single random sample from the population of interest. When sampling without replacement, check that the sample sizes are less than 10% of their corresponding population sizes. All expected counts must be at least 5. The expected count for any cell of a two-way table can be calculated using the formula: expected count $= \frac{\text{(row total)(column total)}}{\text{table total}}$.

3) **DO: If conditions are met, perform calculations.**
   Compute the chi-square test statistic $\chi^2 = \sum \frac{(\text{observed count} - \text{expected count})^2}{\text{expected count}}$ and find the P-value by using a chi-square distribution with df = (number of rows – 1)(number of columns – 1).
   Consider using technology to carry out these calculations.

4) **CONCLUDE: Make a conclusion about the hypothesis in the context of the problem.**
   If the P-value is smaller than the stated significance level, reject H_0. You have convincing evidence in favor of H_a. If the P-value is greater than or equal to the significance level, then you fail to reject the null hypothesis. There is not convincing evidence for H_a.

If you find sufficient evidence to reject the null hypothesis, you should perform a follow up analysis to determine the components that contributed the most to the chi-square statistic. That is, determine which observed categories differed the most from the expected values.

Check for Understanding – Learning Targets 1, 2, 3, 4, and 5

____ *I can state appropriate hypotheses for a test about the relationship between two categorical variables.*

____ *I can calculate expected counts for a test about the relationship between two categorical variables.*

____ *I can check the conditions for a test about the relationship between two categorical variables.*

____ *I can calculate the test statistic and P-value for a test about the relationship between two categorical variable.*

____ *I can perform a chi-square test for independence.*

A government official selects a random sample of 529 households that completed the most recent U.S. census and records the following data regarding the head of household: their highest level of education and whether they made more than $60,000/year.

		Education Level				
		High School	Bachelors	Masters	PhD	Total
Earn	No	54	92	92	14	252
> $60,000	Yes	28	105	116	28	277
	Total	82	197	208	42	529

1. State appropriate hypotheses for performing a chi-square test for independence.

2. Check that the conditions for carrying out the test are met.

3. Calculate the value of the chi-square test statistic and specify the distribution of the chi-square statistic.

4. Calculate the *P*-value. Using a significance level of $\alpha = 0.05$, what conclusion do you draw?

5. If there is convincing evidence that education level is not independent of earnings, perform a follow-up analysis.

Concepts 1, 2, 3, 4, and 6: Chi-Square Test for Homogeneity

To determine whether or not a distribution of a categorical variable differs for two or more populations or treatments, we use a chi-square test for homogeneity. The mechanics of this test are almost the same as those for the test for association. The main difference is that the data come from independent random samples from the population of interest or groups in a randomized experiment. Be sure to state the hypotheses and check the Random condition accordingly.

Check for Understanding – Learning Targets 1, 2, 3, 4, and 6

____ *I can state appropriate hypotheses for a test about the relationship between two categorical variables.*

____ *I can calculate expected counts for a test about the relationship between two categorical variables.*

____ *I can check the conditions for a test about the relationship between two categorical variables.*

____ *I can calculate the test statistic and P-value for a test about the relationship between two categorical variable.*

____ *I can perform a chi-square test for homogeneity.*

At a manufacturing plant, workers produce glass vases in three eight-hour shifts: 12 am to 8 am, 8 am to 4 pm, and 4 pm to 12 am. A quality control manager wants to know if any difference exists in the proportion of defective vases produced during each of the three shifts. The manager randomly selects 100 vases produced from the thousands of vases produced by each of the three shifts for a detailed inspection. Each vase either passes or fails the inspection. Here are the results of the manager's inspection.

		Shift			
		Shift 1	Shift 2	Shift 3	Total
Result	Pass	96	88	92	276
	Fail	4	12	8	24
	Total	100	100	100	300

Do these data provide convincing evidence at the $\alpha = 0.01$ significance level that the distributions of result differ for the three shifts?

(*Need Tech Help?* View the video **Tech Corner 28: Significance Tests for the Relationship Between Two Categorical Variables** at bfwpub.com/TPS7eStrive)

Concept 7 – Distinguishing Between "Independence" and "Homogeneity"

When trying to determine whether the appropriate test is a chi-square test for independence or a chi-square test of homogeneity, consider how the data are produced.

First, some preliminary questions:

1. Do you have data that is classified according to two categorical variables? Data may be displayed in a two-way table, a segmented bar graph, side-by-side bar graph, or mosaic plot.
2. Are you asked to determine if there is convincing evidence of a relationship between the categorical variables?

If the previous two answers are "yes", then ask yourself these final 2 questions:

1. Was a single random sample selected and then classified according to two categorical variables? If so, the appropriate test is a chi-square test of association/independence.
2. Did the data come from independent random samples from distinct populations? Or did the data come from groups of a randomized experiment? If so, the appropriate test is a chi-square test of homogeneity.

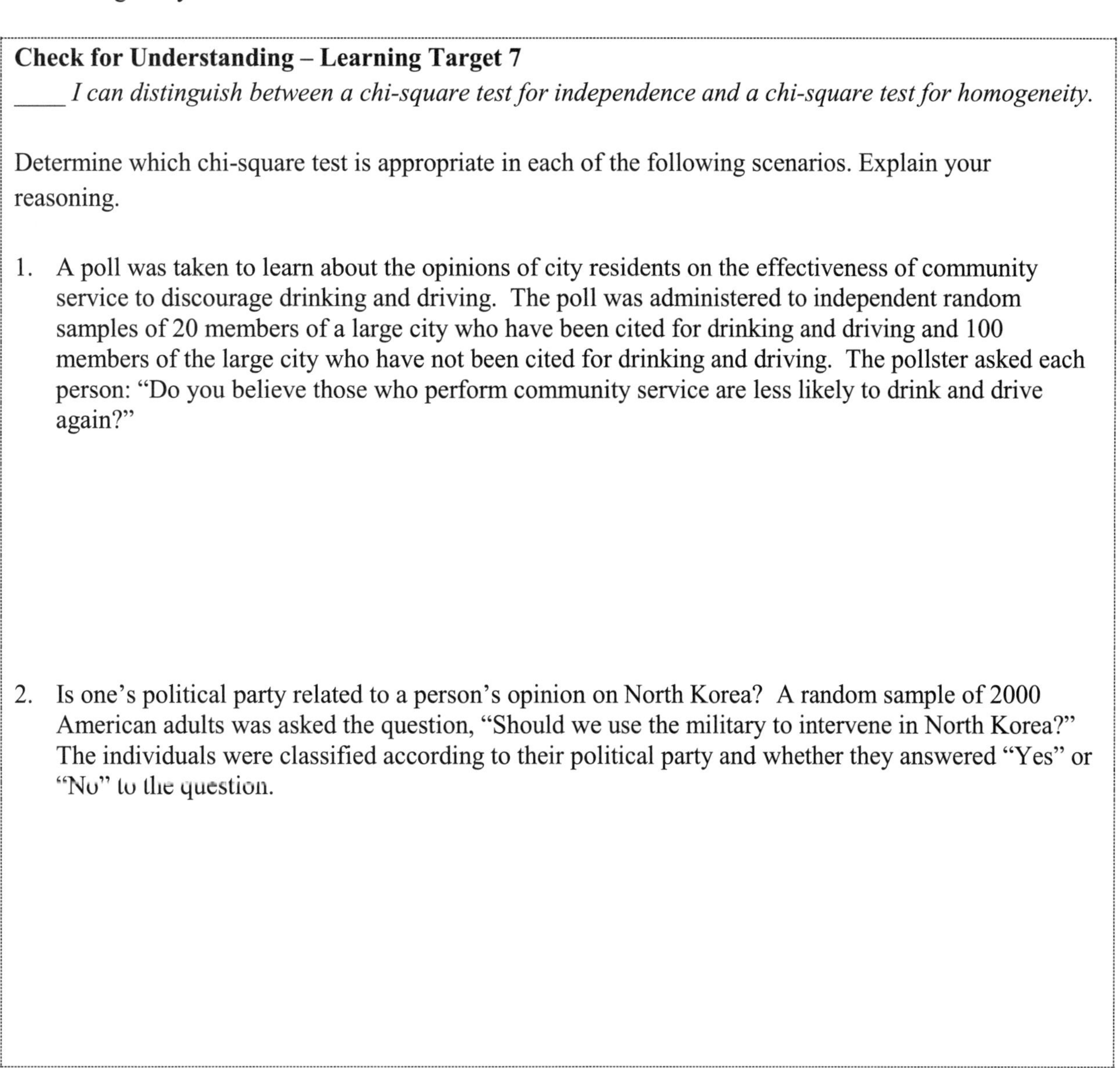

Check for Understanding – Learning Target 7

____ *I can distinguish between a chi-square test for independence and a chi-square test for homogeneity.*

Determine which chi-square test is appropriate in each of the following scenarios. Explain your reasoning.

1. A poll was taken to learn about the opinions of city residents on the effectiveness of community service to discourage drinking and driving. The poll was administered to independent random samples of 20 members of a large city who have been cited for drinking and driving and 100 members of the large city who have not been cited for drinking and driving. The pollster asked each person: "Do you believe those who perform community service are less likely to drink and drive again?"

2. Is one's political party related to a person's opinion on North Korea? A random sample of 2000 American adults was asked the question, "Should we use the military to intervene in North Korea?" The individuals were classified according to their political party and whether they answered "Yes" or "No" to the question.

Unit 8 Summary: Inference for Categorical Data: Chi-Square

In this unit, you learned inference procedures for distributions of categorical data. A chi-square goodness of fit test can be used to determine whether or not an observed distribution of a categorical variable differs from a hypothesized distribution. We can use a test for independence to determine whether or not two categorical variables are related in a population. Finally, when examining the distribution of a single categorical variable in multiple populations or treatments, a test for homogeneity can be used to determine whether the distributions differ. In each test, we use the chi-square test statistic to measure how much the observed counts differ from the expected counts. When the test provides significant evidence against the null hypothesis, we can use a follow-up analysis to determine which component(s) contributed the most to the test statistic.

How well do you understand each of the learning targets?

Learning Target	Got It!	Almost There	Needs Work
I can state appropriate hypotheses for a test about the distribution of a categorical variable.			
I can calculate expected counts for a test about the distribution of a categorical variable.			
I can check the conditions for a test about the distribution of a categorical variable.			
I can calculate the test statistic and *P*- value for a test about the distribution of a categorical variable.			
I can perform a chi-square test for goodness of fit.			
I can state appropriate hypotheses for a test about the relationship between two categorical variables.			
I can calculate expected counts for a test about the relationship between two categorical variables.			
I can check the conditions for a test about the relationship between two categorical variables.			
I can calculate the test statistic and *P*-value for a test about the relationship between two categorical variables.			
I can perform a chi-square test for independence.			
I can perform a chi-square test for homogeneity.			
I can distinguish between a chi-square test for independence and a chi-square test for homogeneity.			

Unit 8 Multiple Choice Practice

Directions. Identify the choice that best completes the statement or answers the question. Check your answers and note your performance when you are finished.

1. To test the effectiveness of your calculator's random number generator, you randomly select 1000 numbers from a standard normal distribution. You classify these 1000 numbers according to whether their values are at most –2, between –2 and 0, between 0 and 2, or at least 2. The results are given in the following table. The expected counts, based on the empirical rule, are given as well.

	At most –2	**Between –2 and 0**	**Between 0 and 2**	**At least 2**
Observed Count	18	492	468	22
Expected Count	25	475	475	25

To test to see if the distribution of observed counts differs significantly from the distribution of expected counts, we can use a χ^2 goodness of fit test. For this test, the test statistic has approximately a χ^2 distribution. How many degrees of freedom does this distribution have?

(A) 3
(B) 4
(C) 7
(D) 999
(E) 1000

2. Refer to problem 1. Which of the following is the component of the χ^2 statistic corresponding to the category "at most −2"?

(A) $\frac{(43)(1000)}{2000}$

(B) $\frac{(43)(25)}{1000}$

(C) $\frac{18}{1000}$

(D) $\frac{(18-25)^2}{25}$

(E) $\frac{(18-25)^2}{18}$

3. Anne wants to know if students and teachers prefer different brands of frozen pizzas. She bakes four dozen pizzas made by each of four manufacturers, which she labels brands A, B, C, and D. She then selects a random sample of 20 students and 30 teachers, gives them one slice of each brand and asks which brand they like best. Here are the results:

	A	B	C	D	Total
Teachers	3	4	6	7	20
Students	12	5	6	7	30
Total	13	9	12	14	50

The appropriate null hypothesis for Anne's question in this problem is:

(A) There is an association between type of person (student/teacher) and preferred frozen pizza.
(B) Type of person (student/teacher) and pizza preference are independent.
(C) The distribution of preferred pizza for is different for each type of person (student/teacher).
(D) The observed count in each cell is equal to the expected count.
(E) There is no difference in the population distribution of pizza preference for students and teachers.

4. Refer to problem 3. Are the conditions for a chi-square test of association/independence met?
   (A) Yes, because the sample size is greater than 30.
   (B) Yes, because a simple random sample was selected.
   (C) No, because the distribution for each gender is different.
   (D) No, because not all of the observed counts are at least 5.
   (E) No, because not all of the expected counts are at least 5.

5. The table provided shows the individual components of the chi-square test of association for a study done on amount of time spent at a computer and whether or not a person wears glasses.

		Wear Glasses?	
		Yes	No
Amount of Screen time	Above Average	8.7	6.3
	Average	0.5	0.3
	Below Average	3.1	2.2

   Which of the following statements is supported by the information in this table?
   (A) Above-average screen time individuals wore glasses much less often than expected.
   (B) Average screen time individuals wore glasses much less often than expected.
   (C) Below-average screen time individuals wore glasses about as often as expected.
   (D) You can't determine this without the original observed counts.
   (E) The chi-square statistic for this test is about 3.5.

6. Refer to problem 5. What are the degrees of freedom of the chi-square test statistic?
   (A) 2
   (B) 3
   (C) 4
   (D) 5
   (E) 6

7. A random sample of 200 Canadian students were asked about their hand dominance and whether they suffer from allergies. Here are the results:

		Allergies?	
		Yes	No
Hand Dominance	Ambidextrous	12	7
	Left-handed	11	9
	Right-handed	95	66

   What can you conclude about the relationship between hand dominance and allergies?
   (A) There is not convincing evidence ($P = 0.13$) to conclude that there is a relationship between hand dominance and allergies.
   (B) There is convincing evidence ($P = 0.87$) to conclude that there is a relationship between hand dominance and allergies.
   (C) There is not convincing evidence ($P = 0.87$) to conclude that there is a relationship between hand dominance and allergies.
   (D) There is not convincing evidence ($P = 0.13$) to conclude that there is a relationship between hand dominance and allergies.
   (E) We cannot perform a chi-square test on these data.

8. Are birthdays equally distributed among the four seasons (winter, spring, summer, fall)? A curious statistics student believes that the birthdays for the students in her school are not evenly distributed among the four seasons. To test the claim, she obtains the birthdays of each student in a random sample of 80 students in his school. Her results are shown in this table.

Season	Winter	Spring	Summer	Fall
Number of birthday	16	21	23	20

Which test is appropriate to determine if the birthdays of all students at the school are not evenly distributed across the four seasons.
(A) Chi-square test for goodness of fit
(B) Chi-square test for independence
(C) Chi-square test for homogeneity
(D) One-sample t test for a mean
(E) One-sample z test for a proportion

9. Refer to problem 8. What is the expected number of students to have a birthday in the summer?
(A) 16 (B) 20 (C) 21 (D) 23 (E) 25

10. Refer to problem 8. What is value of the test statistic of the appropriate test?
(A) $\chi^2 = 1.3$
(B) $\chi^2 = 0$
(C) $\chi^2 = 1.241$
(D) $t = 1.578$
(E) $z = 1.960$

Check your answers below. If you got a question wrong, check to see if you made a simple mistake or if you need to study that concept more. After you check your work, identify the concepts you feel very confident about and note what you will do to learn the concepts in need of more study.

#	Answer	Concept	Right	Wrong	Simple Mistake?	Need to Study More
1	A	Degrees of Freedom				
2	D	Chi-square Components				
3	E	Chi-square test of Homogeneity				
4	E	Conditions for Chi-Square Procedures				
5	D	Follow-up Analysis				
6	A	Degrees of Freedom				
7	C	Chi-square Conclusions				
8	A	Distinguish Between the Chi-Square Tests				
9	B	Calculate an expected count				
10	A	Chi-Square Goodness of Fit Test Statistic				

FRAPPY! Free Response AP® Problem, Yay!

The following problem is modeled after actual Advanced Placement Statistics free response questions. Your task is to generate a complete, concise response in 15 minutes. After you generate your response, read over the solution and scoring guide. Score your response and note what, if anything, you would do differently to increase your own score.

A study was performed to determine whether or not the name of a course had an effect on student registrations. A statistics course in a large state university was given 4 different names in a course catalog. Each name corresponded to the exact same statistics course. A random sample of student registrations was recorded, and the results are given below:

Course Name	Number of Registrations
Statistical Applications	25
Statistical Reasoning	22
Statistical Analysis	30
The Practice of Statistics	40
Total	117

Do these data suggest the name of the course has an effect on student registrations? Conduct an appropriate statistical test to support your conclusion. Use $\alpha = 0.05$.

FRAPPY! Scoring Rubric

The following problem is modeled after actual Advanced Placement Statistics free response questions. Your task is to generate a complete, concise response in 15 minutes. After you generate your response, read over the solution and scoring guide. Score your response and note what, if anything, you would do differently to increase your own score.

Intent of the Question

The primary goals of this question are to assess your ability to use a chi-square test for goodness of fit to (1) state the appropriate hypotheses; (2) identify and compute the appropriate test statistic; (3) make a conclusion in the context of the problem.

Solution

The solution should contain:

- **Test and Hypotheses**: The test must be identified by name or formula as a chi-square goodness-of-fit test and hypotheses must be stated appropriately. H_0: Student registrations do not differ by course name and H_a: Student registrations do differ by course name.
- **Conditions**: Random sample was selected. It is reasonable to believe that 117 < 10% of all registrations for the course for this large state university. All expected counts are 29.25, which is at least 5.
- **Mechanics**: $\chi^2 = 6.38$ and P-value = 0.094 with 3 degrees of freedom.
- **Conclusion**: Because the P-value of $0.094 > \alpha = 0.05$ we fail to reject the null hypothesis. We do not have convincing evidence that the course name has an effect on the number of registrations.

Scoring

Each element scored as essentially correct (E), partially correct (P), or incorrect (I).

"Hypotheses and Name" is essentially correct if the hypotheses are written correctly, and the response correctly identifies the test by name or formula. This part is partially correct if the test isn't identified or if the hypotheses are written incorrectly. Incorrect otherwise.

"Conditions and Mechanics" is essentially correct if the response correctly does at least 4 of the 5 following things: (1) checks the Random condition, (2) checks the 10% condition, (3) checks the Large Counts condition, (4) correct test statistic, (5) correct P-value. This part is partially correct if 3 of those 3 things are done correctly. Incorrect otherwise.

"Conclusion" is essentially correct if the conclusion fails to reject the null hypothesis because the P-value is greater than a significance level of 0.05 and provides an interpretation in context. This part is partially correct if the response fails to justify the decision by comparing the P-value to a significance level OR if the conclusion lacks an interpretation in context. Incorrect otherwise.

Scoring

4 Complete Response
All three parts essentially correct

3 Substantial Response
Two parts essentially correct and one part partially correct

2 Developing Response
Two parts essentially correct and no parts partially correct
One part essentially correct and one or two parts partially correct
Three parts partially correct

1 Minimal Response
One part essentially correct and no parts partially correct
No parts essentially correct and two parts partially correct

My Score:
What I did well:
What I could improve:
What I should remember if I see a problem like this on the AP Exam:

Unit 8 Formula Sheet

Create a one-page summary of important concepts and formulas found in this unit. This will serve as a valuable resource as you prepare for the AP Exam.

Unit 8 Crossword Puzzle

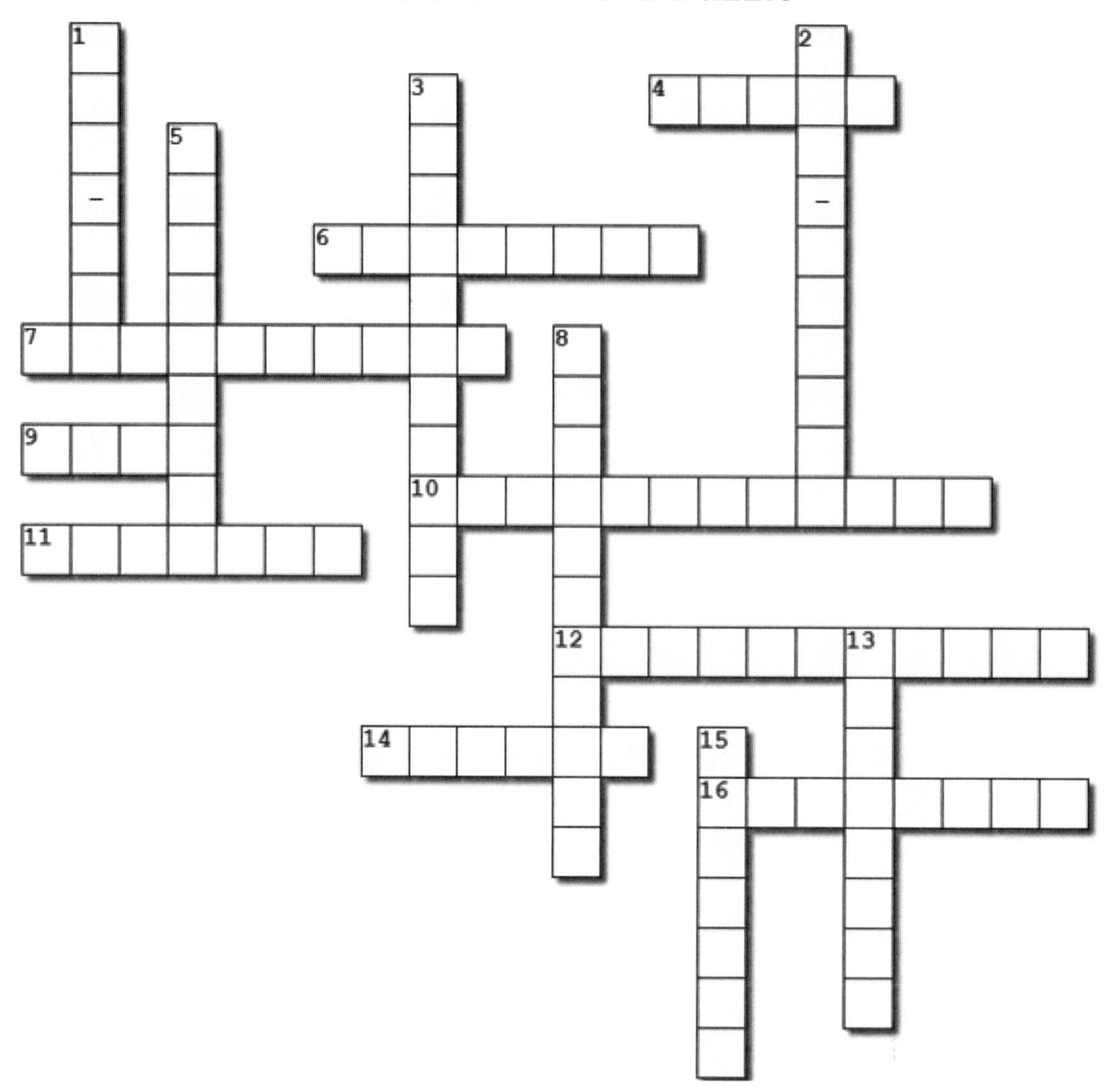

Across:

4. Chi-square distributions are a family of curves that are ____ skewed.
6. The df for a chi-square test of ____ of fit is number of categories – 1.
7. As with any significance test, we begin by stating ____.
9. In any chi-square tests, the expected counts must all be at least ____.
10. A chi-square test for ____ tests that there is no association between two categorical variables in the population of interest.
11. Always state the conclusion in ____.
12. Check the 10% condition when sampling without ____.
14. If the P-value $< \alpha$, you should ____ H_0.
16. The test statistic measures how different the ____ counts are from the expected counts, relative to the expected counts.

Down:

1. When comparing two or more categorical variables over multiple groups, arrange the data in a ___-____ table.
2. A ___-____ test statistic is used to test the hypotheses about distributions of categorical data.
3. Use a chi-square test for ____ to compare the distribution of a single categorical variable for multiple populations or groups.
5. A follow-up analysis involves identifying the ____ that contributed the most to the chi-square test statistic.
8. Chi-square tests allow for inference about ____ variables.
13. ____ counts are often not whole numbers and shouldn't be rounded to a whole number.
15. For a chi-square test of independence, df = (numbers of rows – 1)(number of ____ – 1).

Unit 9: Inference for Quantitative Data: Slopes

"By a small sample, we may judge the whole piece." Miguel de Cervantes (Don Quixote)

UNIT 9 OVERVIEW

In this unit, you will learn how to determine whether or not there is convincing evidence of a relationship between two quantitative variables. You will learn how to construct and interpret a confidence interval for the slope as well as how to perform a significance test for the slope. This final unit will refresh your memory on some of the key concepts of regression while introducing you to some new inference techniques. As you wrap up your studies and begin your preparations for the AP® exam, look back at the learning targets to determine what topics need a little extra review!

Sections in Unit 9
Section 9A: Confidence Intervals for the Slope of a Population Regression Line
Section 9B: Significance Tests for the Slope of a Population Regression Line

PRACTICE FOR MASTERY

Use the following *suggested* guide to the pages and exercises in your text to practice for mastery!
Note: your teacher may assign different problems. Be sure to follow their instructions!

Day	Topics	Read	Do
1	◆ Sampling Distribution of the Sample Slope b ◆ Checking Conditions for a Confidence Interval for β	p. 802-811	**9A:** 1, 3, 5, 19, 23, 25
2	◆ Calculating a Confidence Interval for β ◆ t Interval for the Slope	p. 811-817	**9A:** 7, 9, 13, 15
3	◆ Stating Hypotheses and Checking Conditions for a Test about β ◆ Calculating the Standardized Test Statistic and P-value for a Test about β ◆ t Test for the Slope	p. 825-835	**9B:** 1, 3, 5, 7, 9, 13, 21
4	Unit 9 AP® Statistics Practice Test (p. 847)		Unit 9 Review Exercises (p.845)
5	Unit 9 Test: Celebration of learning!		Cumulative AP® Practice Test 4 (p.849)

Section 9A: Confidence Intervals for the Slope of a Population Regression Line

Section Summary

When you construct a least-squares regression line based on sample data, the line approximates the population (true) regression line. If you take a different sample and construct a least-squares regression line, it is very likely you will end up with a slightly different line. In this section, you will learn about the sampling distribution of the sample slope b so you can perform inference about the true slope of the regression line. As you did with other inference procedures, you will start by learning how to check the conditions for performing inference. Then you will learn how to construct and interpret a confidence interval for the true slope. You will also revisit computer output for a linear regression analysis. It is important that you are able to interpret computer output as a number of AP® exam questions involve standard computer output for regression.

Learning Targets:

____ I can describe the sampling distribution of the sample slope b.
____ I can check the conditions for calculating a confidence interval for the slope β of a population regression line.
____ I can calculate a confidence interval for the slope of a population regression line.
____ I can construct and interpret a t interval for the slope.

Read: Vocabulary and Key Concepts

- When two quantitative variables x and y have a linear relationship, we can model that relationship with the least-squares regression line $\hat{y} =$ __________.
- If the data come from a random sample from the population of interest or from a randomized experiment, we may want to use the ______________ regression line to make an inference about the ________________ regression line.
- A regression line calculated from _______ value in the population is called a _________________ regression line. The equation of a population regression line is $\mu_y = \alpha + \beta x$, where
  - μ_y is the ________ y-value for a given value of x.
  - α is the population y __________.
  - β is the population __________.
- A regression line calculated from ____________ data is called a ________________ regression line. The equation of a sample regression line is $\hat{y} = a + bx$, where
  - $\hat{y}$ is the _____________ y-value or the estimated mean y-value for a given value of x.
  - a is the sample y ___________.
  - b is the sample ___________.

Sampling Distribution of the Sample Slope *b*

- Let b be the ________ of the ______________ regression line in an SRS of n observations (x, y) from a population of size N with regression line $\mu_y = \alpha + \beta x$. Then:
  - The ________ of the sampling distribution of b is $\mu_b =$ ____.
  - When the 10% condition ($n < 0.10N$) is met, the ____________________ of the sampling distribution of b is approximately $\sigma_b = \frac{\sigma}{\sigma_x\sqrt{n}}$, where σ is the ______________ standard deviation of the ____________ and σ_x is the ______________ standard deviation of the explanatory variable, x.
- The value μ_b gives the __________ value of b for all possible samples of a given size from a population. Because $\mu_b = \beta$, we know that the slope b of the sample regression line is an __________ estimator of the slope β of the population regression line.
- The value σ_b measures the ____________ distance between a sample slope b and the population slope β for ________________________ of a given size from a population. Because the sample size appears in the denominator of the formula for σ_b, we know that the sample slope b is less variable when the sample size is larger.
- When we sample _________ replacement, the standard deviation of the sampling distribution of b is exactly $\sigma_b = \frac{\sigma}{\sigma_x\sqrt{n}}$. When we sample ____________ replacement, the observations are not independent and the actual standard deviation of the sampling distribution of b is ____________ than the value given by the formula. However, if the sample size is ____________ 10% of the population size (the 10% condition), the value given by the formula is nearly correct.

Checking Conditions for a Confidence Interval for *β*

Check that the following conditions are met:

- ___________: The form of the relationship between x and y is linear. For each value of x, the mean value of y (denoted by μ_y) falls on the population regression line $\mu_y = \alpha + \beta x$.
- ___________: For each value of x, the distribution of y is approximately normal.
- ___________: For each value of x, the standard deviation of y (denoted by σ) is the same.
- ___________: The data come from a random sample from the population of interest or a randomized experiment.
  - _____: When sampling without replacement, $n < 0.10N$.

How to check the conditions:

- **Linear:** Examine a ______________ to see if the overall pattern is roughly linear. Make sure there are no leftover ____________ patterns in the ___________ plot.

- **Normal:** Make a dotplot, stemplot, histogram, or boxplot of the ______________ and check for no strong ______________ and no __________.
- **Equal SD:** Look at the scatter of the residuals above and below the "residual = 0" line in the ____________ plot. The variability of the residuals in the ______________ direction should be roughly the same from the smallest to the largest x-value.
- **Random:** Ensure that the data came from a ______________ sample from the population of interest or a ________________ experiment.
- **10%:** If sampling is done _____________ replacement, check that the sample size n is less than 10% of the population size N.

Calculating a Confidence Interval for β

- When the conditions are met, the sampling distribution of the sample slope b is approximately _________ with mean $\mu_b = \beta$ and standard deviation $\sigma_b = \frac{\sigma}{\sigma_x \sqrt{n}}$.
- Two issues typically keep us from using this standard deviation formula in practice.
  1. We don't know σ, so we __________ it with s.
  2. We don't know σ_x so we estimate it with s_x.
- Our resulting estimate for the standard deviation of the sampling distribution of b is the _______________________ of the sample slope b: $\sigma_b = \frac{s}{s_x \sqrt{n}}$.
- We interpret s_b like any standard error: It measures how far the ________ slope typically __________ from the _______________ slope if we repeat the data collection process ___________ times.
- When the conditions are met, a C% confidence interval for the slope β of the population regression line is

  Write the formula here: _____________________

  where t^* is the critical value for the t distribution with $n - 2$ degrees of freedom and C% of the area between $-t^*$ and t^*.

Watch: Go to bfwpub.com/TPS7eStrive. Select Unit 9 and watch the following Example videos:

Video	Topic	Done
Unit 9A, Predicting Old Faithful (p. 806)	*Sampling distribution of the sample slope b*	
Unit 9A, Studying ponderosa pines (p. 809)	*Checking conditions for a confidence interval for β*	
Unit 9A, Studying ponderosa pines (p. 813)	*Calculating a confidence interval for β*	
Unit 9A, How much is that truck worth? (p. 814)	*t interval for the slope*	

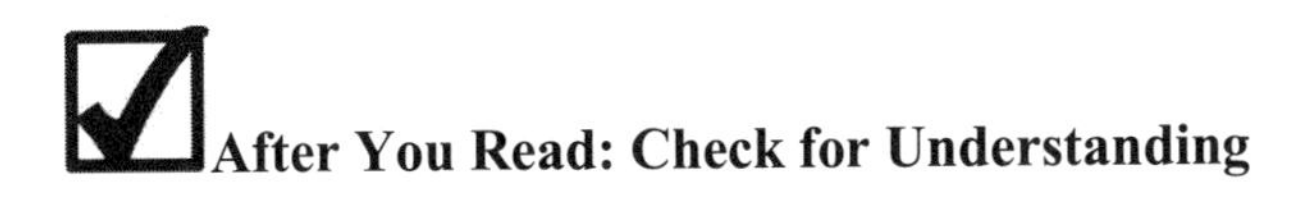

After You Read: Check for Understanding

Concept 1: The Sampling Distribution of b

If we take repeated samples of the same size n from a population with a true regression line $\mu_y = \alpha + \beta x$ and determine the sample regression line $\hat{y} = a + bx$, the sampling distribution of the slope b will have the following characteristics:

1) The shape of the sampling distribution will be approximately normal.
2) The mean of the sampling distribution is $\mu_b = \beta$.
3) The standard deviation of the sampling distribution of b is $\sigma_b = \dfrac{\sigma}{\sigma_x \sqrt{n}}$

   where σ is the population standard deviation of the residuals and σ_x is the population standard deviation of the explanatory variable, x.

Check for Understanding – Learning Target 1

____ *I can describe the sampling distribution of the sample slope b.*

For the population of the 100 most popular dog breeds this year, there is a negative linear relationship between x = average weight (in pounds) and y = average lifespan (in years). The population regression line is $\mu_y = 13.44 - 0.023x$, the standard deviation of the residuals is $\sigma = 1.005$ years, and the standard deviation of average weight is $\sigma_x = 37.62$ pounds.

Imagine selecting a random sample of 5 dog breeds and using the sample data to calculate b = the slope of the sample regression line for predicting y = average lifespan from x = average weight.

1. Calculate and interpret the mean of the sampling distribution of b.

2. Verify that the 10% condition is met. Then calculate and interpret the standard deviation of the sampling distribution of b.

Concept 2: Conditions for Calculating a Confidence Interval for β

If the conditions for performing inference are met, we can use the slope b of the sample regression line $\hat{y} = a + bx$ to estimate the slope β of the population (true) regression line $\mu_y = \alpha + \beta x$. The conditions are:

- **Linear**: The form of the relationship between x and y is linear. For each value of x, the mean value of y (denoted by μ_y) falls on the population regression line $\mu_y = \alpha + \beta x$.
  Is a scatterplot of the sample data linear?
- **Normal**: For each value of x, the distribution of y is approximately normal.
  Does a dotplot of the residuals show no strong skewness and no outliers?
- **Equal SD**: For each value of x, the standard deviation of y (denoted by σ) is the same.
  In the residual plot, is the amount of scatter above and below the line residual = 0 roughly the same from the smallest to largest x-values?
- **Random**: The data come from a random sample from the population of interest or a randomized experiment. *How were the data produced?*
  - **10%**: When sampling without replacement, $n < 0.10N$.
    If sampling without replacement, is the sample size less than 10% of the size of the population?

Check for Understanding – Learning Target 2

____ *I can check the conditions for calculating a confidence interval for the slope β of a population regression line.*

A study by *Consumer Reports* rated 10 randomly selected cereals on a 100-point scale (higher numbers are better) and recorded the number of grams of sugar in each serving. The scatterplot shows the relationship between x = amount of sugar (g) and y = rating for a random sample of 10 cereals, along with the least-squares regression line, the corresponding residual plot, and histogram of residuals. Check if the conditions for constructing a confidence interval for the slope of the population regression line are met.

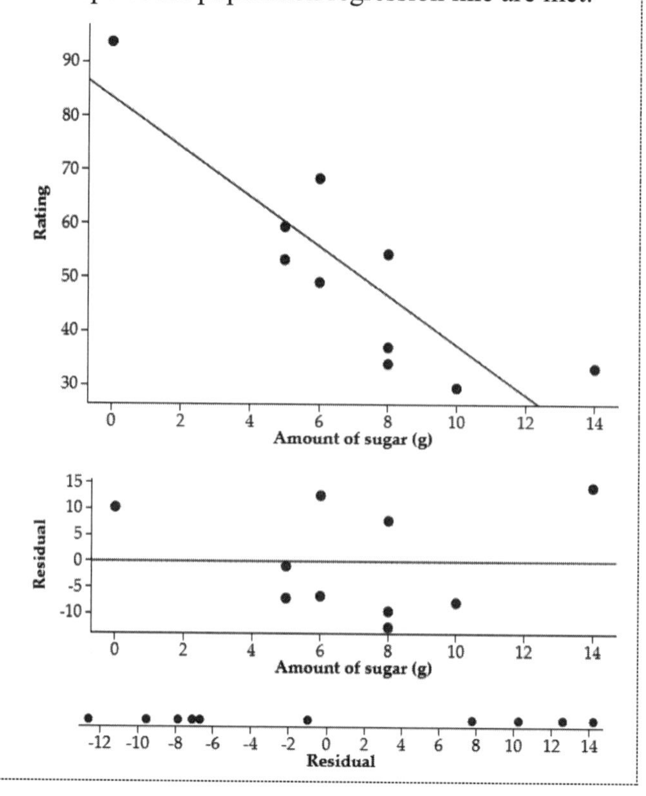

Concepts 3 and 4: Confidence Intervals for β

When we construct a least-squares regression line $\hat{y} = a + bx$ and the conditions noted above are met, b is our estimate of the true slope β. A level C confidence interval for the true slope has the form $b \pm t^*SE_b$ where t^* is the critical value for the t distribution with $n - 2$ degrees of freedom that has area C between $-t^*$ and t^*.

Check for Understanding – Learning Targets 3 and 4

____ *I can calculate a confidence interval for the slope of a population regression line.*

____ *I can construct and interpret a t interval for the slope.*

A study by *Consumer Reports* rated 10 randomly selected cereals on a 100-point scale (higher numbers are better) and recorded the number of grams of sugar in each serving. The data from the study are below:

Sugar	6.0	8.0	5.0	0.0	8.0	10.0	14.0	8.0	6.0	5.0
Rating	68.40	33.98	59.43	93.70	34.38	29.51	33.17	37.04	49.12	53.31

Here is computer output summarizing the relationship between x = amount of sugar (g) and y = rating for these cereals.

```
Predictor             Coef      SE Coef       T          P
Constant             83.4495    7.70749    10.82707    <0.001
Amount of sugar (g)  -4.6065    0.98684    -4.66792    0.00161
S = 10.8103             R-Sq = 73.1%      R-Sq(adj) = 72.6%
```

Construct and interpret a 99% confidence interval for the true slope of the regression line. Assume the conditions for performing inference are met.

Check your calculations by using the data provided and your calculator to compute the confidence interval. (*Need Tech Help?* View the video **Tech Corner 29: Confidence Intervals for the Slope** at bfwpub.com/TPS7eStrive)

Section 9B: Significance Tests for the Slope of a Population Regression Line

Section Summary

You have made it to the final topic in your study of AP® Statistics! You will finish your study of inference by learning how to carry out a significance test for the slope of a population regression line, β. When the conditions for inference are met, not only can we estimate the population slope from b, we can also test whether or not a specified value for β is plausible. The process for testing a claim about a population (true) slope follows the same format used for other significance tests.

Learning Targets:

____ I can state appropriate hypotheses and check the conditions for a test about the slope of a population regression line.

____ I can calculate the standardized test statistic and *P*-value for a test about the slope of a population regression line.

____ I can perform a *t* test for the slope.

Read: Vocabulary and Key Concepts

Stating Hypotheses and Checking Conditions for a Test about β

- The null hypothesis usually says that there is ____ linear association between the explanatory and response variables. The alternative hypothesis can say that there is a ____________ association, a ____________ association, or ___ association (either positive or negative) between the two variables.
- If there is no ___________ association between the *x* and *y* variables, the slope of the population least-squares regression line is ____. Consequently, our usual _______ hypothesis is H_0: $\beta = 0$.
- Put another way, H_0 says that linear regression of *y* on *x* is no better for ____________ *y* than the ______ of the ___________ variable $\bar{y}$.

Conditions for Performing a Significance Test about the Slope of a Population Regression Line:

- **Linear**: The ______ of the relationship between *x* and *y* is __________. For each value of *x*, the mean value of *y* (denoted by μ_y) falls on the _____________ regression line $\mu_y = \alpha + \beta x$.
- **Normal**: For each value of *x*, the distribution of *y* is ___________________________.
- **Equal SD**: For each value of *x*, the ___________________________ of *y* (denoted by σ) is the same.
- **Random**: The data come from a random sample from the ______________________________ or a randomized _______________.
  - **10%**: When _________ without ________________, $n < 0.10N$.

Calculating the Standardized Test Statistic and *P*-Value for a Test about β

- Write the general formula: standardized test statistic = ________________
- Write the specific formula for a test about β: t = ________
- In most cases, we want to test the ____ hypothesis that the ________ of the ______________ regression line is $\beta = 0$. However, sometimes we might use a null value different than ____.
- Regression output from statistical software gives the value of t for a test of H_0: β = ___ by default. If you want to test a null hypothesis other than H_0: β = ___, get the slope and standard error from the output and use the formula to calculate the t statistic.
- When the conditions are met for a test about the slope of a population regression model, we use a t distribution with _______ degrees of freedom to calculate the ________.
- Notice that the *P*-value reported in the computer output is for a _____________ test. To get the correct *P*-value for a _____________ test when there is some evidence for H_a, divide the *P*-value given in the computer output by ____.
- In some contexts, it makes sense to fit a regression model that goes through the point (0, 0). Because there is only _____ parameter to estimate in this case (the ________), the degrees of freedom would be df = _______.

Think About It:

- Testing the null hypothesis H_0: $\beta = 0$ is exactly the same as testing that there is no _____________ between x and y in the ________________ from which we selected our data.
- You can use the test for zero slope to test the hypothesis H_0: $\rho = 0$ of zero _______________ between any two quantitative variables.

Watch: Go to bfwpub.com/TPS7eStrive. Select Unit 9 and watch the following Example videos:

Video	Topic	Done
Unit 9B, Can foot length predict height? (p. 829)	*Stating hypotheses and checking conditions for a test about* β	
Unit 9B, Can foot length predict height for high schoolers? (p. 831)	*Calculating the standardized test statistic and P-value for a test about* β	
Unit 9B, Can infant crying predict aptitude later in life? (p. 832)	*t test for the slope*	

Do: Check for Understanding

Concepts 1, 2, and 3: Significance Test for β

The process for testing a claim about a population (true) slope follows the same format used for other significance tests.

1) **STATE: State the hypotheses, parameter(s), and significance level.**
   The null hypothesis states that the population (true) slope is equal to a particular value (usually zero) while the alternative hypothesis is that the population (true) slope is greater than, less than, or not equal to that value:
   H_0: $\beta = 0$
   H_a: $\beta > 0$ OR $\beta < 0$ OR $\beta \neq 0$
   Define the parameter, β. State a significance level.

2) **PLAN: Identify the appropriate inference method and check the conditions.**
   - **Linear**: The form of the relationship between x and y is linear. For each value of x, the mean value of y (denoted by μ_y) falls on the population regression line $\mu_y = \alpha + \beta x$.
     Is a scatterplot of the sample data linear?
   - **Normal**: For each value of x, the distribution of y is approximately normal.
     Does a dotplot of the residuals show no strong skewness and no outliers?
   - **Equal SD**: For each value of x, the standard deviation of y (denoted by σ) is the same.
     In the residual plot, is the amount of scatter above and below the line residual = 0 roughly the same from the smallest to largest x-values?
   - **Random**: The data come from a random sample from the population of interest or a randomized experiment. *How were the data produced?*
     - **10%**: When sampling without replacement, $n < 0.10N$.
       If sampling without replacement, is the sample size less than 10% of the size of the population?

3) **DO: If the conditions are met, perform calculations.**
   Compute the t test statistic: $t = \dfrac{b - \beta_0}{s_b}$
   Find the P-value by calculating the probability of observing a t statistic at least this extreme in the direction of the alternative hypothesis in a t distribution with $n - 2$ degrees of freedom. If you are given the sample data, you can use your calculator to carry out the calculations.
   (*Need Tech Help?* View the video **Tech Corner 30: Significance Test for the Slope** at bfwpub.com/TPS7eStrive).

4) **CONCLUDE: Make a conclusion about the hypotheses in the context of the problem.**
   If the P-value is smaller than the stated significance level, reject H_0. You have convincing evidence for H_a. If the P-value is larger than the significance level, fail to reject H_0. You do not have convincing evidence for H_a.

As with other inference situations, a significance test can tell us whether or not a claim about the parameter is plausible, while using a confidence interval can give us additional information about its true value.

Check for Understanding – Learning Targets 1, 2, and 3

____ *I can state appropriate hypotheses and check the conditions for a test about the slope of a population regression line.*

____ *I can calculate the standardized test statistic and P-value for a test about the slope of a population regression line.*

____ *I can perform a t test for the slope.*

The study in the previous Check for Understanding (Section 9A) was expanded to include a total of 77 randomly selected cereals. The scatterplot, residual plot, and computer output of the regression analysis are given here. A dotplot of the residuals (not shown here) shows no strong skewness or outliers.

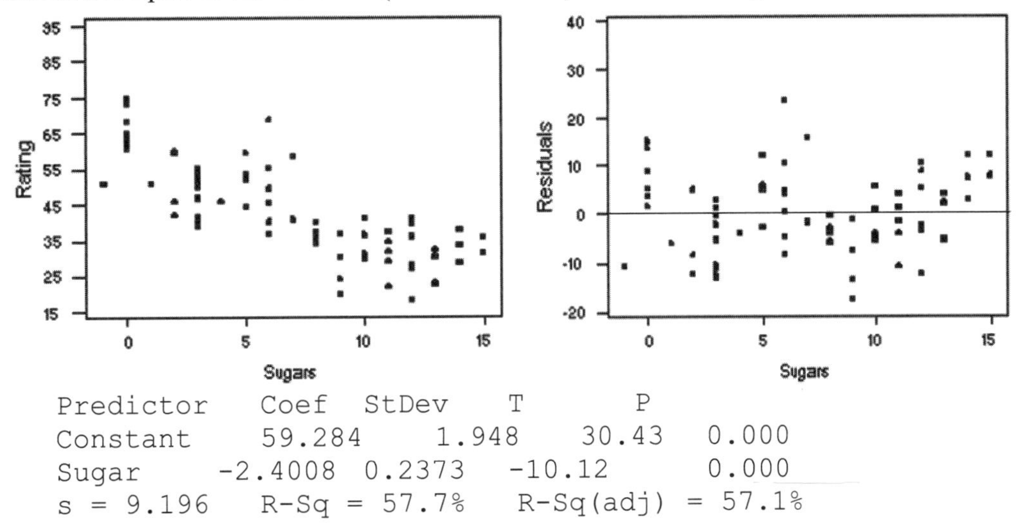

Predictor	Coef	StDev	T	P
Constant	59.284	1.948	30.43	0.000
Sugar	-2.4008	0.2373	-10.12	0.000

s = 9.196   R-Sq = 57.7%   R-Sq(adj) = 57.1%

Do these data provide convincing evidence that the slope of the population regression line relating x = amount of sugar to y = rating is less than zero? Use $\alpha = 0.01$.

Unit 9 Summary: Inference for Quantitative Data: Slopes

In this unit, you learned how to apply your knowledge of inference to linear relationships. When you find a least-squares regression line for a set of sample data, you are constructing a model that approximates the true relationship between x and y. By considering the sampling distribution of b, you can construct a confidence interval for the slope of the true regression line and also test claims about the slope.

How well do you understand each of the learning targets?

Learning Target	Got It!	Almost There	Needs Work
I can describe the sampling distribution of the sample slope b.			
I can check the conditions for calculating a confidence interval for the slope β of a population regression line.			
I can calculate a confidence interval for the slope of a population regression line.			
I can construct and interpret a t interval for the slope.			
I can state appropriate hypotheses and check the conditions for a test about the slope of a population regression line.			
I can calculate the standardized test statistic and P-value for a test about the slope of a population regression line.			
I can perform a t test for the slope.			

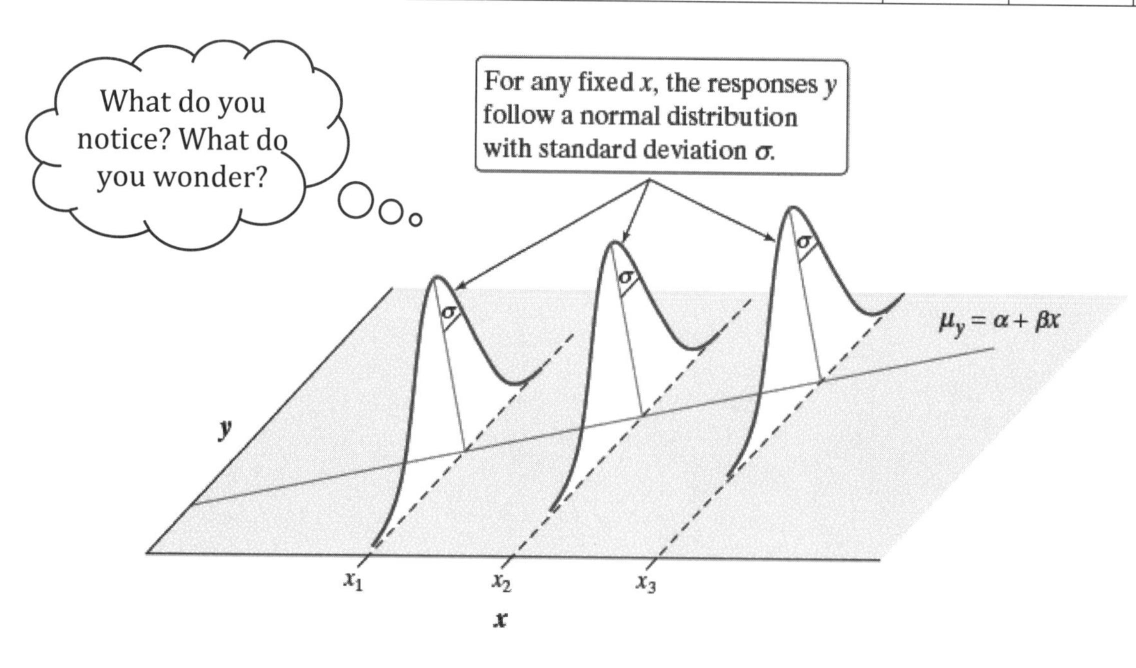

Unit 9 Multiple Choice Practice

Directions. Identify the choice that best completes the statement or answers the question. Check your answers and note your performance when you are finished.

1. A high school guidance counselor wonders if it is possible to predict a student's GPA in their senior year from their GPA in the first marking period of their freshman year. She selects a random sample of 15 seniors from the 348 current seniors and records their full-year GPA in their senior year and first marking period GPA in their freshman year ("Fresh"). A computer regression analysis is given below.

```
Predictor Coef   SE Coef   T   P
Constant  1.6310 0.5328  3.06 0.009
Fresh   0.5304 0.1789  2.96 0.011

S = 0.3558 R-Sq = 40.3% R-Sq(adj) = 35.7%
```

Which of the following is an estimate of the standard error of the slope of the regression line?
(D) 0.1789 (B) 0.3558 (C) 0.5304 (D) 0.5328 (E) 1.6310

2. Refer to Question 1. Here is a residual plot created using the least-squares regression line relating x = first marking period GPA in the freshman year and y = full-year GPA in their senior year.

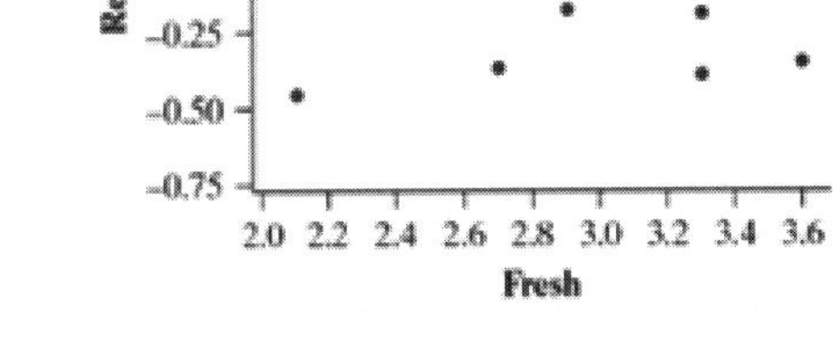

Enough information is provided above to confirm that four conditions for slope inference have been satisfied, but we need a further analysis of the data to confirm that the fifth condition for inference has been satisfied.

Which condition requires further analysis of the data?
(A) Mean Senior GPA is a linear function of Freshman GPA.
(B) The individual observations are independent of each other.
(C) For each value of Freshman GPA, the distribution of Senior GPA is roughly normal.
(D) The standard deviation of Senior GPA is roughly equal for each value of Freshman GPA.
(E) The data come from a random sample from the population of interest.

3. Refer to Questions 1 and 2. Assume the conditions for inference are met. What is the appropriate value of t^* for finding a 90% confidence interval for β, the population slope relating x = first marking period GPA in the freshman year and y = full-year GPA in their senior year for all seniors at this high school?
(A) 1.645 (B) 1.746 (C) 1.753 (D) 1.761 (E) 1.771

4. Refer to Questions 1-3. Which of the following represents a 90% confidence interval for β, the population slope relating x = first marking period GPA in the freshman year and y = full-year GPA in their senior year for all seniors at this high school?
(A) (0.236, 0.825)
(B) (0.692, 2.570)
(C) (0.758, 2.504)
(D) (0.214, 0.847)
(E) (0.687, 2.575)

5. Fitness can be measured by the rate of oxygen consumption (%) during exercise, with more fit people having higher rates. This measurement is quite costly to obtain, and so a study was done to see if this measurement could be predicted from the time it takes (in minutes) to run 1500 meters. Oxygen consumption was measured on a random sample of thirty students, and their finish times on a 1500 m run were recorded. Some of the results of a computer regression analysis are given below.

```
Dependent variable is Oxygen Consumption
R-squared = 72.9%
s = 2.81737 with 30 - 2 = 28 degrees of freedom

Variable      Coefficient    s.e. of Coefficient    P-value
Constant      81.589         3.961                  ≤ 0.0001
Run Time      -3.2088        0.3701                 ≤ 0.0001
```

Which of the following is an appropriate interpretation of the number –3.2088?

(A) For each additional one-minute increase in run time decreases predicted oxygen consumption by 3.2088%, on average.
(B) An increase in oxygen consumption of one-minute decreases predicted run time by 3.2088 minutes, on average.
(C) An increase in run time of one-minute increases predicted oxygen consumption by 3.2088%, on average.
(D) An increase in run time of one-minute is associated with a decrease in predicted oxygen consumption of about 3.2%, on average.
(E) A run time of 10 minutes corresponds to predicted oxygen consumption of 32.088, on average.

6. Refer to Question 5. Which of the following expressions represents the margin of error of a 99% confidence interval for the slope of the population regression line?

(A) $2.576\left(\frac{0.3701}{\sqrt{28}}\right)$ (B) 2.576(0.3701) (C) $2.763\left(\frac{0.3701}{\sqrt{28}}\right)$
(D) $2.763\sqrt{0.3701}$ (E) 2.763(0.3701)

7. A psychologist studies the relationship between a person's annual income and their happiness as rated on a scale of 1-10, where 10 is very happy. He selects a random sample of 15 adults and records their annual income and how happy they are. Assume the conditions for regression inference are met. Is there convincing evidence that happiness increases as annual income increases? To answer this question, test the hypotheses:

(A) $H_0 : \beta = 0$ versus $H_a : \beta > 0$. (B) $H_0 : \beta = 0$ versus $H_a : \beta < 0$.
(C) $H_0 : \beta = 0$ versus $H_a : \beta \neq 0$. (D) $H_0 : \beta > 0$ versus $H_a : \beta = 0$.
(E) $H_0 : \beta = 1$ versus $H_a : \beta > 1$.

8. A history teacher who is concerned about how long her students spend writing papers for her class asks all of her students how long (in minutes) they spent on a recent essay. She then performs a linear regression analysis, with writing time as the explanatory variable and essay score (out of 100 points) as the response variable. The residual plot for the data shows a random scatter of points and that there is no evidence of changes in the standard deviation of essay score for different values of writing time. Nevertheless, her friend who teaches statistics tells her it would not be appropriate to test the null hypothesis that the slope of the population regression line is 0. Why not?

(A) The variable "essay score" is a discrete variable.
(B) She does not have evidence that the relationship between writing time and essay score is linear.
(C) The variable "writing time" is categorical.
(D) There are too many uncontrolled variables, such as student ability and motivation.
(E) It is not appropriate for the teacher to use regression inference when she has information about the entire population.

9. Can we predict annual household electricity costs in a specific region from the number of rooms in the house? Below is a scatterplot of annual electricity costs (in dollars) versus number of rooms for 30 randomly selected houses in Michigan, along with computer output for linear regression of electricity costs on number of rooms.

```
Predictor  Coef SE Coef   T    P
Constant  406.9   164.8 2.47 0.020
Rooms    58.45  24.77 2.36 0.026

S = 246.735 R-Sq = 16.6% R-Sq(adj) = 13.6%
```

If we test the hypotheses $H_0 : \beta = 0$, $H_a : \beta \neq 0$, what is the value of the test statistic?
(A) 0.026 (B) 2.36 (C) 2.47 (D) 24.77 (E) 58.45

10. Refer to Question 9. If we test the hypotheses H_0: $\beta = 0$ vs. H_a: $\beta > 0$ at the $\alpha = 0.05$ level. Which of the following is the appropriate conclusion?
(A) Because the *P*-value of 0.020 is less than α, we reject H_0. There is convincing evidence of a negative linear relationship between annual electricity costs and number of rooms in the population of Michigan homes.
(B) Because the *P*-value of 0.020 is greater than α we fail to reject H_0. We do not have enough evidence to conclude that there is a positive linear relationship between annual electricity costs and number of rooms in the population of Michigan homes.
(C) Because the *P*-value of 0.026 is greater than α, we accept H_0. We have convincing evidence that there is not a linear relationship between annual electricity costs and number of rooms in the population of Michigan homes.
(D) Because the *P*-value of 0.026 is less than α, we accept H_0. We have convincing evidence that there is not a linear relationship between annual electricity costs and number of rooms in the population of Michigan homes.
(E) Because the *P*-value of 0.026 is less than α, we reject H_0. We have convincing evidence of a positive linear relationship between annual electricity costs and number of rooms in the population of Michigan homes.

Check your answers below. If you got a question wrong, check to see if you made a simple mistake or if you need to study that concept more. After you check your work, identify the concepts you feel very confident about and note what you will do to learn the concepts in need of more study.

#	Answer	Concept	Right	Wrong	Simple Mistake?	Need to Study More
1	A	Standard Error of the Slope				
2	C	Conditions for Inference				
3	E	Appropriate Value of t^*				
4	D	CI for β				
5	A	Interpretation of slope				
6	E	Margin of Error				
7	A	Stating the Hypotheses				
8	E	Conditions for Inference				
9	B	Standardized Test Statistic				
10	E	Significance Test from Output				

FRAPPY! Free Response AP® Problem, Yay!

The following problem is modeled after actual Advanced Placement Statistics free response questions. Your task is to generate a complete, concise response in 15 minutes. After you generate your response, read over the solution and scoring guide. Score your response and note what, if anything, you would do differently to increase your own score.

Professor Bready asked a random sample of 10 of the 250 students in her introductory statistics class to record the total amount of time (in hours) they spent studying for a particular test and also recorded the students' scores on the test. She then performed a regression analysis on the data. Below is numerical and graphical output from her computer software.

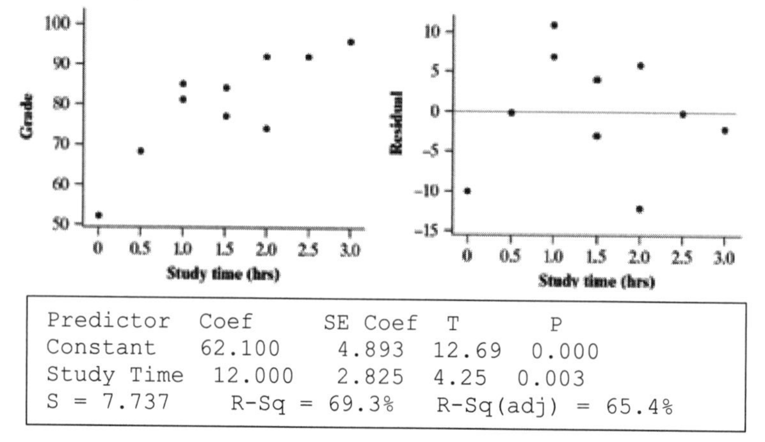

Predictor	Coef	SE Coef	T	P
Constant	62.100	4.893	12.69	0.000
Study Time	12.000	2.825	4.25	0.003

S = 7.737   R-Sq = 69.3%   R-Sq(adj) = 65.4%

(a) Check whether the conditions for regression inference have been met. If you do not have enough information to check a condition, describe what further information would be required.

(b) Assume all conditions for inference are met. Do these data provide convincing evidence at the $\alpha = 0.05$ significance level of a positive linear relationship between study time and test score in the population of all of professor Bready's introductory statistics students?

(c) Do the data establish that studying more will cause an improvement in grades? Why or why not?

FRAPPY! Scoring Rubric

Use the following rubric to score your response. Each part receives a score of "Essentially Correct," "Partially Correct," or "Incorrect." When you have scored your response, reflect on your understanding of the concepts addressed in this problem. If necessary, note what you would do differently on future questions like this to increase your score.

Intent of the Question

The primary goals of this question are to assess your ability to (1) interpret standard computer output; (2) conduct a t test for β; (3) draw a conclusion about cause-and-effect.

Solution

(a) **Linear:** The scatterplot shows a linear relationship between study time and grade. ✓
Normal: We need a graphical display of the distribution of the residuals, such as a dotplot or histogram, in order to see if there is no strong skewness or outliers. ✓
Equal SD: The amount of scatter of points around the residual = 0 line appears to be about the same at the different values of study time. ✓
Random: The 10 students were randomly selected from the population of all students in the introductory statistics class. ✓
10%: 10 < 10% of the 250 students in this introductory statistics class. ✓

(b) **STATE:** H_0: $\beta = 0$, H_a: $\beta \neq 0$ where β = the slope of the population regression line relating y = score to x = study time in the population of all introductory statistics students. Use $\alpha = 0.05$.
PLAN: t test for the slope. Assume the conditions for inference are met. ✓
DO: According to the output, the test statistic is $t = 4.25$ and the two-sided P-value is 0.003.
CONCLUDE: Because the P-value of $0.003 < \alpha = 0.05$, we reject H_0. There is convincing evidence of a positive linear relationship between study time and test score in the population of all professor Bready's introductory statistics students.

(c) No. There was no random assignment of the students to study times because this is not a controlled experiment, so we cannot conclude that studying more will cause an improvement in grades. There may be other variables that are confounded with study time.

Scoring:

Each element scored as essentially correct (E), partially correct (P), or incorrect (I).

Part (a) is Essentially correct if at least 4 of the 5 conditions are correctly checked. Partially correct if 3 of the 5 conditions are correctly checked. Incorrect otherwise. Note: Simply stating a condition is needed does not earn credit. A check mark, "yes" or other notation for each condition being met is required.

Part (b) is Essentially correct if at least 4 of the 5 following items are correctly stated: (1) Hypotheses with parameter defined in context, (2) name of the test, (3) value of the test statistic and P-value, (4) P-value compared to α with correct decision, (5) conclusion in context. Partially correct if 3 of the 5 items are correct. Incorrect otherwise.

Part (c) is Essentially correct if the response indicates "no" with justification (in context) appealing to the fact that an experiment was not conducted. Partially correct if the response indicates "no" but lacks proper justification or context. Incorrect otherwise.

Scoring

4 Complete Response

All three parts essentially correct

3 Substantial Response

Two parts essentially correct and one part partially correct

2 Developing Response

Two parts essentially correct and no parts partially correct
One part essentially correct and one or two parts partially correct
Three parts partially correct

1 Minimal Response

One part essentially correct and no parts partially correct
No parts essentially correct and two parts partially correct

My Score:
What I did well:
What I could improve:
What I should remember if I see a problem like this on the AP Exam:

Unit 9 Formula Sheet

Create a one-page summary of important concepts and formulas found in this unit. This will serve as a valuable resource as you prepare for the AP Exam.

Unit 9 Crossword Puzzle

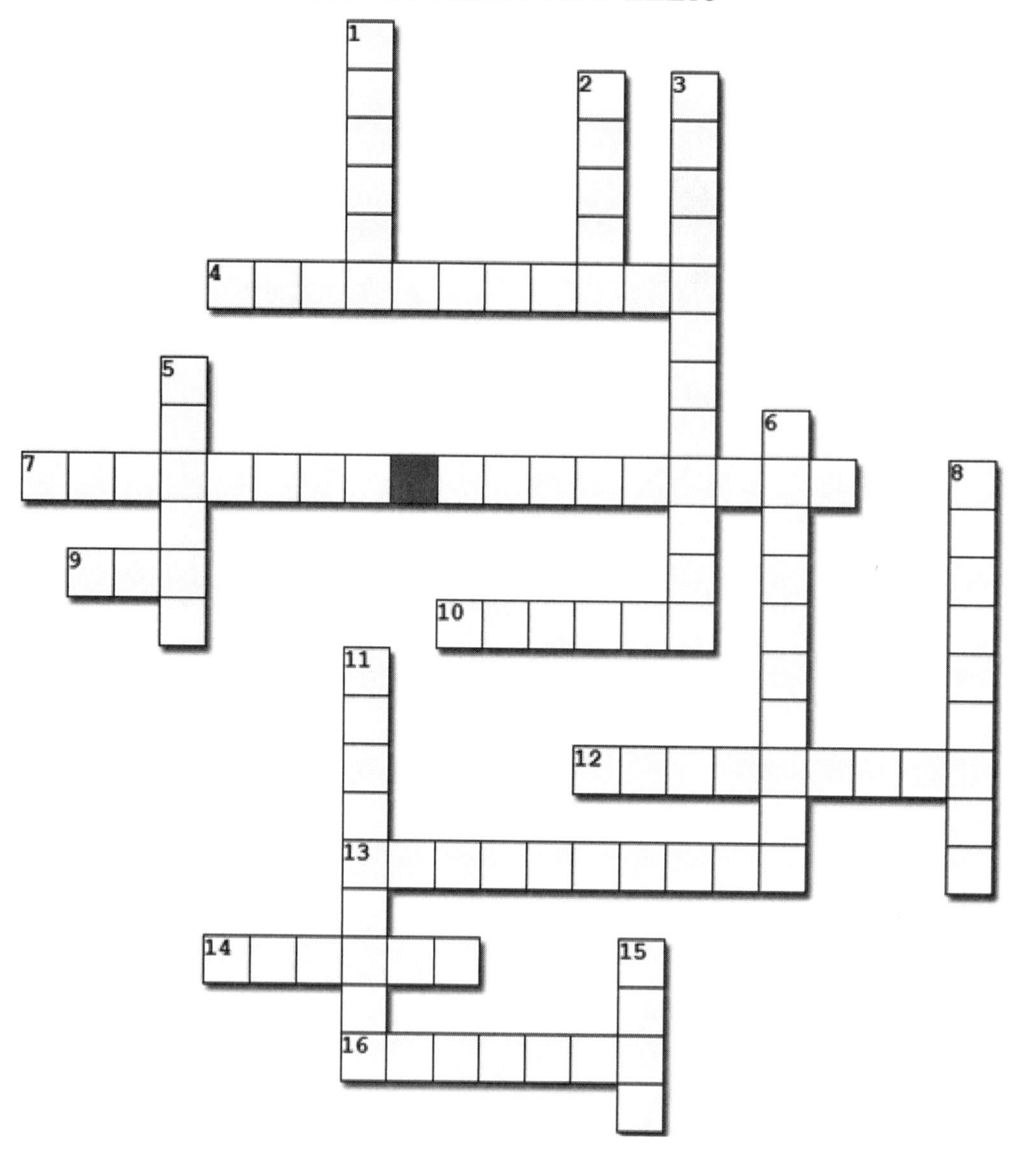

Across:

4. When sampling without ____, make sure that $n < 0.10N$.
7. For each value of x, the ___ of y (denoted by σ) must be the same.
9. When the conditions are met, we use a t distribution with $n -$ ___ (spell the number) degrees of freedom to calculate the P-value.
10. b is the ____ slope.
12. $\hat{y}$ is the ____ y value or the estimated mean y value for a given value of x.
13. A regression line calculated from every value in the population is called a population ____line.
14. The form of the relationship between x and y must be ____.
16. The value σ_b measures the ____ distance between a sample slope b and the population slope for all possible samples of a given size from a population.

Down:

1. For each value of x, the distribution of y must be approximately ____.
2. β is the population ____.
3. When two ____ variables x and y have a linear relationship, we can model that relationship with the least-squares regression line $\hat{y} = a + bx$.
5. The data must come from a ___ sample from the population of interest or a randomized experiment.
6. Use the sample regression line to make an inference about the ____ regression line.
8. The P-value reported in computer output is for a ___-____ test.
11. α is the population y _____.
15. The usual ____ hypothesis in a t test for slope is H_0: $\beta = 0$.

Unit 1, Part I Solutions

Introduction – Smelling Parkinson's Disease

Answers will vary.

Section 1A – Statistics: The Language of Variation

Check for Understanding – Learning Target 1

1. The individuals are the 6 students (James, Jen, DeAnna, Jonathan, Doug, and Sharon).
2. Variables: Class, ACT score, Favorite subject, Height
   Quantitative: ACT score and Height
   Categorical: Class and Favorite Subject.
3. No. Mr. Bush did not select a random sample of students from this school, but simply surveyed 6 students in his statistics class. The students who said yes may not have even been telling the truth! We cannot count on these 6 students to be representative of all students at this school.

Check for Understanding – Learning Target 2

1. Here is a frequency table and relative frequency table:

Frequency Table

Survey response	Frequency
Yes	13
No	8
Not sure	4

Relative Frequency Table

Survey response	Relative Frequency
Yes	13/25 = 0.52 or 52%
No	8/25 = 0.32 or 32%
Not sure	4/25 = 0.16 or 16%

2. 13 students believe in Santa. This is shown in the frequency table.
3. 52% of students surveyed believe in Santa. This is shown in the relative frequency table.

Section 1B – Displaying and Describing Categorical Data

Check for Understanding – Learning Target 1

1. Bar chart is shown below. The most popular coffee-shop is Goodbye Blue (30%), followed by Ugly Mug (20%) and One Mean Bean (20%). The least popular coffee-shop is the national chain.
2. Pie chart is shown below. The pie chart does not show counts. Rather, it displays the percent of the total that preferred each coffee-shop. The pie chart shows that the overwhelming preference is for Goodbye Blue Monday, One Mean Bean, and The Ugly Mug with 70% of residents polled preferring these coffee-shops.

Category Name	Frequency	Relative Frequency
Goodbye Blue	75	30%
Ugly Mug	50	20%
Morning Joe's	38	15.2%
One Mean Bean	50	20%
Brewed own	25	10%
National Chain	12	4.8%
Total	250	100%

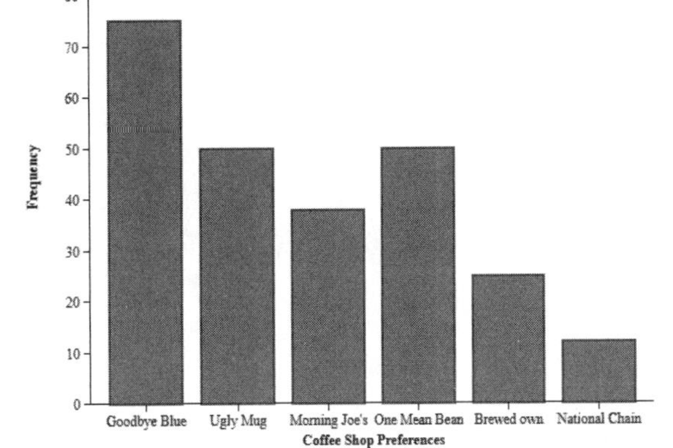

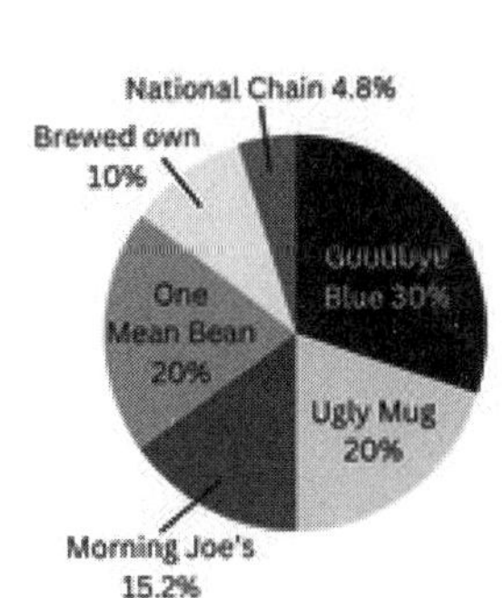

Check for Understanding – Learning Target 2

The side-by-side bar graph shows that as class increases from Freshmen, to Sophomores, to Juniors, to Seniors, the percentage of students who prefer online learning increases and the percentage of students who prefer in-person learning decreases.

Freshmen were most likely to say that they prefer in-person learning (80%), which is twice the relative frequency of seniors of whom only 40% said they prefer in-person learning. Sophomores were second most likely to prefer in-person learning (about 72%), followed by Juniors (about 67%). There was a steep drop-off for seniors, as they are the only group of students who preferred online learning (60%) over in-person learning (40%).

Check for Understanding – Learning Target 3

1. The pictograph is misleading because the areas of the hammers make it look like the number of homes with asphalt shingles is over twice as common as wood shingles, over 10 times as common as ceramic tile, and over 6 times as common as metal roofs. In reality, asphalt shingles are only about 1.5 times as common as wood shingles, about 2.67 times as common as ceramic tile, and about twice as common as metal roofs. Also, the vertical axis should start at 0.
2. No. The U.S. military budget is not over 3 times the size of China's military budget. The vertical axis should start at 0, rather than 45 billion. The U.S. military budget is closer to 2.7 times the size of China's military budget.

Section 1C – Displaying Quantitative Data with Graphs

Check for Understanding – Learning Targets 1 and 2

1.

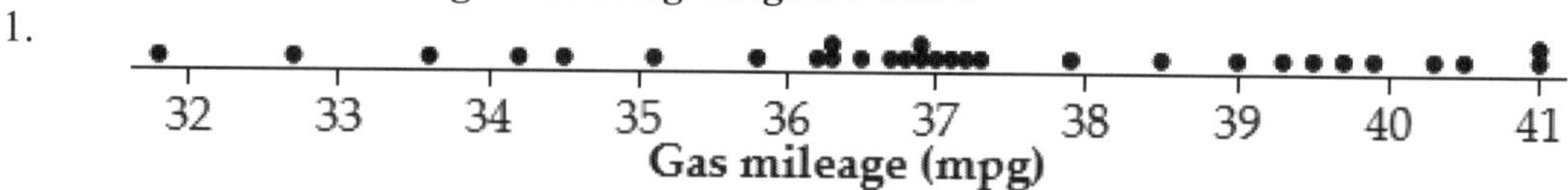

2. The distribution of gas mileage appears to be fairly symmetric with no clear peak and several small gaps.

Check for Understanding – Learning Targets 3 and 4

1. The distribution of Unit 1 exam score is skewed to the left with a peak at 94% and another peak around 100%. There is a gap between the student who scored 70% and the next lowest score of about 78%. The score of 70% is a potential outlier. The center (median) is about 94% and the scores vary from 70% to 100%.
2. The distribution of grade at the end of the semester is also skewed to the left. There is a single peak at 94%. There is a gap from 50% to 57% and another fairly large gap from 60% to about 68%. The three lowest values (50%, 57%, and 60%) are potential outliers. The center (median) is less than that of the Unit 1 exam scores (approximately 89%). The variability in end of semester grades is much greater (50% to 100%) than the variability of Unit 1 exam scores (70% to 100%).

Check for Understanding – Learning Targets 5 and 6

1. $\frac{4}{24} \approx 0.167$ or 16.7%. There are 4 players, out of the 24 in this sample, who are taller than 82 inches.

Stem	Leaves
7	1 2 3
7	5 5 5 5 6 6 8 8 8 8 8 8
8	0 0 1 2 2 3 4 4
8	6

Key:
7 | 1 represents an NBA player who is 71 inches tall.

2. $\frac{8}{50} = 0.16$ or 16% of the cars in this sample get at least 40 miles per gallon.

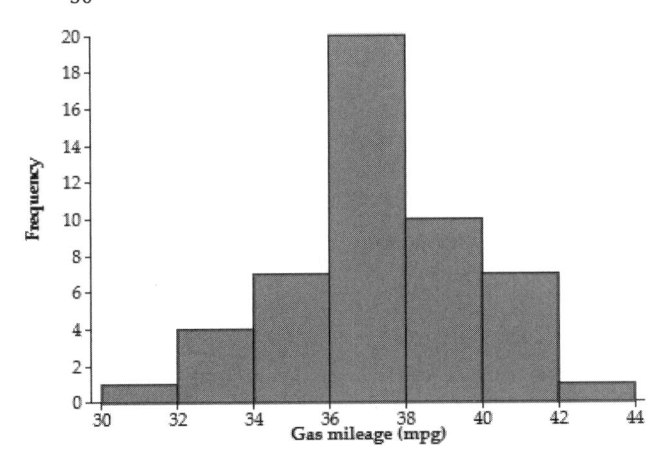

Section 1D – Displaying Quantitative Data with Numbers

Check for Understanding – Learning Targets 1 and 2

1. The stemplot shows the distribution of length of time to complete the logic puzzle is slightly skewed to the right. Therefore, the mean will probably be greater than the median.
2. The average amount of time it took these students to complete the logic puzzle was 32.67 seconds.
3. Median = 29 seconds. This is the "middle" time in the distribution. Half of the students took 29 or more seconds to complete the puzzle and half took 29 or less seconds.

Check for Understanding – Learning Targets 3, 4, and 5

1. Range = 84.6 – 0.6 = 84 million people
2. Standard deviation = 34.866 million people. The populations of these 7 European countries typically vary from the mean by about 34.866 million people.
3. $IQR = Q_3 - Q_1 = 68.2 - 3.7 = 64.5$ million people.

Check for Understanding – Learning Target 6

1. The distribution of home price is right-skewed with potential high outliers so we should use the median to measure the center and the *IQR* to measure the variability of the distribution.
2. The median is in the $200,000 to < $250,000 interval because the median is the 18th sale price in the ordered list. The mean is likely to be greater than $250,000 because the distribution of sale price is skewed to the right and there are potential high outliers.
3. The distribution of flight times is fairly symmetric and has no clear outliers. Therefore, we should use the mean and standard deviation to measure the center and variability of the distribution.

Check for Understanding – Learning Targets 7, 8, and 9

1. min = 170, $Q_1 = 314$, Median = 330, $Q_3 = 344$, max = 374.
2. Low outliers $< 314 - 1.5(344 - 314)$
   Low outliers < 269
   There are two books that had fewer than 269 pages: 170 pages and 242 pages
   High outliers $> 344 + 1.5(344 - 314)$
   High outliers > 389
   There are no high outliers because 389 is greater than the maximum of 374.
3.

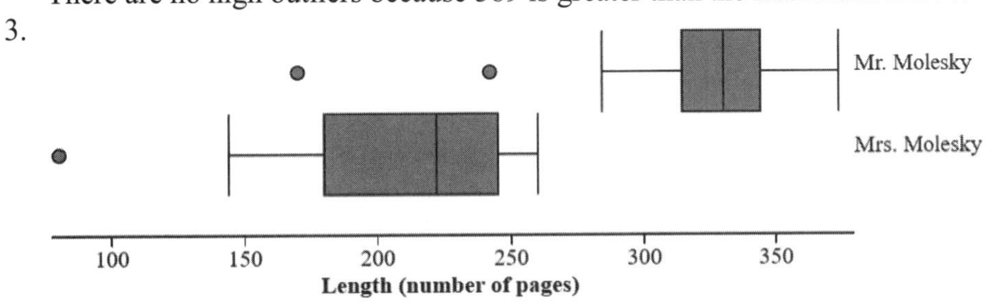

4. **Shape**: The distribution of book length is skewed to the left for both Mr. and Mrs. Molesky.
Outliers: There are two low outliers among the books Mr. Molesky has read and there is one low oulier among the books Mrs. Molesky has read.
Center: The median book length Mr. Molesky has read is 330 pages, which is greater than the median book length of Mrs. Molesky's books of about 222 pages.
Variability: The interquartile range of book length for Mr. Molesky is 30 pages, which is less than the interquartile range of book length for Mrs. Molesky of about 65 pages.

Unit 1, Part I Crossword Puzzle Solutions

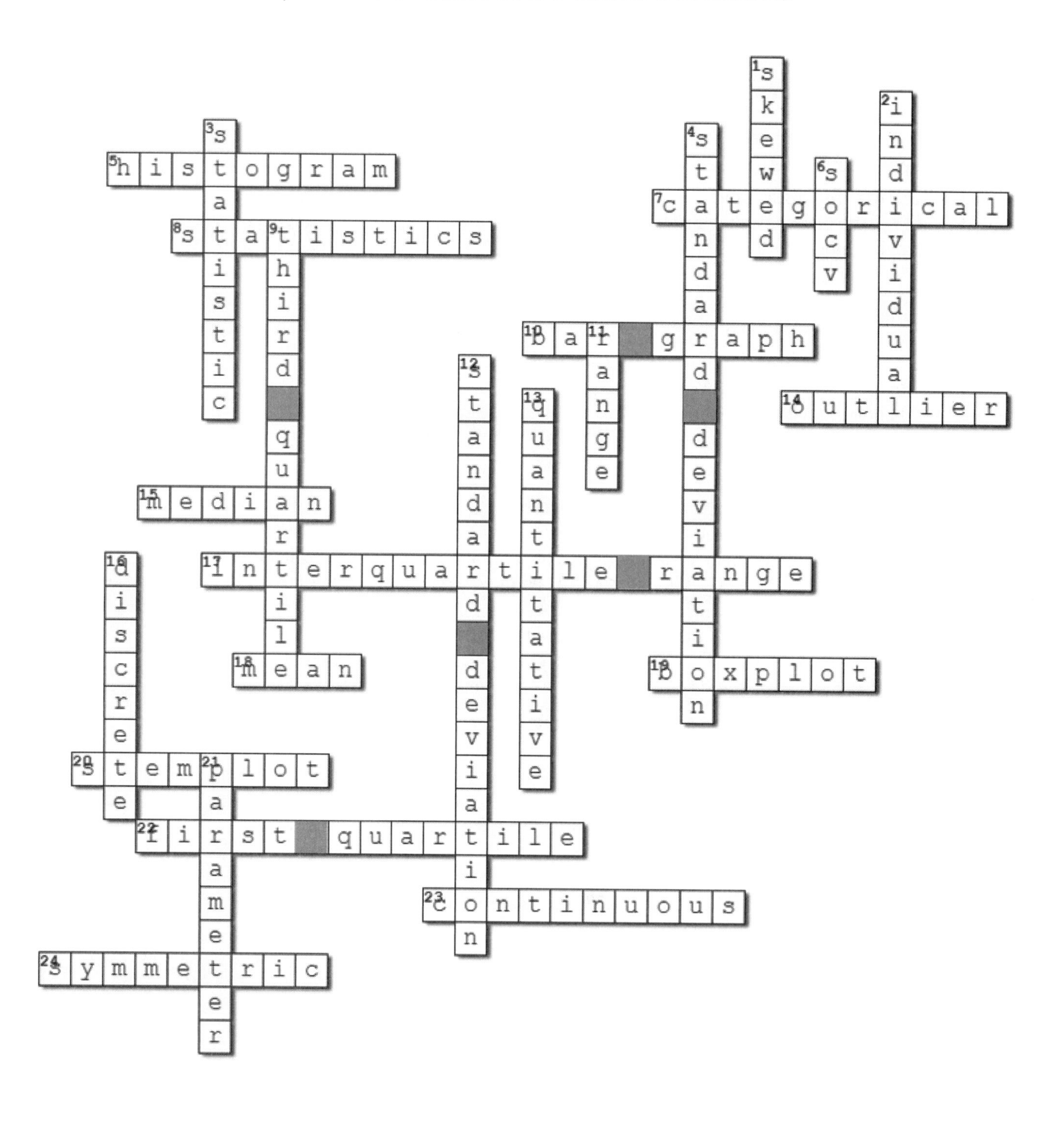

Unit 1, Part II Solutions

Section 1E – Displaying Quantitative Data with Graphs

Check for Understanding – Learning Target 1

1. $\frac{9}{15} = 0.60$, so this contestant is at the 60th percentile in the distribution of number of hot dogs eaten. About 60% of the contestants ate as many or fewer hot dogs as this contestant.
2. $(0.8)(15) = 12$. Because 12 of the values are less than or equal to 45 hot dogs, the 80th percentile is 45 hot dogs.
3. Mele's percentile: $\frac{11}{14} \approx 0.786$, so Mele is at the 78th percentile of the female distribution
   Oji's percentile: $\frac{10}{15} \approx 0.667$, so Oji is at the 66th percentile of the male distribution.
   Even though Oji ate more than twice as many hot dogs as Mele, Mele's performance is more impressive within her division.

Check for Understanding – Learning Targets 2 and 3

1. $z = \frac{113 - 103}{8.9} = 1.124$
   The number of people who watched the 2023 Super Bowl falls 1.124 standard deviations above the mean number of people who watched the Super Bowl over the past 20 years.

2. $-0.112 = \frac{\text{value} - 103}{8.9}$
   $-0.112(8.9) + 103 = \text{value}$
   $102 = \text{value}$
   The number of people who watched the Super Bowl in 2020 is about 102 million people.

3. Super Bowl: $z = 1.124$, World Cup Final: $z = \frac{1.5 - 1}{0.251} = 1.992$
   The World Cup Final had a more impressive viewership because the number of viewers for this event was 1.992 standard deviations above the mean World Cup Final viewership, while the viewership for the Super Bowl was only 1.124 standard deviations above the mean Super Bowl viewership.

Check for Understanding – Learning Target 4

1. Joe Biden's age when he first took office places him at about the 100th percentile of the distribution. This estimate is made by identifying age 78.167 on the horizontal axis, going directly up to the cumulative relative frequency graph, then going directly left to the vertical axis. The cumulative relative frequency is about 100%, which indicates that Joe Biden is at about the 100th percentile.
2. Exact percentile = 46/46 = 100th percentile. All other U.S. presidents were his age or younger. This matches the estimated percentile from Question 1.
3. The median of the distribution is about 57 years old. This estimate is made by identifying 50 on the vertical axis, going directly to the right until coming to the cumulative relative frequency graph, then dropping directly down to the horizontal axis.
4. The median age is in the interval 55 to <60 years old.
   Method 1: There are 46 U.S. presidents, so the median is the average of the ages of the 23rd and 24th oldest U.S. presidents in an ordered list of age when taking office. The cumulative frequency column indicates that 18 presidents were <55 upon taking office and 33 were < 60 upon taking office, so the 23rd and 24th oldest presidents were in this interval.
   Method 2: Examine the cumulative relative frequencies. 39.13% of the presidents were <55 years old and 71.74% were < 60 years old, so the 50th percentile falls in the 55 to < 60 interval.
   This matches the estimation made in Question 3.

Check for Understanding – Learning Target 5

1. The distribution of quiz score is single-peaked and roughly symmetric with no gaps or clear outliers. The mean quiz score is 25.043 points, and the standard deviation of quiz scores is 14.201 points.

2. Adding 5 to each student's score would not change the shape or the variability, but it would increase the center of the distribution. The distribution of adjusted quiz scores would have the same shape as the distribution of original quiz: roughly symmetric. The mean of the adjusted quiz scores would increase by 5 to 30.043 points. The variability of the adjusted quiz scores would remain the same as the original quiz scores. The standard deviation would still be 14.201 points.

3. Doubling each student's score would not change the shape but will double the mean and standard deviation. The distribution of doubled quiz scores would have the same shape as the distribution of original quiz scores: roughly symmetric. The mean of the doubled quiz scores would double to 50.086 points, and the variability would double. The standard deviation of doubled quiz scores would be 28.402 points.

For further investigation: To convince yourself of these rules, make a dotplot of original quiz scores. Remember, the mean of the original quiz scores is 25.04 points with a standard deviation of 14.2 points. Now add 5 points to each student's score. Make a dotplot of the adjusted quiz scores. Recalculate the mean and standard deviation. Compare the shape, center, and variability of adjusted quiz scores to the distribution of original quiz scores. Repeat this process to verify the effect of multiplication of shape, center, and variability.

Section 1F – Normal Distributions

Check for Understanding – Learning Targets 1 and 2

1.

0.52 1.84 3.16 4.48 5.8 7.12 8.44

Amount of trash per day (pounds)

2. $\frac{1}{2}(100\% - 68\%) = 16\%$. About 16% of U.S. residents make more than 5.8 pounds of trash per day.

3. $\frac{1}{2}(95\% - 68\%) = 13.5\%$. About 13.5% of U.S. residents make between 1.84 and 3.16 pounds of trash per day.

Check for Understanding – Learning Target 3

Graphical methods:

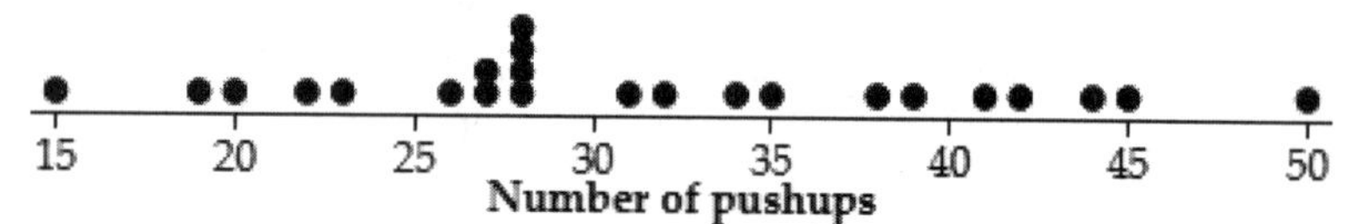

The distribution of the number of push-ups that the 23 students on the JV and High School basketball teams could do is roughly symmetric, single-peaked, and somewhat mound-shaped.

Numerical methods: Mean = 31.39, SD = 9.15

The percentages of values that fall within one, two, and three standard deviations of the mean are:

Lower boundary	Interval	Percent of Observations that fall in this interval	Percent expected to fall in this interval if the data are approximately normal:
Mean ± 1SD	22.24 to 40.54	$\frac{14}{23} \approx 60.9\%$	68%
Mean ± 2SD	13.09 to 49.69	$\frac{22}{23} \approx 95.7\%$	95%
Mean ± 3SD	3.94 to 58.84	$\frac{23}{23} = 100\%$	99.7%

These percentages are close to the 68%, 95%, and 99.7% targets for a normal distribution. The graphical and numerical evidence suggests that the distribution of number of pushups is approximately normal. Keep in mind that you will rarely find a distribution that is perfectly normal.

Check for Understanding – Learning Targets 4 and 5

1. Mean = 14 minutes, SD = 4.5 minutes

2. Using technology: Normalcdf(lower: 25, upper: 1000, mean: 14, SD: 4.5) = 0.0073
   $z = \frac{25-14}{4.5} = 2.444$. Using Table A: The area to the right of $z = 2.44$ is 0.0073.
   About 0.73% of U.S. adults read for personal interest more than 25 minutes per day.

3. Using technology: Normalcdf(lower: –1000, upper: 12, mean: 14, SD: 4.5) = 0.3284
   $z = \frac{12-14}{4.5} = -0.444$. Using Table A: The area to the left of $z = -0.44$ is 0.3300.
   About 32.84% of U.S. adults read for personal interest less than 12 minutes per day.

4. Using technology: Normalcdf(lower: 5, upper: 10, mean: 14, SD: 4.5) = 0.1643
   $z = \frac{5-14}{4.5} = -2$ and $z = \frac{10-14}{4.5} = -0.889$. Using Table A: The area between $z = -2$ and $z = -0.89$ is 0.1639.
   The proportion of U.S. adults who read for personal interest between 5 and 10 minutes per day is about 0.1643.

5. Using technology: invNorm(area:0.75, mean: 14, SD: 4.5) = 17.035 minutes
   Using Table A: 0.75 area to the left → $z = 0.674$
   $$0.674 = \frac{x-14}{4.5}$$
   $0.674(4.5) + 14 = x$
   $x = 17.033$ minutes
   The 75th percentile of the distribution of number of minutes spent reading for personal interest is about 17.035 minutes per day.

Check for Understanding – Learning Target 6

1. Using technology: InvNorm(area: 0.25, mean: 0, SD: 1) = –0.674
   Using Table A: 0.25 area to the left → z = –0.67
   $-0.674 = \frac{10.471-13}{\sigma} \rightarrow -0.674(\sigma) = -2.529 \rightarrow \sigma \approx 3.752$
2. Using technology: InvNorm(area: 0.40, mean: 0, SD: 1) = –0.253
   Using Table A: 0.40 area to the left → –0.25
   $-0.253 = \frac{7.38-\mu}{2.45} \rightarrow -0.253(2.45) - 7.38 = -\mu \rightarrow \mu \approx 8$ robocall

Unit 1, Part II Crossword Puzzle Solutions

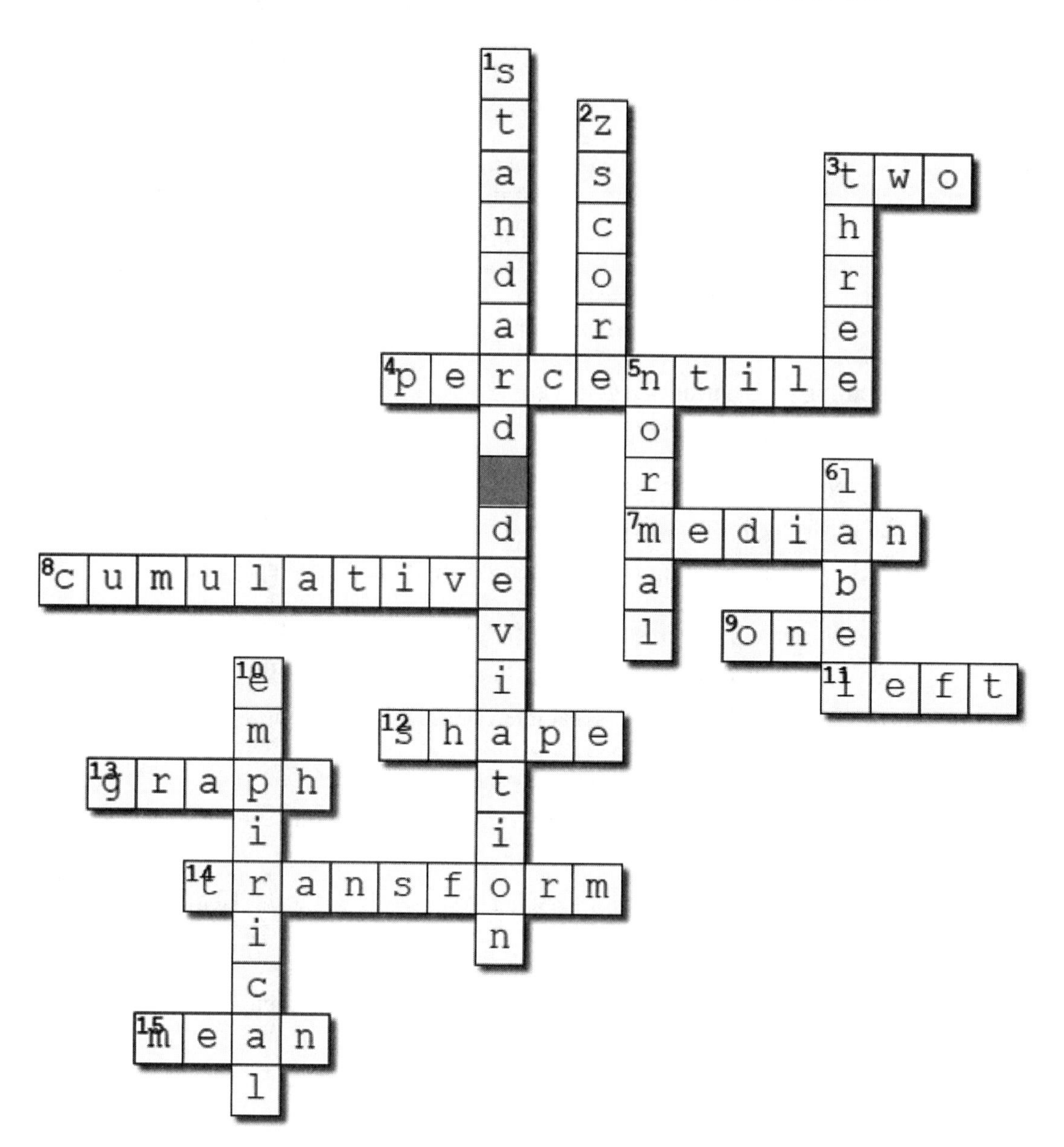

Unit 2 Solutions

Section 2A – Relationships Between Two Categorical Variables

Check for Understanding – Learning Target 1

1. Explanatory: test anxiety — Response: performance on the test
2. Explanatory: volume of hippocampus — Response: verbal retention

Check for Understanding – Learning Target 2

1. $\frac{105}{294} \approx 0.357$ or 35.7%
2. $\frac{88}{294} \approx 0.299$ or 29.9%
3. Age 18 to 29: $\frac{17}{78} \approx 0.218$ or 21.8%

   Age 30 to 59: $\frac{34}{122} \approx 0.279$ or 27.9%

   Age 60+: $\frac{54}{94} \approx 0.574$ or 57.4%

Check for Understanding – Learning Targets 3 and 4

1. Here is a segmented bar graph and a mosaic plot of the data.

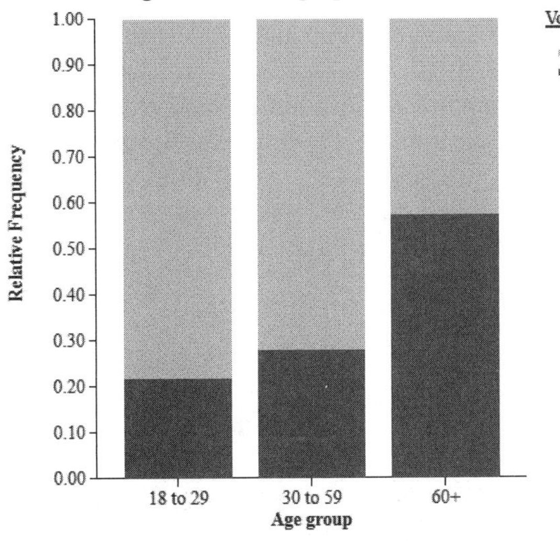

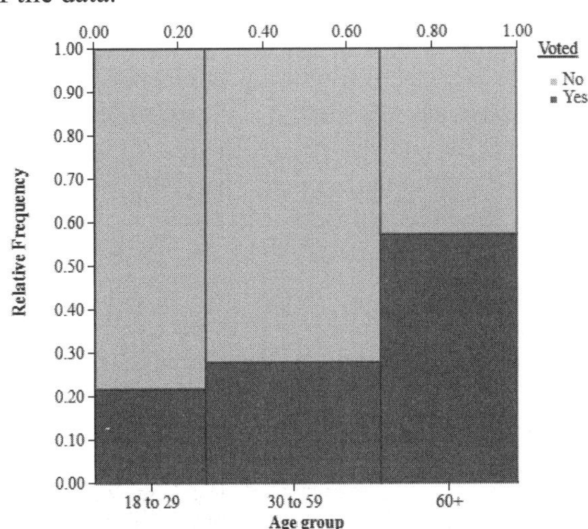

2. There is a clear association between age group and voting for the U.S. adults in this sample. Knowing a person's age group helps us predict whether the individual voted in the most recent midterm election. The youngest individuals (age 18 to 29) were the least likely to vote (21.8%), followed by individuals in age 30 to 59 (27.9%). The oldest individuals (age 60+) were the most likely to vote (57.4%). The mosaic plot reveals that individuals aged 30 to 59 made up 41.5% of the sample, individuals aged 60+ made up 32.0% of the sample, and individuals aged 18 to 29 made up 26.5% of the sample.

Section 2B – Relationships Between Two Quantitative Variables

Check for Understanding – Learning Targets 1 and 2

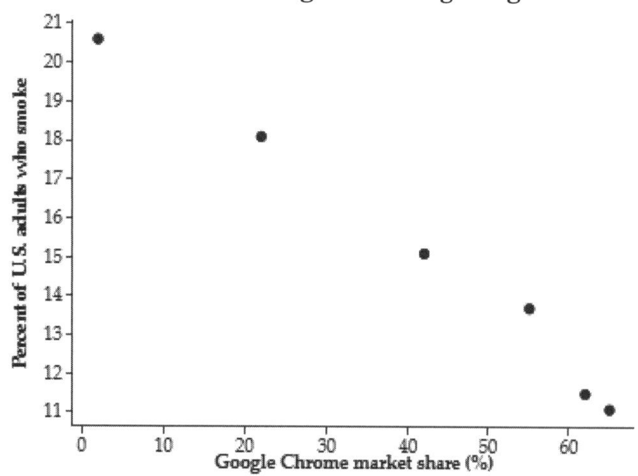

There is a strong, negative, linear relationship between Google Chrome market share and the percent of U.S. adults who smoke. There are no unusual observations.

Check for Understanding – Learning Targets 3, 4, and 5

1. $r = -0.99$. The linear relationship between Google Chrome market share and the percent of U.S. adults who smoke is strong and negative.
2. No. Even though there is a strong correlation between Google Chrome market share and the percent of U.S. adults who smoke, the reporter should not infer causation. It is possible that both variables are changing due to another variable, such advancements in society or increased knowledge. Maybe over time Google chrome has gained popularity due to its usefulness in finding information and also the spread of information has increased awareness about the dangers of smoking.

Section 2C – Linear Regression Models

Check for Understanding – Learning Targets 1, 2, and 3

1. $\hat{y} = 1.5 + 0.25(12) = 4.5$ feet tall. This prediction is reasonable.
2. $\hat{y} = 1.5 + 0.25(40) = 11.5$ feet tall. This is not reasonable. This prediction is an extrapolation.
3. Residual $= y - \hat{y} = 4.25 - 4.5 = -0.25$. Alyse's height is 0.25 feet (or 4 inches) less than the height predicted by the regression model for a 12-year-old.
4. Slope = 0.25. The predicted height increases by 0.25 feet for each additional 1 year of age.
   y intercept = 1.5. The predicted height of a person who is 0 years old is 1.5 feet. The y intercept has meaning in this context, because a person is 0 years old at birth. The line predicts the height of a newborn to be 1.5 feet (18 inches).

Check for Understanding – Learning Targets 4 and 5

1. x = number of gift cards sold and y = number of gift cards redeemed
   $b = 0.92\left(\frac{48}{48.5}\right) = 0.911$
   $\bar{y} = a + b\bar{x}$
   $294 = a + 0.911(355)$
   $a = 294 - 0.911(355) = -29.405$
   The least-squares regression line is $\hat{y} = -29.405 + 0.911x$, where x = the number of gift cards sold and y = the number of gift cards redeemed within 2 years
2. For each additional 1 gift card sold, the predicted number of gift cards redeemed within 2 years increases by 0.911. For each additional 100 gift cards purchased about 91 are predicted to be redeemed within the next 2 years.
3. The least-squares regression line is $\hat{y} = 71.346 - 1.141x$, where x = anxiety score and y = exam score.

Check for Understanding – Learning Targets 6 and 7

1. The residual plot does not show any obvious leftover curved pattern. It appears the linear model is appropriate for making predictions of exam scores.

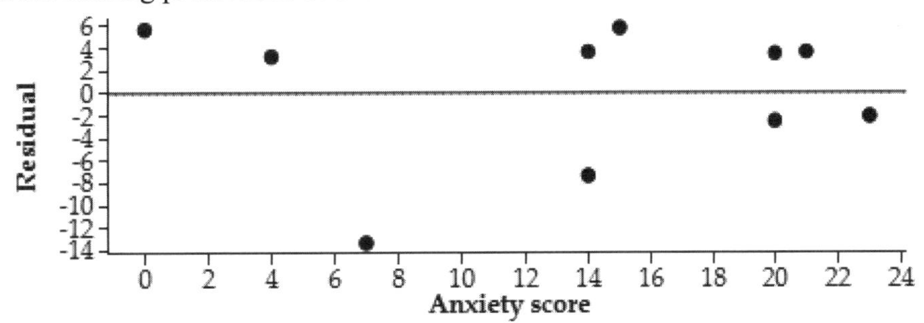

2. *Interpretation of r^2*: Approximately 67% of the variation in exam score is accounted for by the least-squares regression line with x = anxiety score.
   Interpretation of s: The actual exam score is typically about 6.69 points away from the exam score predicated by the least-squares regression line with x = anxiety score.

Section 2D – Analyzing Departures from Linearity

Check for Understanding – Learning Target 1

1. Because the point (64, 38) is far above the regression line and slightly to the left of the mean height:
   - It increases the y intercept slightly and decreases the slope slightly.
   - Makes the values of r and r^2 closer to 0 because it is outside the linear pattern of the other points.
   - It increases the standard deviation of the residuals.
   - Because it has a large residual, the point (64, 38) is an outlier. It is not extreme in the x-direction, so it is not a high-leverage point.

2. Because the player at (77, 210) has an above-average height and below-average weight this point is pulling the line towards itself. This point:
   - Makes the slope of the line closer to zero and increases the value of the y intercept.
   - Makes the values of r and r^2 closer to 0 because it is outside the linear patter of the other points.
   - It increases the standard deviation of the residuals.

Check for Understanding – Learning Target 2

predicted length = $19.47(\text{diameter})^{0.7} + 16.57$
If diameter = 47mm,
predicted length = $19.47(47)^{0.7} + 16.57 = 304.863$ mm.

Check for Understanding – Learning Target 3

1. It is unclear which residual plot has the most random scatter. The model with the greatest value of r^2 is Model 3. Also notice that the scatterplot for Model 3 shows the most linear form.
2. Model 3 is based upon x = ln(police) and y = ln(crime), therefore the equation of the least-squares regression line is: $\ln(\widehat{\text{crimes}}) = 4.886 + 0.5365\ln(\text{police})$.
   For a state with 25,400 police officers,
   $\ln(\widehat{\text{crimes}}) = 4.886 + 0.5365\ln(25.4)$
   $\ln(\widehat{\text{crimes}}) = 6.62144$
   $\widehat{crimes} = e^{6.62144} \approx 751.03$ crimes

Unit 2 Crossword Puzzle Solutions

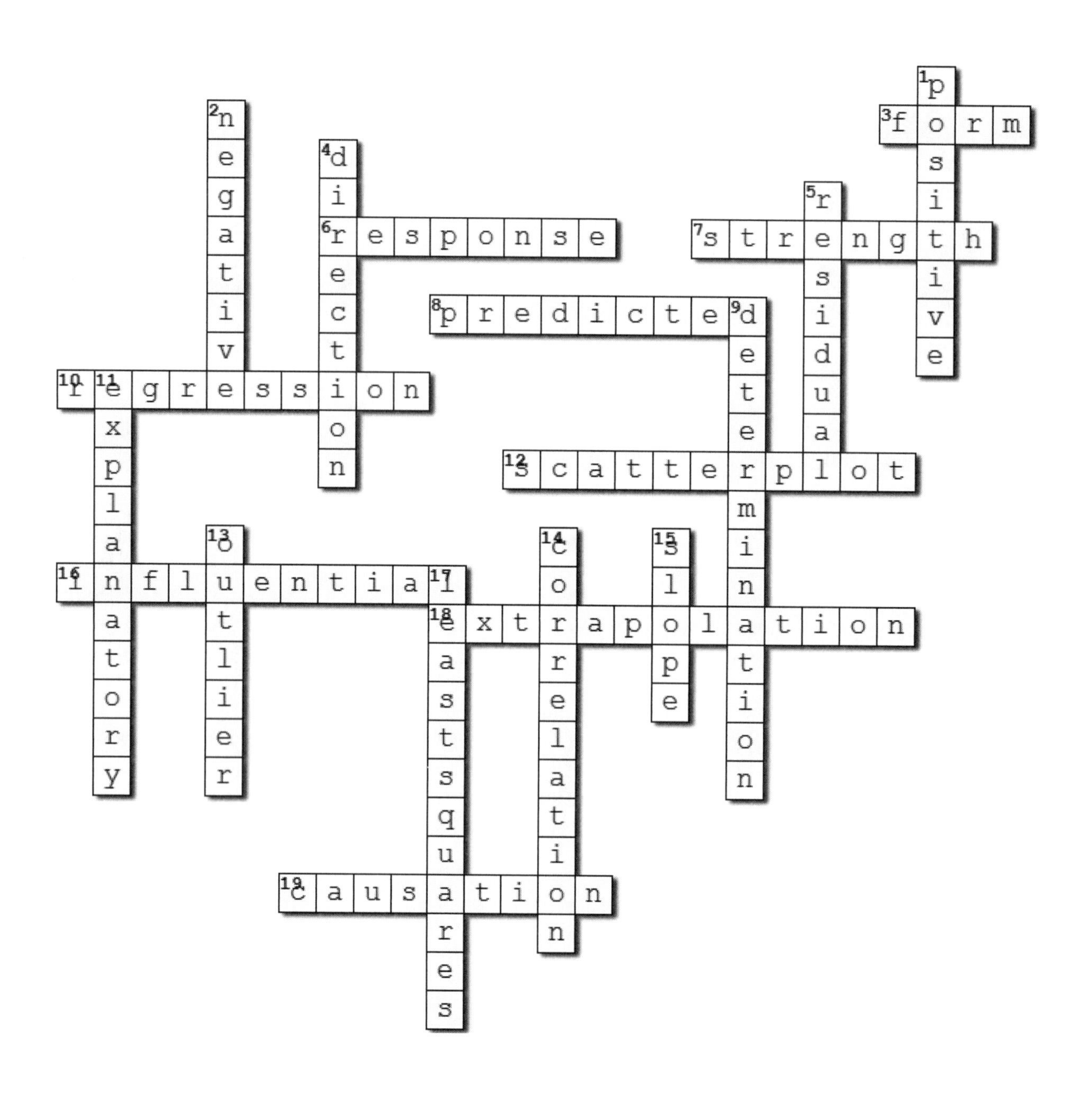

Unit 3 Solutions

Section 3A – Introduction to Data Collection

Check for Understanding – Learning Target 1

1. Population: all sentences and words used in the popular Algebra 1 textbooks.
   Sample: the sentences and words in the 10 randomly selected paragraphs.
2. Population: all registered voters, Sample: the 2845 registered voters who completed the survey.
3. This is a census.

Check for Understanding – Learning Targets 2 and 3

1. This is an observational study since no treatment was imposed on the subjects. Spelling test scores and shoe size were observed. No effort was made to influence either variable.
2. The explanatory variable is shoe size and response variable is the spelling test score.
3. The students were randomly selected from the population of all students at a suburban school district, so we can generalize the results to the population of all students in this suburban school district.
4. No, because this is an observational study and not an experiment. Researchers did not (nor is it possible for them to) randomly assign the students a shoe size (or stretch their feet. Older students would have bigger feet, on average, and would most likely have better spelling test scores.

Section 3B – Sampling and Surveys

Check for Understanding – Learning Target 1

1. Label the vacation destinations 01 to 15 in order of popularity, as provided in the table. Move along a line of random digits from left to right, reading two-digit numbers, until the numbers corresponding to 3 different vacation destinations have been selected (ignoring repeated numbers and the numbers 16-99, 00). Visit the three destinations that correspond with the numbers selected.
2. 19 – skip, 22 – skip, 09 – select, 50 – skip, 34 – skip, 05 – select, 75 – skip, 62 – skip, 87 – skip, 09 – repeat, 96 – skip, 14 – select.
   The three vacation destinations selected are: 09: Yosemite National Park, 05: Wisconsin Dells, 14: Honolulu, HI.

Check for Understanding – Learning Target 2

1. Because opinions may be similar within each political party, but differ across political parties, use political party as strata. Label the Republicans 1 to 49. Use a random number generator to generate 5 different random integers from 1 to 49 and survey the Senators corresponding to the selected integers. Label the Democrats 1 to 51. Use a random number generator to generate 5 different random integers from 1 to 51 and survey the Senators corresponding to the selected integers. Combine these samples.
2. The Senate seating is arranged into 4 arches. Each arch can be used as a cluster because each arch is a small-scale replica of the population because each arch consists of roughly 50% Democrats and 50% Republicans. Label the arches 1 to 4 from the most inward arch to the most outward arch. Use a random number generator to generate one integer from 1 to 4 and survey all Senators in that arch.
3. Because we want to sample 10 Senators from a population of 100 Senators, we should select every 100 / 10 = 10th senator. To do so, use a random number generator to generate a random integer from 1 to 10 to determine which of the first 10 Senators to survey starting in the inner-most arch on the left. Then move across the arches from left to right surveying every 10th Senator thereafter.

Check for Understanding – Learning Target 3
This is an example of voluntary response bias. Those listeners with strong opinions are likely to call in. The poll probably overestimates the true proportion of listeners that are opposed to the increase.

Check for Understanding – Learning Target 4
When asked by a live person, many people may lie about cheating because they don't want to be looked down upon. These results suffer from response bias. The percentage of the population who cheated is likely to be greater than the sample percentage.

Section 3C – Experiments

Check for Understanding – Learning Target 1
It is possible that households who received the app experienced a very rainy year and that the fact that it rained a lot lead to a decrease in water usage (no need to water the plants or wash the car). If both of these things are true, then we would see a relationship between app usage and water consumption even if the app has no effect on water consumption.

Check for Understanding – Learning Targets 2, 3, 4, 5, and 6

1. Experimental units (subjects): Mr. Tyson's students
   Explanatory variable: listening to classical music while studying
   Treatments: listening to classical music; not listening to classical music
   Response variable: test scores
2. The purpose of the control group is to provide a baseline for comparing the effect of classical music on test scores. Otherwise, Mr. Tyson would not be able to determine if an increase in test scores was due to listening to classical music or some other variable, like studying harder.
3. This study could be single-blind but cannot be double-blind. It can be single-blind if Mr. Tyson has a third party carry out the randomization of the students to the treatment groups (classical music or no classical music). That way, Mr. Tyson would not be influenced to score the tests any differently for any particular student because he would not know which students were in which treatment group. The study could not be double-blind because the students will know if they are listening to classical music or not.
4. The 150 students should be randomly assigned to two groups. This could be accomplished by placing all names in a hat, mixing well, and drawing names without replacement until 75 names are selected. The students whose names were drawn will be assigned to listen to classical music while studying. The remaining 75 students will be assigned to listen to no music while studying. Both groups will receive the same instruction from Mr. Tyson. All students will take the same assessment and their average results will be compared.
5. All students are given the same instruction from Mr. Tyson and all students take the same test. Although students are told to study for no more than 45 minutes, there still will be variability in study time because some students may study for 20 minutes, 45 minutes, or not study at all. The playlist is only controlled for the treatment group, not for all students.
   It is important to keep these variables the same because if students received different instruction or took different tests we would not know if the difference in score was due to the classical music or the instruction / test.

Check for Understanding – Learning Targets 7 and 8

1. Answers may vary. Suppose some of Mr. Tyson's students have a very high class rank (top 20%) and others do not have a high class rank. In order to ensure that not all of the students who have a very high class rank are assigned to listen to classical music (which could happen through random assignment), he should block by class rank. He should block all students who are in the top 20% together and all students who are not in the top 20% should be blocked separately. Then, he should randomly assign half of the students in each block to listen to classical music while studying and the other half should study in silence. Then all students should take the same assessment and the results within each block should be compared.
2. Answers may vary. Mr. Tyson could pair the students according to their class rank so that the students with the two highest class ranks form a pair, and the next two form a pair, and so on. Within each pair, he can flip a coin to decide which student in each pair will study listening to classical music and which student will study with no music. For each student, record the test score. Compare the average test score for those who did and did not listen to classical music.

Check for Understanding – Learning Targets 9 and 10

1. No. The treatments were not randomly assigned to the participants. All participants used electronic devices to record their talking patterns. Also, the results were not statistically significant.
2. This study used volunteers, so the largest population we can generalize the results of this study to are all volunteers like the ones in this study.
3. The difference in the number of words spoken by the men and women in this experiment can be plausibly attributed to chance variation in the two groups.

Unit 3 Crossword Puzzle Solutions

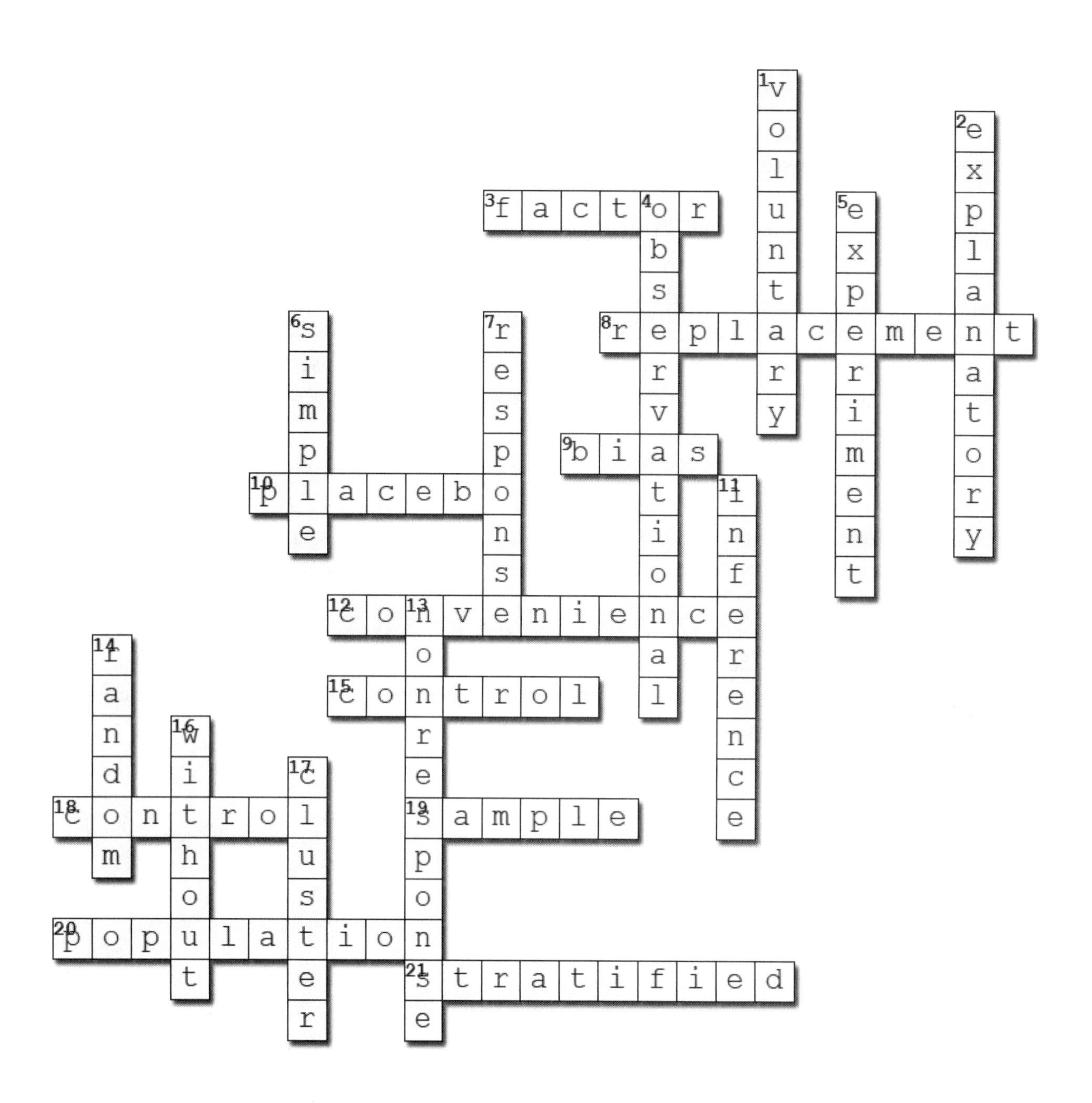

Unit 4, Part I Solutions

Section 4A: Randomness, Probability, and Simulation

Check for Understanding – Learning Target 1

1. If you were to repeatedly select ducks with replacement from a well-mixed duck pond, you would draw a duck with a red dot on the bottom about 10% of the time.
2. No. Although we would expect to select a duck with a red dot approximately 10 times out of 100, we are not guaranteed to do so *exactly* 10 times.

Check for Understanding – Learning Target 2

1. Let 1 to 95 represent a passenger showing up for the flight and 96 to 99 and 100 represent a "no show." Generate 205 random integers from 1 to 100, allowing repeats, to simulate the response of showing up or not showing up for 205 passengers. Count the number of integers that are from 1 to 95. Record the number of simulated passengers who show up.
2. In one simulated trial, 202 passengers showed up for the flight. This flight was overbooked.
3. Because 3 of the 100 dots are greater than 200, there is a $\frac{3}{100} \approx 0.03$ probability that the flight will be overbooked.

Section 4B: Probability Rules

Check for Understanding – Learning Targets 1, 2, and 3

1. There are 10 possible outcomes. Each outcome has a probability of $\frac{1}{10}$.
2. $S = \{S, T, A, I, C\}$. $P(S) = 0.3$, $P(T) = 0.3$, $P(A) = 0.1$, $P(I) = 0.2$, $P(C) = 0.1$.
3. $P(V) = P(A) + P(I) = 0.1 + 0.2 = 0.3$
4. $P(V^C) = 1 - 0.3 = 0.7$. This is the probability that a randomly selected card does not have a vowel.

Check for Understanding – Learning Targets 4 and 5

1. Use a two-way table to display the sample space.

	B	B^C	Total
W	165	23	188
W^C	84	17	101
Total	249	40	289

2. 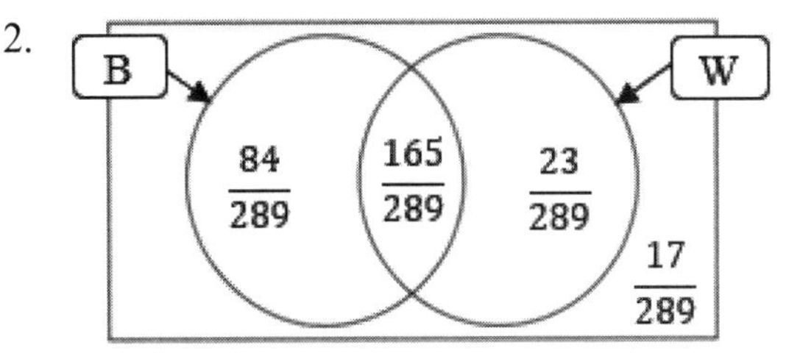

3. $P(\text{B or W}) = P(\text{B}) + P(\text{W}) - P(\text{B and W}) = \frac{249}{289} + \frac{188}{289} - \frac{165}{289} = \frac{272}{289} \approx 0.941.$

4. The complement of P(B or W) = Ben does not play and the Steelers do not win.
   $P(B^C \text{ and } W^C) = \frac{17}{289} \approx 0.059.$

Section 4C: Conditional Probability and Independent Events

Check for Understanding – Learning Targets 1 and 2

1. $P(\text{survived}) = \frac{148}{406} \approx 0.365$
2. $P(\text{survived} \mid \text{child}) = \frac{19}{100} = 0.19$
3. No, because $P(\text{survived}) \neq P(\text{survived} \mid \text{child})$. Knowing the randomly selected passenger is a child makes it much less likely they survived.

Check for Understanding – Learning Targets 3 and 4

1. $P(\text{athlete and athlete}) = \frac{10}{25} \cdot \frac{9}{24} = 0.15$
2. Here is the tree diagram.

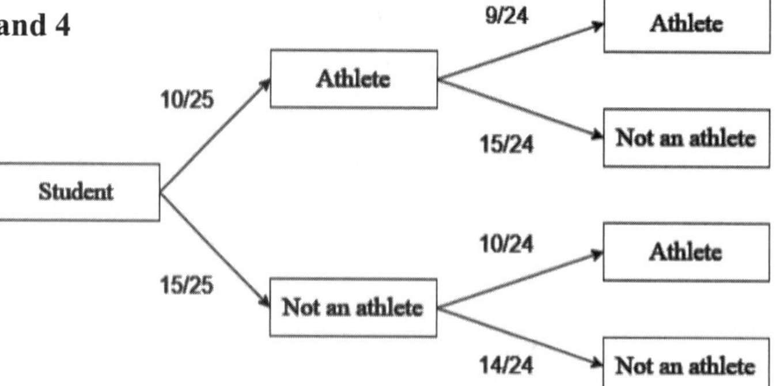

3. There are 2 branches that lead to an athlete winning the second homework pass.
   $P(\text{athlete won } 2^{nd} \text{ homework pass})$
   $= \frac{10}{25} \cdot \frac{9}{24} + \frac{15}{25} \cdot \frac{10}{24} = 0.40.$
4. Of the two branches that lead to an athlete winning the second homework pass (denominator = answer from Question 3), the top branch is the one for which the first pass also went to an athlete (numerator = answer from Question 1).
   $P(1^{st} \text{ pass went to an athlete} \mid 2^{nd} \text{ pass went to an athlete}) = \frac{0.15}{0.40} = 0.375$

Check for Understanding – Learning Targets 5 and 6

1. $P(\text{Roll five 6's}) = \frac{1}{6} \cdot \frac{1}{6} \cdot \frac{1}{6} \cdot \frac{1}{6} \cdot \frac{1}{6} = 0.0001286$
2. $P(\text{Get a Yahtzee}) = 6(0.0001286) = 0.0007716$
3. Yes! A and B are independent because $(0.60)(0.75) = 0.45$ which equals $P(A \text{ and } B)$.
4. No. The first and second foul shots are dependent events. If a player misses the first shot, they do not get to attempt a second shot, so $P(\text{make } 2^{nd} \text{ shot} \mid \text{miss } 1^{st} \text{ shot}) = 0$.

Unit 4, Part I Crossword Puzzle Solutions

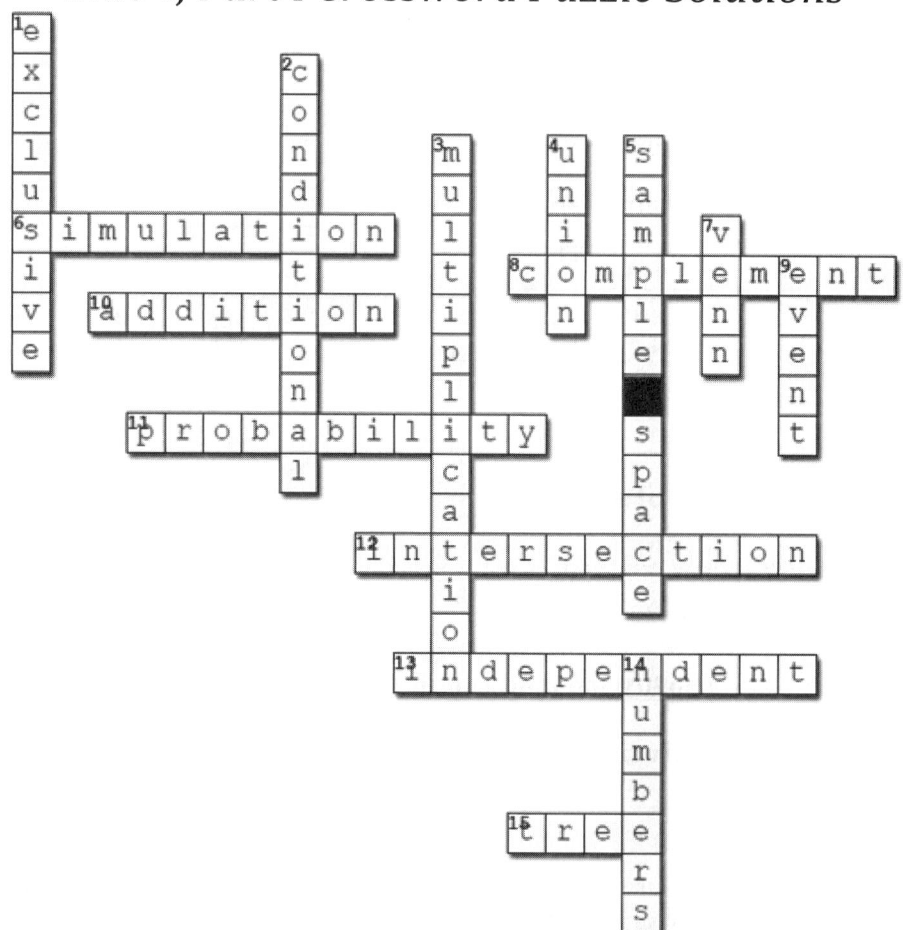

Unit 4, Part II Solutions

Section 4D: Introduction to Discrete Random Variables

Check for Understanding – Learning Targets 1 and 2

1. Here are the sums along with the probability distribution table and probability distribution histogram.

		First roll			
		1	2	3	4
	1	2	3	4	5
Second	2	3	4	5	6
roll	3	4	5	6	7
	4	5	6	7	8

X	2	3	4	5	6	7	8
$P(X)$	1/16	2/16	3/16	4/16	3/16	2/16	1/16

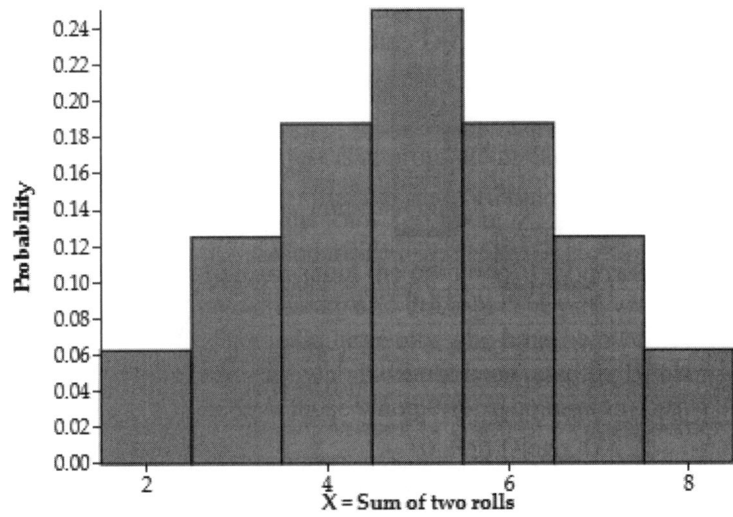

The probability distribution for the sum of two rolls is symmetric with a single peak at $X = 5$.

2. Yes, we should be surprised. In 10 rolls we would only expect to see a sum of 3 or less about once or twice because $10\left(\frac{1}{16}+\frac{2}{16}\right) = 1.875$.

Check for Understanding – Learning Targets 3 and 4

1. $\mu_Y = 0(0.155) + 1(0.295) + 2(0.243) + 3(0.233) + 4(0.074) = 1.776$ goals.
If many, many high school hockey games are randomly selected, the average number of goals will be about 1.776 goals.

2. The median number of goals is 2 goals because this is the smallest value for which the cumulative probability is at least 0.50.

Goals	0	1	2	3	4
Probability	0.155	0.295	0.243	0.233	0.074
Cumulative Probability	0.155	0.450	0.693	0.926	1

3. $\sigma_S = \sqrt{(0-1.776)^2(0.155)+(1-1.776)^2(0.295)+\cdots+(4-1.776)^2(0.074)} = \sqrt{1.394} =$ 1.181 goals.
   If many, many high school hockey games are randomly selected, the number of goals will typically vary from the mean of 1.776 goals by about 1.181 goals.

Section 4E: Transforming and Combining Random Variables

Check for Understanding – Learning Target 1

1. The shape of the probability distribution of *Y* is the same as the shape of the probability distribution of *X*, skewed to the right with a single peak.
2. $\mu_Y = 1.50(\mu_X) - 2 = 1.50(1.1) - 2 = -\$0.35$.
3. $\sigma_Y = 1.5\sigma_X = 1.5(0.943) = \$1.415$.

Check for Understanding – Learning Targets 2 and 3

1. $\mu_{\text{total}} = 1.4 + 1.2 + 0.9 + 1 = 4.5$ min
   $\sigma_{\text{total}} = \sqrt{0.1^2 + 0.4^2 + 0.8^2 + 0.7^2} = \sqrt{1.3} \approx 1.14$min
2. $\mu_{\text{Doug-Ann}} = 1 - 1.4 = -0.4$ min
   If many, many days are randomly selected, Doug will be faster than Ann by about 0.4 min, on average.
   $\sigma_{\text{Doug-Ann}} = \sqrt{0.7^2 + 0.1^2} \approx 0.7071$
   If many, many days are randomly selected, the difference (Doug – Ann) in time needed to check their homework will typically vary by 0.7071 min from the mean difference of 0.4 min.

Section 4F: Binomial and Geometric Random Variables

Check for Understanding – Learning Targets 1 and 2

1. *X* is a binomial random variable with $n = 8$ and $p = 0.1$ because the following conditions are met:
   Binary? "Success" = you are selected, "Failure" = you are not selected
   Independent? Knowing whether you are selected in one trial tells you nothing about whether you will be selected in another trial because the selected card will be replaced each time and the cards will be mixed well.
   Number? $n = 8$
   Same probability? $p = 0.10$
2. Using the formula: $1 - P(X = 0) = 1 - \binom{8}{0}(0.10)^0(0.90)^8 = 0.5695$
   Using technology: $1 - P(X = 0) = 1 -$ binompdf(n: 8, p: 0.10, x: 0) = 0.5695

Check for Understanding – Learning Target 3

1. The Independent condition is not met because the 500 students are selected without replacement from the population of all students in the U.S.
2. It is appropriate to treat the individual observations as independent even through the Independent condition is not met because the 10% condition is met. The sample size of $n = 500 < 10\%$ of the population of all students in the U.S.
3. X is approximately a binomial random variable with $n = 500$ and $p = 0.72$ because the following conditions are met: Binary? "Success" = positive rating. "Failure" = not a positive rating. Independent? The students are selected without replacement, but the 10% condition is met. Number? $n = 500$. Same probability? $p = 0.72$.
4. $\mu_X = np = 500(0.72) = 360$
5. $\sigma_X = \sqrt{np(1-p)} = \sqrt{500(0.72)(0.28)} \approx 10.04$

Check for Understanding – Learning Targets 4 and 5

1. X has a geometric distribution with $p = 0.2$ because the following conditions are met:
   - There are two outcomes (Success: ring or Failure: no ring).
   - Each box is independent, meaning, knowing one box has a ring would give no information about whether or not any other box has a ring.
   - The probability of a ring in any given box is 0.2.
   - We are interested in counting the number of boxes until a ring is found.
2. Using the formula: $P(X = 7) = 0.8^6(0.2) = 0.0524$
   Using technology: $P(X = 7)$ = geometpdf(p: 0.2, x: 7) = 0.0524
3. $P(X < 4) = P(X = 1) + P(X = 2) + P(X = 3)$
   Using the formula: $P(X < 4) = (0.2) + (0.8)(0.2) + (0.8)^2(0.2) = 0.488$
   Using technology: $P(X < 4)$ = geometcdf(p: 0.2, x: 3) = 0.488
4. $E(X) = \frac{1}{0.2} = 5$ boxes.
5. $\sigma_X = \frac{\sqrt{1-p}}{p} = \frac{\sqrt{1-0.2}}{0.2} = 4.472$
   If many, many cereal boxes were randomly selected, the number of boxes it would take to get the first ring would typically vary by about 4.472 boxes from the mean of 5 boxes.

Unit 4, Part II Crossword Puzzle Solutions

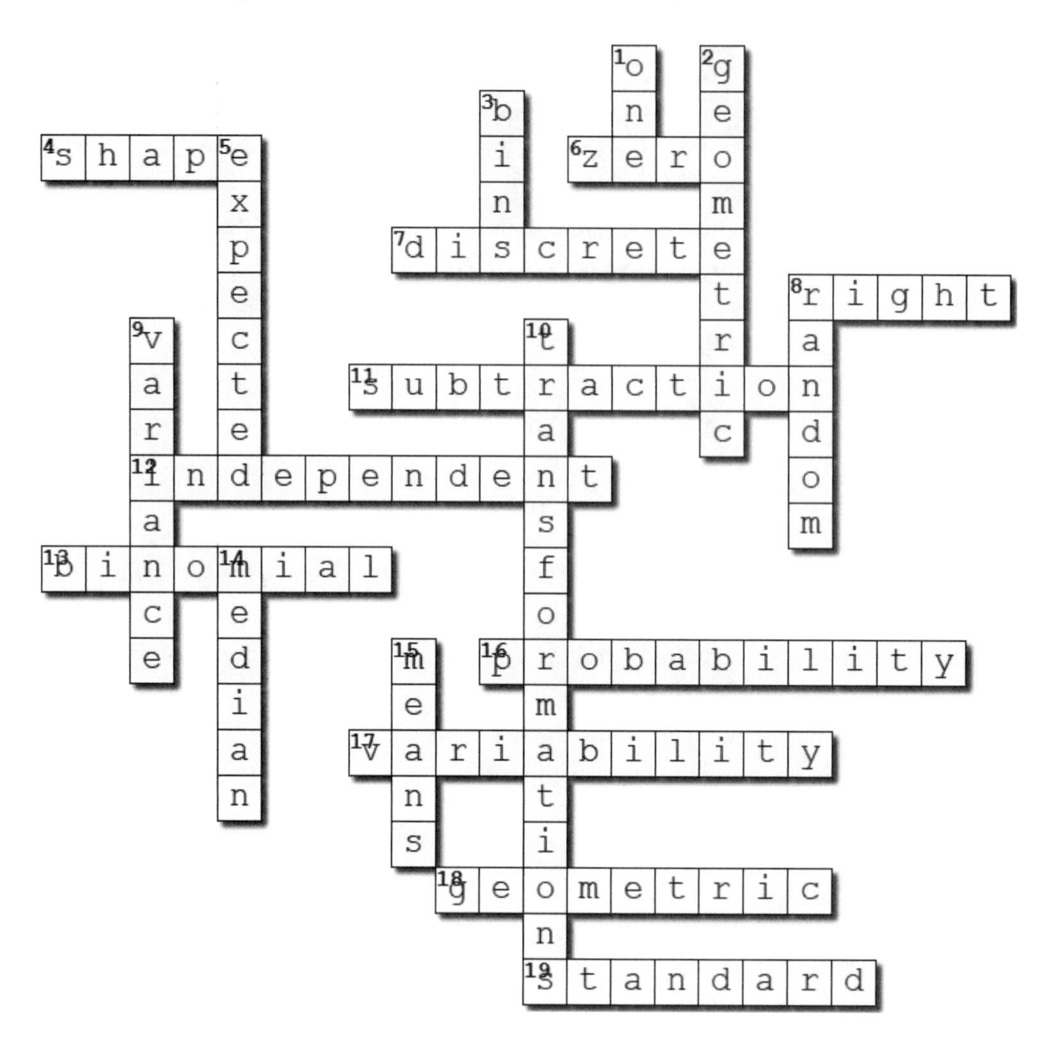

Unit 5 Solutions

Section 5A: Normal Distributions, Revisited

Check for Understanding – Learning Targets 1 and 2

1. Let X = the amount of time it takes Mr. Molesky to complete a particular course in his favorite game of video golf. The distribution of X is approximately normal with a mean of 110 minutes and a standard deviation of 10 minutes. We want to find $P(X < 95)$.
   $$z = \frac{95 - 110}{10} = -1.50$$
   Using technology: normalcdf(lower: –1000, upper: –1.5, mean: 0, SD: 1) = 0.0668.
   Using technology: normalcdf(lower: –1000, upper: 95, mean: 110, SD: 10) = 0.0668.
   Using Table A: 0.0668.
   There is a 0.0668 probability that Mr. Molesky will complete the course in less than 95 minutes.

2. We want to find the 90th percentile of the distribution of X.
   Using technology: invNorm(area: 0.90, mean: 0, SD: 1) = 1.282.
   Using technology: invNorm(area: 0.90, mean: 110, SD: 10) = 122.816
   Using Table A: $z = 1.28$
   $$1.282 = \frac{x - 110}{10} \rightarrow x = (1.282)(10) + 110 \rightarrow x = 122.82$$
   The 90th percentile of Mr. Molesky's course completion times is 122.816 minutes.

3. Let $D = X - Y$ = the difference in course completion time for Mr. Molesky and Ms. Liberty.
   $\mu_D = \mu_X - \mu_Y = 110 - 100 = 10$ minutes. (Ms. Liberty wins by 10 minutes, on average)
   $\sigma_D = \sqrt{\sigma_X^2 + \sigma_Y^2} = \sqrt{10^2 + 8^2} = 12.806$ minutes
   D is normally distributed with mean 10 minutes and SD 12.806 minutes.
   We want to find $P(\text{Molesky} - \text{Liberty} < 0) = P(D < 0)$.
   $$z = \frac{0 - 10}{12.806} = -0.781$$
   Using technology: normalcdf(lower: –1000, upper: –0.781, mean: 0, SD: 1) = 0.2174.
   Using technology: normalcdf(lower: –1000, upper: 0, mean: 10, SD: 12.806) = 0. 2174.
   Using Table A: 0.2177.
   There is a 0.2174 probability that Mr. Molesky will finish his game before Ms. Liberty.

Section 5B: What is a Sampling Distribution?

Check for Understanding – Learning Target 1

1. *Population*: All adults with high blood pressure
   Parameter: μ = mean reduction in blood pressure for all adults who take the new drug
   Sample: The 500 adults who participated in the study.
   Statistic: $\bar{x}$ = the mean reduction of blood pressure for the 500 adults who participated in the study.

2. *Population*: All 16- to 24-year old drivers.
   Parameter: p = proportion of all 16- to 24-year old drivers who text while driving.
   Sample: The 1500 drivers aged 16 to 24 who were surveyed.
   Statistic: $\hat{p}$ = the proportion of the sample who indicated they text while driving = 0.12.
   The sample proportion is not trustworthy. The population proportion is likely greater than the sample proportion due to response bias. (Liar, liar pants on fire.)

Check for Understanding – Learning Target 2

1. Here is the distribution of cost for the population of all power banks available at this store.

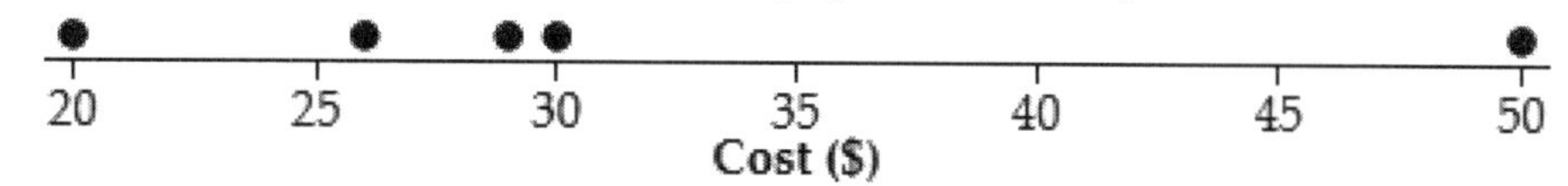

2. Here is the distribution of cost for one possible sample of $n = 2$ power banks.

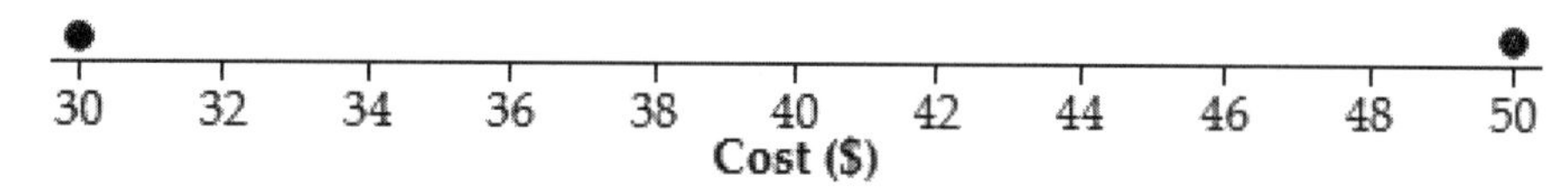

3. Here are the 10 possible samples of size $n = 2$ from this population along with the sample means.

Sample	26, 20	26, 30	26, 29	26, 50	20, 30	20, 29	20, 50	30, 29	30, 50	29, 50
Mean	23	28	27.5	38	25	24.5	35	29.5	40	39.5

Here is the sampling distribution:

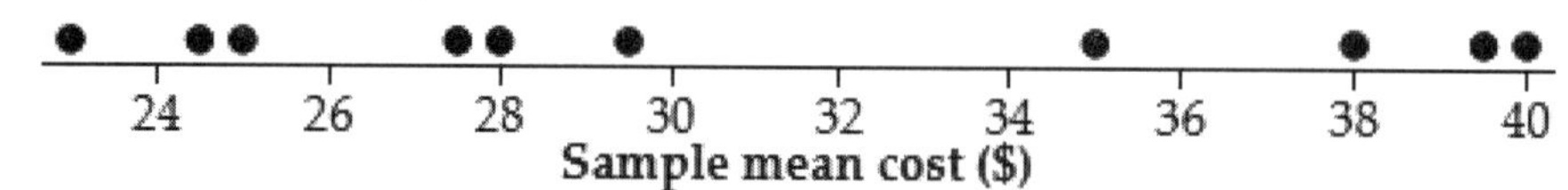

Check for Understanding – Learning Target 3

1. Population = all inexpensive bathroom scales produced by this manufacturer

2. The manufacturer claims $\mu = 150$ pounds = the true population mean weight these scales would display for a 150 pound weight.

3. The simulated sampling distribution tells us what sample means we might expect to see when selecting 12 scales from a population for which $\mu = 150$ pounds and $\sigma = 2$ pounds. If the manufacturer is telling the truth we would most likely obtain a sample mean in the ballpark of 148.75 to 151.25 pounds, because it is fairly unlikely that a random sample of 12 scales would produce a sample mean that is outside of that interval of weights.

4. It is surprising to get $\bar{x}$ that is greater than or equal to 151.5 pounds by chance alone when $\mu = 150$ pounds. Therefore, there is convincing evidence that the manufacturer is not correct about the accuracy of its bathroom scales.

Check for Understanding – Learning Targets 4 and 5

1. No. It is unlikely the population mean amount of change per locker for all U.S. high school students is \$0.42. Different samples of size 50 would produce different means.

2. A random sample of 200 students would be more likely to give an estimate closer to the value of the population parameter, because larger random samples tend to produce more precise estimates than smaller random samples.

Section 5C: Sample Proportions

Check for Understanding – Learning Targets 1 and 2

1. $\mu_{\hat{p}} = 0.05$. If you selected all possible samples of 250 potatoes from the truckload and calculated the sample proportion that are unacceptable for each sample, the sample proportions would have an average value of 0.05.

2. It is safe to assume that $n =$ 250 potatoes is less than 10% of all potatoes in the truckload.

   $$\sigma_{\hat{p}} = \sqrt{\frac{0.05(1-0.05)}{250}} = 0.0138.$$

   If you selected all possible samples of 250 potatoes from the truckload and calculated the sample proportion that are unacceptable for each sample, the sample proportions would typically vary from the population proportion of 0.05 by about 0.0138.

3. Because $np = 250(0.05) = 12.5 \geq 10$ and $n(1-p) = 250(1-0.05) = 237.5 \geq 10$, the sampling distribution of $\hat{p}$ is approximately normal.

Check for Understanding – Learning Target 3

We want to find $P(\hat{p} > 0.75)$ where $\hat{p}$ is the sample proportion of students who use social media.

- *Shape:* Approximately normal because $np = 1000(0.70) = 700 \geq 10$ and $n(1-p) = 1000(1-0.70) = 300 \geq 10$.
- *Center:* $\mu_{\hat{p}} = p = 0.70$
- *Variability:* $\sigma_{\hat{p}} = \sqrt{\frac{0.70(1-0.70)}{1000}} = 0.0145$

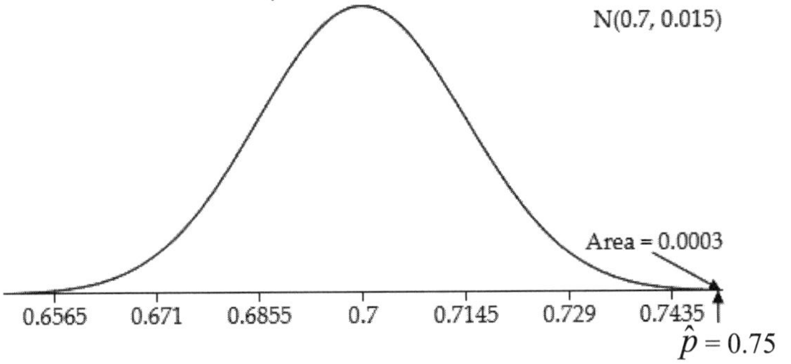

N(0.7, 0.015)

Area = 0.0003

0.6565 0.671 0.6855 0.7 0.7145 0.729 0.7435

$\hat{p} = 0.75$

Sample proportion of U.S. high school students who have social media

$$z = \frac{0.75-0.70}{0.0145} = 3.448$$

Using technology: normalcdf(lower: 3.448, upper: 1000, mean: 0, SD: 1) = 0.0003

Using technology: normalcdf(lower: 0.75, upper: 1000, mean: 0.70, SD: 0.0145) = 0.0003

Using Table A: 1 – 0.9997 = 0.0003

There is a 0.0003 probability that more than 75% of the students in the sample use social media.

Check for Understanding – Learning Target 4

1. The shape of the sampling distribution is approximately normal because
   $n_W p_W = (50)(0.45) = 22.5 \geq 10$, $n_W(1 - p_W) = (50)(1 - 0.45) = 27.5 \geq 10$,
   $n_E p_E = (50)(0.4) = 20 \geq 10$, and $n_E(1 - p_E) = (50)(1 - 0.4) = 30 \geq 10$.

2. The mean of the sampling distribution is $\mu_{\hat{p}_1 - \hat{p}_2} = p_1 - p_2 = 0.45 - 0.40 = 0.05$.

   In all possible independent random samples of 50 students from East High and 50 students from West High, the resulting differences in the sample proportions $(\hat{p}_1 - \hat{p}_2)$ of students who play sports has an average of 0.05.

3. Because 50 < 10% of all students at West High (it's a large high school) and 50 < 10% of all students at East High (it is also a large high school), and the two samples are independent,

   $\sigma_{\hat{p}_W - \hat{p}_E} = \sqrt{\frac{0.45(1-0.45)}{50} + \frac{0.4(1-0.4)}{50}} \approx 0.0987$.

4. $P(\hat{p}_E > \hat{p}_W)$ is equivalent to $P(\hat{p}_W < \hat{p}_E) = P(\hat{p}_W - \hat{p}_E < 0)$.
   Based on a normal distribution with a mean of 0.05 and a standard deviation of 0.0987:
   $P(\hat{p}_W - \hat{p}_E < 0)$ = Normalcdf(lower: –1000, upper: 0, mean: 0.05, SD: 0.0987) ≈ 0.3062.

Section 5D: Sample Means

Check for Understanding – Learning Targets 1 and 2

1. $\mu_{\bar{x}} = 6.8$ absences
   If you selected all possible samples of 40 students from this high school and calculated the sample mean number of absences for each sample, the sample means would have an average value of 6.8 absences.

2. Because we can assume that $n = 40$ is less than 10% of all students at their large high school,
   $\sigma_{\bar{x}} = \frac{6.8}{\sqrt{40}} = 1.075$ absences
   If you selected all possible samples of 40 students from the high school and calculated the sample mean number of absences for each sample, the sample means would typically vary from the population mean of 6.8 absences by about 1.075 absences.

3. Because $n = 40 \geq 30$, the sampling distribution of $\bar{x}$ will be approximately normal.

Check for Understanding – Learning Target 3

Let $\bar{x}$ = the mean blood cholesterol level for a random sample of 250 men.
The distribution of $\bar{x}$ is approximately normal because $n = 250 \geq 30$.
$\mu_{\bar{x}} = \mu =$ 188 mg/dl
$\sigma_{\bar{x}} = \frac{\sigma}{\sqrt{n}} = \frac{41}{\sqrt{250}} = 2.593$ mg/dl because it is reasonable to assume that 250 < 10% of all men.
We want to find $P(\bar{x} > 193)$.

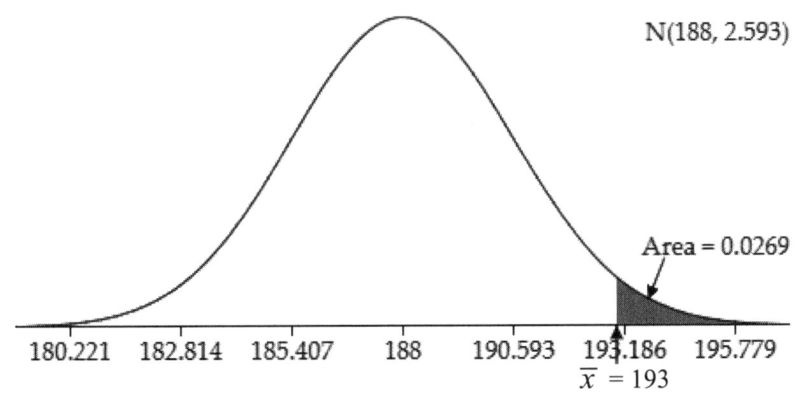

$z = \frac{193-188}{2.593} = 1.928$

Using technology: normalcdf(lower: 1.928, upper: 1000, mean: 0, SD: 1) = 0.0269.

Using technology: normalcdf(lower: 193, upper: 1000, mean: 188, SD: 2.593) = 0.0269.

Using Table A: 1 – 0.9732 = 0.0268.

There is about a 0.0269 probability that the sample mean blood cholesterol level will be greater than 193 mg/dl.

Check for Understanding – Learning Target 4

1. The shape of the sampling distribution of $\bar{x}_W - \bar{x}_E$ is approximately normal because both sample sizes are large ($n_W = 50 \geq 30$and $n_E = 50 \geq 30$).

2. The mean of the sampling distribution of $\bar{x}_W - \bar{x}_E$ is 95 – 55 = 40 minutes.

   In all possible independent random samples of 50 students from East High and 50 students from West High, the resulting differences in sample mean amount of time spent on homework ($\bar{x}_W - \bar{x}_E$) has an average of 40 minutes.

3. Because the samples are independent, and because we can assume 50 < 10% of all students at West High (it's a large high school) and 50 < 10% of all students at East High (it is also a large high school), $\sigma_{\bar{x}_W - \bar{x}_E} = \sqrt{\frac{25^2}{50} + \frac{35^2}{50}}$ = 6.083 minutes.

4. $P(\bar{x}_E > \bar{x}_W)$ is equivalent to $P(\bar{x}_W < \bar{x}_E) = P(\bar{x}_W - \bar{x}_E < 0)$.
   Based on a normal distribution with a mean of 40 and a standard deviation of 6.083:
   $P(\bar{x}_W - \bar{x}_E < 0)$ = Normalcdf(lower: –1000, upper: 0, mean: 40, SD: 6.083) ≈ 0.
   It is nearly impossible to select two random samples from these schools (each of size 50) for which the sample mean amount of time spent on homework for the East High sample is greater than that of the West High sample.

Unit 5 Crossword Puzzle Solutions

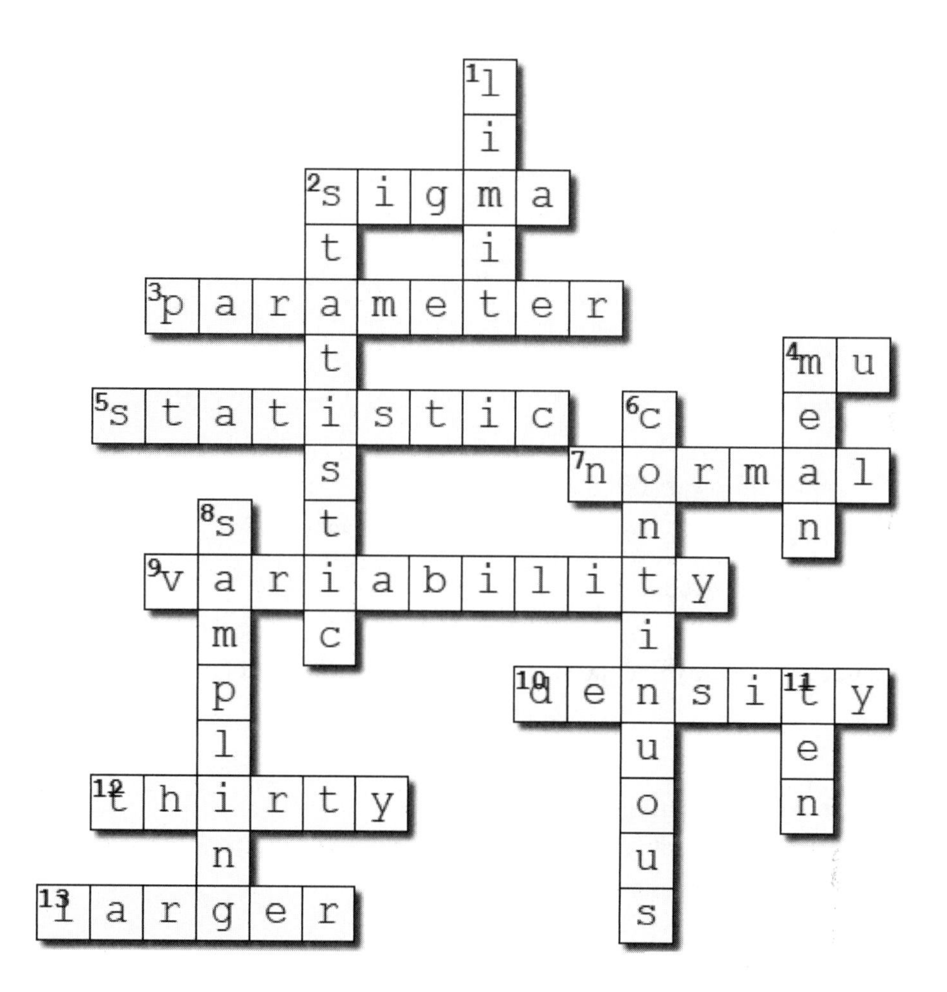

Unit 6, Part I Solutions

Section 6A: Confidence Intervals: The Basics

Check for Understanding – Learning Targets 1, 2, and 3

1. We are 90% confident the interval from 19.10 ounces to 20.74 ounces captures the true mean contents of a "20 oz." bottle of water.

2. If we were to select many random samples of fifty "20 oz." water bottles and construct a confidence interval for the mean contents using each sample, 90% of the intervals would capture the population mean content of all "20 oz." water bottles.

3. Because there are values in the interval from 19.10 to 20.74 that are greater than or equal to 20, there is not convincing evidence that the mean contents of all "20 oz." water bottles is less than 20 ounces.

Section 6B: Confidence Intervals: The Basics

Check for Understanding – Learning Targets 1, 2, and 3

State: 90% CI for p = the proportion of all adults who cannot name a single freedom that is protected by the First Amendment.

Plan: One-sample z interval for p.

- Random: Random sample of 300 adults. ✓
- 10%: 300 < 10% of all adults. ✓
- Large Counts: The number of successes (126) and failures (174) are both at least 10. ✓

Do: $\hat{p} = \frac{126}{300} = 0.42$

$$0.42 \pm 1.645\sqrt{\frac{0.42(0.58)}{300}}$$

$= 0.42 \pm 0.047$

$= (0.373, 0.467)$

Using technology: (0.373, 0.467)

Conclude: We are 90% confident the interval from 0.373 to 0.467 captures p = the proportion of all adults who cannot name a single freedom that is protected by the First Amendment.

Note: They are freedom of press, religion, speech, assembly, and the right to petition the government.

Check for Understanding – Learning Targets 4 and 5

1. The confidence interval would be narrower because decreasing the confidence level from 90% to 80% decreases the margin of error.

2. The confidence interval would be wider because decreasing the sample size from 300 to 100 increases the margin of error when everything else stays the same.

3. $1.960\sqrt{\frac{0.5(0.5)}{n}} \leq 0.02 \Rightarrow \sqrt{\frac{0.5(0.5)}{n}} \leq 0.0102 \Rightarrow \frac{0.25}{n} \leq 0.000104$

   $n \geq 2401$ adults

   We would need a sample size of at least 2401 adults to estimate the proportion of all adults who cannot name a single freedom that is protected by the First Amendment with 95% confidence and a margin of error of at most 2%.

Section 6C: Significance Tests: The Basics

Check for Understanding – Learning Target 1

H_0: $p = 1/6$ vs. H_a: $p > 1/6$ where p = the true proportion of rolls that will result in a 6. Use $\alpha = 0.05$.

Check for Understanding – Learning Targets 2 and 3

1. If the null hypothesis is true, the proportion of rolls that will result in a six will be about $\frac{1}{6} \approx 0.167$.

2. Assuming the die is still fair (Mrs. Chauvet failed at "loading" the die and p is still 1/6), there is less than a 0.001 probability of observing 38 or more sixes in 100 rolls of the die.

3. Yes. Because the P-value of less than $0.001 < \alpha = 0.01$, we reject H_0. We have convincing evidence that the proportion of rolls that will result in a six is greater than 1/6.

Section 6D: Significance Tests for a Population Proportion

Check for Understanding – Learning Targets 1, 2, and 3

State: We want to test H_0: $p = 0.05$ versus H_a: $p < 0.05$, where p = the proportion of all Alaskan moose hunters who have had someone whisper in their ear while moose hunting using $\alpha = 0.01$.

Plan: One-sample z test for a proportion

- Random: We have a random sample of 250 Alaskan moose hunters. ✓
- 10%: 250 < 10% of the large population of Alaskan moose hunters. ✓
- Large Counts: $250(0.05) = 12.5 \geq 10$ and $500(0.95) = 237.5 \geq 10$. ✓

Do: $\hat{p} = \frac{4}{250} = 0.016$. $z = \frac{0.016-0.05}{\sqrt{\frac{0.05(1-0.05)}{250}}} = -2.467$

P-value:
Using technology: $z = -2.467$, *P*-value = 0.0068
Using Table A: Area to the left of $z = -2.47$ is 0.0068.

Conclude: Because the *P*-value of $0.0068 < \alpha = 0.01$, we reject H_0. There is convincing evidence that the proportion of all Alaskan moose hunters who have had someone whisper in their ear while moose hunting is less than 0.05. This provides convincing evidence to allow for the removal of the law in Alaska.

Check for Understanding – Learning Target 4

1. *Type I error*: The researcher finds convincing evidence that less than 5% of all Alaskan moose hunters have had someone whisper in their ear while moose hunting, when the population proportion is really 0.05.
   Consequence: The lawmakers will remove the law from the Alaskan law books, when they should leave it in the books.

2. *Type II error*: The researcher does not find convincing evidence that less than 5% of all Alaskan moose hunters have had someone whisper in their ear while moose hunting, when the population proportion really is less than 0.05.
   Consequence: The Alaskan lawmakers keep the law, wasting law enforcement's time and effort when there are more important matters.

Check for Understanding – Learning Target 5

1. If the true proportion of processors that use more than 70 watts of power in the shipment is $p = 0.04$, there is a 0.12 probability that the quality control engineer will find convincing evidence for H_a: $p > 0.03$.
2. Decrease. Decreasing the significance level from 0.05 to 0.01 makes it harder to reject H_0 when H_a is true.
3. Increase. Increasing the sample size from 30 to 50 gives more information about the population proportion p.
4. Increase. It is easier to detect a bigger difference (0.02 versus 0.01) between the null and alternative parameter values.

Unit 6, Part I Crossword Puzzle Solutions

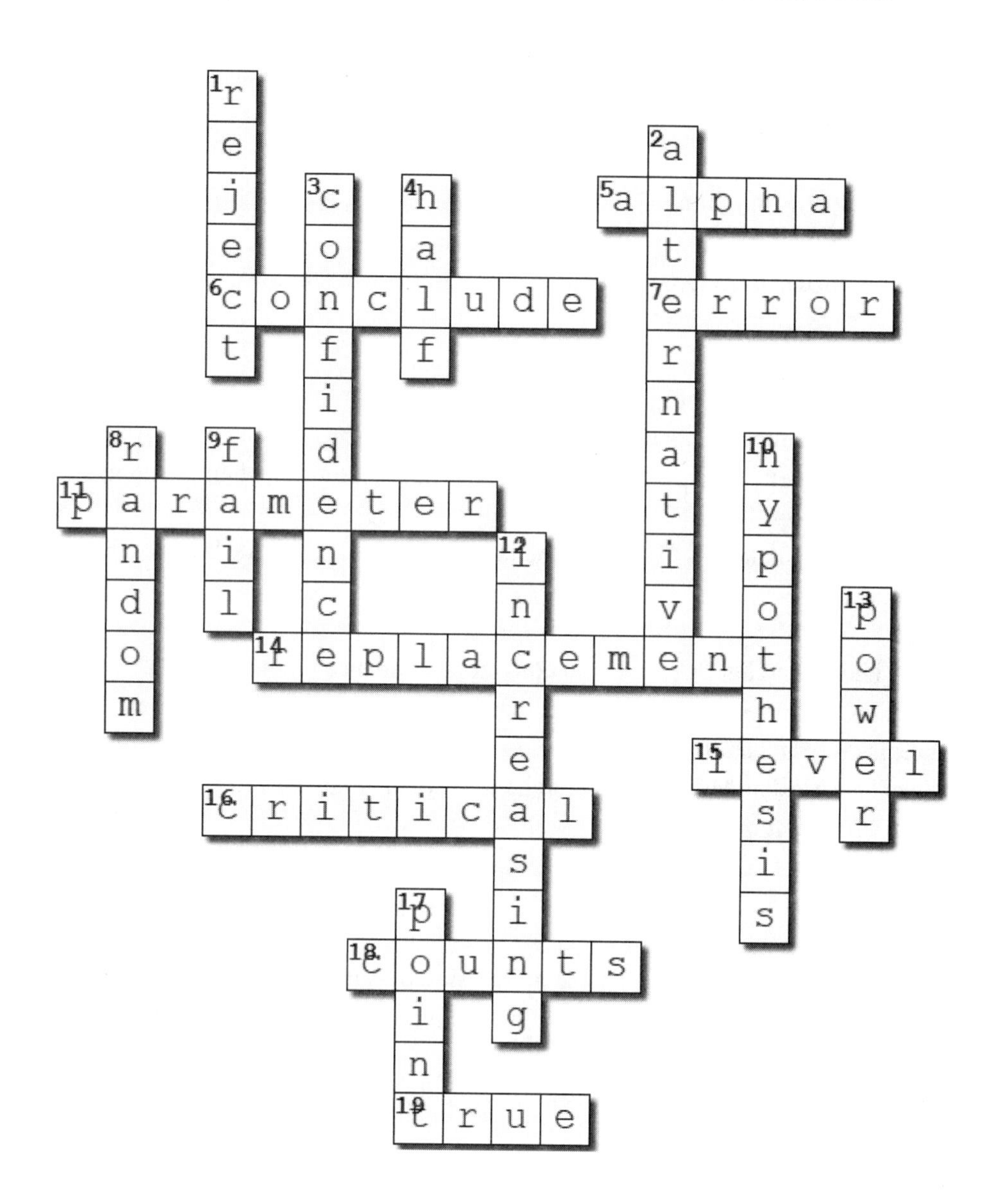

Unit 6, Part II: Solutions

Section 6E: Confidence Intervals for a Difference in Population Proportions

Check for Understanding – Learning Targets 1, 2, and 3

State: 95% confidence interval for $p_{2000} - p_{2020}$ where p_{2000}= the proportion of all U.S. adults who smoked in 2000 and p_{2020}= the proportion of all U.S. adults who smoked in 2020.

Plan: Two-sample *z* interval for a difference in proportions

- Random: We have independent random samples of 1500 U.S. adults in 2000 and 2000 U.S. adults in 2020. ✓
- 10%: 1500 < 10% of all U.S. adults in 2000 and 2000 < 10% of all U.S. adults in 2020. ✓
- Large Counts: 551, 1500 – 551 = 949, 352, 2000 – 352 = 1648 are all ≥ 10. ✓

Do: $\hat{p}_{2000} = \frac{551}{1500} = 0.367$ and $\hat{p}_{2020} = \frac{352}{2000} = 0.176$

95% CI: $(0.367 - 0.176) \pm 1.960\sqrt{\frac{0.367(1-0.367)}{1500} + \frac{0.176(1-0.176)}{2000}} = (0.1618, 0.2209)$

Conclude: We are 95% confident the interval from 0.1618 to 0.2209 captures the difference (2000 – 2020) in the proportions of all U.S. adults who smoked in 2000 and in 2020. Since all values contained in this interval are greater than 0, we have convincing evidence that the proportion of U.S. adults who smoked in 2000 is greater than in the proportion of U.S. adults who smoked in 2020.

Section 6F: Significance Tests for a Difference in Population Proportions

Check for Understanding – Learning Targets 1, 2, 3, and 4

State:

H_0: $p_H = p_G$

H_a: $p_H \neq p_G$

where p_H= the proportion of all healthcare providers who smoke and p_G= the proportion of the general population of adults who smoke. Use $\alpha = 0.05$.

Plan: Two-sample z test for a difference in proportions

- Random: We have two independent random samples of 150 healthcare providers and 120 adults from the general population. ✓
- 10%: 150 < 10% of all healthcare providers and 120 < 10% of all adults in the general population. ✓
- Large Counts: $\hat{p}_C = \frac{30+28}{150+120} = \frac{58}{270} = 0.215$

  $n_H\hat{p}_C = 150(0.215) = 32.25$

  $n_H(1 - \hat{p}_C) = 150(1 - 0.215) = 117.75$

  $n_G\hat{p} = 120(0.215) = 25.8$

  $n_G(1 - \hat{p}_C) = 120(1 - 0.215) = 94.2$

  The expected counts are all at least 10. ✓

Do: $\hat{p}_H = \frac{30}{150} = 0.200$ and $\hat{p}_G = \frac{28}{120} = 0.233$

$$z = \frac{0.200 - 0.233}{\sqrt{(0.215)(1 - 0.215)\left(\frac{1}{150} + \frac{1}{120}\right)}} = -0.663$$

P-value:
Using technology: $z = -0.663$, *P*-value = 0.5075
Using Table A: 0.2546(2) = 0.5092

Conclude:
Because the *P*-value of 0.5075 > $\alpha = 0.05$, we fail to reject the null hypothesis. We do not have convincing evidence to conclude that the proportion of all healthcare providers who smoke is different than the proportion of the general population of adults who smoke.

Unit 6, Part II Crossword Puzzle Solutions

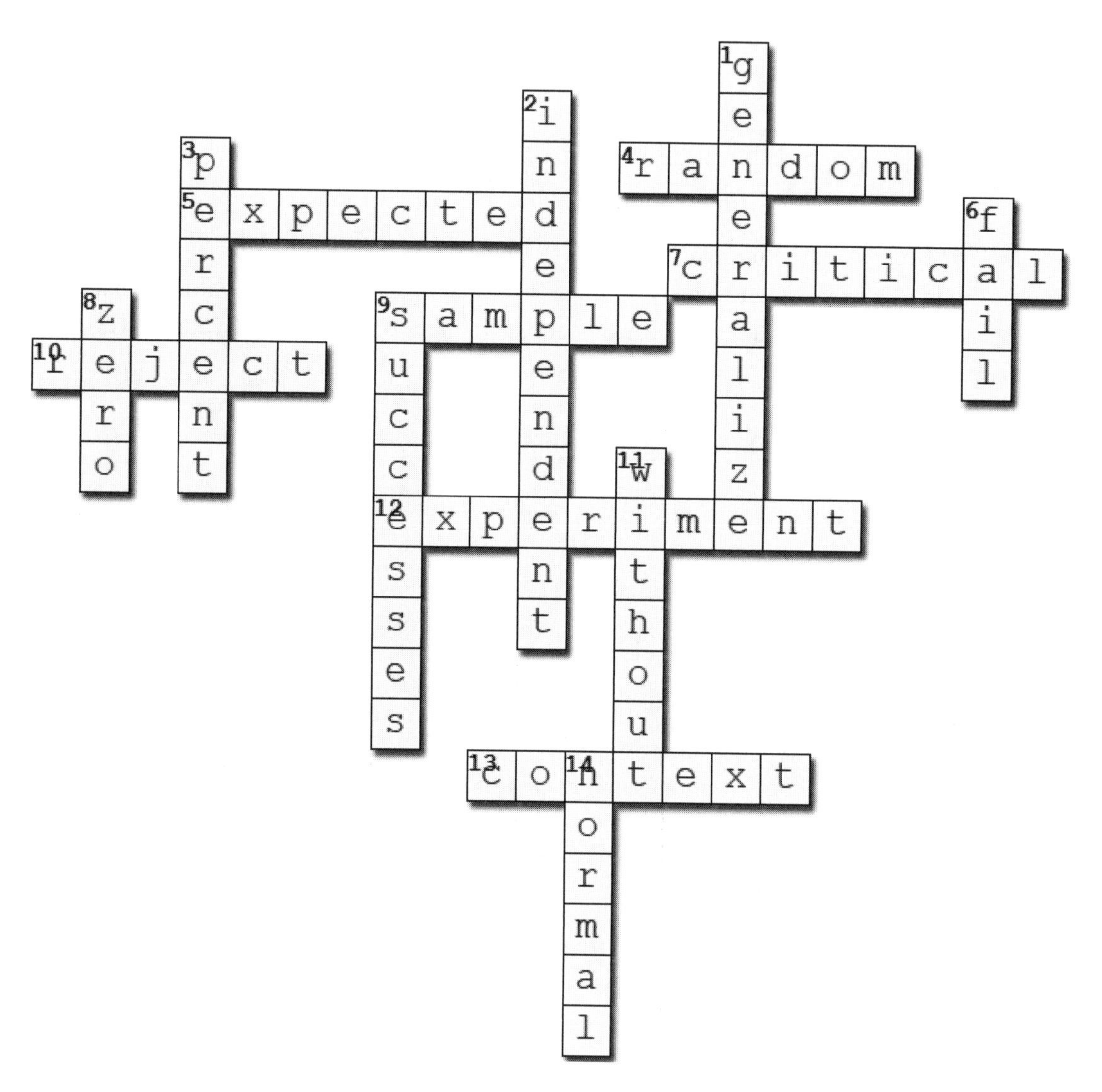

Unit 7 Solutions

Section 7A: Confidence Intervals for a Population Mean or Mean Difference

Check for Understanding – Learning Targets 1 and 2

1. $t^* = 1.330$
2. $t^* = 1.984$
3. $t^* = 2.756$
4. Random: Random sample of 28 college students. ✓
   10%: 28 < 10% of all college students. ✓
   Normal/Large Sample: The sample size is small. ($n = 28 < 30$) and the dotplot is strongly skewed to the right with some possible high outliers. – Condition not met!

Check for Understanding – Learning Targets 3 and 4

State: 95% CI for μ = the true mean amount of sugar for all soft drinks from this manufacturer.
Plan: One-sample t interval for a mean

- Random: A random sample of 8 soft drinks was selected. ✓
- 10%: 8 < 10% of all soft drinks from this manufacturer. ✓
- Normal/Large Sample: The shape of the distribution of the population is not provided and the sample size is small ($n = 8 < 30$). We must look at the distribution of the sample. The dotplot shows no strong skewness and no outliers. ✓

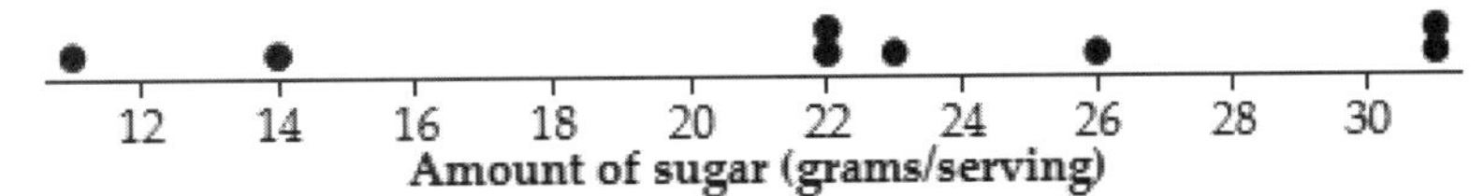

Do: $\bar{x} = 22.5$, $s_x = 7.191$, $n = 8$
With 95% confidence and df = 8 – 1 = 7, $t^* = 2.365$

$$22.5 \pm 2.365 \frac{7.191}{\sqrt{8}}$$

(16.487, 28.513)
Using technology: (16.488, 28.512) with df = 7.
Conclude: We are 95% confident the interval from 16.488 g/serving to 28.512 g/serving captures the mean amount of sugar for all soft drinks produced by this manufacturer.

Check for Understanding – Learning Target 5

State: 99% CI for μ_{Diff} = the true mean difference (Caffeine – Placebo) in the number of presses for subjects like the ones in this study.
Plan: Paired t interval for a mean difference

- Random: Paired data come from 2 groups in a randomized experiment. ✓
- Normal/Large Sample: Because the shape of the distribution of the population is unknown and the number of differences is small ($n = 11 < 30$), we will look at the distribution of the 11 differences. Here are the differences:

Subject		1	2	3	4	5	6	7	8	9	10	11
# of Presses	Caffeine	251	284	300	321	240	294	377	345	303	340	408
	Placebo	201	262	283	290	259	291	354	346	283	361	411
Difference	Caffeine – Placebo	50	22	17	31	–19	3	23	–1	20	–21	–3

Here is a dotplot of the differences:

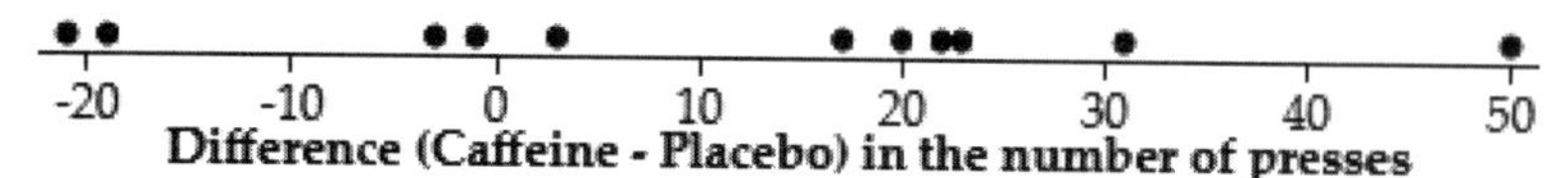

The distribution of the difference (Caffeine – Placebo) in the number of presses shows no strong skewness and contains no outliers. ✓

Do: $\bar{x}_{\text{Diff}} = 11.091$, $s_{\text{Diff}} = 21.520$, $n_{\text{Diff}} = 11$

With 99% confidence and df = 11 – 1 = 10, $t^* = 3.169$

$$11.091 \pm 3.169\left(\frac{21.520}{\sqrt{11}}\right)$$

(–9.471, 31.653)

Using technology: (–9.473, 31.654) with df = 10

Conclude: I am 99% confident that the interval from –9.473 to 31.654 captures the true mean difference (Caffeine – Placebo) in the number of presses for students like the ones in this study.

Section 7B: Significance Tests for a Population Mean or Mean Difference

Check for Understanding – Learning Targets 1, 2, and 3

State: We want to test H_0: $\mu = 8.9$ versus H_a: $\mu \neq 8.9$ where μ = the mean ratio for all humerus bone fossils at the Minnesota Iron Range site, using $\alpha = 0.05$.

Plan: One-sample *t* test for a mean

- Random: The researchers are willing to view the humerus bones they found as a random sample of all such humerus bones. ✓
- 10%: 41 < 10% of the large population of all humerus bones in this area. ✓
- Normal/Large Sample: $n = 41 \geq 30$ ✓

Do: $\bar{x} = 9.269$, $s_x = 1.199$, $n = 41$

$$t = \frac{9.269 - 8.9}{\frac{1.199}{\sqrt{41}}} = 1.971$$

P-value = 0.0557 with df = 41 – 1 = 40.

Using technology: $t = 1.970$, *P*-value = 0.0558, df = 40

Conclude: Because the *P*-value of $0.0558 > \alpha = 0.05$, we fail to reject the null hypothesis. There is not convincing evidence to suggest that the mean ratio for all humerus bones of the species Moleskius Primatium is different than 8.9.

Check for Understanding – Learning Target 4

State: H_0: $\mu_{\text{Diff}} = 0$ versus H_a: $\mu_{\text{Diff}} > 0$

μ_{Diff} = the true mean difference (Caffeine – Placebo) in the number of presses for subjects like the ones in this study.

Plan: Paired *t* test for a mean difference.

- Random: The treatment order was randomly assigned to the subjects. ✓
- Normal/Large Sample: Because the shape of the distribution of the population is unknown and the number of differences is small ($n = 11 < 30$), we will look at the distribution of the 11 differences. Here are the differences:

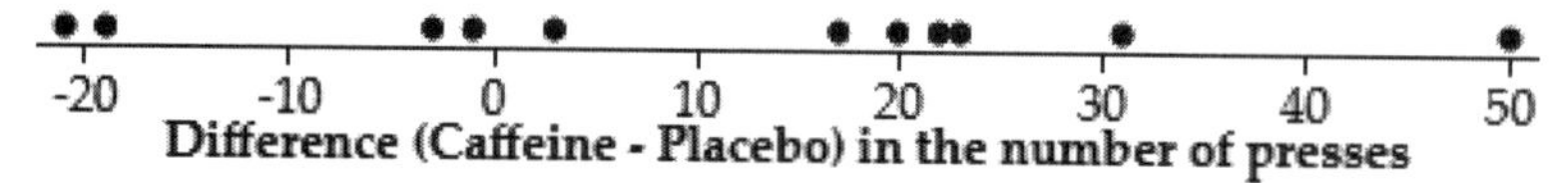

The distribution of the difference (Caffeine – Placebo) in the number of presses shows no strong skewness and contains no outliers. ✓

Do: $\bar{x}_{\text{Diff}} = 11.091$, $s_{\text{Diff}} = 21.520$, $n_{\text{Diff}} = 11$

$$t = \frac{11.091 - 0}{\frac{21.520}{\sqrt{11}}} = 1.709 \qquad P\text{-value} = 0.0591 \text{ with df} = 11 - 1 = 10$$

Using technology: $t = 1.709$, P-value = 0.0591, df = 10

Conclude: Because the P-value of $0.0591 > \alpha = 0.05$, we fail to reject H_0. There is not convincing evidence that the true mean difference (Caffeine – Placebo) in the number of presses for subjects like these is greater than 0.

Section 7C: Confidence Intervals for a Difference in Population Means

Check for Understanding – Learning Targets 1, 2, and 3

State: 95% CI for $\mu_1 - \mu_2$ = the true difference in mean weight loss where μ_1 = the mean weight loss for the new diet and μ_2 = the mean weight loss for the current diet.

Plan: Two-sample t interval for a difference in means

- Random: Patients were randomly assigned to the new diet or current diet. ✓
- Normal / Large Samples: $n_1 = 100$ and $n_2 = 100$ are both at least 30. ✓

Do: $(9.3 - 7.4) \pm 1.990\sqrt{\frac{4.7^2}{100} + \frac{4^2}{100}} = (0.672,\ 3.128)$ with df = 99 (using df = 80).

Using technology: (0.683, 3.117), df = 193.06

Conclude: We are 95% confident the interval from 0.683 to 3.117 captures the true difference in mean weight loss for the two diets for subjects like the ones in this study. Because all values contained in this interval are positive, we have evidence to suggest the new diet is more effective than the current diet.

Section 7D: Significance Tests for a Difference in Population Means

State: We want to test $H_0: \mu_1 - \mu_2 = 0$ versus $H_a: \mu_1 - \mu_2 > 0$, where μ_1 = the mean length of all country songs and μ_2 = the mean length of all pop songs using $\alpha = 0.05$.

Plan: Two-sample t-test for $\mu_1 - \mu_2$.

- Random: Random sample of 9 country songs and 8 pop songs. ✓
- 10%: 9 < 10% of all country songs and 8 < 10% of all pop songs. ✓
- Normal/Large Sample: The sample sizes are small ($n_1 = 9 < 30$ and $n_2 = 8 < 30$), but the dotplots show no strong skewness and no outliers. ✓

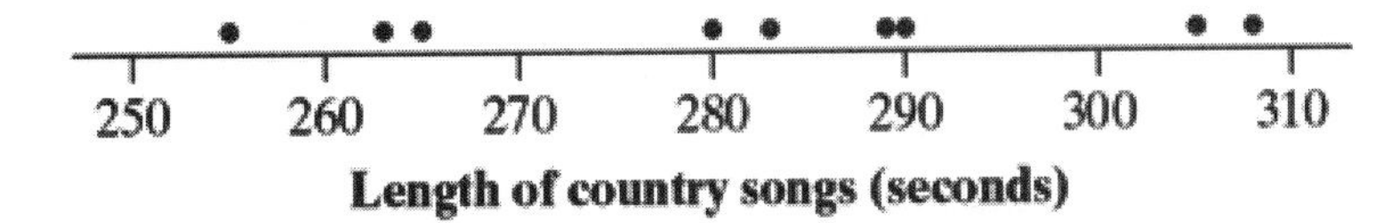

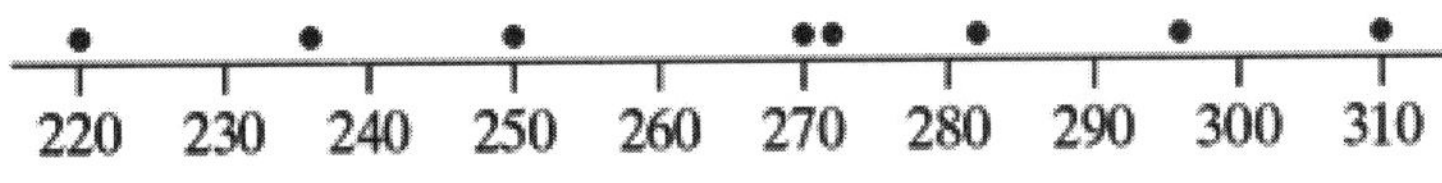

Do: $\bar{x}_1 = 282$, $s_1 = 18.378$, $n_1 = 9$, $\bar{x}_2 = 267.375$, $s_2 = 30.199$, and $n_2 = 8$

$$t = \frac{282 - 267.375}{\sqrt{\frac{18.378^2}{9} + \frac{30.199^2}{8}}} = 1.188$$

P-value:
Using technology: $t = 1.188$, *P*-value = 0.1296, df = 11.296.

Conclude: Because the *P*-value of $0.1296 > \alpha = 0.05$, we fail to reject H_0. We do not have convincing evidence that the mean length of all country songs is greater than the mean length of all pop songs.

Check for Understanding – Learning Target 4

1. Two-sample *t* procedures. The data come from two groups in a randomized experiment, where one group listens to Mozart for 10 minutes a day and the other group does not listen to Mozart.
2. Two-sample *t* procedures. The data come from independent random samples of drive-thru customers and customers who order inside.
3. Paired *t* procedures. The data come from two measurements of the same variable (number of steps) for each dog.

Unit 7 Crossword Puzzle Solutions

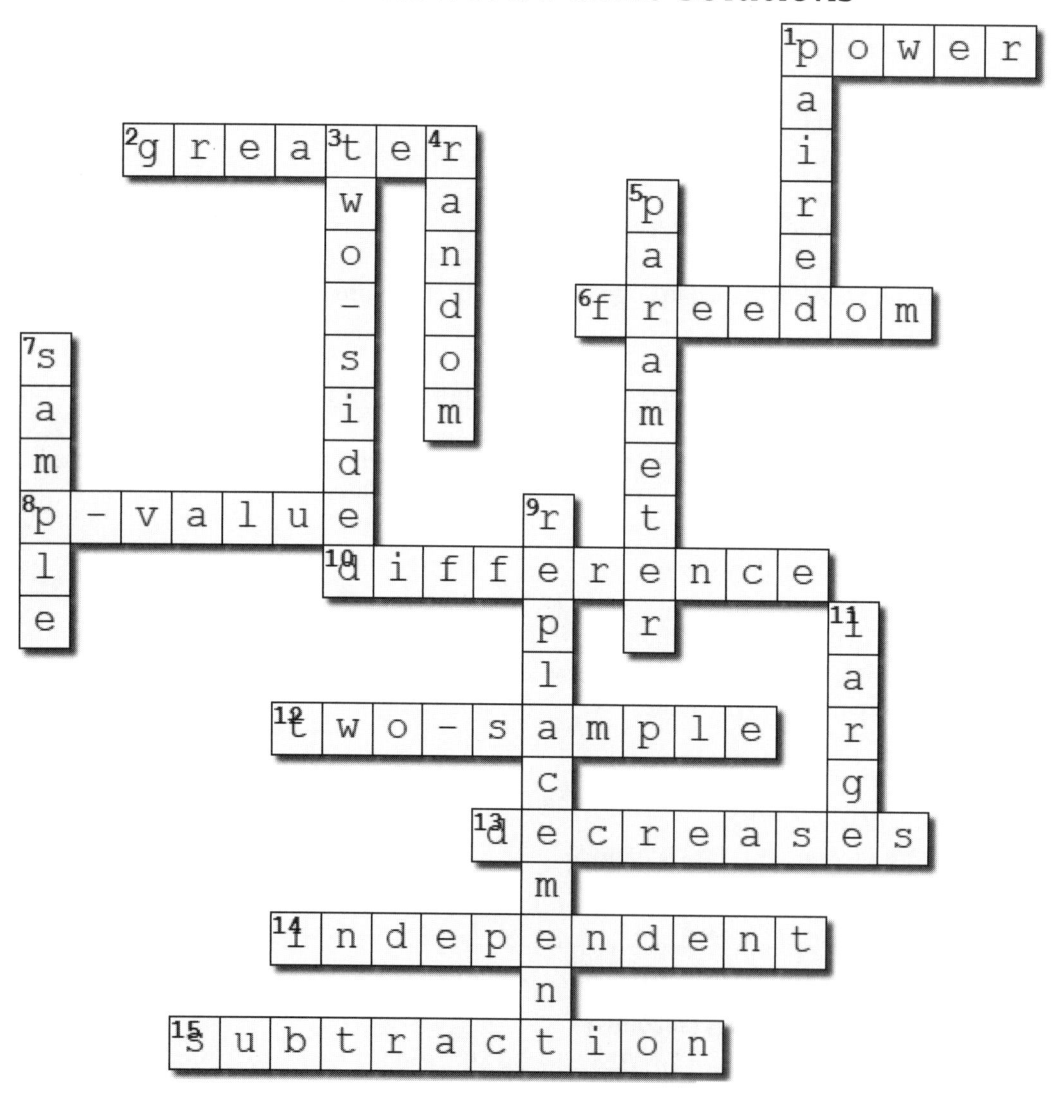

Unit 8 Solutions

8A: Chi-Square Tests for Goodness of Fit

Check for Understanding - Learning Targets 1, 2, 3, 4, and 5

State:

H_0: The distribution of die outcome is in the ratio 1:1:1:1:2:2.
H_a: The distribution of die outcome is not in the ratio 1:1:1:1:2:2.
OR
H_0: $p_1 = 0.125, p_2 = 0.125, p_3 = 0.125, p_4 = 0.125, p_5 = 0.25, p_6 = 0.25$
H_a: At least two of these proportions differ from the values stated by the null hypothesis where p_{outcome} = the population proportion of die rolls for each outcome.

Plan: Chi-Square Test for Goodness of Fit

- Random: We have a random sample of 400 die rolls. ✓
- 10%: Not needed, because we did not sample without replacement. ✓
- Large Counts: The expected counts in each category: Ones: 400(0.125) = 50, Twos: 400(0.125) = 50, Threes: 400(0.125) = 50, Fours: 400(0.125) = 50, Fives: 400(0.25) = 100, Sixes: 400(0.25) = 100 are all at least 5. ✓

Do:

$$\chi^2 = \frac{(42-50)^2}{50} + \frac{(55-50)^2}{50} + \frac{(48-50)^2}{50} + \frac{(57-50)^2}{50} + \frac{(94-100)^2}{100} + \frac{(104-100)^2}{100}$$

$\chi^2 = 3.36$

P-value: df = 6 – 1 = 5

Using technology: $\chi^2 = 3.36$, P-value = 0.6447, df = 5.

Using Table C: The P-value is greater than 0.25.

Conclude: Because the P-value of $0.6447 > \alpha = 0.05$, we fail to reject the null hypothesis. We do not have convincing evidence to conclude that the predicted 1:1:1:1:2:2 ratio is incorrect.

Section 8B: Chi-Square Tests for Independence or Homogeneity

Check for Understanding – Learning Targets 1, 2, 3, 4, and 5

1. H_0: There is no association between education level and earning more than $60,000/year in the population of all U.S. heads of household. H_a: There is an association between education level and earning more than $60,000/year in the population of all U.S. heads of household.
2. *Random*: The data come from a random sample of 529 heads of household. ✓
   10%: $n = 529$ is less than 10% of all U.S. heads of household. ✓
   Large Counts: The expected counts (39.06, 93.84, 99.09, 20.01, 42.94, 103.16, 108.91, and 21.99) are all at least 5. ✓
3. The test statistic is $\chi^2 = 15.391$. The test statistic has a chi-square distribution with (4 – 1)(2 – 1) = 3 degrees of freedom.
4. The P-value is 0.002. Because the P-value of $0.002 < \alpha = 0.05$, we reject H_0. There is convincing evidence of an association between education level and earning more than $60,000/year in the population of all U.S. heads of household.
5. The components that contribute the most to the significance of this test come from the category of individuals whose highest level of education is high school (Contributions: No: 5.7114, Yes: 5.1980). Among individuals whose highest education is high school, a much larger proportion earn $60,000 or less and a much smaller proportion earn more than $60,000 than would be expected if education level and earning more than $60,000 were independent.

Check for Understanding – Learning Targets 1, 2, 3, 4, and 6

State: H_0: The proportion of vases that are defective is the same for all three shifts.
H_a: The proportion of vases that are defective is not the same for all three shifts. Use $\alpha = 0.05$.

Plan: Chi-square test for Homogeneity
Random: The data come from independent random samples of vases from each of the 3 shifts. ✓
10%: $n_1 = 100 < 10\%$ of all vases produced by shift 1, $n_2 = 100 < 10\%$ of all vases produced by shift 2, and $n_3 = 100 < 10\%$ of all vases produced by shift 3. ✓
Large Counts: The expected counts (92, 92, 92, 8, 8, and 8) are all at least 5. ✓

Do: $\chi^2 = 4.348$, P-value = 0.1137, df = 2

Conclude: Because the P-value of $0.1137 > \alpha = 0.01$, we fail to reject H_0. There is not convincing evidence that the proportion of defective vases is not the same for the three shifts.

Check for Understanding – Learning Target 7

1. Chi-square test for homogeneity. The data come from independent random samples of 20 members of the community who have been cited for drinking and driving and 100 members of the community who have not been cited for drinking and driving.
2. Chi-square test for independence. The data come from a single random sample of 2000 American adults who are classified according to two categorical variables.

Unit 8 Crossword Puzzle Solutions

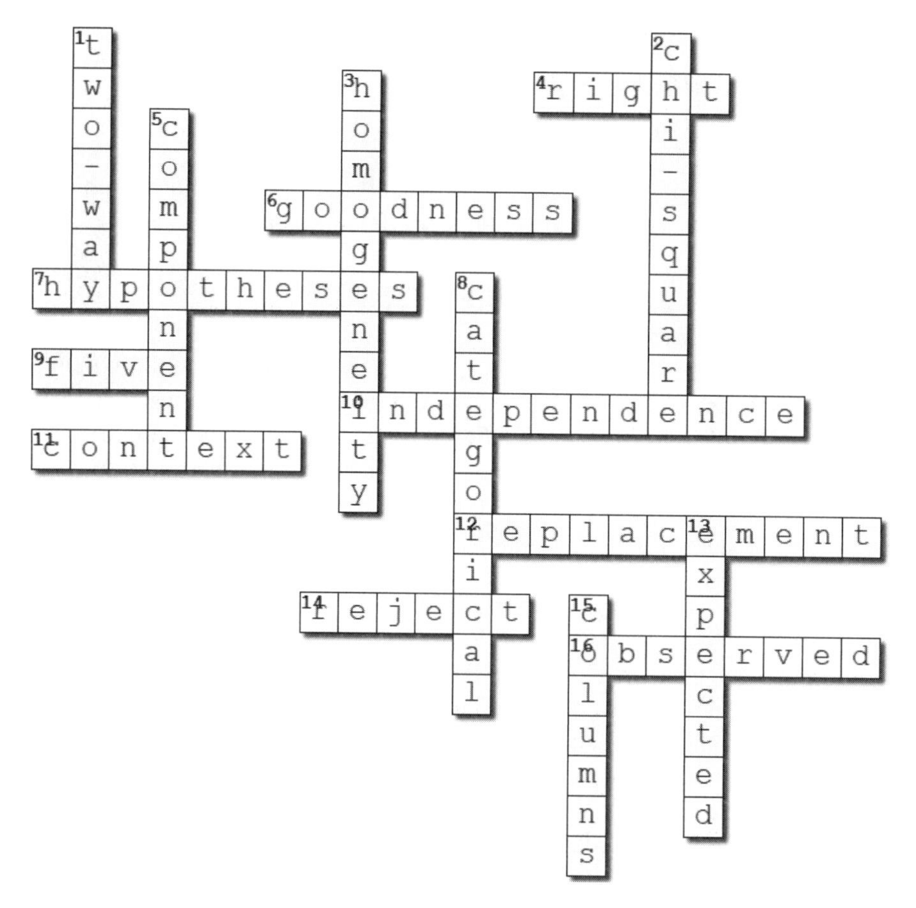

Unit 9 Solutions

Section 9A: Confidence Intervals for the Slope of a Population Regression Line

Check for Understanding – Learning Target 1

1. $\mu_b = -0.023$
   If you selected all possible samples of 5 dog breeds from the list of the 100 most popular dog breeds this year and calculated the slope of the least-squares regression line for predicting average lifespan (in years) from the average weight (in pounds) for each sample, the sample slopes would have an average value of –0.023.
2. Because 5 < 10% of the 100 most popular dog breeds this year, $\sigma_b = \frac{1.005}{37.62\sqrt{5}} = 0.012$.
   If you selected all possible samples of 5 dog breeds from the list of the 100 most popular dog breeds this year and calculated the slope of the least-squares regression line for predicting average lifespan (in years) from the average weight (in pounds) for each sample, the sample slopes would typically vary from the population slope of –0.023 by about 0.012.

Check for Understanding – Learning Target 2

- Linear: The scatterplot shows a linear relationship between amount of sugar and rating, and there is no leftover curved pattern in the residual plot. ✓
- Normal: The dotplot of the residuals shows no strong skewness or outliers. ✓
- Equal SD: In the residual plot, we do not see a clear < pattern or > pattern. ✓
- Random: Random sample of 10 cereals. ✓
  - 10%: 10 < 10% of all cereals. ✓

Check for Understanding – Learning Targets 3 and 4

STATE: 99% CI for β = the slope of the population regression line relating y = rating to x = amount of sugar (g) for all cereals.
PLAN: t interval for the slope – Assume the conditions for inference are met. ✓
DO: With 99% confidence and df = 10 – 2 = 8, $t^* = 3.355$
$-4.6065 \pm 3.355(0.98684)$
$= -4.6065 \pm 3.3108$
$= (-7.9173, -1.2957)$
CONCLUDE: We are 99% confident that the interval from −7.9173 to −1.2957 captures the slope of the population regression line relating y = rating to x = amount of sugar for all cereals.

Section 9B: Significance Tests for the Slope of a Population Regression Line

Check for Understanding – Learning Targets 1, 2, and 3

STATE: We want to test H_0: $\beta = 0$ versus H_a: $\beta < 0$ where β = the slope of the population regression line relating y = rating to x = amount of sugar (g) for all cereals. Use $\alpha = 0.01$
PLAN: t test for the slope

- Linear: The scatterplot shows a linear relationship between amount of sugar and rating, and there is no leftover curved pattern in the residual plot. ✓
- Normal: The dotplot of the residuals shows no strong skewness or outliers. ✓
- Equal SD: In the residual plot, we do not see a clear < pattern or > pattern. ✓
- Random: Random sample of 10 cereals. ✓
  - 10%: 77 < 10% of all cereals. ✓

DO: $t = -10.12$, P-value = 0.000 / 2 = 0 using df = 77 – 2 = 75
CONCLUDE: Because the P-value of approximately $0 < \alpha = 0.01$, we reject H_0. These data provide convincing evidence that the slope β of the population regression line relating x = amount of sugar to y = rating is less than 0.

Unit 9 Crossword Puzzle Solutions

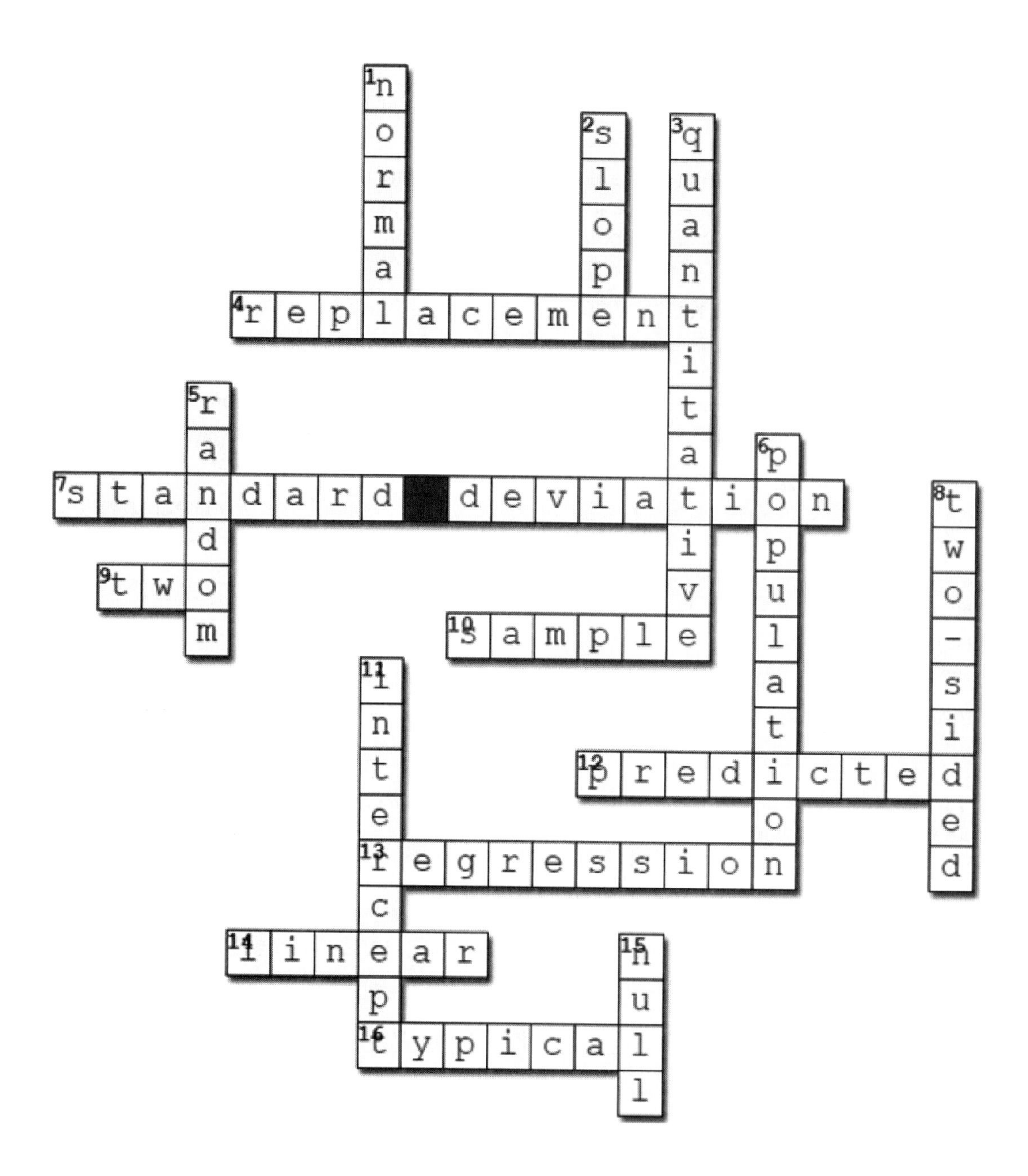

Preparing for the AP® Statistics Examination

After studying statistics all year, you will have a chance to take the AP® Statistics Exam in May. Not only will you be able to apply all that you have learned, but you will also have an opportunity to earn college credit for your efforts! If you earn a "passing score" on the exam, you may be eligible to receive AP® credit at a college or university. This means you will have demonstrated a level of knowledge equivalent to that of students completing an introductory statistics course. You may gain credit hours, advanced placement in a course sequence, and possibly save money on tuition! For these reasons, it is to your advantage to do your best on the exam. Hopefully this guide has helped you understand the concepts in AP® Statistics. Not only is a strong conceptual understanding necessary, but you must also plan and prepare for the exam itself. This section is designed to help you get ready for the exam. Best of luck in your studies!

The number of students taking the AP® Statistics Exam has been increasing in recent years. In 2023, about 243,000 students took the exam and earned scores between 1 and 5.

Score	Qualification	Translation
1	No recommendation	No credit, but you're still better off after having taken the course!
2	Possibly qualified	Credit? Probably not.
3	Qualified	Maybe college credit
4	Well qualified	Most likely college credit
5	Extremely well qualified	Statistics rock star…woohoo!

Of the 243,000 students who took the exam in 2023, 60% earned a score of 3 or higher and have a good chance of earning college credit. And just think, the majority of students taking the exam probably didn't prepare as much as you did! After working through this Strive Guide along with your textbook, you are in a good position to earn a high score on the exam!

Exam Prep Sections
1. Sample Schedule
2. The AP® Statistics Course Outline
3. The AP® Exam Format
4. Test Taking Tips
5. TPS7e AP® Exam Tips by Unit
6. Planning Your Exam Preparation and Review
7. Practice Exams

Section 1 – Sample Schedule

This section presents a general outline to help you prepare and review for the AP® Statistics Exam. Every student has their own routine and method of studying. Consider these suggestions as you work through the course and create a plan to maximize your chance of earning a 3 or higher on the exam.

1. ***At the start of your AP Statistics course***

   During the first few weeks of the school year, familiarize yourself with the course outline and AP® Statistics Exam format. It has been said that any student can hit a target that is clear and holds still for them. The course outline clearly indicates the topics you will study, and the exam format is always the same from year to year. Both the course outline and exam format are presented in the next few sections. Additional information on the exam can be found on the College Board's AP® Central website.

2. ***During your AP Statistics course***

   As you work through each unit, read the textbook and do all assigned exercises. Use this guide to help organize your notes, define key terms, and practice important concepts. Each unit in this guide includes practice multiple choice and free-response questions. Try all of them and note any concepts that give you difficulty.

3. ***Six weeks before the exam***

   About six weeks before the exam, begin planning for your review and preparation. If you are currently taking more than one AP® course, make sure you understand when each exam will be given and plan accordingly.

4. ***Four weeks before the exam***

   In early April, you should be wrapping up your studies in the course. This is a good time to attempt a practice test. Your teacher may provide one and/or you may want to take one of the practice exams in this book. Be sure to note any concepts that give you difficulty so you can tailor your preparation and review for the exam. There are still several weeks of class left and plenty of time to review, practice, and solidify your understanding of the key concepts in the course.

5. ***The week before the exam***

   The week before the exam, you should be done with your studies and should practice, practice, practice! Take this time to refresh your memory on the key concepts and review the topics that gave you the most trouble during the year. Study the flashcards that are found at the back of this book. Take another practice test to get used to the exam format. Allow 90 minutes for the multiple-choice section and 90 minutes for the free-response section. Use only the formulas and tables provided on the actual exam.

6. ***The day of the exam***

   You have prepared and reviewed as much as possible and are ready for the exam! Make sure you get a good night's sleep, eat a good breakfast the morning of the exam, and check that your calculator is working and has new batteries. Bring several pencils, a clean eraser, and arrive at your exam site early. Good luck!

Section 2 – The AP® Statistics Course Outline

The course outline for AP® Statistics is provided below. This outline lists the topics covered in the course and the percentage of the exam devoted to that material. Additional information on the course outline can be found on the College Board's AP® Central website.

Exam Weighting for the Multiple-Choice Section of the AP Exam

Units	Exam Weighting
1: Exploring One-Variable Data	15–23%
2: Exploring Two-Variable Data	5–7%
3: Collecting Data	12–15%
4: Probability, Random Variables, and Probability Distributions	10–20%
5: Sampling Distributions	7–12%
6: Inference for Categorical Data: Proportions	12–15%
7: Inference for Quantitative Data: Means	10–18%
8: Inference for Categorical Data: Chi-Square	2–5%
9: Inference for Quantitative Data: Slopes	2–5%

Section 3 – The AP® Statistics Exam Format

The AP® Statistics Exam is divided into two sections. The first section consists of 40 multiple-choice questions and the second section consists of 6 free-response questions. Each section counts for 50% of the total exam score. The number of questions you will be asked from each section of the topic outline corresponds to the percentages provided in the AP® Statistics Course outline. On the free-response section, you will most likely be asked at least one question about exploratory data analysis, at least one about probability, at least one about sampling or experimental design, and at least one about inference. Question 6 on the free-response section is considered an "investigative task" that will stretch your skills beyond what you have learned in the course and is worth almost twice that of each of the other 5 questions. Even though this question most commonly deals with an unfamiliar topic, you should be able to provide a reasonable answer based on your preparation.

The Multiple-Choice Section – 50% of the total score

You will have 90 minutes to complete the 40-question multiple-choice section of the exam. This gives you roughly about 2 minutes to answer each question. Each question has five answer choices (A-E), only one of which is correct. Each correct answer earns you one point, while each question answered incorrectly (or left blank) earns you no points. It is in your best interest to answer every question, even if you need to make an educated guess! The questions will cover all of the topics of the course and may require some calculation. You are allowed to use a graphing calculator for this section. Some may require interpreting a graph or reflecting on a given situation. Don't feel it is necessary to answer the questions in order. A good strategy is to answer all of the ones you feel are easy and then go back to the ones that might take a little more time. Be sure to keep an eye on the clock!

The Free-Response Section – 50% of the total score

The second section of the exam is made up of 6 free-response questions including one "investigative task." You will have 90 minutes to complete this section of the exam. You should allow about 25 minutes for the investigative task, leaving just over 12 minutes for each of the other questions. The first 5 free-response questions make up 37.5% of your total score and the investigative task makes up 12.5% of your total score.

You are also allowed to use a graphing calculator in this section. The free-response questions are designed to measure your statistical reasoning and communication skills and are graded on the following 0-4 scale.

4 = Complete Response {NO statistical errors and clear communication}
3 = Substantial Response {Minor statistical error/omission or fuzzy communication}
2 = Developing Response {Important statistical error/omission or poor communication}
1 = Minimal Response {A "glimmer" of statistical knowledge related to the problem}
0 = Inadequate Response {Statistically dangerous to self and others}

Each problem is graded holistically by the AP® Statistics Readers, meaning your entire response to the problem and all its parts is considered before a score is assigned. Be sure to keep an eye on the clock so you can provide at least a basic response to each question!

Section 4 – Test Taking Tips

Once you have mastered all of the concepts and have built up your statistical communication skills, you are ready to begin reviewing for the actual exam. The following tips were written by your textbook's author, Daren Starnes, and a former AP® Statistics teacher, Sanderson Smith, and are used with permission.

General Advice

Relax, and take time to think! Remember that everyone else taking the exam is in a situation identical to yours. Realize that the problems will probably look considerably more complicated than those you have encountered in other math courses. That's because a statistics course is, necessarily, a "wordy" course.

Read each question carefully before you begin working. This is especially important for problems with multiple parts or lengthy introductions. Underline key words, phrases, and information as you read the questions.

Look at graphs and displays carefully. For graphs, note carefully what is represented on the axes, and be aware of number scale. Some questions that provide tables of numbers and graphs relating to the numbers can be answered simply by "reading" the graphs.

About graphing calculator use: As noted throughout this guide, your graphing calculator is meant to be a tool and is to be used sparingly on some exam questions. Your brain is meant to be your primary tool. Don't waste time punching numbers into your calculator unless you're sure it is necessary. Entering lists of numbers into a calculator can be time-consuming, and certainly doesn't represent a display of statistical intelligence. Do not write directions for calculator button-pushing on the exam and avoid calculator syntax, such as *normalcdf* or *1-PropZTest.*

Multiple-choice questions:

- Examine the question carefully. What statistical topic is being tested? What is the purpose of the question?
- Read carefully. After deciding on an answer, make sure you haven't made a careless mistake or an incorrect assumption.
- If an answer choice seems "too obvious," think about it. If it's so obvious to you, it's probably obvious to others, and chances are good that it is not the correct response.
- Since there is no penalty for a wrong answer or skipped problem, it is to your advantage to attempt every question or make an educated guess, if necessary.

Free-response questions:

- Do not feel it necessary to work through these problems in order. Question 1 is meant to be straightforward, so you may want to start with it. Then move to another problem that you feel confident about. Whatever you do, don't run out of time before you get to Question 6. This Investigative Task counts almost twice as much as any other question.
- Read each question carefully, sentence by sentence, and underline key words or phrases.
- Decide what statistical concept/idea is being tested. This will help you choose a proper approach to solving the problem.
- You don't have to answer a free-response question in paragraph form. Sometimes an organized set of bullet points or an algebraic process is preferable. NEVER leave "bald answers" or "just numbers" though!
- ALWAYS answer each question in context.
- The amount of space provided on the free-response questions does not necessarily indicate how much you should write.
- If you cannot get an answer to part of a question, make up a plausible answer to use in the remaining parts of the problem.

On problems where you have to produce a graph:

- Label and scale your axes! Do not copy a calculator screen verbatim onto the exam.
- Don't refer to a graph on your calculator that you haven't drawn. Transfer it to the exam paper. Remember, the person grading your exam can't see your calculator!

Communicate your thinking clearly.

- Organize your thoughts before you write, just as you would for an English paper.
- Write neatly. The AP® Readers cannot score your solution if they can't read your writing!
- Write efficiently. Say what needs to be said, and move on. Don't ramble.
- The burden of communication is on you. Don't leave it to the reader to make inferences.
- When you finish writing your answer, look back. Does the answer make sense? Did you address the context of the problem?

Follow directions. If a problem asks you to "explain" or "justify," then be sure to do so.

- Don't "cast a wide net" by writing down everything you know, because you will be graded on everything you write. If part of your answer is wrong, you will be penalized.
- Don't give parallel solutions. Decide on the best path for your answer, and follow it through to the logical conclusion. Providing multiple solutions to a single question is generally not to your advantage. You will be graded on the lesser of the two solutions. Put another way, if one of your solutions is correct and another is incorrect, your response will be scored "incorrect."

Remember that your exam preparation begins on the first day of your AP® Statistics class. Keep in mind the following advice throughout the year.

- READ your statistics book. Most AP® Statistics Exam questions start with a paragraph that describes the context of the problem. You need to be able to pick out important statistical cues. The only way you will learn to do that is through hands-on experience.
- PRACTICE writing about your statistical thinking. Your success on the AP® Statistics Exam depends not only on how well you can "do" the statistics, but also on how well you explain your reasoning.
- WORK as many problems as you can in the weeks leading up to the exam.

Section 5 – Updated TPS6e AP® Exam Tips by Unit

These AP® Statistics Exam tips will look familiar to you from reading "The Practice of Statistics, 7e" text. We have included them here as a valuable review as you prepare for the AP® Statistics Exam

Unit 1: Exploring One-Variable Data

- If you learn to distinguish categorical from quantitative variables now, it will pay big rewards later. You will be expected to analyze categorical and quantitative variables correctly on the AP® exam.
- When comparing distributions of quantitative data, it's not enough just to list values for the center and spread of each distribution. You have to explicitly *compare* these values, using words like "greater than," "less than," or "about the same as."
- If you're asked to make a graph on a free-response question, be sure to label and scale the axes. Unless your calculator shows labels and scaling, don't just transfer a calculator screen shot to your paper.
- You may be asked to determine whether a quantitative data set has any outliers. Be prepared to state and use the rule for identifying outliers.
- Use statistical terms carefully and correctly of the AP® exam. Don't say "mean" if you really mean "median." Range is a single number; so are Q_1, Q_3, and *IQR*. Avoid the use of language like "the outlier *skews* the mean." Skewed is a shape. If you misuse a term, expect to lose some credit.
- Avoid using calculator speak to show your work on normal calculations. To get full credit, take the time to draw and label a normal distribution, label the boundary value(s), and shade the area of interest. Show your work when performing calculations!

Unit 2: Exploring Two-Variable Quantitative Data

- If you are asked to make a scatterplot on a free-response question, be sure to label and scale both axes. Don't copy an unlabeled calculator graph directly onto your paper.
- If you're asked to interpret correlation, make sure the relationship displayed in the scatterplot is fairly linear. Then say, "The linear relationship between [*x* context] and [*y* context] is [strength description] and [direction description]."
- When displaying the equation of a least-squares regression line, the calculator will report the slope and intercept with much more precision than we need. However, there is no firm rule for how many decimal places to show for answers on the AP® exam. Our advice: Decide how much to round based on the context of the problem you are working on.

- Students often have a hard time interpreting the value of r^2 on AP® exam questions. They frequently leave out key words in the definition. Our advice: Treat this as a fill-in-the-blank exercise. Write "________% of the variation in [response variable name] is accounted for by the linear model relating [response variable name] to [explanatory variable name]."

Unit 3: Collecting Data

- If you're asked to describe how the design of a study leads to bias, you're expected to identify the *direction* of the bias. Suppose you were asked, "Explain how using a convenience sample of students in your statistics class to estimate the proportion of all high school students who own a graphing calculator could result in bias." You might respond, "This sample would probably include a much higher proportion of students with a graphing calculator than in the population at large. That is, this method would probably lead to an overestimate of the actual population proportion."
- If you are asked to identify a possible confounding variable in a given setting, you are expected to explain how the variable you choose (1) is associated with the explanatory variable and (2) affects the response variable.
- If you are asked to describe the design of an experiment on the AP® exam, you won't get full credit for a diagram. You are expected to describe how the treatments are assigned to the experimental units and to clearly state what will be measured or compared. Some students prefer to start with a diagram and then add a few sentences. Others choose to skip the diagram and put their entire response in narrative form.
- Don't mix the language of experiments and the language of sample surveys or other observational studies. You will lose credit for saying things like "use a randomized block design to select the sample for this survey" or "this experiment suffers from nonresponse because some subjects dropped out during the study."

Unit 4: Probability, Random Variables, and Probability Distributions

- On the AP® exam, you may be asked to describe how you will perform a simulation using rows of random digits. If so, provide a clear enough description of your simulation process for the reader to get the same results you did from *only* your written explanation.
- Many probability problems involve simple computations that you can do on your calculator. It may be tempting to just write down your final answer without showing the supporting work. Don't do it! A "naked answer", even if it's correct, will usually earn you no credit on a free response question.
- If the mean of a random variable should have a non-integer value, but you report it as an integer, your answer will be marked as incorrect.
- When showing your work on a free response question, you must include more than a calculator command. Writing normalcdf(68, 70, 64, 2.7) will *not* earn you full credit for a normal calculation. At a minimum you must indicate what each of those calculator inputs represents. Better yet, sketch and label a normal curve to show what you're calculating.
- Don't rely on "calculator speak" when showing your work on free-response questions. Writing binompdf(5, 0.25, 3) = 0.08789 will *not* earn you full credit for a binomial probability calculation. At the very least, you must indicate what each of those calculator inputs represents. For example, "binompdf(n: 5, p: 0.25, x: 3)."

Unit 5: Sampling Distributions

- Terminology matters. Don't say "sample distribution" when you mean sampling distribution. You will lose credit on free-response questions for misusing statistical terms.
- Notation matters. The symbols $\hat{p}$, $\overline{x}$, p, μ, σ, $\mu_{\hat{p}}$, $\sigma_{\hat{p}}$, $\mu_{\overline{x}}$, and $\sigma_{\overline{x}}$ all have specific and different meanings. Either use notation correctly—or don't use it at all. You can expect to lose credit if you use incorrect notation.

Unit 6: Inference for Categorical Data: Proportions

- On a given problem, you may be asked to interpret the confidence interval, the confidence level, or both. Be sure you understand the difference: the confidence level describes the long-run capture rate of the method, and the confidence interval gives a set of plausible values for the parameter.
- If a free-response question asks you to construct and interpret a confidence interval, you are expected to do the entire four-step process. That includes clearly defining the parameter and checking conditions.
- You may use your calculator to compute a confidence interval for one proportion on the AP® exam. But there's a risk involved. If you just give the calculator answer with no work, you'll get either full credit for the "Do" step (if the interval is correct) or no credit (if it's wrong). We recommend showing the calculation with the appropriate formula and then checking with your calculator. If you opt for the calculator-only method, be sure to name the procedure (ex. one-sample *z* interval for a proportion) and to give the interval (ex. 0.514 to 0.607).
- If a question of the AP® exam asks you to calculate a confidence interval, all the conditions should be met. However, you are still required to state the conditions and show evidence that they are met.
- The formula for the two-sample z interval for $p_1 - p_2$ often leads to calculation errors by students. As a result, we recommend using the calculator's 2-PropZInt feature to compute the confidence interval of the AP® Exam. Be sure to name the procedure (two-proportion *z* interval) and to give the interval (0.076, 0.143) as part of the "Do" step.
- The conclusion to a significance test should always include three components: (1) an explicit comparison of the *P*-value to a stated significance level (2) a decision about the null hypothesis: reject or fail to reject H_0, and (3) a statement in the context of the problem about whether or not there is convincing evidence for H_a.
- When a significance test leads to a fail to reject H_0 decision, be sure to interpret the results as "we don't have enough evidence to conclude H_a." Saying anything that sounds like you believe H_0 is (or might be) true will lead to a loss of credit. Don't write text-message-type responses, like "FTR H_0."
- You can use your calculator to carry out the mechanics of a significance test on the AP® exam. But there's a risk involved. If you just give the calculator answer with no work, and one or more of your values is incorrect, you will probably get no credit for the "Do" step. We recommend doing the calculation with the appropriate formula and then checking with your calculator. If you opt for the calculator-only method, be sure to name the procedure (one-proportion *z* test or two-proportion *z* test) and to report the test statistic ($z = 1.15$) and *P*-value (0.1243).

Unit 7: Inference for Quantitative Data: Means

- It is not enough just to make a graph of the data on your calculator when assessing normality. You must *sketch* the graph on your paper to receive credit. Don't expect to receive credit for describing a graph that you made on your calculator but didn't put on paper. You don't have to draw multiple graphs – any appropriate graph will do.
- You may use your calculator to compute a confidence interval for one mean on the AP® exam. But there's a risk involved. If you just give the calculator answer with no work, you'll get either full credit for the "Do" step (if the interval is correct) or no credit (if it's wrong). We recommend showing the calculation with the appropriate formula and then checking with your calculator. If you opt for the calculator-only method, be sure to name the procedure (ex. one-sample *t* interval for μ) and to give the interval (ex. 514 to 607).
- The formula for the two-sample *t* interval for $\mu_1 - \mu_2$ often leads to calculation errors by students. As a result, we recommend using the calculator's 2-SampTInt feature to compute the confidence interval of the AP® Exam. Be sure to name the procedure (two-sample *t* interval) and to give the interval (3.9362, 17.724) and df: 55.728 as part of the "Do" step.
- It is not enough just to make a graph of the data on your calculator when assessing Normality. You must *sketch* the graph on your paper to receive credit. You don't have to draw multiple graphs – any appropriate graph will do.
- Remember: if you just give calculator results with no work, and one or more values are wrong, you probably won't get any credit for the "Do" step. We recommend doing the calculation with the appropriate formula and then checking with your calculator. If you opt for the calculator-only method, name the procedure (*t* test), and report the test statistic: $t = -0.94$, degrees of freedom: df = 14, and *P*-value: 0.1809.
- The formula for the two-sample *t* statistic for $\mu_1 - \mu_2$ often leads to calculation errors by students. As a result, we recommend using the calculator's 2-SampTTest feature to compute the confidence interval of the AP® Exam. Be sure to name the procedure (two-sample *t* test) and to report the test statistic: $t = 1.600$, *P*-value: 0.0644, and df: 15.59 as part of the "Do" step.

Unit 8: Inference for Categorical Data: Chi-Square

- You can use your calculator to carry out the mechanics of a significance test on the AP® exam. But there's a risk involved. If you just give the calculator answer with no work, and one or more of your values are incorrect, you will probably get no credit for the "Do" step. We recommend writing out the first few terms of the chi-square calculation followed by "...". This approach might help you earn partial credit if you enter a number incorrectly. Be sure to name the procedure (χ^2GOF-Test, χ^2-Test for homogeneity, or χ^2-Test for association/independence) and to report the test statistic: $\chi^2 = 11.2$, degrees of freedom: df = 3, and *P*-value: 0.011.
- In the "Do" step, you are not required to show every term in the chi-square statistic. Writing the first few terms of the sum and the last term, separated by ellipsis, is considered as "showing work." We suggest that you do this and then let your calculator tackle the computations.
- If you have trouble distinguishing the two types of chi-square tests for two-way tables, you're better off just saying "chi-square test" than choosing the wrong type. Better yet, learn to tell the difference!

Unit 9: Inference for Quantitative Data: Slopes

- The formula for the t interval for the slope of a population (true) regression line often leads to calculation errors by students. As a result, we recommend using the calculator's LinRegTInt feature to compute the confidence interval of the AP® Exam. Be sure to name the procedure (t interval for slope) and to give the interval (–0.217, –0.108) and df = 14 as part of the "Do" step.
- When you see a list of data values on an exam question, don't just start typing the data into your calculator. Read the question first. Often, additional information is provided that makes it unnecessary for you to enter the data at all. This can save you valuable time on the exam.

Section 6 – Planning Your Exam Preparation and Review

This book has been designed to help you identify your statistical areas of strength and areas in which you need improvement. If you have been working through the checks-for-understanding, multiple-choice questions, FRAPPYs, and vocabulary puzzles for each unit, you should have an idea which topics need additional study or review.

The practice tests that are included in the section of the book are designed to help you get familiar with the format of the exam as well as check your understanding of the key concepts in the course. Plan on taking both of these tests as part of your preparation and review.

Allow yourself 90 minutes for each section of the test and be sure to check your answers in the provided keys. For each question you missed, determine whether it was a simple mistake or whether you need to go back and study that topic again. After you complete the tests, continue practicing problems before the exam date.

The best preparation for the exam (other than having a solid understanding of statistics) is to practice as many multiple-choice and free-response questions as possible. Ask your teacher or refer to the College Board's AP Central website for additional resources to help you with this!

During the week before the exam, spend some time memorizing the flash cards that are found at the end of this book. These flash cards cover the most commonly tested concepts on the exam and are a great way to help maximize your score.

Section 7 – Practice Exams

Use the following two practice exams to help you prepare for the AP® Statistics Exam. Allow yourself 90 minutes for the multiple-choice questions, take a break, and allow yourself 90 minutes for the free-response section. When you finish, correct your exam and note your areas in need of improvement. Answers for each exam are included in the answer key.

PRACTICE EXAM #1

SECTION I

Time—1 hour and 30 minutes
Number of questions—40
Percent of total score—50

Directions. Identify the choice that best completes the statement or answers the question. Check your answers and note your performance when you are finished.

1. A statistics class wants to estimate the average number of tattoos worn by motorcyclists in their city. At a rally attended by many motorcyclists in their city, the class randomly selects 30 rally attendees and asks each attendee how many tattoos they wear. Which of the following describes the sample and the population to which it would be most reasonable for the class to generalize the results?

   (A) The sample is the 30 selected rally attendees, and the population is all the motorcyclists in their city.
   (B) The sample is the 30 selected rally attendees, and the population is all rally attendees.
   (C) The sample is all rally attendees, and the population is all motorcyclists in their city.
   (D) The sample is all rally attendees, and the population is the 30 rally attendees.
   (E) The sample is the average number of tattoos that all rally attendees have, and the population is all rally attendees.

2. Researchers of an agricultural station wanted to test the yields of six different varieties of corn. The station had four large fields available in four different parts of the county. The researchers divided each field into six sections then randomly assigned one variety of corn seed to each section in that field. This procedure was done for each field. At the end of the growing season, the corn was harvested, and the yield (measured in tons per acre) was compared. The difference in mean yield of the varieties of corn seed was found to be statistically significant. Based on the design of this study, which statement is true?

   (A) The variety of corn was the cause of the difference in mean yield because the seeds were randomly assigned to sections.
   (B) The variety of corn was the cause of the difference in mean yield because the seeds were randomly selected.
   (C) The fields were the cause of the difference in mean yield because the varieties of corn were assigned at random to the fields.
   (D) The fields were the cause of the difference in mean yield because the study used replication.
   (E) No conclusion can be drawn because climate and soil conditions are confounding variables.

3. A large high school offers AP Statistics and AP Calculus. Among the seniors in this school, 65% take AP Statistics, 45% take AP Calculus, and 30% take both. If a senior from this school is randomly selected, what is the probability they are in AP Statistics or AP Calculus, but not both?

(A) 35%
(B) 50%
(C) 70%
(D) 75%
(E) 80%

4. An avid football fan wonders if there is an association between the number of penalties a team commits and the number of points they score. The fan collects data on the number of penalties and the total points scored in the first three weeks of a recent National Football League (NFL) season for all 32 teams. The scatterplot below shows the results of the study, and the least-squares regression line is included on the graph.

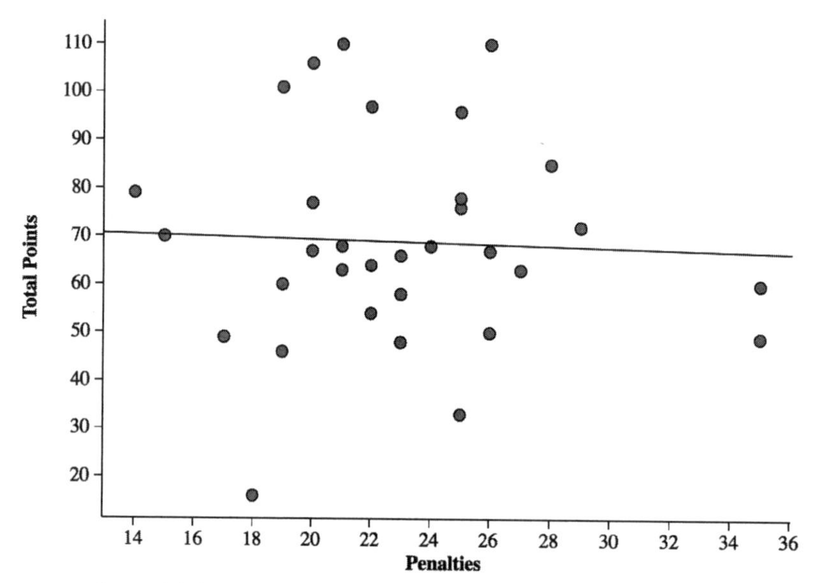

Based on the scatterplot, which statement is **FALSE**?

(A) The point at (35, 60) is an outlier in regression.
(B) The residual plot of this data would show random scatter.
(C) There is a weak association between penalties and total points among these NFL teams.
(D) There is a slightly negative association between penalties and total points among these NFL teams.
(E) The point at (35, 50) has higher leverage than the point at (18, 18).

5. Traffic engineers studied the traffic patterns of two busy intersections on opposite sides of town at rush hour. At the first intersection, the average number of cars waiting to turn left was 17 with a standard deviation of 4 cars. At the second intersection, the average number of cars waiting to turn left was 25 cars with a standard deviation of 7 cars. Assume that the number of cars waiting to turn left at each intersection is independent. What is the standard deviation of the total number of cars waiting to make a left turn at the two intersections?
   (A) $\sqrt{11}$
   (B) $\sqrt{33}$
   (C) $\sqrt{65}$
   (D) 11
   (E) 13

6. Adiflu is a drug to treat the flu and is effective in reducing symptoms in 70% of flu cases but results in unpleasant side effects for 40% of individuals who take it. Suppose a random sample of 10 people with the flu take Adiflu. What is the probability that fewer than 3 experience unpleasant side effects?

   (A) 0.0016
   (B) 0.0106
   (C) 0.1209
   (D) 0.1673
   (E) 0.3823

7. A pharmaceutical company has developed a new medication to lower a person's cholesterol. The advertisement for the new drug cites the results of an experiment comparing the drug with a placebo, stating that the cholesterol level had been lowered by an average of 13.2 mg/l with a P-value < 0.01. Which one of the following best explains the meaning of the P-value?

   (A) There was a less than 1% difference in the mean cholesterol levels between those taking the placebo and those taking the new medication using the same experimental method.
   (B) Less than 1% of the people who were given the new cholesterol-reducing drug experienced any serious side effects.
   (C) Assuming that the placebo is equally as effective as the medication, there is less than a 1% probability of obtaining a difference of 13.2 mg/l or greater by chance alone.
   (D) Less than 1% of the people who were given the experimental drug experienced a drop of 13.2 mg/l when compared to the placebo group.
   (E) The difference in the mean cholesterol levels between those taking the placebo and those taking the medication was not significant at the 1% level.

8. A simple random sample of 100 batteries is selected from a process that produces batteries with a mean lifetime of 32 hours and a standard deviation of 3 hours. Thus, the standard deviation of the sampling distribution of the sample mean, $\sigma_{\bar{x}}$ is equal to 0.3 hour. If the sample size had been 400, how would the value of $\sigma_{\bar{x}}$ change?

(A) It is one-fourth as large as when $n = 100$.
(B) It is one-half as large as when $n = 100$.
(C) It is twice as large as when $n = 100$.
(D) It is four times as large as when $n = 100$.
(E) The value of $\sigma_{\bar{x}}$ does not change.

9. A poll of 83 AP Statistics students asked, "How many hours per week do you spend studying statistics?" The results are summarized in the computer output below.

Variable	N	Mean	Median	TrMean		StDev
Hours	83	8.5	7.2	7.38		4.9
	Minimum		Maximum	Q1	Q3	
Hours	0.00		25.00	5	12	

Suppose an outlier is a value located 2 or more standard deviations away from the mean. What is the smallest value that is considered an outlier according to this rule?

(A) 14
(B) 17.2
(C) 18.5
(D) 23
(E) 24.5

10. A child psychologist claims the mean attention span for a 1-year-old is 12 seconds. An advocacy group claims that exposure to a popular cartoon results in a decrease in the mean attention span among the population of 1-year old children. To investigate this claim, which of the following hypotheses would be appropriate?

(A) H_0: $\mu < 12$ versus H_a: $\mu = 12$
(B) H_0: $\mu > 12$ versus H_a: $\mu = 12$
(C) H_0: $\mu = 12$ versus H_a: $\mu < 12$
(D) H_0: $\mu = 12$ versus H_a: $\mu > 12$
(E) H_0: $\mu = 12$ versus H_a: $\mu \neq 12$

11. Prospective salespeople for a large tech company are being offered a sales training program before they start working for the company. Previous data indicate that 25% of salespeople who did not participate in the program sold over $50,000 per month. To determine whether the sales-training program is effective, a random sample of 40 new salespeople are given the training. The company is interested in assessing whether the sales program significantly increases the proportion, p, of new salespeople with over $50,000 in monthly sales. A considerable investment in time and money would be used to conduct the training program. The company wants to test H_0: $p = 0.25$ versus H_a: $p > 0.25$. Which of the following describes a Type II error?

(A) The company finds convincing evidence that the proportion of new salespeople who will sell over $50,000 per month is greater than 25%, when in fact, the population proportion is greater than 25%.
(B) The company finds convincing evidence that the proportion of new salespeople who will sell over $50,000 per month is greater than 25%, when in fact, the population proportion is not greater than 25%.
(C) The company does not find convincing evidence that the proportion of new salespeople who will sell over $50,000 per month is greater than 25%, when in fact, the population proportion is greater than 25%.
(D) The company does not find convincing evidence that the proportion of new salespeople who will sell over $50,000 per month is greater than 25%, when in fact, the proportion is not greater than 25%.
(E) The company finds convincing evidence that the proportion of new salespeople who will sell over $50,000 per month is 25%, when in fact, the population proportion is greater than 25%.

12. Researchers wanted to know if playing soft music improved students' performance on tests. Twelve students worked a complicated paper-and-pencil maze while listening to music and again while not listening to music. The order of the treatments was randomly assigned, and the mazes were completed one week apart. The time to complete each maze (in seconds) was recorded and the results are shown in the following table.

Subject	1	2	3	4	5	6	7	8	9	10	11	12
Time with music	83	79	73	76	75	79	74	67	80	69	71	63
Time without music	89	82	85	82	81	90	75	69	74	82	78	68

Which of the following inference procedures is appropriate to determine if there is a significant difference in average time to complete the maze?

(A) A chi-square test for independence.
(B) A two-sample test for a difference in proportions.
(C) A t test on the slope of the regression line.
(D) A two-sample t test for a difference in means.
(E) A matched pairs t-test for a mean difference.

13. A cereal manufacturer is packaging collectible cards from a popular movie in each box of cereal. One card, a holographic card of a memorable scene, is rarer than the others and occurs in only 3% of boxes. The manufacturer places cards in boxes at random. Which of the following represents the probability that the fifth box you purchase will contain your first holographic card?

(A) $\binom{5}{1}(0.97)^4(0.03)^1$
(B) $(0.97)^4(0.03)^1$
(C) $(0.97)^1(0.03)^4$
(D) $1-(0.97)^5$
(E) $1-(0.03)^5$

14. The histogram below shows the distribution of age for 300 members of a pickleball association. Based on the histogram, which of the following is closest to the interquartile range of the distribution?

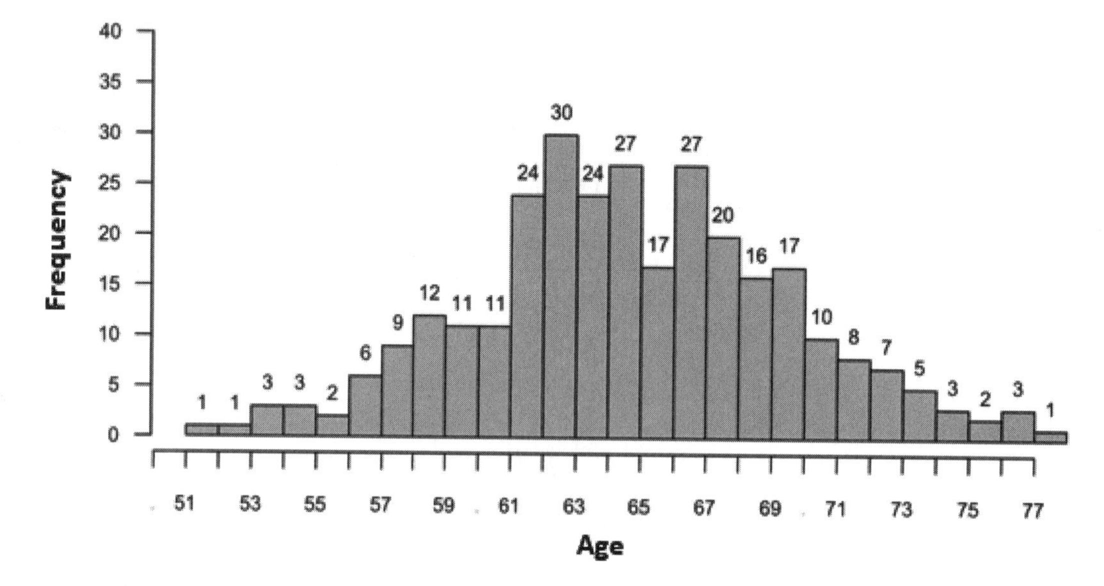

(A) 2
(B) 6
(C) 9
(D) 12
(E) 27

15. A study tested a claim that a "gas pill" boosts fuel economy (measured in miles per gallon) when added to a tank of gas. Ten randomly selected new cars were used in the study and would be run over an identical course twice, once with the additive and once without the additive. The order of the two treatments (pill and no pill) was randomized. Each car's tank was filled with standard gasoline and one of the treatments applied. The car's tank was completely drained, then refilled with the same standard gasoline and the process was repeated for the other treatment. Professional drivers drove the same course at identical speeds and did not know which treatment was used. The fuel economy (mpg) for each car was calculated for both treatments. A 95% confidence interval for the true mean difference (pill – no pill) in fuel economy is given by (–0.13, 1.73).

Based on the information, which one of the following statements is TRUE?

(A) Ninety-five percent of the time, this confidence interval will contain 0.
(B) A two-sample *t*-test should have been used instead of a paired *t* test.
(C) The true mean mileage is greater for cars without the additive than those with the additive because the mean difference is more likely to be positive.
(D) The sample size is too small to draw any conclusions.
(E) Because the interval contains 0, we cannot conclude that using the gas pill made a difference in fuel economy (mpg).

16. A shipping department of a large company claims they have 5 shipments returned per day, on average. The finance department believes more than 5 shipments are returned per day, on average. The finance department selects a random sample of 100 days and records how many returned shipments there were each day. With all conditions for inference met, a significance test of H_0: $\mu = 5$ versus H_a: $\mu > 5$, where μ = the mean number of shipments returned for all days resulted in a *P*-value of 0.106.

For a significance level of $\alpha = 0.05$, which of the following is a correct conclusion?

(A) The *P*-value is less than 0.05, so the null hypothesis is rejected. There is convincing statistical evidence that more than 5 shipments are returned per day, on average.
(B) The *P*-value is less than 0.05, so the null hypothesis is not rejected. There is not convincing statistical evidence that less than 5 shipments are returned per day, on average.
(C) The *P*-value is greater than 0.05, so the null hypothesis is rejected. There is not convincing statistical evidence that more than 5 shipments are returned per day, on average.
(D) The *P*-value is greater than 0.05, so the null hypothesis is not rejected. There is convincing statistical evidence that 5 shipments are returned per day, on average.
(E) The *P*-value is greater than 0.05, so the null hypothesis is not rejected. There is not convincing statistical evidence that more than 5 shipments are returned per day, on average.

17. Eight people who suffer from hay fever volunteer to test a new medication that will relieve their symptoms. The last names of the volunteers are:

(1) Rodriguez, (2) Liu, (3) Brown, (4) Kim, (5) Harris, (6) Munoz, (7) Klein, (8) Scott

Four of the volunteers will receive the new medication, while the other four will receive a placebo as part of a double-blind experiment. Starting at the left of the list of random numbers below, and reading from left to right, assign four people to be given the medication.

07119 97336 71048 08178 77233 13916 47564 81056 97025 85977 29372

The four people assigned to be given the medication are

(A) Rodriguez, Liu, Brown, Klein
(B) Liu, Harris, Klein, Scott
(C) Rodriguez, Brown, Munoz, Klein
(D) Rodriguez, Brown, Klein, Scott
(E) Rodriguez, Kim, Klein, Scott

18. A farmer randomly selected 46 sheep from a very large herd to construct a 95% confidence interval to estimate the mean weight of wool (in pounds) from sheep in the herd. The interval obtained was (23.7, 33.6). If the farmer had used a 90% confidence interval instead, the 90% confidence interval will be

(A) Narrower and would involve a larger risk of being incorrect.
(B) Narrower and would involve a smaller risk of being incorrect.
(C) Narrower and would involve the same risk of being incorrect.
(D) Wider and would involve a larger risk of being incorrect.
(E) Wider and would involve a smaller risk of being incorrect.

19. Company A has 500 employees and Company B has 5000 employees. Union negotiators want to compare the salary distributions for the two companies. Which one of the following would be the most useful for accomplishing this comparison?

(A) Dotplots for A and B drawn on the same scale.
(B) Back-to-back stemplots for A and B.
(C) Two frequency histograms for A and B drawn on the same scale.
(D) Two relative-frequency histograms for A and B drawn on the same scale.
(E) A scatterplot of A versus B.

20. Most flights at airports leave at their posted departure time. The frequency table below summarizes the times in the last month at a major airport where flights have been delayed past their posted departure time. The delays occurred for a variety of reasons—mechanical problems with the plane, congestion on the tarmac, the previous flight at a gate was late in leaving, weather, etc.

Delay Time	Frequency
Less than 10 minutes	11
At least 10 but less than 20 minutes	37
At least 20 but less than 30 minutes	33
At least 30 but less than 40 minutes	30
At least 40 but less than 50 minutes	26
At least 50 but less than 60 minutes	14
At least 60 but less than 70 minutes	6
At least 70 minutes	3

Which one of the following represents possible values for the median and mean delay times for flights from this major airport?

(A) Median = 27 minutes and mean = 24 minutes
(B) Median = 29 minutes and mean = 29 minutes
(C) Median = 29 minutes and mean = 32 minutes
(D) Median = 31 minutes and mean = 35 minutes
(E) Median = 35 minutes and mean = 39 minutes

21. A national newspaper manager wants to estimate the true proportion of all people who believe that life exists elsewhere in the universe. Which of the following intervals contains the minimum number of people that should be sampled to estimate, with 95% confidence, the true proportion of those who believe that life exists elsewhere within 0.04 of the true proportion?

(A) Less than 500
(B) 501 to 550
(C) 551 to 600
(D) 601 to 650
(E) More than 650

22. Consumers frequently complain that there is a large variation in prices charged by different pharmacies for the same medication. A survey of a large random sample of pharmacies in a major city revealed that the price charged for one bottle of 50 tablets of a popular pain reliever were approximately normally distributed. The price of $10.48 for this bottle was at the 90th percentile and 20% of the bottles cost less than $8.61. What are the approximate mean and standard deviation of the price distribution for bottles containing 50 tablets of the pain reliever in this major city?

(A) $\mu = \$9.35$ and $\sigma = \$0.88$
(B) $\mu = \$9.41$ and $\sigma = \$2.67$
(C) $\mu = \$9.55$ and $\sigma = \$0.88$
(D) $\mu = \$9.55$ and $\sigma = \$1.70$
(E) $\mu = \$9.73$ and $\sigma = \$1.70$

23. Fertilizers often contain nitrogen, phosphate, and potassium to encourage fruit production. A study of the effect of various amounts of nitrogen on the yield, measured in pounds per plant, of 10 plots of tomato plants resulted in the following computer output:

Predictor	COEF	SE COEF	T	P
Constant	10.100	0.7973	12.67	0.000
Nitrogen	1.150	0.1879	6.12	0.000
S = 0.8404	R-Sq = 82.4%		R-Sq(adj) = 80.2%	

Which of the following would represent a 99% confidence interval to estimate the true slope of the regression line of yield on nitrogen? Assume all conditions for inference are met.

(A) $1.150 \pm 3.355(0.1879)$
(B) $1.150 \pm 3.250(0.1879)$
(C) $1.150 \pm 3.355(0.8404)$
(D) $1.150 \pm 3.250\left(\frac{0.1879}{\sqrt{10}}\right)$
(E) $1.150 \pm 3.355\left(\frac{0.1879}{\sqrt{10}}\right)$

24. Flossing helps prevent tooth decay and gum disease, yet reports suggest millions of Americans never floss. You are planning a sample survey to determine the proportion of the students at your high school who never floss. Which of the following will result in the largest margin of error?

(A) A sample size of 200 and a confidence level of 95%
(B) A sample size of 200 and a confidence level of 96%
(C) A sample size of 200 and a confidence level of 99%
(D) A sample size of 300 and a confidence level of 96%
(E) A sample size of 300 and a confidence level of 99%

25. A statistics class selected a random sample of students at their large high school to find the proportion of those who claimed to be vegetarians. They found 12 out of the 150 students questioned were vegetarians. Another statistics class in a different large high school selected a random sample of the students at their school and found that 10 out of 90 claimed to be vegetarians. Which of the following represents the approximate 90% confidence interval for the difference between the proportions of students of the two schools who would claim to be vegetarians?

(A) $-0.031 \pm 1.645\left(\frac{(0.08)(0.92)+(0.111)(0.889)}{\sqrt{150+90}}\right)$
(B) $-0.031 \pm 1.645\left(\frac{(0.08)(0.92)}{\sqrt{150}} + \frac{(0.111)(0.889)}{\sqrt{90}}\right)$
(C) $-0.031 \pm 1.645\left(\sqrt{\frac{(0.08)(0.92)}{150} + \frac{(0.111)(0.889)}{90}}\right)$
(D) $-0.031 \pm 1.960\left(\frac{(0.08)(0.92)}{\sqrt{150}} + \frac{(0.111)(0.889)}{\sqrt{90}}\right)$
(E) $-0.031 \pm 1.960\left(\sqrt{\frac{(0.08)(0.92)}{150} + \frac{(0.111)(0.889)}{90}}\right)$

26. A large sales company recruits many graduating students from universities for its workforce. Thirty percent of those hired for management positions come from private universities and colleges and the rest from public colleges and universities. It is very expensive and time-consuming to train new managers, so the company is examining its retention rate (those still working for the company after six years) of these hires. Over the past six years, 35% of the managers who were hired from private schools had left for other jobs, while 20% of those from public schools had done so. What is the probability that a randomly selected person, who left the company within the past six years, was hired from a private university or college?

(A) 0.105
(B) 0.4286
(C) 0.2308
(D) 0.35
(E) 0.2450

27. In a recent social research poll, 4800 randomly selected adults aged 18 to 26 were asked to record the number of hours they spent watching television and the number of hours they spent watching videos online during a typical week. Parallel boxplots of the data are given below.

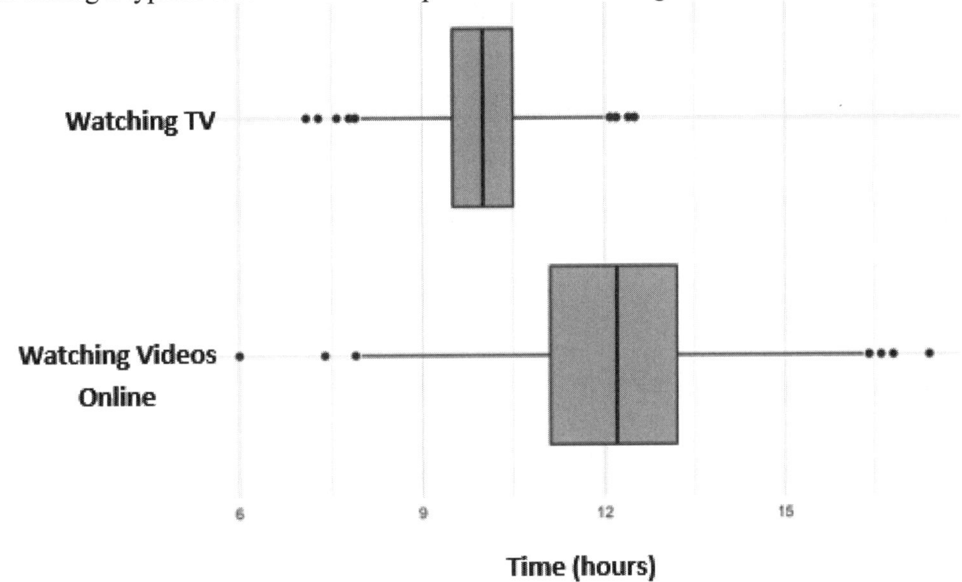

Based on the boxplots, which one of the following statements is **FALSE**?

(A) The range of the distribution of time spent watching videos online is greater than the range of the distribution of time spent watching television.
(B) On average, these adults spent more time watching videos online than watching television.
(C) The median time spent watching videos online is greater than the median time spent watching television.
(D) The *IQR* of the distribution of time spent watching videos online is less than the *IQR* of the distribution of time spent watching TV.
(E) Both distributions have at least one outlier.

28. Three distributions are shown below. One is the population distribution, one is the approximate sampling distribution of means for samples of size $n = 5$ from the population, and one is the approximate sampling distribution of means for samples of size 20 from the population. If the distributions are placed in this order: population, sampling distribution of means for $n = 5$, and sampling distribution of means for $n = 20$, which of the following gives the correct order?

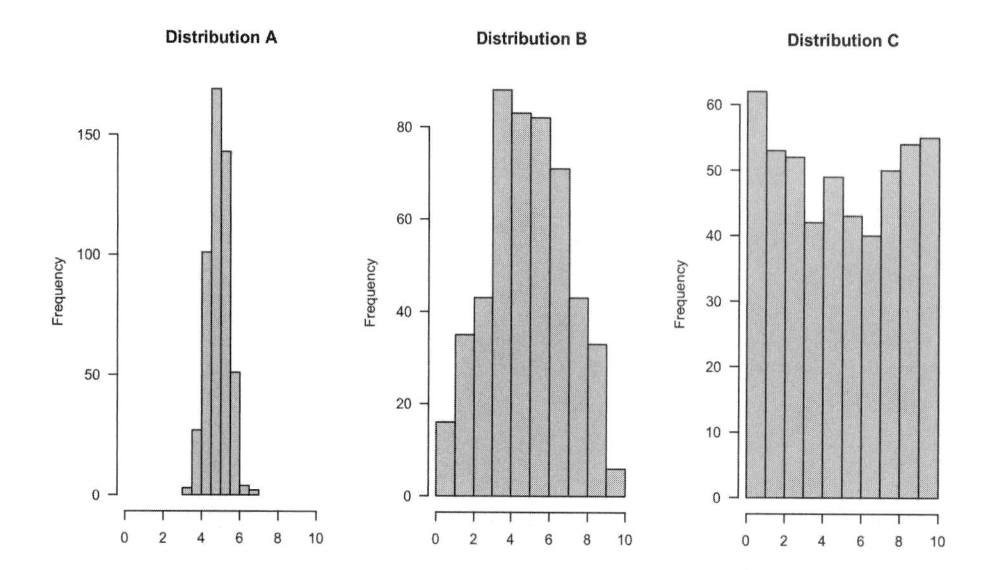

(A) A, B, C
(B) B, A, C
(C) B, C, A
(D) C, A, B
(E) C, B, A

29. In a dice game a player is paid \$5 for rolling a 3 or a 5 with a single fair die. The player will pay \$2 if any other number is rolled. If a person plays the game 30 times, what is the approximate probability that the person will win at least \$15?

(A) 0.0030
(B) 0.0643
(C) 0.2767
(D) 0.3085
(E) 0.3910

30. A study is conducted to determine the effectiveness of mental exercises on short-term memory. A randomized comparative experiment is conducted, and the number of items recalled on a short-term memory test is recorded for 25 individuals who did not practice mental exercises and 35 individuals who did. The individuals that did not do any mental exercises recalled 12.5 items on average with a variance of 7.29. The individuals who practiced mental exercises recalled 16.0 items on average with a variance of 6.25. Assuming the conditions for inference are met, which expression shows the correct standard error for a two-sample *t* interval for a difference in means?

(A) $\sqrt{7.29 + 6.25}$
(B) $\sqrt{7.29 - 6.25}$
(C) $\sqrt{\frac{7.29}{35} + \frac{6.25}{25}}$
(D) $\sqrt{\frac{7.29}{35} - \frac{6.25}{25}}$
(E) $\sqrt{\frac{7.29^2}{35} + \frac{6.25^2}{25}}$

31. In a random sample of older patients at a large medical practice, the age of a patient and a measure of that patient's hearing loss were recorded. For the patients in the sample, the correlation between age and hearing loss was found to be 0.70. Which of the following would be a correct statement if the age of a patient were used to predict the amount of hearing loss for a patient?

(A) About 49% of the variation in age can be explained by the least-squares regression line using hearing loss as the explanatory variable.
(B) About 49% of the variation in hearing loss can be explained by the least-squares regression line using age as the explanatory variable.
(C) About 70% of the variation in hearing loss can be explained by the least-squares regression line using age as the explanatory variable.
(D) About 70% of a person's hearing loss can be explained by age, according to the regression line relating hearing loss and age.
(E) About 70% of the time, age will correctly predict the amount of hearing loss.

32. A study performed by a psychologist determined that a person's score on a resiliency test is linearly related to the person's score on an inventory measuring the person's overall health. The equation of the least-squares regression line is $\widehat{resiliency} = -49 + 1.8(health)$. What is the residual for an individual with an overall health score of 110 and a resiliency score of 140?

(A) –30
(B) –9
(C) 9
(D) 30
(E) 149

33. There is an ever-increasing number of sources for getting the daily news—traditional newspapers, online, radio, nightly television newscasts, comedy shows, cell phones, podcasts, etc. In a recent telephone survey, 3204 randomly selected adults were asked to cite their primary source of daily news. Four in 10 adults said that they read a newspaper, either in print or online, almost every day. A 98% confidence interval to estimate the proportion of all adults who read either a print newspaper or its online equivalent for their daily news is given by (0.38, 0.42). Which of the following is a correct interpretation of the confidence level?

(A) Ninety-eight percent of all samples of this size would yield a confidence interval of (0.38, 0.42).
(B) There is a 98% chance that the true proportion of readers who would read either a print newspaper or its online equivalent for their daily news is in the interval (0.38, 0.42).
(C) The procedure used to generate this interval will capture the proportion of all adults who read either a print newspaper or its online equivalent for their daily news in 98% of all possible random samples of 10 adults.
(D) Ninety-eight percent of all the samples of size 3204 lie in the confidence interval (0.38, 0.42).
(E) There is a 98% chance that a randomly selected reader is one of the 40% who would read a print newspaper or its online equivalent for their daily news.

34. A school counselor believes that students who enroll in a fine arts class (like band, orchestra, or choir) are more likely to participate in a fine arts activity (such as the school musical, play, or speech). The counselor collects data from a random sample of high school students in a large city. The table below summarizes the results.

		Class			
		Band	Orchestra	Choir	Total
Activity	One-Act Play	42	21	5	68
	Musical	120	59	10	189
	Speech	20	10	15	45
	Total	182	90	30	302

What is the expected count for the cell for the musical and orchestra?

(A) 59
(B) $\frac{(59)(90)}{302}$
(C) $\frac{(189)(59)}{302}$
(D) $\frac{(189)(90)}{302}$
(E) $\frac{(59)(302)}{189}$

35. A tax assessor developed a model to predict the value (in thousands of dollars) of a home based on the square footage of the home. The computer output is provided below.

Variable	COEF	SE	T	P
CONSTANT	35.70032	6.9533214	13.049	0.0237
SQFEET	0.14875	0.0030725	4.014	0.0017

Based on the information, which of the following is the best interpretation for the slope of the least-squares regression line?

(A) Each additional square-foot increase in the size of the house will increase the predicted home value by approximately \$35.70.
(B) Each additional square-foot increase in the size of the house will increase the predicted home value by approximately \$14.88.
(C) Each additional square-foot increase in the size of the house will increase the predicted home value by approximately \$148.75.
(D) Each additional 1000 square-foot increase in the size of the house will increase the predicted home value by approximately \$35,700.32.
(E) Each additional 1000 square-foot increase in the size of the house will increase the predicted home value by approximately \$0.14875.

36. Emily, a superintendent of a school district, is interested in discovering student opinions about the quality of extracurricular activities, such as clubs and athletics, offered at the various campuses. The district has 2 high schools, 4 middle schools, and 10 elementary schools. To obtain a sample of district students, Emily first selects a random sample of students from each high school, then from each middle school, then from each elementary school. What type of sampling procedure did she implement?

(A) A simple random sample
(B) A stratified random sample
(C) A cluster sample
(D) A systematic random sample
(E) A convenience sample

37. A community member is surprised by how many people in the neighborhood park in their driveway rather than their garage. The table summarizes the number of vehicles parked in each driveway.

Number of Vehicles	0	1	2	3
Proportion of Homes	0.10	0.25	0.50	0.15

What is the average number of vehicles parked in driveways for homes in this community?

(A) 1.5
(B) 1.7
(C) 2.5
(D) 2.7
(E) 3

38. A city manager is interested in how a certain program helped change the city's distribution of household income. The manager selects a random sample of 1758 households and obtains their household income information. The results were compared to a similar poll of 1494 randomly selected households in 2020. The table below summarizes the data for household income.

Household Income	**Current**	**2020**
Under $30,000	196	184
$30,000 – under $50,000	414	316
$50,000 – under $75,000	488	382
At least $75,000	660	612

A chi-square test was performed using the hypotheses H_0: The distribution of household income in 2020 is the same as it is currently in this city versus H_a: The distribution of household income in 2020 is not the same as it is currently in this city. The test resulted in a *P*-value of 0.0760. Which of the following conclusions is correct?

(A) At the 5% level of significance, there is convincing evidence that the distribution of household income has not changed since 2020.
(B) At the 5% level of significance, there is convincing evidence that the distribution of household income has changed since 2020.
(C) At the 5% level of significance, there is convincing evidence to conclude that an association exists between the year and the distribution of household income.
(D) At the 10% level of significance, there is convincing evidence that the distribution of household income has not changed since 2020.
(E) At the 10% level of significance, there is convincing evidence that the distribution of household income has changed since 2020.

39. A pharmaceutical company wants to test two new acid reflux medications, A and B, to see which is effective in relieving the reflux pain in the shortest amount of time. The researchers plan to give the two medications in three dosages: 50 mg, 100 mg, and 150 mg. They have 120 volunteers available for the study. For this part of the study, researchers think that the weight of the patient might play a role in the incidence of reflux disease. Volunteers are separated into 20 groups—the six heaviest, the next six heaviest, and so on, down to the six lightest. Within each group of six, three are randomly assigned to medication A and three to medication B. Within each medication group of three, one is randomly assigned to each dosage. Which one of the following statements best describes this experiment?

(A) There are two treatments, participants were blocked on weight, and the response variable is time until pain relief is achieved.
(B) There are two treatments, participants were blocked on dosage, and the response variable is time until pain relief is achieved.
(C) There are three treatments, participants were blocked on medication A and B, and the response variable is proportion of participants who achieved pain relief.
(D) There are six treatments, participants were blocked on weight, and the response variable is time until pain relief is achieved.
(E) There are six treatments, participants were blocked on weight, and the response variable is proportion of participants who achieved pain relief.

40. Researchers plan to conduct a study of cell phone usage among middle school students. They will select a simple random sample of students from a large middle school and record the number of hours that each of the student spends on their cell phone each week. The researchers will report the sample mean as a point estimate for the population mean. Which of the following statements is correct for the sample mean as a point estimator?

(A) A sample of size 30 will produce a more biased point estimator than a sample of size 60.
(B) A sample of size 30 will produce a less biased point estimator than a sample size of 60.
(C) A sample of size 30 will produce more variable point estimator than a sample of size 60.
(D) A sample of size 30 will produce less variable point estimator than a sample of size 60.
(E) A sample of size 30 will produce a biased point estimator, but a sample size of 60 will produce an unbiased point estimator.

PRACTICE EXAM #1

SECTION II
Part A
Questions 1-5
Spend about 1 hour and 5 minutes on this part of the exam.
Percent of Section II score—75

Directions: Show all your work. Indicate clearly the methods you use, because you will be scored on the correctness of your methods as well as on the accuracy and completeness of your results and explanations.

1. Radon is a radioactive gas formed by the natural radioactive decay of uranium in rock, soil, and water. It occurs throughout the United States but is more prevalent in some areas than others. Problems can occur when the gas accumulates in homes, especially basements, and exposure carries some health risks. A team of builders recently constructed 400 homes in a large development, where half of the homes used a new radon-resistant construction material, and the other used the current radon-resistant construction material. The builders are concerned about whether radon levels are within acceptable levels. If the levels are not within acceptable levels, the builders will have to install radon removal systems in each home that has higher than accepted levels. The dotplots below show the distribution of radon level in picocuries per liter of air (pCi/L).

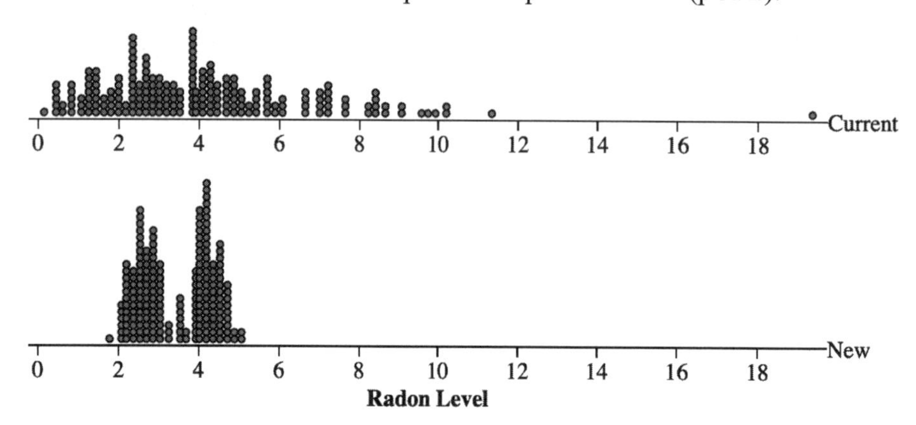

(a) Compare the distributions of radon level between homes built with the new construction material and homes built with the old construction material.

(b) Based on the dotplots, what is one benefit for homes that use the current construction material?

(c) Based on the dotplots, what is one benefit for homes that use the new construction material?

2. A graduate student who is studying dentistry is interested in comparing three different types of toothbrushes. The student recruited 60 volunteers to participate in the study. The volunteers will be divided evenly among three groups, where each group will use one of the three types of toothbrushes twice a day. The student will measure the amount of plaque on each subject's teeth before and after the treatment. Plaque is a substance that grows on teeth that causes cavities and gum disease.

(a) Describe a process that the student can use to randomly assign the treatments to the subjects in this study.

(b) Why is it important that the student implements random assignment in this experiment?

(c) If the student finds statistically significant evidence that the subjects in one treatment group have less plaque than the other groups, on average, would the student be able to conclude that the toothbrush caused the reduction of plaque? Explain your answer.

3. A candy manufacturer is marketing a gift box containing four chocolate creams and five pieces of fudge. The manufacturing process for each candy and each empty box is designed with the specifications shown in the table below.

	Mean (oz.)	**Standard Deviation (oz.)**
One chocolate cream	3	0.2
One piece of fudge	4	0.3
The empty box	3	0.1

The weights of the chocolate creams, the pieces of fudge, and the empty boxes are independent, and each distribution of weight is approximately normal.

(a) What are the mean and standard deviation of the weight of the full box of candy?

(b) What is the probability that a randomly selected full box of candy will weigh less than 34 ounces?

(c) What is the probability that at least one of three randomly selected full boxes of candy will weigh less than 34 ounces?

4. A gluten free diet is a diet that excludes certain grains, such as wheat, barley, and rye. Certain health problems require individuals to eat a gluten-free diet. A group of researchers reviewed a database with information from large-scale national surveys used to investigate the health and diet of U.S. adults. In 2020, the researchers found that 0.5% of 2105 randomly selected survey participants were eating a gluten free diet. By 2024, they found that 2% of 1757 randomly selected survey participants were on a gluten free diet.

   Do the data provide convincing statistical evidence, at the level of $\alpha = 0.05$, that there has been an increase in the proportion of U.S. adults who eat a gluten free diet?

5. A journalist is writing an article about what movie genre—horror, action/adventure, or comedy—that teen and adult moviegoers prefer. The journalist selects a systematic random sample of 80 moviegoers at the local movie theater during a day when over 1600 moviegoers attended the theater. The segmented bar graph below shows the distribution of movie genre preference of the people included in the sample.

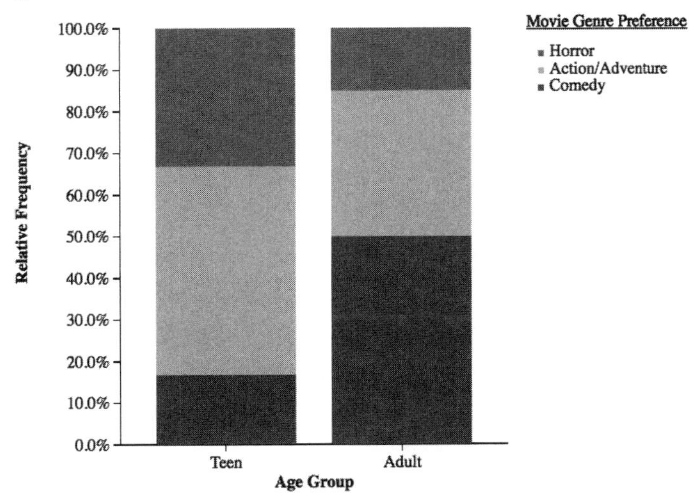

(a) Describe the association between movie genre preference and age group.

(b) The journalist used a systematic random sample to collect the 80 responses. Describe how this process may have been implemented in this context.

(c) The journalist's sample consisted of 60 teens and 20 adults. Use the following grid to sketch a mosaic plot of movie genre preference among teens and adults based on the graph in part (a).

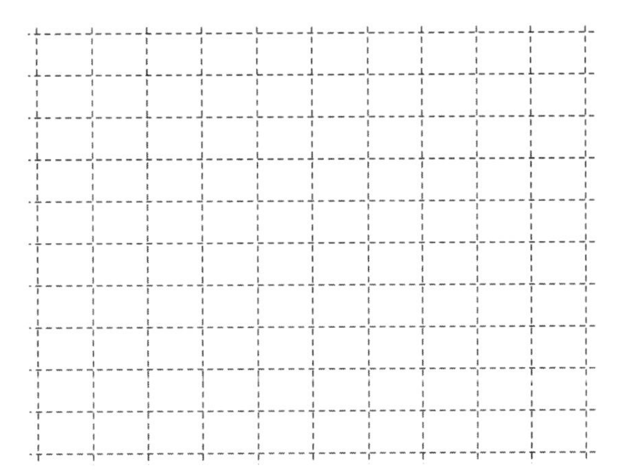

(d) What does the mosaic plot reveal about the distribution of movie genre preference among teens and adults that the segmented bar graph did not?

SECTION II
Part B
Question 6
Spend about 25 minutes on this part of the exam.
Percent of Section II score—25

Directions: Show all your work. Indicate clearly the methods you use, because you will be scored on the correctness of your methods as well as on the accuracy and completeness of your results and explanations.

6. In a certain town, a river frequently floods. An insurance agent believes there is a strong association between the amount of claims for damage done to houses and businesses and the distance the building is from the river. The agent records the total amount in claims, in millions of dollars, and the distance of the building to the river. The following scatterplot shows the distribution total amount in claims and the distance, in miles, the home or business is from the river. Computer output is also provided.

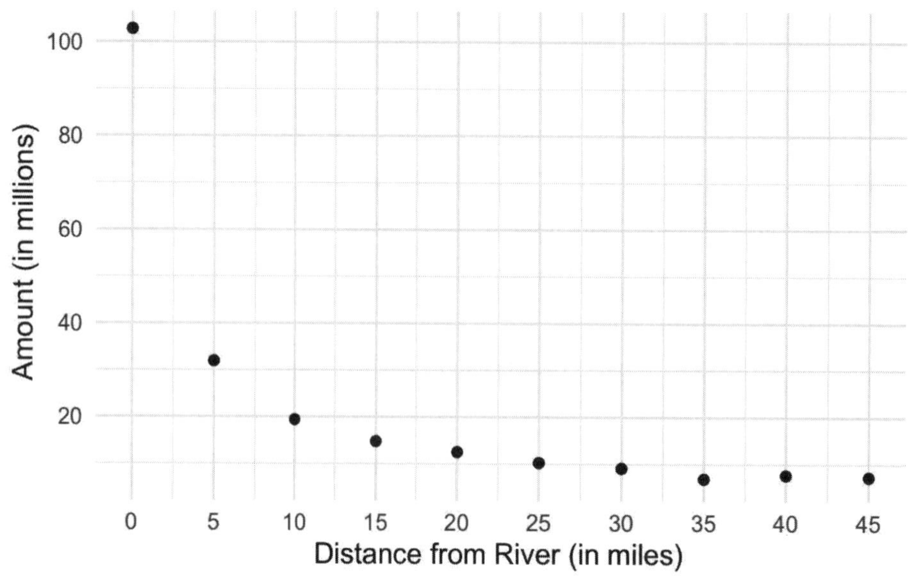

Variable	COEF	SE	T	P
Constant	52.551	13.124	4.004	0.0039
Distance	–1.346	0.492	–2.738	0.0255

S = 22.33 R-SQ = 0.4837 R-SQ(adj) = 0.4192

(a) Based on the scatterplot, is there evidence for the agent's belief? Explain your reasoning.

(b) Based on the values of the coefficient of determination and the correlation coefficient, is there evidence for the agent's belief? Justify your answer.

The agent decides to use a different statistic, called r_s, to assess the strength of the association. To calculate this statistic, the agent must follow the following steps.

1) Assign ranks to the distances. The distance closest to the river will be rank 1, the 2nd closest is rank 2, etc.
2) Assign ranks to the total amount. The largest amount will be rank 10, the 2nd largest is rank 9, etc.
3) Find the difference between the ranks (distance – amount).
4) Calculate $r_s = 1 - \frac{6\Sigma(d_i^2)}{n(n^2-1)}$, where n is the number of ordered pairs and $\Sigma(d_i^2)$ is the sum of the squared differences in ranks.

(c) Complete the rank (distance), rank (amount), and difference in ranks columns in the table below.

Distance	Total Amount	Rank (Distance)	Rank (Amount)	Difference in Ranks
0-4	102.8	1	10	–9
5-9	31.9	2	9	–7
10-14	19.4	3	8	–5
15-19	14.8	4	7	–3
20-24	12.5	5	6	–1
25-29	10.3			
30-34	9.1			
35-39	6.9			
40-44	7.7			
45-49	7.3			

(d) Calculate $\Sigma(d_i^2)$ by squaring each of the differences and then by finding the sum of those squared differences.

(e) Use the formula above to calculate the value of r_s.

(f) Based on your answer in part (e), what is the benefit of using this new statistic compared to the correlation coefficient calculated in part (b)? Explain your reasoning.

Answer Key for Practice Exam 1
Part I Multiple Choice Solutions

<table>
<tr><th>Problem</th><th>Answer</th><th>Explanation</th></tr>
<tr><td>1</td><td>B</td><td>The sample is the 30 selected individuals at the rally. The population is all those at the rally. Although the students want to estimate the average number of tattoos that motorcyclists in the city have, they did not select their sample from the entire city, but only those at the rally.</td></tr>
<tr><td>2</td><td>A</td><td>Because the seeds were randomly assigned to sections, we can infer that the variety of seed was the cause of the statistically significant difference in means.</td></tr>
<tr><td>3</td><td>B</td><td><table><tr><th></th><th>Stats</th><th>Not Stats</th></tr><tr><td>Calc</td><td>0.30</td><td>0.15</td></tr><tr><td>Not Calc</td><td>0.35</td><td>0.20</td></tr></table>

P(Calc and not Stat OR Stat and not Calc) = 0.35 + 0.15 = 0.50</td></tr>
<tr><td>4</td><td>A</td><td>To be an outlier in regression, the point should have a large residual compared to the other points. The point at (35, 60) has a small residual, so it is not an outlier in regression.</td></tr>
<tr><td>5</td><td>C</td><td>Because the number of cars at the two intersections are independent, we add the variances then take the square root to find the standard deviation of the total number of cars waiting to make a left turn: $\sigma_T = \sqrt{4^2 + 7^2} = \sqrt{65}$</td></tr>
<tr><td>6</td><td>D</td><td>Let X = the number who experience unpleasant side effects.

We need $P(X < 3)$, which is equal to $P(X = 0) + P(X = 1) + P(X = 2)$

$= {}_{10}C_0(0.40)^0(0.60)^{10} + {}_{10}C_1(0.40)^1(0.60)^9 + {}_{10}C_2(0.40)^2(0.60)^8 = 0.1673$

With technology: binomcdf(n = 10, p = 0.40, x = 2) = 0.1673</td></tr>
<tr><td>7</td><td>C</td><td>The P-value is the probability of obtaining a sample statistic at least as extreme as that found in the experiment (a difference of 13.2), given that the placebo is equally as effective as the medication.</td></tr>
<tr><td>8</td><td>B</td><td>Because $\sigma_{\bar{x}} = \frac{\sigma}{\sqrt{n}}$, multiplying the sample size by 4 would result in $\sigma_{\bar{x}} = \frac{\sigma}{\sqrt{4n}} = \frac{\sigma}{2\sqrt{n}} = \frac{1}{2}\left(\frac{\sigma}{\sqrt{n}}\right)$. Therefore, quadrupling the sample size will cut the standard deviation of the sampling distribution in half.</td></tr>
</table>

9	C	$8.5 - 2(4.9) = -1.3$ $8.5 + 2(4.9) = 18.3$ Any observation greater than 18.3 would be considered outliers. There are no observations less than –1.3. Among the choices provided, the smallest value that is considered an outlier is 18.5.
10	C	The null hypothesis always includes an equal sign. The one-sided alternative hypothesis, in this case, is less than 12 seconds.
11	C	A Type II error occurs when we fail to reject the null hypothesis when the alternative hypothesis is true. In this case, we make a Type II error if we fail to find evidence that $p > 0.25$, when in reality, $p > 0.25$.
12	E	There are two measurements on each subject (time with music versus time without music). Each subject serves as his or her own control. This is a matched pairs experiment.
13	B	This is a geometric distribution with $P(\text{success}) = 0.03$. Because the fifth box is the "first success", we have four failures and then a success. $P(X = 5) = (0.97)^4(0.03)^1$.
14	B	The interquartile range is the difference between the third and first quartiles: $IQR = Q_3 - Q_1$. With 300 members, the first quartile is at the age of the $300(0.25) = 75^{th}$ member and the third quartile is at the age of the $300(0.75) = 225^{th}$ member. To find the age of the 75th member, count from the left: 1+1+3+3+2+6+9+12+11+11 = 59 members are less than 61-year-old and 1+1+3+3+2+6+9+12+11+11+24 = 83 members are less than 62 years old. The age of the 75th member is approximately 61 years old. $Q_1 \approx 61$. To find the age of the 225th member, count 75 members from the right. 1+3+2+3+5+7+8+10+17+16 = 72 members are 68 and older and 1+3+2+3+5+7+8+10+17+16+20 = 92 members are 67 and older. The age of the 225th member is approximately 67 years old. $Q_3 \approx 67$. $IQR = 67 - 61 = 6$ years.

15	E	(A) False. This confidence interval either contains 0 or it does not. (B) False. The data are paired. (C) False. Even though the mean difference is positive, the confidence interval contains negative values that are plausible for the true mean difference in gas mileage based on the sample data. (D) False. A sample of size 10 can be sufficient to make a conclusion. (E) True. Because the confidence interval contains zero, there is no statistically significant difference between the additive and the non-additive. We cannot conclude that the pill is effective in improving gas mileage.
16	E	Because the *P*-value is greater than the significance level ($0.106 > 0.05$), we fail to reject the null hypothesis. We do not have convincing evidence for the alternative hypothesis (H_a: $\mu > 5$). The finance department does not have convincing statistical evidence that the average number of returned shipments is more than 5 per day.
17	C	The subjects are labeled as 1, 2, 3, 4, 5, 6, 7, 8. Ignore repeats because subjects were not used more than once. The random numbers chosen are 0 7 1 1 9 9 7 3 3 6 resulting in the selection of 7, 1, 3, 6. This corresponds to Klein, Rodriguez, Brown, and Munoz.
18	A	Decreasing the confidence level would make the confidence interval narrower because the critical value (z^*) would be smaller. A smaller confidence level leads to a larger risk of being incorrect.
19	D	The number of data points for the two companies would not lend itself to easily using dotplots or stemplots. Because you have only one variable, a scatterplot is inappropriate. Because the companies are quite different in size, a relative-frequency histogram would let you display the percent of workers in each salary group for comparison purposes.
20	C	There are 160 observations in the data set. The median would be the value between the 80th and 81st data points. These both fall in the "at least 20 but less than 30 minutes" interval resulting in a median of at least 20 but less than 30 minutes. This narrows the choices to (A), (B), or (C). In addition, the distribution of waiting times is skewed toward the higher numbers, causing the mean to be greater than the median. This eliminates choices (A) and (B).
21	D	Solve for n in $1.96\sqrt{\frac{(0.5)(0.5)}{n}} \leq 0.04$ $n \geq 600.25$ Rounding up to the nearest integer, the minimum number for the sample is 601 people. The interval that contains this answer is 601 to 650.

22	A	From the z-table or by using a calculator, InvNorm(0.20) = –0.84 and InvNorm(0.90) = 1.28 $$z = \frac{x-\mu}{\sigma} \Rightarrow -0.84 = \frac{8.61-\mu}{\sigma} \text{ and } 1.28 = \frac{10.48-\mu}{\sigma}$$ Solving the system: $$\begin{cases} -0.84\sigma = 8.61-\mu \\ 1.28\sigma = 10.48-\mu \end{cases}$$ $\mu = \$9.35$ and $\sigma = \$0.88$
23	A	For $n = 10$, there are $n - 2$ degrees of freedom in regression on slope. Therefore, with a 99% confidence level, df = 8 and $t^* = 3.355$. The amount of nitrogen in the fertilizer is the explanatory variable, so the slope coefficient is 1.150 and the standard error is 0.1879, giving $1.150 \pm 3.355(0.1879)$.
24	C	The margin of error is defined by $z^*\sqrt{\frac{\hat{p}(1-\hat{p})}{n}}$. Assuming that $\hat{p}$ remains the same, a higher level of confidence yields a larger margin of error. A larger sample size yields a smaller margin of error. So, a higher level of confidence and a smaller sample size yields the largest margin of error.
25	C	p_1 = true proportion who claim to be vegetarians at school #1 with sample proportion of 12 / 150 = 0.08. p_2 = true proportion who claim to be vegetarians at school #2 with sample proportion of 10 / 90 = 0.111. CI: $-0.031 \pm 1.645\left(\sqrt{\frac{(0.08)(0.92)}{150}+\frac{(0.111)(0.889)}{90}}\right)$
26	B	$$P(\text{private} \mid \text{left}) = \frac{P(\text{private} \cap \text{left the company})}{P(\text{left the company})} = \frac{(0.3)(0.35)}{(0.3)(0.35)+(0.7)(0.2)}$$ $= 0.4286$
27	D	The distribution of hours spent surfing the Web exhibits more variation (has a larger *IQR*) than does the distribution of hours spent watching TV.
28	E	From the Central Limit Theorem (CLT), we know that the larger the sample size, the approximate sampling distribution of the sample mean becomes closer to approximately normal. We also know that $\sigma_{\bar{x}} = \frac{\sigma}{\sqrt{n}}$, so the distribution of sample means will have less variability than the population distribution, and the larger the sample size, the less variability there will be in the distribution of sample means. Histogram C is the original population, Histogram B is sample size 5, and Histogram A is the most normal-like of the three and has the smallest variability.

29	E	The mean (expected value) on each roll is: $E(\text{winnings}) = (\$5)\left(\frac{1}{3}\right) + (-\$2)\left(\frac{2}{3}\right) = \$\frac{1}{3}$ with a standard deviation of: $\sigma(\text{winnings}) = \sqrt{\left(5-\frac{1}{3}\right)^2\left(\frac{1}{3}\right) + \left(-2-\frac{1}{3}\right)^2\left(\frac{2}{3}\right)} = \$3.30$ For 30 turns, the expected winnings are $E(total) = 30\left(\frac{1}{3}\right) = \$10$. The standard deviation of the 30 turns is $\sigma(\text{total}) = \sqrt{30(3.30)^2} = \$18.07$. $P(\text{total} > \$15) = P\left(z > \frac{\$15 - \$10}{\$18.07}\right) = P(z > 0.277) \approx 0.3910$
30	C	The standard error of the difference in sample means is given by $\sqrt{\frac{s_1^2}{n_1} + \frac{s_2^2}{n_2}} = \sqrt{\frac{6.25}{25} + \frac{7.29}{35}}$
31	B	You are asked to interpret the coefficient of determination, given by r^2. Because $r = 0.7$, $r^2 = 0.49$. The coefficient of determination, r^2, is the percent of variation in the y-variable that can be explained by the regression line, using the x-variable as the predictor. In this case, the y-variable is the amount of hearing loss and the x-variable is the age of the patient.
32	B	Predicted resiliency = – 49 + 1.8(110) = 149. Residual = actual y – predicted y = 140 – 149 = –9.
33	C	This question is asking for an interpretation of the confidence level, not the interval. The 98% level indicates how often, on average, the procedure will produce an interval that will capture the true population parameter of interest.
34	D	The expected counts for each cell are determined by $\frac{\text{(row total)(column total)}}{\text{table total}}$

35	C	The equation is $\widehat{\text{home value}} = 35.70032 + 0.14875(\text{sqfeet})$. The slope is 0.14875. Every increase of one square foot in house size leads to an increase in the predicted home value of 0.14875($1000) = $148.75.
36	B	Because the superintendent selected a random sample from each type of school to form the overall sample, this sampling method is a stratified random sample. Each type of school is a stratum and individuals within each school may have similar characteristics (i.e., similar opinions about extracurricular activities), making them homogeneous groupings.
37	B	0(0.10) + 1(0.25) + 2(0.50) + 3(0.15) = 1.7
38	E	Because the *P*-value of $0.0760 > \alpha = 0.05$, we would fail to reject H_0 at the 5% level and not find convincing evidence, eliminating choices A, B, and C. Because the *P*-value of $0.0760 < \alpha = 0.10$, we would reject H_0 at the 10% level. Choice (E) rejects H_0 and concludes that a change has occurred.
39	D	The treatments are the two types of medication, each at three different dosages, for a total of six treatments. Subjects are blocked into groups of six according to their weight. Because the company is testing which medication is effective in the least amount of time, the response variable is the time until pain relief is achieved.
40	C	Because a random sample was selected, the sample mean will be an unbiased estimator, *regardless* of the size of the sample, which eliminates options A, B, and E. A smaller sample size will yield a more variable point estimator and a larger sample size will yield less variable point estimator. In other words, a larger sample size n ($60 > 30$) will make the standard deviation of the sample mean decrease, given the formula $\sigma_{\bar{x}} = \frac{\sigma}{\sqrt{n}}$.

Answer Key & Textbook Mapping
Multiple Choice Practice Exam 1

Multiple Choice Question	Answer	Learning Objective	Skill	CED Unit	Textbook Section
1	B	DAT-2.A	1.C	1	3A
2	A	VAR-3.E	4.B	3	3C
3	B	VAR-4.E	3.A	4	4B
4	A	DAT-1.A	2.A	2	2B
5	C	VAR-5.E	3.B	4	4E
6	D	UNC-3.B	3.A	4	4F
7	C	DAT-3.E	4.B	7	7B
8	B	UNC-3.Q	3.B	5	5D
9	C	UNC-1.K	4.B	1	1D
10	C	VAR-7.C	1.F	7	7B
11	C	UNC-5.A	1.B	6	6D
12	E	VAR-7.B	1.E	7	7B
13	B	UNC-3.E	3.A	4	4F
14	B	UNC-1.J	2.C	1	1D
15	E	UNC-4.AA	4.D	7	7B
16	E	DAT-3.F	4.E	7	7B
17	C	VAR-3.C	1.C	3	3C
18	A	UNC-4.U	4.A	7	7A
19	D	UNC-1.N	2.D	1	1C
20	C	UNC-1.M	2.A	1	1D
21	C	UNC-4.C	3.D	6	6B

22	A	VAR-2.B	3.A	1	1F
23	D	UNC-4.AF	3.D	9	9A
24	C	UNC-4.H	4.A	6	6B
25	C	UNC-4.K	3.D	6	6E
26	B	VAR-4.D	3.A	4	4C
27	D	UNC-1.N	2.D	1	1C
28	E	UNC-3.R	3.C	5	5D
29	E	UNC-5.C	3.B	4	4F
30	C	UNC-3.T	3.B	7	7D
31	B	DAT-1.G	2.C	2	2C
32	B	DAT-1.E	2.B	2	2C
33	C	UNC-4.F	4.B	6	6A
34	D	VAR-8.H	3.A	8	8B
35	C	DAT-1.H	4.B	2	2C
36	B	DAT-2.C	1.C	3	3B
37	D	VAR-5.C	3.B	4	4D
38	E	DAT-3.J	4.E	8	8B
39	D	DAT-2.A, VAR-3.C, VAR-3.A	1.C	3	3C
40	C	UNC-3.V	4.B	5	5B

Part II Free-Response Section Solutions and Scoring

Each of the 6 Free-Response Questions is scored using the following scale:

4 Complete Response

All three parts essentially correct

3 Substantial Response

Two parts essentially correct and one part partially correct.

2 Developing Response

Two parts essentially correct and no parts partially correct

One part essentially correct and one or two parts partially correct

Three parts partially correct

1 Minimal Response

One part essentially correct and no parts partially correct

No parts essentially correct and two parts partially correct

0 Incorrect Response

No parts essentially correct and one part partially correct

No parts essentially correct and no parts partially correct

Question 1:
Intent of the Question

The primary goals of this question are to assess your ability to (1) compare 2 distributions of data; (2) draw conclusions about the populations by identifying unique benefits in each population.

Solution

(a) The radon level in homes built with the new construction material has less variability than homes with the current construction material. The centers of both distributions of radon level are roughly the same ($\approx$ 3 pCi/L). The shape of the distribution of radon level in homes with the current construction material is right skewed with at least one high outlier, but the shape of the distribution of radon level in homes with the new construction material is bimodal and roughly symmetric with no outliers.

(b) One benefit of using the current construction material is that a larger proportion of homes had radon levels less than 2 pCi/L. In homes with the new material, only 1/200 homes had a radon level less than 2 pCi/L. If the acceptable level of radon is less than 2 pCi/L, then all but one of the homes with the new material would fail by that standard, while some of the homes with the current material would pass.

(c) One benefit of using the new construction material is that all homes have radon levels of about 5 pCi/L or less, whereas a large proportion of homes using the current material have radon levels more than 6 pCi/L. If 5 pCi/L or less is the acceptable level, all the homes using the new material would pass, but several homes using the current material would fail by that standard.

Scoring:

Each section scored as essentially correct (E), partially correct (P), or incorrect (I).

Part (a): Section 1 is essentially correct if response satisfies all 3 components: (1) statement about the shape for each material, (2) explicit comparison of the centers with words or symbols, and (3) context. Partially correct if the response satisfies 2 of the 3 components OR if the communication is weak. Incorrect otherwise.

Part (a): Section 2 is essentially correct if response satisfies both: (1) explicit comparison of the variability with words or symbols; (2) statement of outliers only for current material. Partially correct if the response satisfies 1 of the 2 components OR if the communication is weak. Incorrect otherwise.

Parts (b) and (c): Section 3 is essentially correct if response satisfies both: (1) unique statement about the benefit for current material not also true for new material (2) unique statement about the benefit for new material not also true for current material. Partially correct if the response satisfies 1 of the 2 components OR if the communication of unique benefits is weak. Incorrect otherwise.

Question 2:

Intent of the Question

The primary goals of this question are to assess your ability to (1) describe how to conduct random assignment for an experiment; (2) describe a benefit of random assignment; (3) evaluate causation for a statistically significant result.

Solution

(a) On equally sized slips of paper, write "type I toothbrush" on 20 slips, "type II toothbrush" on 20 slips, and "type III toothbrush" on 20 slips. Put the 60 slips into a hat and mix them well. Have each volunteer select a slip of paper (and not replace it). The slip they select will be the treatment group they are in.

(b) There are a variety of variables that the experimenter may not be able to control for, such as diet, overall teeth health, genetics, etc. To make roughly equivalent treatment groups at the beginning of the experiment, the student will randomly assign subjects into groups. It is hoped that this randomization will distribute the variety of uncontrolled variables equally among the three groups so any observed differences are due to the treatment, not an extraneous variable.

(c) Yes, if a statistically significant result is found, the student can infer the toothbrush type caused the reduction in plaque because this was an experiment in which the volunteers were randomly assigned to the treatments (type of toothbrush).

Scoring:

Each part is scored as essentially correct (E), partially correct (P), or incorrect (I).

Part (a) is essentially correct if the response describes an appropriate method of random assignment to the 3 equal groups. This method should include labeling the individuals and employing a sufficient means of random assignment that could be repeated by someone knowledgeable in statistics and should avoid duplicate assignment. Note: If appropriate, mixing slips of paper well must be explicitly stated. Partially correct if random assignment is described correctly, but the description does not provide sufficient detail for implementation. Incorrect, otherwise.

Part (b) is essentially correct if the response states a unique benefit of random assignment AND appeals to at least 2 of: (1) establishing treatment groups that are roughly equivalent; (2) to distribute uncontrolled variables roughly equally among the groups; (3) to ensure any observed differences are due to the treatments. Partially correct if the response states a unique benefit of random assignment but appeals to only 1 of the 3 components OR if the communication is weak. Incorrect otherwise.

Part (c) is essentially correct if response answers yes AND justifies answer appealing to both the experiment and random assignment of treatments. Partially correct if response answers yes AND appeals only to either the experiment or to the random assignment of treatments OR if the communication is weak. Incorrect otherwise.

Question 3:

Intent of the Question

The primary goals of this question are to assess your ability to (1) calculate the mean and standard deviation for the sum of independent random variables; (2) use a sampling distribution to calculate a probability; (3) calculate a probability for a binomial distribution.

Solution

(a) Let W = the weight of the box plus the chocolate creams plus the fudge pieces.

$\mu_W = 4(3) + 5(4) + 1(3) = 35$ ounces

$\sigma_W = \sqrt{4(0.2)^2 + 5(0.3)^2 + 1(0.1)^2} = 0.79$ ounce

(b) $P(W < 34) = P\left(z < \dfrac{34-35}{0.79}\right) = P(z < -1.27) = 0.1020$

With Table A: $z = -1.27$, P-value = 0.1020

With technology: normalcdf(lower : -1000, upper: 34, mean: 35, SD: 0.79) = 0.1028

The probability that a randomly selected box of candy will weigh less than 34 ounces ≈ 0.1028.

(c) Let X = number of boxes that weigh less than 34 ounces.

$P(X \geq 1) = 1 - P(X = 0) = 1 - (1 - 0.1020)^3 = 0.2758$

$P(X \geq 1) = 1 - P(X = 0) = 1 - (1 - 0.1028)^3 = 0.2778.$

Scoring:

Each part is scored as essentially correct (E), partially correct (P), or incorrect (I).

Part (a) is essentially correct if the response correctly includes at least 4 of the 5 components: (1) calculates the mean weight; (2) calculates the standard deviation; (3) provides supporting work for at least one calculation; (4) provides units for at least one parameter; (5) variable is defined and any symbols used for the parameters are appropriate. Partially correct if the response satisfies 2 or 3 of the 5 components. Incorrect, otherwise.

Part (b) is essentially correct if the response sets up and performs a correct probability calculation with all parameters clearly identified. Part (b) is partially correct if the response includes a correctly set up calculation but fails to calculate the correct value OR sets up an incorrect, but plausible, calculation and carries it through correctly OR sets up an appropriate calculation that does not clearly identify the parameters. Incorrect, otherwise.

Part (c) is essentially correct if the response sets up and performs a correct probability calculation with all parameters clearly identified. Part (c) is partially correct if the response includes a correctly set up calculation but fails to calculate the correct value OR sets up an incorrect, but plausible, calculation and carries it through correctly OR sets up an appropriate calculation that does not clearly identify the parameters. Incorrect, otherwise.

Question 4:

Intent of the Question

The primary goals of this question are to assess your ability to (1) identify an appropriate significance test; (2) conduct a 2-sample z-test for the difference in proportions; (3) use test results to draw an appropriate conclusion to answer a question.

Solution:

If conditions are met, we will perform a two-sample z test for a difference in proportions whose hypotheses are:

$$H_0: p_{2020} = p_{2024}$$
$$H_a: p_{2020} < p_{2024}$$

p_{2020} = population proportion of U.S. adults who eat a gluten free diet in 2020
p_{2024} = population proportion of U.S. adults who eat a gluten free diet in 2024

Conditions for inference:

1. Independent random samples: The samples in 2020 and in 2024 were independent and selected at random.
    - 10%: Because these samples were selected without replacement, we will check the 10% condition. There were at least 21,050 U.S. adults in 2020 and 17,570 U.S. adults in 2024.
2. Large Counts: The pooled proportion is $\hat{p}_C = \frac{(0.005*2105)+(0.02*1757)}{2105+1757}$ which equals 0.011824.

   Because (2105)(0.011824) = 24.9 and (2105)(1 – 0.011824) = 2080.1 are both at least than 10 and (1757)(0.011824) = 20.8 and (1757)(1 – 0.011824) = 1736.2 are both at least than 10, the sampling distribution of the difference in proportions is approximately normal.

All conditions are met. We will use a two-sample z test for a difference in proportions.

$$z = \frac{\hat{p}_1 - \hat{p}_2}{\sqrt{(\hat{p}_c)(1-\hat{p}_c)\left(\frac{1}{n_1}+\frac{1}{n_2}\right)}}$$

where $\hat{p}_c = 0.011824$

$$z = \frac{0.005 - 0.02}{\sqrt{(0.011824)(0.988176)\left(\frac{1}{2105}+\frac{1}{1757}\right)}} = -4.294$$

P-value = $P(z < -4.294) = 8.78 \times 10^{-6} \approx 0$

Because the P-value of $0 < 0.05$, we reject the null hypothesis. There is convincing statistical evidence to conclude that the proportion of U.S. adults who eat gluten free diets increased from 2020 to 2024 (that is, the proportion in 2020 was less than the proportion in 2024).

Scoring:

Each section scored as essentially correct (E), partially correct (P), or incorrect (I).

Section (1) Section 1 is essentially correct if response satisfies all 4 components: (1) identifies a 2-sample z-test for the difference in proportions by name or formula; (2) states hypotheses using a difference of 2 proportions (3) states the correct equality for the null hypothesis and a correct inequality for the alternative hypothesis; (4) defines or provides sufficient context to identify the parameters. Partially correct if the response satisfies 3 of the 4 components. Incorrect otherwise.

Section (2) Section 2 is essentially correct if response satisfies all 5 components: (1) states independent random samples were selected; (2) shows the 10% condition is satisfied for each of the independent samples; (3) shows the normality/large counts condition is satisfied using the pooled proportion with supporting work for all 4 counts and a comparison to an appropriate count; (4) correctly reports the test statistic for the named test; (5) correctly reports the P-value, consistent with the stated alternative hypothesis and reported test statistic. Partially correct if the response satisfies 2 or 3 or 4 of the 5 components. Incorrect otherwise. Note: To earn component (1), the response must clearly reference 2 independent samples. Components (2) and (3) can be earned with supporting evidence AND an explicit statement of each condition being met or using check marks.

Section (3) Section 3 is essentially correct if response satisfies all 3 components: (1) provides correct comparison of the P-value to the stated alpha (2) provides a correct decision (reject/fail to reject null hypothesis or convincing evidence exists or does not exist for alternative hypothesis) based on the comparison in component one; (3) states a conclusion in context, consistent with, and in terms of, the alternative hypothesis using non-deterministic language. Partially correct if the response satisfies 2 of the 3 components. Incorrect otherwise.

Question 5:
Intent of the Question

The primary goals of this question are to assess your ability to (1) describe the association between 2 independent categorical variables based on a segmented bar graph; (2) describe how to conduct a systematic random sample; (3) sketch and interpret a mosaic plot.

Solution:

(a) The segmented bar graph illustrates an association between age and movie genre preference. Adults in this sample are more likely to prefer comedy (about 3 times as much as teens), whereas teens are about twice as likely (32%) to prefer horror than adults (15%). Teens also prefer action/adventure more than adults.

(b) The journalist selected the sample from 1600 individuals at the movie theater on one day. For a systematic sample, the journalist would pick a random starting point between 1 and 80, such as $k = 5$, and begin with the 5th person that arrives at the movie ticket window and survey that person. Then the journalist could select (for example) every twentieth person for the fixed, periodic interval. This continues until the journalist has a total of 80 survey participants.

(c) Because there were 60 teens and 20 adults that were surveyed, the horizontal widths of the bars will account for this difference in a mosaic plot. The heights of the bars and the percentages for each movie genre preference should be identical to the original segmented bar graph, but the width of the teen bar should be 3 times as large as the adult bar (because 60 is 3 times as large as 20).

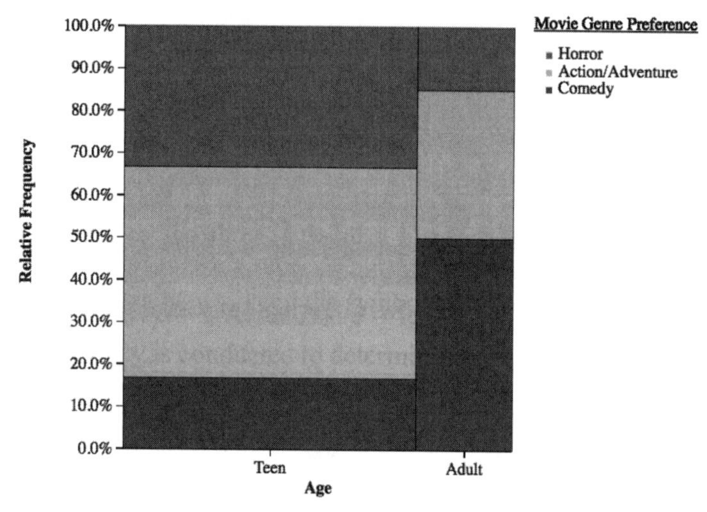

(d) The mosaic plot reveals that a higher proportion of teens responded to the survey than adults. This fact is hidden in the segmented bar graph.

Scoring:

Each section scored as essentially correct (E), partially correct (P), or incorrect (I).

Part (a): Section 1 is essentially correct if the response states an association exists between age and genre AND states at least 2 correct direct comparisons (either similarities or differences, or one of each) between age and genre. Partially correct if an association is identified with only 1 correct direct comparison (either similarity or difference) OR if the communication is weak. Incorrect otherwise. Note: Incorrect if response states a correlation exists between the variables.

Part (b): Section 2 is essentially correct if the response includes all 4 components: (1) the sample was selected from the 1600 moviegoers on one day; (2) the *k*th person was randomly selected with $1 \leq k \leq 20$ (or another reasonable value) for the 1st member of the sample; (3) every 20th person after the kth was selected (or another reasonable fixed interval); (4) stop when the 80th individual was selected. Partially correct if the response satisfies 2 or 3 of the 4 components, OR if the communication is weak. Incorrect otherwise.

Parts (c) and (d): Section 3 is essentially correct if the response includes all 5 components: (1) the width of the bar for teens is approximately 3 times that for adults; (2) the heights of the 2 bars should be the same; (3) percentages graphed in each bar should match the original graph; (4) both the horizontal (age) and vertical (relative frequency) axes are labeled AND a key is provided; (5) a unique observation made from the mosaic plot that cannot be identified in the segmented bar graph. Partially correct if the response satisfies 3 or 4 of the components. Incorrect otherwise.

Question 6:
Intent of the Question

The primary goals of this question are to assess your ability to (1) describe an association from a scatterplot; (2) interpret computer output for a linear regression; (3) calculate the correlation given the coefficient of determination; (4) interpret the coefficient of determination and the correlation for a linear regression; (5) calculate a new statistic; (6) identify the benefit of using the new statistic for a non-linear association.

Solution:

(a) Yes, there appears to be a strong, negative, curved association between distance from the river and the amount of insurance claims.

(b) No. The coefficient of determination (*r*-squared) is 0.4837 and the correlation coefficient is $\pm\sqrt{0.4837} = -0.6955$. (The sign is negative since the slope, -1.346, is negative.) The coefficient of determination accounts for only 50% of the variability in amount and the correlation coefficient is not very high, so there doesn't appear to be evidence of a strong association. These values would indicate a more "moderate" (not strong) association.

(c) To complete the table:

Distance	Total Amount	Rank (Distance)	Rank (Amount)	Difference in Ranks
0-4	102.8	1	10	-9
5-9	31.9	2	9	-7
10-14	19.4	3	8	-5
15-19	14.8	4	7	-3
20-24	12.5	5	6	-1
25-29	10.3	**6**	**5**	**1**
30-34	9.1	**7**	**4**	**3**
35-39	6.9	**8**	**1**	**7**
40-44	7.7	**9**	**3**	**6**
45-49	7.3	**10**	**2**	**8**

(d) $\Sigma(d^2) = (-9)^2 + (-7)^2 + (-5)^2 + (-3)^2 + (-1)^2 + (1)^2 + (3)^2 + (7)^2 + (6)^2 + (8)^2 = 324$

(e) $r_s = 1 - \frac{6(324)}{10(10^2-1)} = -0.9636$

(f) The correlation in part (e) is -0.9636, which is much stronger than the correlation in (b): -0.6955. Based on this new statistic for correlation, the association now appears to be much stronger than using the correlation from part (b). This new statistic gives stronger evidence of the insurance agent's belief that there is a strong association between amount and distance from the river.

Scoring:

Each section scored as essentially correct (E), partially correct (P), or incorrect (I).

Parts (a) and (b): Section 1 is essentially correct if the response includes all 5 components: (1) answers yes in part (a); (2) supports answer to part (a) by stating a strong, curved (nonlinear) association exists between distance from river and amount of insurance claims; (3) calculates the correlation with the negative sign; (4) answers no in part (b); (5) supports answer to part (b) based on both r^2 and r. Partially correct if the response satisfies 3 or 4 of the 5 components OR if the communication is weak. Incorrect otherwise. Note: Component (2) does not require direction for credit.

Parts (c), (d), and (e): Section 2 is essentially correct if the response includes all 4 components: (1) table completed with correct values; (2) correct value in part (d) for values in table; (3) correct value in part (e) for prior calculations; (4) supporting work for part (e). Partially correct if the response satisfies 2 or 3 of the 4 components. Incorrect otherwise. Note: Overlook minor arithmetic errors.

Part (f): Section 3 is essentially correct if the response includes all 4 components: (1) explicit comparison with words or symbol between the correlations reported in parts (b) and (e); (2) identifies the stronger correlation based on calculated results; (3) selects the appropriate statistic; (4) justifies choice of statistic selected. Partially correct if the response satisfies 2 or 3 of the components OR if the communication is weak. Incorrect otherwise.

Practice Exam 1 Scoring Worksheet

Section I: Multiple Choice

_______________ × 1.2500 = ___________

Number Correct (out of 40)

Weighted Section I Score (Do not round)

Section II: Free Response

Question 1 _________ × 1.8750 = _________
(out of 4) (Do not round)

Question 2 _________ × 1.8750 = _________
(out of 4) (Do not round)

Question 3 _________ × 1.8750 = _________
(out of 4) (Do not round)

Question 4 _________ × 1.8750 = _________
(out of 4) (Do not round)

Question 5 _________ × 1.8750 = _________
(out of 4) (Do not round)

Question 6 _________ × 3.1250 = _________
(out of 4) (Do not round)

Composite Score

___________ + ___________ = ___________

Weighted Section I Score

Weighted Section II Score

Composite Score (Round to nearest whole number)

AP Score Conversion Chart

Composite Score Range	AP Score
73-100	5
59-72	4
44-58	3
32-43	2
0-31	1

Textbook Mapping
Free Response Practice Exam 1

FRQ Part	Learning Objective	Skill	CED Unit	Textbook Section	SKILL #
1a	UNC-1.N	2.D	1	1C	2
1b	UNC-1.N	2.D, 4.A	1	1C	4
1c	UNC-1.N	2.D, 4.A	1	1C	4
2a	VAR-3.C	1.C	3	3C	1
2b	VAR-3.C	1.C	3	3C	1
2c	VAR-3.E	4.B	3	3C	4
3a	VAR-5.E	3.B	4	4E	3
3b	VAR-2.B	3.A	4	4E	3
3c	UNC-3.B	3.A	4	4E	3
4-i	VAR-6.H	1.F	6	6F	1
4-ii	VAR-6.J	4.C	6	6F	4
4-iii	VAR-6.K	3.E	6	6F	3
4-iv	DAT-3.D	4.E	6	6F	4
5a	UNC-1.P	2.D	2	2A	2
5b	DAT-2.C	1.C	3	3B	1
5c	UNC-1.P	2.D	2	2A	2
5d	UNC-1.E	2.D	1	2A	2
6a	DAT-1.A	2.A	2	2B	2
6b	DAT-1.C, DAT-1.G	2.C, 4.B	2	2B, 2C	2, 4
6c	DAT-1.B	2.C	2	2B	2
6d	DAT-1.B	2.C	2	2B	2
6e	DAT-1.B	2.C	2	2B	2
6f	DAT-1.C	4.B	2	2B	4

PRACTICE EXAM #2

SECTION I

Time—1 hour and 30 minutes
Number of questions—40
Percent of total score—50

Directions. Identify the choice that best completes the statement or answers the question. Check your answers and note your performance when you are finished.

1. Which of the following situations would be easiest to explore using a census?

   (A) You wish to know the proportion of teachers in the state that have a master's degree.
   (B) You want to know the average amount of time spent on homework by students in your high school over the semester.
   (C) You want to know the proportion of homes in a city that have wireless Internet access.
   (D) You want to know the proportion of trees in a large state forest that are infected with Dutch Elm disease.
   (E) You want to know the average height of the twenty students in your history class.

2. It is known that 15% of the seniors in a large high school enter military service upon graduation. If a group of 20 seniors are randomly selected from this school, what is the probability of observing at most one senior who will be entering military service?

   (A) $20(0.15)^1(0.85)^{19}$
   (B) $1 - 20(0.15)^1(0.85)^{19}$
   (C) $(0.85)^{20} + 20(0.15)^1(0.85)^{19}$
   (D) $(0.85)^{20}$
   (E) $1 - (0.85)^{20}$

3. To evaluate the yield per acre of three different varieties of corn, 12 one-acre plots of land will be used. One variety of corn will be randomly assigned to 4 of the plots, the second variety will be randomly assigned to 4 other plots, and the third variety to the 4 remaining plots of land. At the end of the growing season, the yield per acre will be evaluated. Which one of the following is a TRUE statement?

   (A) The type of land is being used as a blocking factor.
   (B) The treatments in this experiment are the plots of land.
   (C) The experimental units are the varieties of corn.
   (D) The randomization process will reduce the variation that exists due to the plots of land.
   (E) Replication of corn variety on four plots of land allows us to measure variability across plots of land.

4. Pediatricians were interested in determining if there is an association between how frequently a child snores per week and sleep-disordered breathing (whether the child has nocturnal coughing or asthma). The results of a survey of 1048 randomly selected children aged 2 to 5 are given below.

	Snored four or more times per week	**Snored three or fewer times per week**	Total
Had nocturnal coughing	59	255	314
Had asthma	22	222	244
Did not have asthma or nocturnal coughing	115	375	490
Total	196	852	1048

The hypotheses that the pediatricians used were given as:

H_0: There is no association between frequency of snoring and the prevalence of sleep-disordered breathing among children aged 2 to 5.
H_a: An association does exist between frequency of snoring and the prevalence of sleep-disordered breathing among children aged 2 to 5.

If the test statistic for this procedure is $\chi^2 = 22.38$, which one of the following is a correct conclusion at the 1% level of significance?

(A) Because the *P*-value is greater than 0.01, an association does exist between frequency of snoring and the prevalence of sleep-disordered breathing among children aged 2 to 5.
(B) Because the *P*-value is less than 0.01, there is no association between frequency of snoring and the prevalence of sleep-disordered breathing among children aged 2 to 5.
(C) Because the *P*-value is greater than 0.01, there is convincing statistical evidence that there is no association between frequency of snoring and the prevalence of sleep-disordered breathing among children aged 2 to 5.
(D) Because the *P*-value is less than 0.01, there is convincing statistical evidence of an association between frequency of snoring and the prevalence of sleep-disordered breathing among children aged 2 to 5.
(E) Because the test statistic is greater than 0.01, there is not convincing statistical evidence of an association between frequency of snoring and the prevalence of sleep-disordered breathing among children aged 2 to 5.

5. A study of two popular sleep-aids measured the mean number of hours slept for 200 individuals. In the group of 200 volunteer subjects, 100 were randomly assigned to be administered Drug 1 and the other 100 were administered Drug 2. A two-sample *t* test was performed on the difference in the mean sleep times experienced by the subjects. The *P*-value was 0.12. Which of the following is a correct interpretation of the *P*-value?

   (A) Approximately 12% of the subjects did not sleep at all.
   (B) There was a 12% difference between the mean sleep times of the two groups.
   (C) We would expect to see a difference in the mean sleep times at least as extreme as the observed difference about 12% of the time.
   (D) Assuming both drugs are equally effective, we would expect to see a difference in the mean sleep times at least as extreme as the observed difference in about 12% of repeated administrations due only to chance.
   (E) Assuming both drugs are not equally effective, we would expect to see a difference in the mean sleep times at least as extreme as the observed difference about 12% of the time.

6. Biologists have collected data relating the age (in years) of one variety of oak tree to its height (in feet). A scatterplot for 20 of these trees is given below.

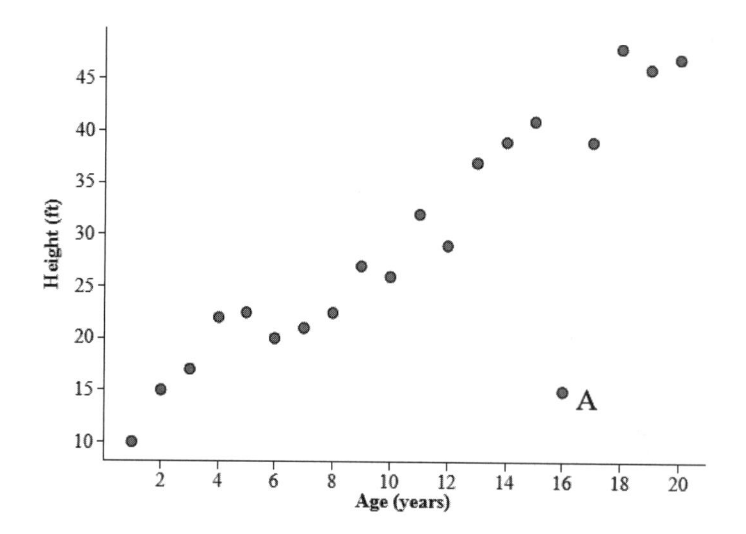

   The point labeled A was graphed at an incorrect height and will be moved to the correct height of 40 feet. When point A is moved, which one of the following statements will be TRUE?

   (A) The slope of the least-squares regression line will decrease, and the correlation will increase.
   (B) The slope of the least-squares regression line will decrease, and the correlation will decrease.
   (C) The slope of the least-squares regression line will increase, and the correlation will increase.
   (D) The slope of the least-squares regression line will increase, and the correlation will decrease.
   (E) The slope of the least-squares regression line and the correlation will remain the same.

7. A school principal is interested in calculating a 96% confidence interval for the true mean number of days students are absent during the school year. The attendance office lists the number of days absent for each student during the month of December. Which of the following is the best reason the principal cannot construct the confidence interval?

(A) A confidence interval cannot be constructed unless all data for the population are known.
(B) The critical value for 96% is not listed in the table, therefore it cannot be calculated.
(C) The attendance office should report the average for the month, not the number for each individual student.
(D) Because most students would not miss any days, the average will be too small to construct a confidence interval.
(E) The number of absences in December may not be representative of the rest of the months during the year. The principal should select a random sample of students and record the number of days absent during the school year for each student in the sample.

8. A study of 156 people addicted to drugs in Amsterdam recorded how often each subject had recently injected drugs and whether the person was infected with HIV, the virus that causes AIDS. Here is a two-way table of the number of people in each condition:

		HIV Positive	HIV Negative
Injected Drugs?	Daily	32	45
	Less than Daily	20	18
	Never	18	23

If a person is chosen at random from this group, what is the probability that they test positive for HIV, given that they inject daily?

(A) 0.21
(B) 0.42
(C) 0.46
(D) 0.54
(E) 0.58

9. Suppose the mean rate of return on the common stocks in a large, diversified investment portfolio was 12% last year. If the rates of return of the stocks within the portfolio are approximately normally distributed and a rate of return of 15% represents the 80th percentile, what is the approximate standard deviation of the rates of return in the portfolio?

(A) 1.79%
(B) 2.8%
(C) 3.57%
(D) 3.98%
(E) 4.1%

10. A sports physiologist wants to compare the effects of two different exercise machines, A and B, on the flexibility of gymnasts. There are 20 gymnasts available for the study. All gymnasts have been given a flexibility rating before the experiment starts. A similar test will be given at the end of the experiment. Which one of the following methods of assigning gymnasts to machines would be optimal to assess the difference of the change in flexibility due to the machines?

(A) Randomly assign a number (1 through 20) to each gymnast. Match the gymnasts by consecutive number (i.e., #1 with #2, #3 with #4, etc.). Then assign the odd-numbered gymnast in each pair to machine A and the even-numbered gymnast to machine B.
(B) Have all the gymnasts use both machines. A coin flip will determine which machine is used first.
(C) Flip a coin for each gymnast to assign them into the "heads" or "tails" group until one group has 10 gymnasts. The remaining unassigned gymnasts will fill the other group. Flip the coin again to assign the heads or tails group to use machine A and the other group to use machine B.
(D) Match the gymnast with the highest flexibility rating with the gymnast with the next highest flexibility rating to form the first pair. Match the gymnasts with the next two highest flexibility ratings to form the second pair, etc. Flip a coin for each pair to assign that pair of gymnasts to one of the two machines.
(E) Match the gymnast with the highest flexibility rating with the gymnast with the next highest flexibility rating to form the first pair. Match the gymnasts with the next two highest flexibility ratings to form the second pair, etc. For each pair, flip a coin to assign one member of the pair to machine A and the other will be assigned to machine B.

11. According to the Insurance Institute for Highway Safety in 2017, 58% of 18-year-olds admitted to texting while they are driving. In a large city in the Midwest, a random sample of 250 18-year-old drivers found that 160 of them had admitted to texting while driving. If the texting while driving rate of 18-year-old drivers in this Midwestern city is the same as the national rate, which of the following represents the probability of getting a random sample of 250 18-year-old drivers whose texting rate is greater than 64%?

(A) $P\left(z > \frac{0.64-0.58}{\sqrt{\frac{(0.58)(0.42)}{250}}}\right)$

(B) $P\left(z > \frac{0.64-0.58}{\sqrt{\frac{(0.64)(0.36)}{250}}}\right)$

(C) $P\left(z > \frac{0.58-0.64}{\sqrt{\frac{(0.64)(0.36)}{250}}}\right)$

(D) $\binom{250}{160}(0.58)^{160}(0.42)^{90}$

(E) $\binom{250}{160}(0.64)^{160}(0.36)^{90}$

12. A poll was conducted to determine the level of support within a large suburban school district for the construction of a new football stadium. Of the 655 people who were surveyed, 57% said they support spending money to construct the stadium. The poll had a margin of error of 4%. Which is the correct interpretation for this margin of error?

(A) About 4% of the school district residents polled refused to respond to the question.
(B) We would expect, at most, a 4% difference between the proportion of support in the sample and the true proportion of support in the district.
(C) We would expect, at least, a 4% difference between the proportion of support in the sample and the true proportion of support in the district.
(D) If we conducted repeated samples from this population, the results would vary by no more than 4% from the current sample result of 57%.
(E) If the poll had sampled more people, the proportion of district residents who would have said they supported the stadium would have been about 4% higher.

13. A company that ships vitamins claims that 97% of such shipments arrive on time. Let the random variable A represent the number of shipments that arrive on time in 50 randomly selected shipments. Random variable A follows a binomial distribution with a mean of 48.5 and a standard deviation of 1.21. Which of the following is the best interpretation of the mean?

(A) Every random sample of 50 shipments will have 48.5 shipments that arrive on time.
(B) Every random sample of 50 shipments will have 48.5 shipments that do not arrive on time.
(C) On average, the company ships 48.5 shipments until one is late.
(D) For all possible shipments of size 50, the average number of on time shipments is equal to 48.5.
(E) For all possible shipments of size 50, the average number of not on time shipments is equal to 48.5.

14. In a recent year, there were more than 116,000 men, women, and children in the United States who were awaiting transplants of a variety of organs such as livers, hearts, and kidneys. A national organ donor organization is trying to estimate the proportion of all people willing to donate their organs after their death to help transplant recipients. Which of the following choices is the smallest sample size that would be required to ensure a margin of error of at most 3 percent for a 98% confidence interval estimate of the proportion of all people in the U.S. who would be willing to donate their organs?

(A) 1075
(B) 1180
(C) 1510
(D) 1740
(E) 1845

15. A researcher is interested in the effect of energy drinks on a person's physical actions. The number of times each person can tap on a surface per minute was measured before and after they consumed a popular energy drink. The results are below:

Subject	1	2	3	4	5	6	7	8	9	10	$\bar{x}$	s_x
Before Drink	105	99	93	96	95	99	94	87	100	89	95.7	5.36
After Drink	103	102	105	102	101	110	95	89	102	98	100.7	5.70
Difference	2	–3	–12	–6	–6	–11	–1	–2	–2	–9	–5	4.59

The conditions for inference were checked and verified. Which one of the following represents the standardized test statistic for a test of $H_0: \mu_{B-A}$, where μ_{B-A} is the true mean difference (before drink minus after drink) in the number of taps for subjects like the ones in this study?

(A) $t = \dfrac{95.7-100.7}{\sqrt{\dfrac{5.36}{10}+\dfrac{5.70}{10}}}$

(B) $t = \dfrac{95.7-100.7}{\sqrt{\dfrac{5.36^2}{10}+\dfrac{5.70^2}{10}}}$

(C) $t = \dfrac{95.7-100.7}{\sqrt{\dfrac{5.36^2}{9}+\dfrac{5.70^2}{9}}}$

(D) $t = \dfrac{-5-0}{\dfrac{4.59}{\sqrt{10}}}$

(E) $t = \dfrac{-5-0}{\dfrac{4.59}{\sqrt{9}}}$

16. The number of goals scored per game in a full season for a professional soccer league is strongly skewed to the right with a mean of 2.3 goals and a standard deviation of 3.7 goals. An SRS of size n = 15 is selected from the population and the sample mean is computed. This process is repeated a total of 375 times. Which one of the following best describes the shape, center and variability of the resulting distribution of sample means?

(A) Skewed to the right with a mean of 2.3 goals and a standard deviation of 3.7 goals.
(B) Skewed to the right with a mean of 2.3 goals and a standard deviation of 0.955 goal.
(C) Approximately normal with a mean of 2.3 goals and a standard deviation of 3.7 goals.
(D) Approximately normal with a mean of 2.3 goals and a standard deviation of 0.955 goal.
(E) Approximately normal with a mean of 2.3 goals and a standard deviation of 0.191 goal.

17. The enrollment rate of graduating high school seniors accepted into a random sample of public and private colleges and universities was compared. The enrollment rate is the percent of students who were offered admission and chose to enroll at that school. The side-by-side boxplots summarize the distribution of enrollment rate for the sample of colleges and universities.

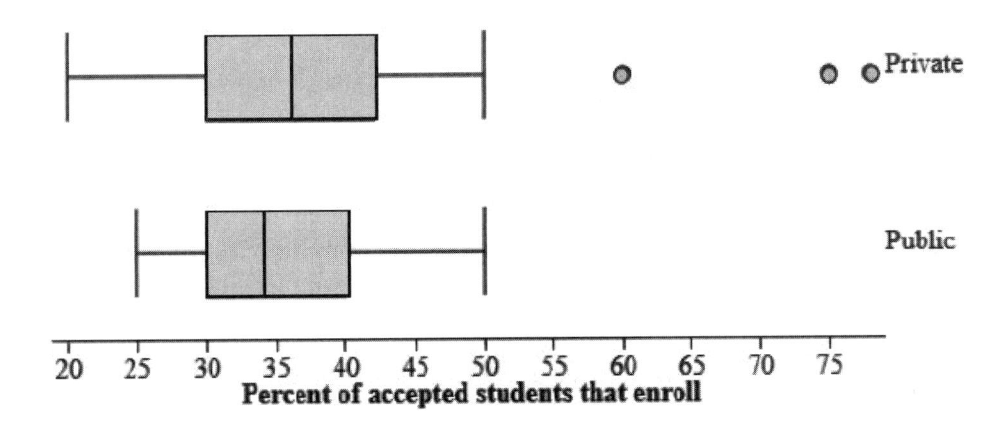

Which one of the following statements is TRUE about the distribution of enrollment rate?

(A) The *IQR* of the distribution of enrollment rate at private colleges and universities is less than the *IQR* of the distribution of enrollment rate at public colleges and universities.
(B) The median of the distribution of enrollment rate at private colleges and universities is less than the median of the distribution of enrollment rate at public colleges and universities.
(C) The mean of the distribution of enrollment rate at private colleges and universities is greater than the mean of the distribution of enrollment rate at public colleges and universities.
(D) The distribution of enrollment rate at private colleges and universities is skewed to the left.
(E) The third quartile for the distribution of enrollment rate at public colleges and universities is greater than the third quartile of the distribution of enrollment rate at private colleges and universities.

18. The Department of Traffic Safety (DTS) wants to reduce the number of drivers who speed on a certain stretch of road. To reduce repeat offenders, the DTS will randomly select 5 drivers, without replacement, from a population of 50 drivers convicted of speeding to assess the effectiveness of a new "safe driving" program. To select a random sample, the DTS will label the drivers 01, 02, 03, ..., 50 and use the following line from a random number table:

22368 46573 25595 85393 30995 89198 27982 53401 93965 34095 52666 19174

Which one of the following represents the sample of 5, starting from the left side of the table?

(A) 22, 36, 8, 46, 32
(B) 22, 36, 84, 65, 73
(C) 22, 36, 25, 30, 27
(D) 22, 36, 46, 32, 39
(E) 22, 23, 36, 46, 32

19. The following frequency table summarizes the final exam scores for 150 students in a statistics class at a large university.

Interval	Frequency
90–99	18
80–89	27
70–79	44
60–69	37
50–59	19
40–49	5

Which of the following is closest to the percentile for a score of 79 in this distribution?

(A) 25
(B) 30
(C) 70
(D) 75
(E) 88

20. A family owns two pizza restaurants in a town, one located on Valley Parkway and the other on Grand Avenue. The Valley Parkway restaurant sells 70% of their pizzas, and the Grand Avenue location sells the rest. At the Valley Parkway location, 40% of all pizzas sold are deep-dish. At the Grand Avenue location, 30% of all pizzas sold are deep-dish. If a customer who purchases a pizza at one of these locations is randomly selected, what is the approximate probability that the customer purchased a deep-dish pizza?

(A) 0.175
(B) 0.33
(C) 0.35
(D) 0.37
(E) 0.70

21. A car salesperson receives a commission of $300 for each car she sells. The number of cars she sells on a randomly selected day follows the probability distribution given in the table below.

Cars Sold	0	1	2	3	4
Probability	0.15	0.25	0.35	0.20	0.05

Based on this distribution, what are the mean and standard deviation of the daily commission that she should receive?

(A) Mean = $525 and standard deviation = $87.46
(B) Mean = $525 and standard deviation = $326.92
(C) Mean = $525 and standard deviation = $335.41
(D) Mean = $600 and standard deviation = $335.41
(E) Mean = $600 and standard deviation = $474.34

22. A company provides insurance coverage for automobile drivers at a fixed yearly premium. The company loses money on high-risk drivers. The company earns a profit on low-risk drivers. The company will not accept any new high-risk drivers for the next three months. Each time a potential customer applies for insurance the company must decide to insure them or not based on the following hypotheses.

H_0: The driver is low-risk.
H_a: The driver is high-risk.

Which of the following represents a Type II error and its consequence for the company?

(A) The company decides that the driver is high-risk, but was in fact, low-risk. The company misses an opportunity to make a profit.
(B) The company decides that the driver is high-risk, but was in fact, low-risk. The company may end up paying more money than it collects for the premium.
(C) The company decides that the driver is low-risk, but was in fact, high-risk. The company may end up paying more money than it collects for the premium.
(D) The company decides that the driver is low-risk, but was in fact, high-risk. The company makes a profit.
(E) The company decides that the driver is high-risk, and the driver is high-risk. The company avoids losing money.

23. A study on dieting recorded the lean body mass (in kilograms) and resting metabolic rate (in kcal/day) for a random sample of 12 women. The output shown in the table is from a least-squares regression to predict resting metabolic rate given lean body mass.

Term	Coef	SE Coef
Constant	201.162	181.701
Lean Body Mass	24.026	4.174

Suppose a woman has a lean body mass of 40 kg and a resting metabolic rate of 1000 kcal/day. Based on the residual, does the regression model overestimate or underestimate the resting metabolic rate for this woman?

(A) Underestimate, because the residual is positive.
(B) Underestimate, because the residual is negative.
(C) Overestimate, because the residual is positive.
(D) Overestimate, because the residual is negative.
(E) Neither, because the residual is 0.

24. A spring-loaded launcher is used to propel a ping pong ball for a game at a school carnival. If the ball lands in a bucket, the participant wins a prize. In testing the launcher to determine where the bucket should be placed, 20 test shots are taken. The mean distance traveled is 64.5 inches with a standard deviation of 2.1 inches. The conditions for inference were checked and verified. Which one of the following is the approximate 98% confidence interval estimate for μ, the true mean distance the ping pong ball would travel for all shots taken?

(A) $64.5 \pm 2.539\left(\frac{2.1}{\sqrt{19}}\right)$
(B) $64.5 \pm 2.539\left(\frac{2.1}{\sqrt{20}}\right)$
(C) $64.5 \pm 2.528\left(\sqrt{\frac{2.1}{20}}\right)$
(D) $64.5 \pm 2.528\left(\frac{2.1}{\sqrt{19}}\right)$
(E) $64.5 \pm 2.528\left(\frac{2.1}{\sqrt{20}}\right)$

25. Automobile engineers want to study the effect of a new automatic transmission on mileage. They decided on a specific make and model. They randomly selected 25 cars of this make and model that had the old transmission and selected a separate random sample 25 cars of this make and model that had the new model of transmission. A significance test on the difference between the means was conducted. Which one of the following is a condition necessary to conduct this test?

(A) The sample sizes must be at least 10 times the size of their respective populations.
(B) The distribution of the difference in population means is approximately normal.
(C) The expected number of successes and failures in each group must be at least 10.
(D) The distribution of mileage showed no strong skewness or outliers for either sample.
(E) The data come from two groups in a randomized comparative experiment.

26. Which of the following would be classified as a discrete quantitative variable?

(A) The heights of 10 randomly selected students in middle school.
(B) The 40-yard dash times of 50 randomly selected professional athletes.
(C) The weights of the first five dogs that arrive at a dog park.
(D) The favorite color of a group of students in an art studio.
(E) The number of recyclable cans found in trash cans at a local park on a randomly selected day.

27. In which of the following should the random variable X not be modeled with a geometric distribution?

(A) According to an Internet search, 68% of U.S. households own a pet. Let X represent the number of randomly selected U.S. households surveyed to find one with a pet.
(B) Suppose that at a certain stop light 40% of drivers text while waiting for a green light. Let X represent the number of randomly selected drivers at this light to find one that is texting.
(C) Alex struggles to throw a ball into the air and catch it. He catches only about 10% of throws. Let X represent the number of throws Alex makes to catch one.
(D) In a classroom of 25 students, 20% are blonde. Students will be selected, one at a time without replacement, and the student's hair color will be recorded. Let X represent the number of students selected to find one blonde student.
(E) A marketing company claims that 10% of people on a website will click on an ad. Let X represent the number of people that are randomly selected to find one person who clicked on the ad.

28. An insurance salesperson sells an average of 3.1 policies per week with a standard deviation of 1.1 policies. The salesperson is paid $300 per week plus a commission of $250 for each policy sold. What are the mean and standard deviation, respectively, of the salesperson's total weekly pay?

(A) $1705.00, $605.00
(B) $1705.00, $275.00
(C) $1075.00, $775.00
(D) $1075.00, $302.50
(E) $1075.00, $275.00

29. A group of friends believe there is a difference between the popularity of four teachers (1, 2, 3, and 4) at their school. They select a random sample of students and ask who the most popular teacher is from a list of these four teachers. Their results are shown in the table below.

Teacher	1	2	3	4
Votes	12	11	15	18

Which of the following would be the most appropriate inference procedure to test the student's claim?

(A) A chi-square test for independence.
(B) A chi-square test for goodness of fit.
(C) A two-sample t test for a difference in means.
(D) A one-sample t test for a mean.
(E) A t test for the slope.

30. A fast-food restaurant manager is interested in how many items are sold from 9 PM to 6 AM. The following dotplot shows the number of items sold each day over a 66-day period during these times.

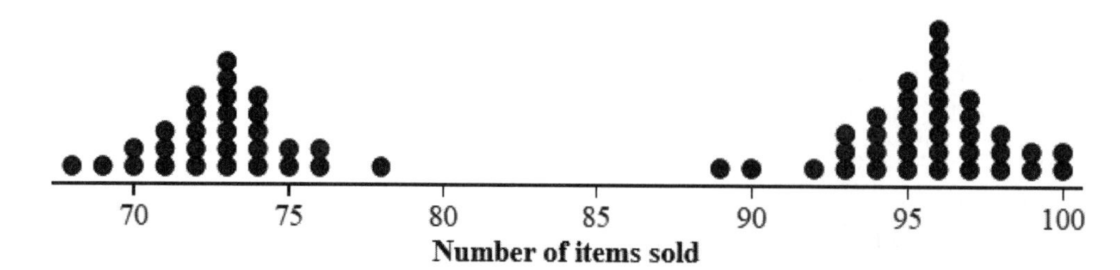

Which of the following is the best description of the distribution of the number of items sold?

(A) Approximately normal
(B) Roughly uniform without a gap
(C) Roughly uniform with a gap
(D) Bimodal without a gap
(E) Bimodal with a gap

31. A biologist is comparing the lengths of two varieties of bowerbird, the MacGregor's bowerbird and the Vogelkop bowerbird. The length of the MacGregor's bowerbird is approximately normal with mean 26 cm. The length of the Vogelkop bowerbird is approximately normal with mean 28 cm. Twelve birds of each variety will be randomly selected and the lengths will be recorded. Let $\bar{x}_M$ represent the sample mean length of the 12 MacGregor's bowerbirds and let $\bar{x}_V$ represent the sample mean length of the 12 Vogelkop bowerbirds.

Which of the following is the best interpretation of $P(\bar{x}_M - \bar{x}_V > 4) = 0.17$?

(A) The probability that the lengths of all 12 MacGregor's bowerbirds will exceed the lengths of all 12 Vogelkop bowerbirds by more than 4 cm is 0.17.
(B) The probability that the lengths of all 12 MacGregor's bowerbirds will exceed the lengths of all 12 Vogelkop bowerbirds by more than 0.17 cm is 4.
(C) The probability that the mean length of all 12 MacGregor's bowerbirds will exceed the mean length of all 12 Vogelkop bowerbirds by more than 4 cm is 0.17.
(D) The probability that the lengths of all MacGregor's bowerbirds will exceed the lengths of all Vogelkop bowerbirds by more than 4 cm is 0.17.
(E) The probability that the mean length of all 12 MacGregor's bowerbirds will exceed the mean length of all 12 Vogelkop bowerbirds is 0.17.

32. In which of the following distributions is the mean most likely greater than the median?

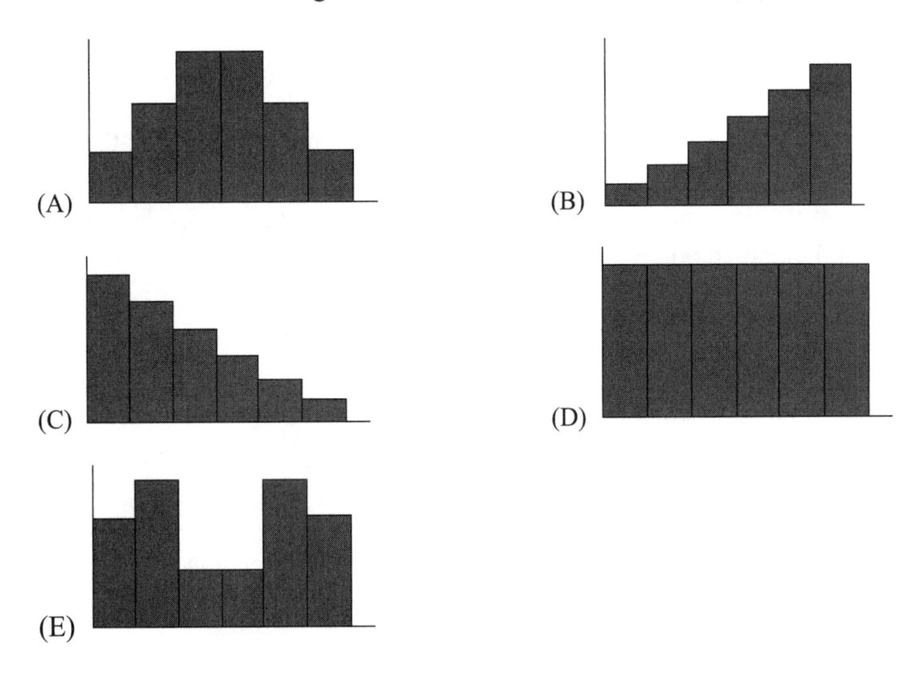

33. At a large university, a random sample of 20 students will be asked, "How many times per week do you eat breakfast?" The sample mean $\bar{x}$ will be computed to estimate the population mean μ. Is $\bar{x}$ an unbiased estimator of μ?

(A) No, because the sample mean $\bar{x}$ does not always equal the population mean μ.
(B) No, because a sample size of 20 is not large enough to assume the sampling distribution of $\bar{x}$ is approximately Normal.
(C) Yes, because the wording of the question is not biased.
(D) Yes, because the expected value of $\bar{x}$ based upon a random sample is equal to μ.
(E) Yes, because with a sample size of 20 the variability of the sample mean $\bar{x}$ will be small.

34. To determine the effectiveness of speed-boosting a computer by "overclocking" its processor, researchers randomly selected 16 computers of various makes and models and paired each of them with a second identical make and model. For each pair, it was randomly decided which of the two would be "overclocked" and which would be unmodified. The 95% confidence interval for the mean difference in speeds ("overclocked" – unmodified) was (1.3, 17.3) MHz. Assuming the conditions for inference are met, what can we conclude?

(A) The mean difference in speeds will be greater than 1.3 MHz approximately 95% of the time.
(B) The unmodified computers are anywhere from 1.3 to 17.3 MHz faster than those that were overclocked 95% of the time.
(C) Because the interval only contains positive values, there is a significant increase in mean speed from unmodified to "overclocked" computers.
(D) 95% of the differences between speeds will be between 1.3 and 17.3 MHz.
(E) A two-sample t test should have been used instead of an interval.

35. A newspaper manager wants to determine the level of support in a large town regarding the establishment of a city-wide residential recycling program. Which one of the following would represent a method of obtaining a stratified random sample?

(A) Randomly select eight residential blocks in the town and ask everyone who lives on those blocks if they would be in favor of a city-wide residential recycling program.
(B) Randomly select one of the first four people who enters City Hall, and every fourth person who enters thereafter until the desired number of people is selected.
(C) Select a random sample of people from the town phone directory.
(D) Select a random sample of residents from each of the northwest, northeast, southwest, and southeast quadrants of the city.
(E) Number the residents of the town using the latest census data. Use a random number generator to pick the sample.

36. A biologist in the Northeast has gathered data on a random sample of 22 white tailed deer in New England and performed a regression analysis on the weight of the white-tailed deer and their length. Some of the output is given below.

Regression Analysis: Weight versus Length

Predictor	Coef	SE Coef	T	P
Constant	-441.39	29.91	-14.76	0.000
Length	10.3382	0.4825	21.42	0.000

S = 53.7777 R-Sq = 76.5% R-Sq(adj) = 76.3%

The conditions for inference were checked and verified. Which of the following would represent a 95% confidence interval to estimate the population slope of the regression line relating white-tailed deer weight to length?

(A) $-441.39 \pm 2.086(29.91)$

(B) $-441.39 \pm 2.080(29.91)$

(C) $10.3382 \pm 2.086(0.4825)$

(D) $10.3382 \pm 2.086\left(\frac{0.4825}{\sqrt{22}}\right)$

(E) $10.3382 \pm 2.080(0.4825)$

37. A simple random sample of 50 adults is surveyed to estimate the proportion of all adults who visit the dentist at least once per year, and a confidence interval for the proportion is constructed. Suppose the researcher had surveyed a random sample of 450 adults instead and got the same sample proportion. How would the width of the confidence interval based on the 450 adults compare to the width of the interval for the 50 adults?

(A) The width would be about one-ninth the width of the original interval.
(B) The width would be about one-third the width of the original interval.
(C) The width would be the same as the width of the original interval.
(D) The width would be about three times the width of the original interval.
(E) The width would be about nine times the width of the original interval.

38. A biologist has gathered sample data about a population of bears in the forests of the northeast. The distribution of the weights (in pounds) of the sample of bears and their sex is given below.

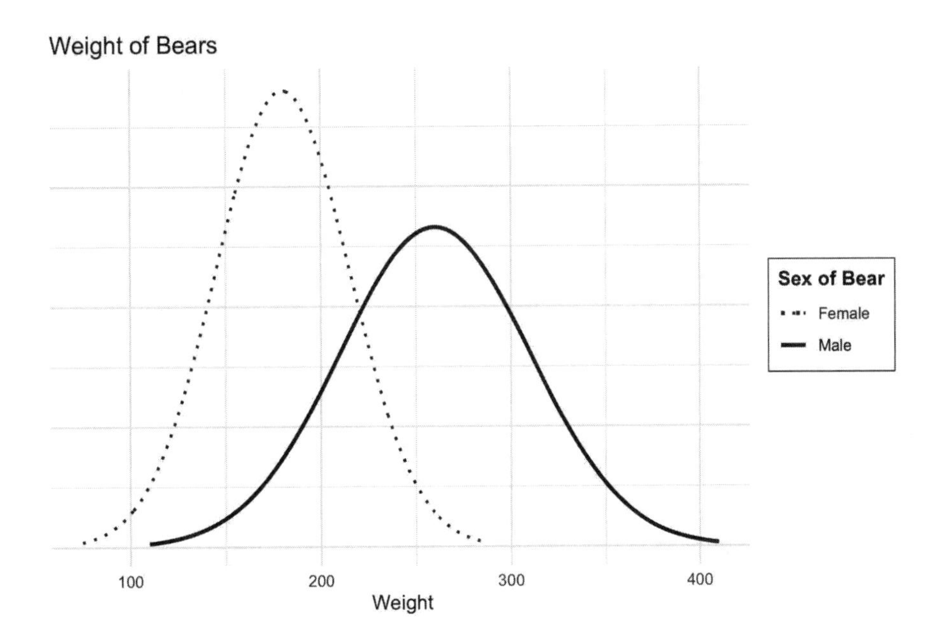

Based on the plot, which statement below is TRUE?

(A) Because the distributions overlap, there is not much difference between male and female bears.
(B) The female bears have a greater mean weight than the male bears and also exhibit more variability in those weights.
(C) The female bears have a greater mean weight than the male bears and also exhibit less variability in those weights.
(D) The male bears have a greater mean weight than the female bears and also exhibit more variability in those weights.
(E) The male bears have a greater mean weight than the female bears and also exhibit less variability in those weights.

39. A professional football analyst is using Elo (a measure of the strength of a team) to predict the team's probability of making the playoffs. The analyst selects a random sample of professional football teams and records the team's Elo and the probability that the team will make the playoffs. A scatterplot (not shown) shows a roughly linear form. Below is the output from a least-squares regression analysis examining the linear relationship between Elo and the projected probability the team will make the playoffs. Which one of the following is the correct value and corresponding interpretation for the correlation?

Predictor	Coef	SE Coef	T	P
Constant	-344.5883	40.1511	-8.582	0.000
Elo	0.2543	0.0267	9.540	0.000
S = 14.78	R-Sq = 75.21%		R-Sq(adj) = 74.38%	

(A) The correlation is 0.7521, and 75.21% of the variation in a team's probability of making the playoffs can be explained by its Elo.
(B) The correlation is 0.7521. There is a strong, positive, linear relationship between a team's probability of making the playoffs and its Elo.
(C) The correlation is 0.867, and 86.7% of the variation in a team's probability of making the playoffs can be explained by its Elo.
(D) The correlation is –0.867. There is a strong, negative, linear relationship between a team's probability of making the playoffs and its Elo.
(E) The correlation is 0.867. There is a strong, positive, linear relationship between a team's probability of making the playoffs and its Elo.

40. Echinacea is widely used as an herbal remedy for the common cold, and researchers were interested in its effectiveness. In a double-blind experiment, healthy volunteers agreed to be exposed to common-cold-causing rhinovirus type 39 and have their symptoms monitored. The volunteers were randomly assigned to take either a placebo or an echinacea supplement daily for 5 days following exposure to the virus. Among the 103 volunteers taking a placebo, 88 developed a cold, whereas 75 of 116 subjects taking echinacea developed a cold. Which of the following represents the 95% confidence interval for the difference in the proportion of individuals like these who would develop a cold after viral exposure when taking echinacea versus when taking a placebo?

(A) $(0.647 - 0.854) \pm 1.645\sqrt{(0.744)(0.256)\left(\frac{1}{116}+\frac{1}{103}\right)}$

(B) $(0.647 - 0.854) \pm 1.645\sqrt{\frac{(0.647)(0.353)}{116}+\frac{(0.854)(0.146)}{103}}$

(C) $(0.647 - 0.854) \pm 1.960\sqrt{\left(\frac{75}{116}\right)\left(\frac{88}{103}\right)\left(\frac{1}{116}+\frac{1}{103}\right)}$

(D) $(0.647 - 0.854) \pm 1.960\sqrt{\frac{(0.647)(0.353)}{116}+\frac{(0.854)(0.146)}{103}}$

(E) $(0.647 - 0.854) \pm 1.960\sqrt{(0.744)(0.256)\left(\frac{1}{116}+\frac{1}{103}\right)}$

PRACTICE EXAM #2

SECTION II
Part A
Questions 1-5
Spend about 1 hour and 5 minutes on this part of the exam.
Percent of Section II score—75

Directions: Show all your work. Indicate clearly the methods you use, because you will be scored on the correctness of your methods as well as on the accuracy and completeness of your results and explanations.

1. A high school student is interested in how politicians use social media to convey their message and policies. The table and statistics below summarize the distribution of the number of social media posts each day over a year for the student's favorite politician.

Number of Posts	10-12	13-15	16-18	19-21	22-24	25-27	28-30	31-33
Frequency	7	35	86	112	75	35	14	1

Summary Statistics

n	Mean	SD	Min	Q_1	Med	Q_3	Max
365	20.1	3.9	10	18	20	23	33

(a) Use the following grid to sketch a histogram of the number of social media posts each day over a year.

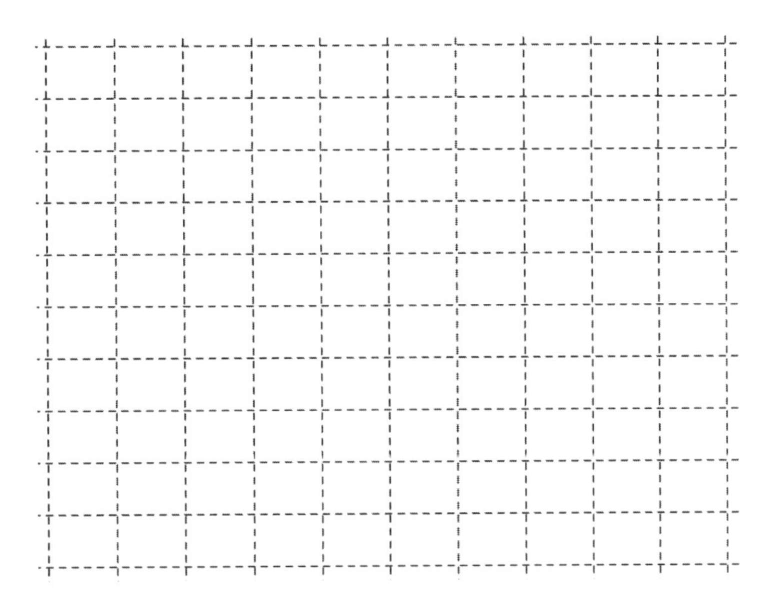

(b) Find <u>and</u> interpret the z-score of the minimum.

(c) Use the two standard deviations rule to determine how many data points are outliers in the distribution of the number of posts each day over a year. Justify your answer.

The student constructs a boxplot to visualize the distribution of the number of posts each day over a year.

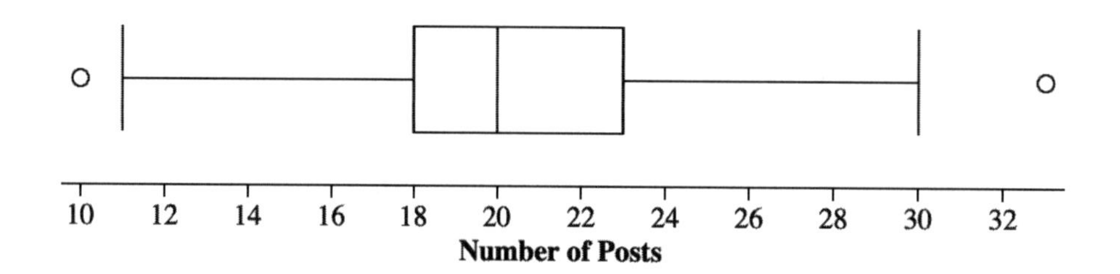

(d) Is the boxplot consistent with the answer from part (c)? Explain your reasoning.

2. A group of medical researchers believe that swimming regularly lessens the chances of developing depression among high school seniors. One hundred–twenty seniors will be recruited and classified by their level of social media usage: low, moderate, high. The experiment will consist of assigning the seniors into one of four groups: no swimming, swimming 15 minutes per day, swimming 30 minutes per day, or swimming 45 minutes per day. A randomized block design is planned, with blocking by classification of social media usage. The researchers will have the students take an inventory that assesses depression level before and after the experiment.

   (a) What is the statistical advantage of blocking by the classification of social media usage?

   (b) Why is it important to randomize the amount of swimming time to students rather than allowing each student to choose the amount of time?

   (c) Explain how the design of the experiment will address replication. What is the benefit of the replication?

3. In a certain board game, players roll three fair, six-sided dice on their turn. Two of the dice are white and have faces labeled one to six. The third die is red and has two sides with a picture of a bear and four sides with pictures of frogs. There are two problematic events in this game: rolling a sum of 7 on the white dice and rolling a bear on the red die.

   (a) Suppose one of the players rolls the three dice.

      i) What is the probability a bear appears on the red die?

      ii) What is the probability that the sum of the faces on the two white dice is 7?

      iii) What is the probability that the two white dice show a sum of 7 and the red die shows a bear?

   (b) Find the probability that the first bear rolled occurs after four turns.

   Let the random variable R be defined as the number of turns until a bear is rolled.

   (c) Describe the distribution of R.

4. An instructor teaches an online statistics class with 650 students. Some students have expressed concern about the amount of video content that they are required to watch to pass the class. The instructor decides that if more than 75% of the class responds that there is too much video content, the instructor will remove 10% of the video content; otherwise, the instructor will keep the course requirements the same. The instructor randomly selects 60 students from the class and 50 respond that there is too much video content.

   (a) Is there convincing evidence, at the 0.05 significance level, that the students' concern is justified? Complete the appropriate inference procedure to support your answer.

   (b) Based on your conclusion from part (a), which of the two errors, Type I or Type II, could have been made? Describe the consequence of the error in context.

5. A group who researches the use of tools powered by artificial intelligence (AI) conducted a national survey among high schoolers. The researchers took a random sample, recorded the grade level of each student and asked whether they had used AI to assist in completing their schoolwork in the past month. The following table shows the counts of responses among three grade levels.

	Sophomores	Juniors	Seniors	Total
Yes	63	270	481	814
No	156	82	295	533
Total	219	352	776	1347

(a) Suppose one teen is randomly selected from this sample. A researcher claims that if a senior is selected, the likelihood that they used an AI tool is more than that if a sophomore is selected that they used an AI tool and more than if a junior is selected that they used an AI tool in the past month because the seniors had the most of the three grade levels who used AI tools in the sample. Is the researcher's claim correct? Explain your answer.

(b) Consider the values in the table.

(i) Construct a segmented bar graph of the relative frequencies based on the information in the table.

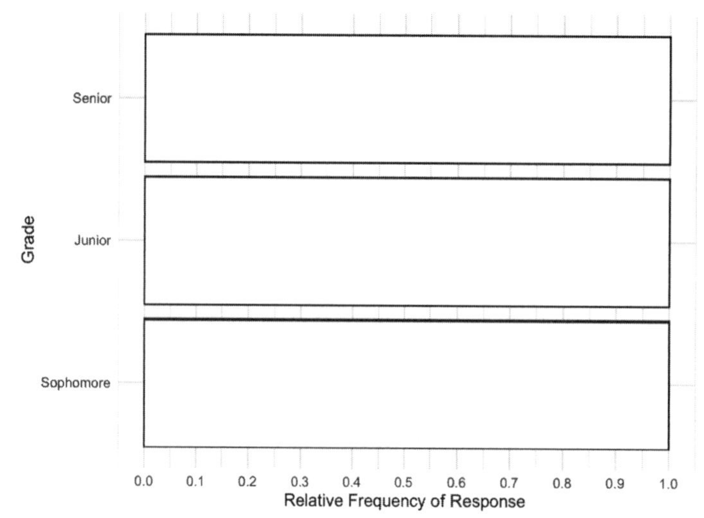

(ii) Which grade level had the smallest proportion of students who used an AI-powered tool in the past month? Determine the value of the proportion.

(c) Consider the inference procedure that is appropriate for investigating whether there is a difference in the proportions of all students in these three grade levels who used AI-powered tools in the past month.

(i) Identify the appropriate inference procedure.

(ii) Identify the hypotheses of the test.

SECTION II

Part B
Question 6
Spend about 25 minutes on this part of the exam.
Percent of Section II score—25

Directions: Show all your work. Indicate clearly the methods you use, because you will be scored on the correctness of your methods as well as on the accuracy and completeness of your results and explanations.

6. Educational researchers are interested in measuring the role that home environment plays in academic achievement. The researchers are concerned that person-to-person genetic differences may influence academic achievement. To address this concern, the researchers will use identical twins in this study. The twin sets, who were each adopted at a very early age, were identified because at the time of adoption, each child was placed in a different home—one of the children had been placed in a home where academics were emphasized and the other had been placed in a home where academics were not emphasized. The 12 twin sets used in the study were randomly selected from among those sets that shared this characteristic.

   An academic achievement test was given to each child when they were 8 years old. The data are given below.

Set of Twins	Academic	Nonacademic	Difference = Academic – Nonacademic
1	88	73	15
2	76	70	6
3	86	88	–2
4	94	85	9
5	70	65	5
6	65	57	8
7	78	78	0
8	52	55	–3
9	77	73	4
10	61	51	10
11	78	98	–20
12	84	72	12

   (a) Researchers were interested in testing the hypothesis that the mean difference in academic achievement is greater for twins when one twin was raised in a home where academics were emphasized and the other twin was raised in a home where academics were not emphasized. Write the appropriate hypotheses.

   (b) What condition for inference would not be met? Explain.

Given that one of the necessary inference conditions was not met, the researchers asked a statistician about the appropriate significance test to use. The statistician explained how to calculate a test statistic in this situation. The steps are described below.

1. Calculate the differences for each of the 12 pairs of twins.
2. Find the absolute value of each difference.
3. List the absolute values of the differences in increasing order, shown below.

0, 2, 3, 4, 5, 6, 8, 9, 10, 12, 15, 20

4. Assign ranks: 0 is the lowest ranked number, which is given the rank of 1, 2 is the next lowest ranked number given a rank of 2, etc.
5. If the difference had a negative sign in the 4th column, assign a negative in the rank-with-sign column. For example, the difference of 2 was originally negative. Assign a negative sign to its rank with sign.

Set of Twins	Academic	Nonacademic	Difference = Academic – Nonacademic	Absolute value of difference	Rank	Rank With Sign
1	88	73	15	15		
2	76	70	6	6	6	6
3	86	88	–2	2	2	–2
4	94	85	9	9	8	8
5	70	65	5	5	5	5
6	65	57	8	8	7	7
7	78	78	0	0	1	1
8	52	55	–3	3	3	–3
9	77	73	4	4	4	4
10	61	51	10	10		
11	78	98	–20	20		
12	84	72	12	12		

(c) Complete the "rank" and "rank with sign" columns in the table above.

(d) Let T_- represent the sum of all the ranks that have negative signs in the last column.

Let T_+ represent the sum of all the ranks that have positive signs in the last column.

The test statistic T is defined as the smaller of $|T_-|$ or T_+. The statistical test will find convincing evidence for the alternative hypothesis if the value of the test statistic or less occurs infrequently by random chance alone.

Find the value of T.

(e) The statistician used software to generate 1000 values of T under the assumption that the null hypothesis of no difference in the achievement scores between the twins in each pair is correct. The results are shown in the histogram below.

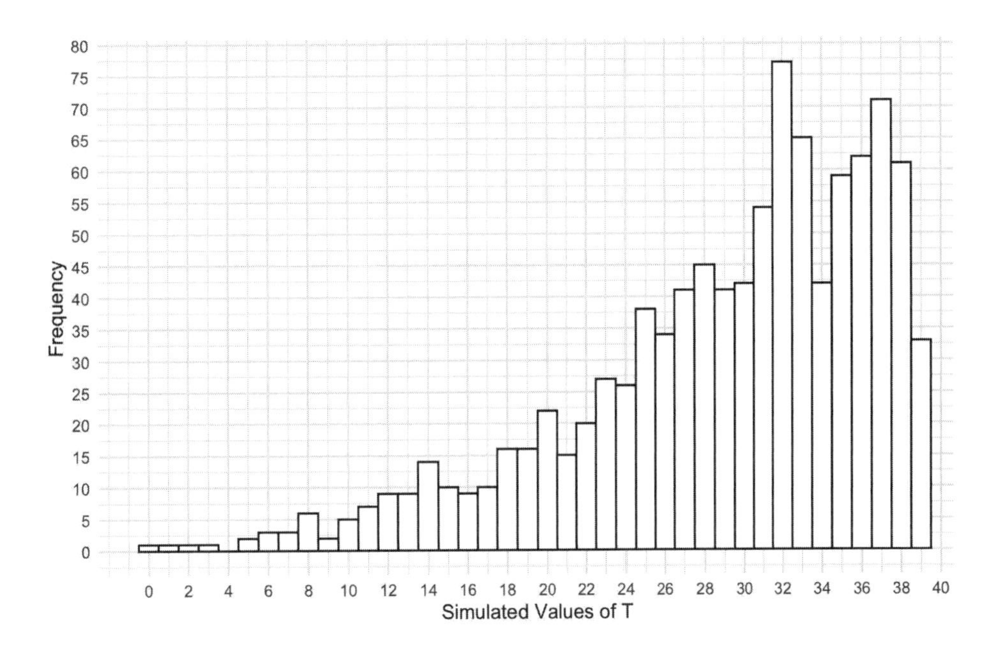

Use the value of T calculated in part (d) and the simulated values of the statistic in the histogram above to determine if the observed data provide convincing statistical evidence of a mean difference in academic achievement between twin pairs where one was raised in a home where academics were emphasized and the other was raised in a home where academics were not emphasized. Explain your reasoning.

Answer Key for Practice Exam 2
Part I Multiple Choice Section Solutions

Problem	Answer	Explanation
1	E	A census involves measuring every individual in the population. It would be easiest to record the heights of twenty students in your history class. Options A, C, and D would have very large populations, making it nearly impossible to survey every individual. Option B would be difficult to measure over the entire semester for every student in the high school.
2	C	This is a binomial distribution with $n = 20$ and $p = 0.15$. $P(X \leq 1) = P(X = 0) + P(X = 1)$ $= \binom{20}{0}(0.15)^0(0.85)^{20} + \binom{20}{1}(0.15)^1(0.85)^{19}$
3	E	This is a completely randomized experiment with three treatments—varieties of corn. The experimental units are the plots of land. Randomization is used to create approximately equivalent groups. It does not eliminate any variability that exists in the plots of land. Having four plots of land for each treatment allows the researcher to measure the variability of crop yield due to being planted in different types of soil.
4	D	With a chi-square value of 22.38 with degrees of freedom $(3 - 1)(2 - 1) = 2$, the *P*-value is $P(\chi^2 \geq 22.38) = 0.0000138$ Because the *P*-value is less than 0.01, the null hypothesis is rejected. There is convincing evidence to conclude that an association exists between frequency of snoring and the prevalence of asthma or nocturnal coughing among children aged 2 to 5.
5	D	The *P*-value is the probability of getting a sample result (a test statistic) at least as extreme as the observed result you have, given that the null hypothesis is true.
6	C	Moving point A will increase the slope because point A acts as an influential point by drawing the regression line toward itself. Increasing the *y*-value of point A will increase the slope of the least-squares regression line. In addition, the correlation would increase because point A will now fall in the same linear pattern formed by the rest of the data points.
7	E	One of the conditions is that the sample must be a random sample from the population of interest.

8	B	$P(\text{infected} \mid \text{injects daily}) = \dfrac{P(\text{infected} \cap \text{injects daily})}{P(\text{injects daily})} = \dfrac{\frac{32}{156}}{\frac{77}{156}} = \dfrac{32}{77} = 0.416$
9	C	Using a calculator or a table; $z = \text{invnorm}(0.80) = 0.84$ $z = \dfrac{x-\mu}{\sigma} \Rightarrow 0.84 = \dfrac{0.15-0.12}{\sigma}$ $\sigma = \dfrac{0.03}{0.84} = 0.0357$
10	E	You are checking for the improvement in flexibility for each gymnast. There is likely a lot of variability between gymnasts so you would want some way to reduce the variability in the estimate. Therefore, a matched-pairs design is best, which means either choice (B) (where each gymnast uses both machines) or choice (E) (where each gymnast uses only one machine). Having a gymnast use both machines is not ideal, because the residual benefits of one machine might carry over to using the other machine. The researchers wouldn't know which of the two machines was the real contributor to improved flexibility. Therefore, choice (E) is the best approach.
11	A	$\hat{p} = \dfrac{160}{250} = 0.64$ $P\left(z > \dfrac{\hat{p}-p}{\sqrt{\dfrac{p(1-p)}{n}}}\right) = P\left(z > \dfrac{0.64-0.58}{\sqrt{\dfrac{(0.58)(0.42)}{250}}}\right)$
12	B	The margin of error is the largest extent to which the sample statistic you calculated and the unknown population parameter would most likely differ. A 4% margin of error would mean that the difference between the sample proportion and the population proportion is at most 4%.
13	D	The mean of the binomial random variable A is the average number of shipments that arrive on time in many, many sets of shipments each of size 50. It does not mean that all samples of size 50 will have exactly 48.5 on-time shipments.

14	C	To find the sample size needed to estimate a population proportion within a specified margin of error at a given confidence level, use $E = z^* \sqrt{\frac{p(1-p)}{n}}$, where $p = 0.5$ and the z^* critical value for a 98% confidence interval for a proportion is 2.326. $2.326\sqrt{\frac{(0.5)(0.5)}{n}} \leq 0.03,\ n \geq \left(\frac{2.326\sqrt{(0.5)(0.5)}}{0.03}\right)^2 = 1502.85 \rightarrow 1503$ The smallest sample size of at least 1503 among the choices is 1510.
15	D	The question asks for the difference in the number of times a person can tap a surface before and after consuming the energy drink. This is a paired data analysis, because there are two measurements on each subject. The appropriate test is a paired (matched) t–test for a mean difference. $t = \frac{\bar{x}_d - \mu}{\frac{s_d}{\sqrt{n}}} = \frac{-5 - 0}{\frac{4.59}{\sqrt{10}}}$
16	B	The mean of the sampling distribution of a sample mean is equal to the mean of the population. Since the sample size is small ($n = 15$), the shape of the sampling distribution will still be somewhat skewed to the right. The standard deviation of the sampling distribution is approximately $\sigma_{\bar{x}} = \frac{\sigma}{\sqrt{n}} = \frac{3.7}{\sqrt{15}} = 0.955$ goal.
17	C	The median for private institutions is slightly greater than that of public institutions. There are no outliers for the public institutions, but this type (public) has a slight skew toward the larger numbers. Therefore, the mean for the public institutions will be just a bit above its median. Because the distribution of enrollment rates for the private institutions has several high outliers (all higher than the maximum of 50 for the distribution of public schools), this will raise the mean for private schools substantially above its median (which is already slightly greater than the median for public institutions), making the mean for private colleges and universities greater than the mean for public colleges and universities.
18	C	22 368 46573 25595 85393 30995 89198 27982 53401 The numbers selected are 22, 36, 25, 30, and 27.
19	C	A score of 79 lies at $\frac{44+37+19+5}{150} = \frac{105}{150} = 0.70$ or about the 70th percentile.
20	D	P(deep-dish) = (0.70)(0.40) + (0.30)(0.30) = 0.28 + 0.09 = 0.37

21	B	Let X = the commission that this salesperson receives on a randomly selected day $E(X) = 0(0.15) + 300(0.25) + 600(0.35) + 900(0.20) + 1200(0.05) = \$525.00$ $\sigma_X = \sqrt{0.15(0-525)^2 + 0.25(300-525)^2 + \cdots + 0.05(1200-525)^2} = \sqrt{106875} = \$326.92$
22	C	A Type II error is committed when we fail to reject a null hypothesis that is false. In this case the company fails to reject the null hypothesis (that the driver is low-risk) but it is false. The company then enrolls a high-risk driver who may cost them money.
23	D	Residual = Actual resting metabolic rate – predicted resting metabolic rate The actual resting metabolic rate is 1000. The predicted resting metabolic rate is 201.162 + 24.026(40) = 1162.202. Residual = 1000 – 1162.202 = –162.202 kcal/day Because the residual is negative, the regression model overestimates the resting metabolic rate for this woman.
24	B	A 98% confidence interval for a mean is given by $\bar{x} \pm t^* \left(\frac{s_x}{\sqrt{n}}\right)$. For this situation, df = 19 and t^* = 2.539 (from the table or technology). $C.I. = 64.5 \pm 2.539\left(\frac{2.1}{\sqrt{20}}\right)$
25	D	The required conditions to carry out a two-sample t test for a difference in population means are: (1) Random, (2) 10% condition, (3) Normal/Large Sample. The Random condition is met because we have two independent random samples of cars that were selected. The 10% condition is met, because the population is not finite. The condition that needs to be checked is the Normal/Large Sample condition because the sample sizes are small (25 < 30). However, the distribution of mileage showed no strong skewness or outliers.
26	E	A discrete variable can take on a countable number of values. Heights, times, and weights are all continuous (not discrete) variables. Favorite color is a categorical, not quantitative, variable. The number of cans is a countable value.
27	D	A geometric setting occurs when we repeat a chance process and count the number of trials until the first success. Answer D is not a geometric setting because there are a fixed number of trials (25 students).

28	E	Let X = number of policies sold each week and let Y = salesperson's weekly pay $Y = 300 + 250X.$ $E(Y) = 300 + 250(3.1) = \$1075$ $SD(Y)$
29	B	Are popularity votes equally distributed across teachers? This question would be answered with a chi-square goodness-of-fit test.
30	E	There are two distinct clusters with respective peaks (at 73 and at 96), so the shape of the distribution is bimodal. There is also a large gap with no values present from the high 70s to the high 80s.
31	C	There is a 17% probability that the average length of the 12 MacGregor's bowerbirds will be more than 4 cm greater than the average length of the Vogelkop bowerbirds.
32	C	A distribution that is skewed to the right will most likely have a mean that is greater than the median.
33	D	The sample mean is an unbiased estimator of the population mean when a random sample is selected because the mean of the sampling distribution of $\overline{x}$ is equal to the population mean μ.
34	C	This is a matched pairs interval, and we are interested in whether "overclocking" improves the speed. Because the interval contains only positive values, there is a significant increase in mean speed from unmodified to "overclocked" computers.
35	D	A stratified random sampling process would involve taking the population of interest and separating it into clearly identifiable, homogeneous subgroups and then drawing a random sample from each subgroup. One way to think of this is "some from all strata". Choice (A) is a cluster sample, choice (B) is a systematic random sample, choices (C) and (E) are simple random samples.

36	C	The confidence interval for a slope is given by $b \pm t^*(SE_b)$ for df = $n - 2 =$ 20. InvT(area: 0.025, df: 20) = 2.086. From the computer output and technology or the *t*-table, the 95% confidence interval is given by $10.3382 \pm 2.086(0.4825)$.
37	B	Because $\sigma_{\hat{p}} = \sqrt{\frac{p(1-p)}{n}} = \frac{\sqrt{p(1-p)}}{\sqrt{n}}$, multiplying the sample size by 9 would change the denominator of the standard deviation of the sampling distribution of the proportion to $\sqrt{9n} = 3\sqrt{n}$, This is equivalent to multiplying the denominator by a factor of 3, or dividing the standard deviation by 3.
38	D	The distribution of weights for male bears is wider than that for female bears, which means the male weights are more variable than are female weights. Because both distributions are relatively symmetric, their peaks are at the means. The mean for males is about 260 pounds and is about 180 pounds for females.
39	E	Because a team's probability of making the playoffs increases as the team's Elo increases (the slope is positive) thus $r = +\sqrt{r^2} = +\sqrt{0.7521} = +0.867$, so there is a strong, positive, linear relationship between a team's probability of making the playoffs and its Elo.
40	D	We only combine proportions for two-sample *z*-tests (not intervals) for a difference in proportions, so answers A, C, and E are incorrect. As a two-sample *z* interval for a difference in proportions where $z^* = 1.960$, $\hat{p}_{Echinacea} = \frac{75}{116} = 0.647$ and $\hat{p}_{Placebo} = \frac{88}{103} = 0.854$

Answer Key & Textbook Mapping
Multiple Choice Practice Exam 2

Multiple Choice Question	Answer	Learning Objective	Skill	CED Unit	Textbook Section
1	E	DAT-2.C	1.C	3	3A
2	C	UNC-3.B	3.A	4	4F
3	E	VAR-3.B	1.B	3	3C
4	D	DAT-3.L	4.E	8	8B
5	D	DAT-3.G	4.B	7	7D
6	C	DAT-1.I	2.A	2	2B
7	E	UNC-4.P	4.C	7	7B
8	B	VAR-4.D	3.A	4	4C
9	C	VAR-2.B	3.A	1	1F
10	E	VAR-3.C	1.C	3	3C
11	A	VAR-6.G	3.E	6	6D
12	B	UNC-4.C	3.D	6	6B
13	D	UNC-3.D	4.B	4	4F
14	C	UNC-4.C	3.D	6	6B
15	D	VAR-7.I	3.E	7	7B
16	B	UNC-3.Q, UNC-3.R	3.B, 3.C	5	5D
17	C	UNC-1.N	2.D	1	1C
18	C	DAT-2.C	1.C	3	3B
19	C	UNC-1.I	2.C	1	1E
20	D	VAR-4.E	3.A	4	4C
21	B	VAR-5.C	3.B	4	4D

22	C	UNC-5.A	1.B	6	6D
23	D	DAT-1.E	2.B	2	2C
24	B	UNC-4.Q	3.D	7	7A
25	D	VAR-7.H	4.C	7	7D
26	E	UNC.1.F	2.A	1	4D
27	D	UNC-3.G	4.B	4	4F
28	E	VAR-5.E	3.B	4	1E, 4E
29	B	VAR-8.C	1.E	8	8A
30	E	UNC-1.H	2.A	1	1C
31	C	UNC-3.V	4.B	5	5D
32	C	UNC-1.M	2.A	1	1D
33	D	UNC-3.I	4.B	5	5B
34	C	UNC-4.AA	4.D	7	7C
35	D	DAT-2.C	1.C	3	3B
36	C	UNC-4.AF	3.D	9	9A
37	B	UNC-4.H	4.A	6	6B
38	D	UNC-1.N	2.D	1	1C
39	E	DAT-1.B, DAT-1.G	2.C	2	2C
40	D	UNC-4.K	3.D	6	6E

Part II Free-Response Section Solutions and Scoring

Each of the 6 Free-Response Questions is scored using the following scale:

4 Complete Response

All three parts essentially correct

3 Substantial Response

Two parts essentially correct and one part partially correct.

2 Developing Response

Two parts essentially correct and no parts partially correct

One part essentially correct and one or two parts partially correct

Three parts partially correct

1 Minimal Response

One part essentially correct and no parts partially correct

No parts essentially correct and two parts partially correct

0 Incorrect Response

No parts essentially correct and one part partially correct

No parts essentially correct and no parts partially correct

Part II Free-Response Section Solutions

Question 1:
Intent of the Question

The primary goals of this question are to assess your ability to (1) construct a histogram; (2) calculate and interpret a z-score; (3) use the two standard deviation rule to identify outliers; (4) interpret a boxplot.

Solution

(a)

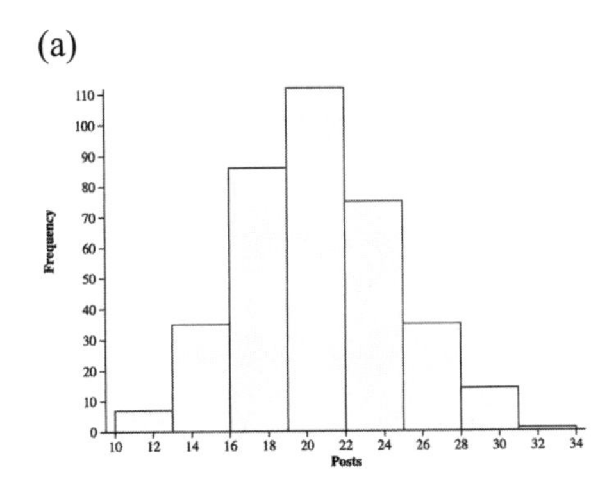

(b) The z-score of the minimum is $z = \frac{10-20.1}{3.9} = -2.59$. The lowest number of posts in a single day is 2.59 standard deviations below the mean number of posts.

(c) Any value greater than $20.1 + 2(3.9) = 27.9$ is considered a high outlier and any value less than $20.1 - 2(3.9) = 12.3$ is considered a low outlier. In other words, any day with more than 27.9 posts or less than 12.3 posts will be counted as an outlier. From the frequency table, there are $14 + 1 = 15$ days with at least 28 posts and there are 7 days with 12 or fewer posts. There are $15 + 7 = 22$ total outliers.

(d) The boxplot is not consistent with the answer from part (c). In part (c), there were 22 total outliers identified, but the boxplot shows outliers only at 10 and 33 posts for at least 2, but at most 7 outliers (ten posts occurred from 1 to 6 days and 33 posts only once). The 2(SD) rule is more sensitive to extreme values and identifies a much larger number of outliers compared to the 1.5(IQR) rule that the boxplot uses.

Scoring:

Each section scored as essentially correct (E), partially correct (P), or incorrect (I).

Parts (a) and (b): Section 1 is essentially correct if response correctly satisfies all 6 components: (1) histogram is graphed, (2) axes labeled (3) appropriate scaling (4) the z-score calculated, (5) supporting work for the z-score; (6) interpretation of the z-score in context. Partially correct if the response satisfies 3, 4, or 5 of the 6 components OR if the communication is weak. Incorrect otherwise.

Part (c): Section 2 is essentially correct if response correctly satisfies all 5 components: (1) calculates both the upper and lower bounds using the 2–standard deviation rule; (2) shows supporting work for the calculation of at least one of the bounds; (3) provides a direct comparison between the minimum and the lower bound; (4) provides a direct comparison between the maximum and the upper bound; (5) links an appropriate statement of the existence of outliers to the comparisons. Partially correct if the response satisfies 3 or 4 of the 5 components OR if the communication is weak. Incorrect otherwise.

Part (d): Section 3 is essentially correct if response correctly satisfies all 4 of the following components: (1) responds no; (2) states a plausible number of outliers shown in the boxplot; (3) provides a direct comparison between the number of outliers reported in part (c) to the number of outliers identified in the boxplot; (4) directly links response (component 1) to the comparison (component 3) using a term such as because. Partially correct if the response satisfies 2 or 3 of the 4 components OR if the communication is weak. Incorrect otherwise.

Question 2:
Intent of the Question

The primary goals of this question are to assess your ability to (1) identify a statistical advantage of blocking; (2) identify the importance of randomizing treatments in an experiment; (3) identify replication in an experimental design; (4) identify a benefit of replication.

Solution
Question 2:

(a) Blocking ensures that at the beginning of the experiment the subjects within each block are as similar as possible to each other with respect to the blocking variable. In this situation, the response variable is a student's change in depression score. Seniors who use social media at increasingly higher levels may tend to have increasingly higher rates of depression, so the variability in the change in depression scores across seniors should be smaller within each classification group than it would be for all seniors combined. Having smaller variability in responses makes it easier to detect a difference between the four treatments, if it exists.

(b) Randomization is used to reduce or eliminate the effect of confounding variables that might be related to the explanatory variable (in this case, swimming time) and might be associated with differences in the response (depression score). If seniors are allowed to choose which amount of time to swim, it's possible that seniors who choose to swim a longer time might be different from those seniors who choose not to swim at all, or only 15 minutes and those differences might be related to an individual's depression. For example, seniors who are more depressed might choose not to swim at all or to swim very little and those who are less depressed might choose to swim for a longer amount of time.

(c) The design addresses replication by assigning multiple seniors in each classification to each treatment (amount of swim time). Replication is important to estimate the natural variability in depression scores among seniors in each different social media classification. The estimate of natural variability is needed so the mean depression scores for each treatment group can be compared.

Scoring:

Each part scored as essentially correct (E), partially correct (P), or incorrect (I).

Part (a) is essentially correct if response correctly satisfies 4 or 5 of the following components: (1) states members of each block are likely to be similar to others in the same block with respect to the response variable; (2) states that subjects with different social media use may tend to have different rates of depression; (3) provides a specific example of how social media use could be linked to depression level; (4) explicitly addresses the impact of blocking on the variability of the response variable; (5) answers in context. Partially correct if the response satisfies 2 or 3 of the 5 components OR if the communication is weak. Incorrect otherwise.

Part (b) is essentially correct if response correctly satisfies all 5 components: (1) states a substantial benefit for random assignment; (2) stated benefit addresses potential confounding; (3) stated benefit is in context; (4) states a detriment of not randomly assigning treatments; (5) stated detriment is in context. Partially correct if the response satisfies 3 or 4 of the 5 components OR if the communication is weak. Incorrect otherwise.

Part (c) is essentially correct if response correctly satisfies all 4 of the following components: (1) addresses assigning multiple subjects to each treatment; (2) addresses variability in response variable linked to the explanatory variable; (3) in context; (4) allows for comparison of response variable between treatments. Partially correct if the response satisfies 2 or 3 of the 4 components OR if the communication is weak. Incorrect otherwise.

Question 3:

Intent of the Question

The primary goals of this question are to assess your ability to (1) calculate probabilities; (2) calculate a geometric probability; (3) describe a geometric distribution.

Solution

(a-i) The probability that a bear appears on the red die is $P(\text{bear}) = \frac{2}{6} = \frac{1}{3} = 0.333$.

(a-ii) There are (6)(6) = 36 total possibilities in the sample space of rolling two six-sided dice. There are six possible outcomes for a sum of 7: (1, 6), (2, 5), (3, 4), (4, 3), (5, 2), and (6, 1). The probability that the sum of the faces on the two white dice is 7 is $P(\text{sum of } 7) = \frac{6}{36} = \frac{1}{6} = 0.167$.

(a-iii) The probability of both problematic events occurring on one roll is $P(\text{bear and sum of } 7) = P(\text{bear}) * P(\text{sum of } 7) = \left(\frac{1}{3}\right)\left(\frac{1}{6}\right) = \frac{1}{18} = 0.056$, because the events are independent.

(b) The probability that the first bear rolled occurs after four turns is $P(\text{first bear after 4th turn}) = P(\text{not on the first, second, third or fourth turns})$

When the first bear occurs	Work	Answer
On the first turn	$\left(\frac{1}{3}\right)$	0.333
On the second turn	$\left(\frac{2}{3}\right)\left(\frac{1}{3}\right)=\frac{2}{9}$	0.222
On the third turn	$\left(\frac{2}{3}\right)^2\left(\frac{1}{3}\right)=\frac{4}{27}$	0.148
On the fourth turn	$\left(\frac{2}{3}\right)^3\left(\frac{1}{3}\right)=\frac{8}{81}$	0.099

$P(\text{after the fourth turn}) = 1 - (0.333 + 0.222 + 0.148 + 0.099) = 0.198.$

Alternate solution with technology: 1 – geometcdf($p = \frac{1}{3}$, x-value = 4) = 0.1975.

(c) R is a geometric random variable. Each trial has two possible outcomes (bear or not), the trials are independent (die rolling), we are counting the number of turns until the first success, and the probability of success remains constant at $p = \frac{1}{3}$. The conditions are satisfied to determine a geometric distribution.

This geometric distribution has a center of $\mu = \frac{1}{p} = \frac{1}{1/3} = 3$, a standard deviation of $\sigma = \frac{\sqrt{1-p}}{p} = \frac{\sqrt{1-\frac{1}{3}}}{\frac{1}{3}} = 2.45$ and the shape is right-skewed.

Scoring:

Each part scored as essentially correct (E), partially correct (P), or incorrect (I).

Part (a) is essentially correct if the response correctly satisfies all 4 components: (1) correct probability in (a)–i; (2) correct probability in (a)–ii; (3) correct probability in (a)–iii; (4) supporting work for at least two calculations. Partially correct if the response satisfies 2 or 3 of the 4 components OR if the supporting work is weak. Incorrect otherwise.

Part (b) is essentially correct if the response correctly satisfies both: (1) correct probability based on probabilities reported in part (a); (2) shows supporting work. Partially correct if the response reports the complement of the correct probability with supporting work OR has weak supporting work OR if the response provides the probability for the first roll of a bear on the 4th attempt or on the 5th attempt WITH clear supporting work. Incorrect otherwise. Note: The second component can be earned with clear calculations, correct formula, or a complete technology statement.

Part (c) is essentially correct if the response correctly satisfies all 5 of the following components: (1) calculates the mean as the center; (2) calculates the standard deviation for variability; (3) states the shape of the distribution; (4) classifies R as geometric; (5) shows at least two of the conditions for a geometric distribution are satisfied. Partially correct if the response satisfies 3 or 4 of the 5 components OR if the communication is weak. Incorrect otherwise. Note: The fourth component can be earned with words, formula, or a complete technology statement.

Question 4:

Intent of the Question

The primary goals of this question are to assess your ability to (1) identify the need for and to conduct a one sample *z*-test for a proportion; (2) identify a potential error; (3) state a consequence of the potential error.

Solution:

(a) **STATE:** We want to test:

$H_0: p = 0.75$

$H_0: p > 0.75$

where *p* = the unknown proportion of the entire class that would respond that there is too much video content in the course. Use $\alpha = 0.05$.

PLAN: One-sample *z* test for a population proportion.

- Random: Random sample of 60 students from the course. ✓
    - 10%: 60 < 10% of the population of 650 total students in the course. ✓
- Large counts: $np_0 = (60)\left(\frac{50}{60}\right) = 50 \geq 10$ and $n(1 - p_0) = (60)\left(1 - \frac{50}{60}\right) \geq 10$ ✓

DO: $\hat{p} = \frac{50}{60} = 0.833$

- $z = \frac{0.833 - 0.75}{\sqrt{\frac{(0.75)(0.25)}{60}}} = 1.485$
- *P*-value

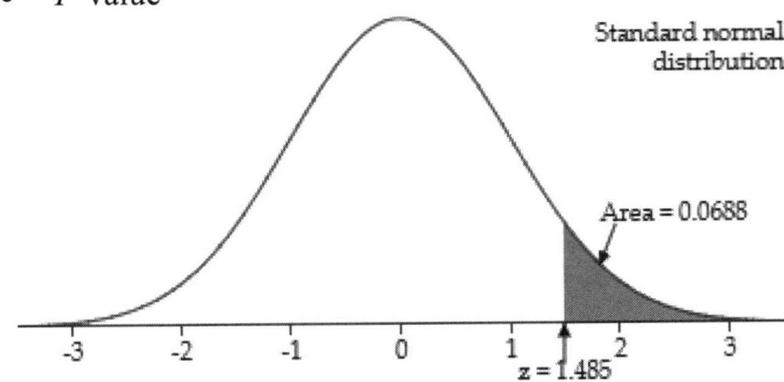

P(*z* > 1.485) = Normalcdf(lower: 1.485, upper: 1000, mean: 0, SD: 1) = 0.069

Using technology: *z* = 1.491, *P*-value = 0.068

Using Table A: Using *z* = 1.49, *P*-value = 1 – 0.9319 = 0.068.

CONCLUDE: Because the *P*-value of $0.068 > \alpha = 0.05$, we fail to reject H_0. There is not convincing evidence that the proportion of all students in the course who would respond that there is too much video content is greater than 0.75.

(b) Type II error, where we fail to reject H_0, when H_a is true. A consequence of this error is that more than 75% of students in the course believe there is too much video content, but the instructor decides to leave the amount the same rather than reducing it by 10%. Students will be upset and may negatively rate the instructor for the heavy workload.

Scoring:

Each section scored as essentially correct (E), partially correct (P), or incorrect (I).

Section (1) Section 1 is essentially correct if response satisfies all 5 components: (1) identifies a 1-sample z-test for a proportion by name or formula; (2) states hypotheses for a population proportion; (3) states the correct equality for the null hypothesis and a correct inequality for the alternative hypothesis; (4) defines or provides sufficient context to identify the parameter; (5) identifies the level of significance. Partially correct if the response satisfies 3 or 4 of the 5 components OR if the communication is weak. Incorrect otherwise.

Section (2) Section 2 is essentially correct if response correctly satisfies all 5 components: (1) states a random sample was conducted; (2) explains how the 10% condition is satisfied; (3) shows the Large Counts condition is satisfied with supporting work comparing both counts to an appropriate value; (4) correctly reports the test statistic for the named test; (5) correctly reports the P-value, consistent with the stated alternative hypothesis and reported test statistic. Partially correct if the response satisfies 3 or 4 of the 5 components. Incorrect otherwise. Note: Components (2) and (3) can only be earned with supporting evidence AND an explicit statement of each condition being met or using check marks.

Section (3) Section 3 is essentially correct if response correctly satisfies all 5 components: (1) provides correct comparison of the P-value to the stated alpha; (2) provides a correct decision (reject/fail to reject null hypothesis or convincing evidence exists or does not exist for alternative hypothesis) based on the comparison in component one; (3) states a conclusion in context, consistent with, and in terms of, the alternative hypothesis using non-deterministic language; (4) identifies the correct potential error type for reported conclusion; (5) provides an appropriate consequence of the potential error stated. Partially correct if the response satisfies 3 or 4 of the 5 components OR if the communication is weak. Incorrect otherwise.

Question 5:
Intent of the Question

The primary goals of this question are to assess your ability to (1) calculate conditional probabilities; (2) compare probabilities to counts; (3) create a segmented bar graph; (4) identify the appropriate test procedure; (5) state appropriate hypotheses.

Solution:

(a) $P(\text{Yes} \mid \text{Sophomore}) = 63 / 219 = 0.288$
$P(\text{Yes} \mid \text{Junior}) = 270 / 352 = 0.767$
$P(\text{Yes} \mid \text{Senior}) = 481 / 776 = 0.620$

No, the researcher's claim is not correct. Although a higher count of seniors said "yes" to the question, a higher proportion of juniors used an AI-powered tool in the past month than seniors ($0.767 > 0.620$).

(b-i)

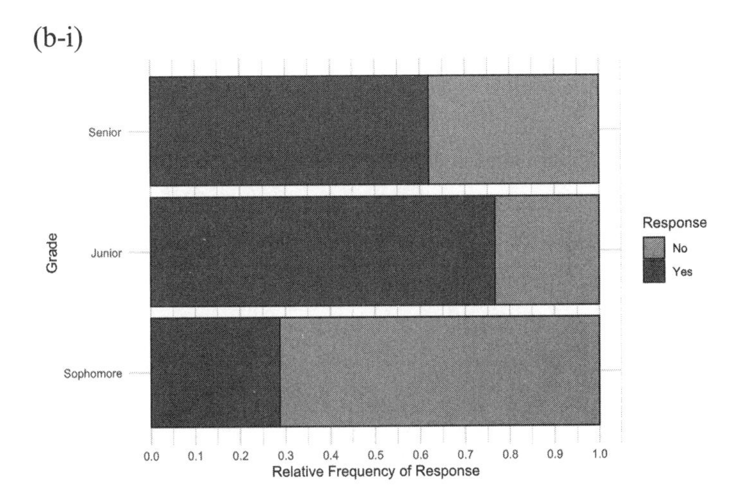

(b-ii) The grade level with the smallest proportion of student using AI was sophomores with 63/219 = 0.288.

(c-i) A chi-square test for association

(c-ii) H_0: There is no association between grade level and whether the student uses AI-powered tools to assist on homework in the past month.

H_a: There is an association between grade level and whether the student uses AI-powered tools to assist on homework in the past month.

Scoring:

Each part scored as essentially correct (E), partially correct (P), or incorrect (I).

Part (a) is essentially correct if the response correctly satisfies all 6 components: (1) correct probability for yes given sophomore (2) correct probability for yes given junior; (3) correct probability for yes given senior; (4) supporting work for at least two calculations; (5) answers the claim is not correct; (6) justifies answer. Partially correct if the response satisfies 3 – 5 of the 6 components OR if the response provides correct calculations with supporting work for an incorrect but reasonable interpretation of part (a) with consistent answer to the question OR if the supporting work is weak. Incorrect otherwise.

Part (b) is essentially correct if the response correctly satisfies all 3 components: (1) provides a clear key; (2) divides the bars accurately, (3) follows the shading shown in the reported key. Partially correct if the response correctly satisfies only 2 of the 3 components. Incorrect otherwise.

Part (c) is essentially correct if the response correctly satisfies all 4 components: (1) identifies a Chi-square test; (2) for association; (3) states an appropriate null hypothesis; (4) states an appropriate alternative hypothesis. Partially correct if the response satisfies 2 or 3 of the 4 components OR if an incorrect test given with appropriate hypotheses for the stated test procedure. Incorrect otherwise. Note: component 2 can be earned with association OR independence OR dependence.

Question 6:

Intent of the Question

The primary goals of this question are to assess your ability to correctly and completely (1) write appropriate hypotheses when given paired data; (2) check appropriate conditions for a paired *t*-test; (3) calculate, then record ranks and signed ranks in a table; (4) calculate a new, unfamiliar test statistic; (5) use a histogram from a simulation to draw an appropriate conclusion; (6) justify the conclusion.

Solution:

(a) The researchers have paired data, so the appropriate hypotheses for this test are H_0: $\mu_d = 0$ and $H_a : \mu_d > 0$, where μ_d = population mean difference (emphasized – not emphasized) in achievement scores between twin pairs where one was raised in a home where academics were emphasized, and the other was raised in a home where academics were not emphasized.

(b) Based on the boxplot and dotplot below, the condition of normality appears to be violated because the sample size is small (12) <u>and</u> there is an outlier among the differences in scores. We should <u>not</u> assume the sampling distribution of the sample mean difference is approximately normal, so it is not appropriate to use a *t* test for paired data.

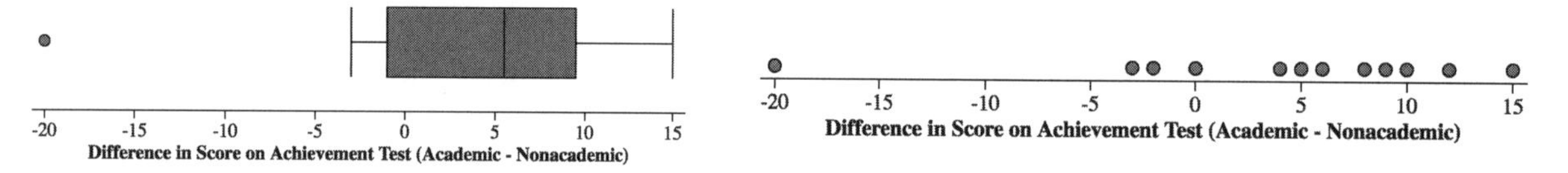

(c) The ranks and ranks with signs are shown in the table below.

Difference = Academic –Nonacademic	Absolute value of difference	Rank	Rank With Sign
15	15	**11**	**11**
6	6	6	6
2	2	2	–2
9	9	8	8
5	5	5	5
8	8	7	7
0	0	1	1
–3	3	3	–3
4	4	4	4
10	10	**9**	**9**
–20	20	**12**	**–12**
12	12	**10**	**10**

(d) T_- : The sum of all the ranks that have negative signs is $-2 + -3 + -12 = -17$

T_+ : The sum of all the ranks that have positive signs is $1 + 4 + 5 + 6 + 7 + 8 + 9 + 10 + 11 = 61$.

Since $|-17| = 17 < 61$, the value of the test statistic T is 17.

(e) As indicated in part (d), to find convincing evidence for the alternative hypothesis, we count how often values of T of 17 or less occur by random chance when the null hypothesis is true. (For example, the strongest evidence for H_a would be if all differences were positive, which is when $T_- = 0$ and therefore with the test statistic $T = 0$.) From the histogram, we determine the heights of each bar that is 17 or less and find a sum: $10 + 9 + 10 + 14 + 9 + 9 + 7 + 5 + 2 + 6 + 3 + 3 + 2 + 0 + 1 + 1 + 1 + 1 = 93$. Out of 1000 trials, 17 or less occurred 93 times, giving an approximate P-value of $93/1000 = 0.093$. At the 5% significance level, we fail to reject the null hypothesis. We do not find convincing statistical evidence of a mean difference in academic achievement between twin pairs where one was raised in a home where academics were emphasized, and the other was raised in a home where academics were not emphasized.

Scoring:

Each section scored as essentially correct (E), partially correct (P), or incorrect (I).

Parts (a) and (b): Section 1 is essentially correct if response correctly satisfies 6 or 7 of the following components anywhere in parts (a) and (b): (1) states hypotheses for a paired *t*-test of a mean difference in words or symbol; (2) defines or provides sufficient detail to identify the parameter with context; (3) states a correct equality for the null hypothesis; (4) states a correct inequality for the alternative hypothesis; (5) creates an appropriate display of the sample differences with sufficient labeling; (6) identifies the normality condition as not being met; (7) justifies why the normality condition is not met by appealing to BOTH the small sample size AND the lack of normality of the sample differences. Partially correct if the response satisfies 4 – 5 of the 7 components or if weak communication. Incorrect otherwise. Notes: Ignore any other reasonable condition references in part (b). Component 2 does not need to reference a *t*-test but cannot be satisfied if any incorrect test is identified.

Parts (c) and (d): Section 2 is essentially correct if the response correctly includes all 4 components: (1) table fully completed with correct values including signs; (2) correct values in part (d) for both T_- and T_+; (3) correct value in part (d) for T; (4) supporting work for part (d). Partially correct if the response satisfies 2 or 3 of the 4 components. Incorrect otherwise. Note: Overlook minor arithmetic errors.

Part (e): Section 3 is essentially correct if the response correctly includes all 5 components: (1) counts the number of simulated values of T that are less than or equal to the value of T reported in part (d); (2) divides the count reported by the number of trials in the simulation; (3) explicitly compares the proportion (or percentage) reported to an appropriate level of significance; (4) states an appropriate decision about the null hypothesis (or alternative hypothesis) for the test linked to the comparison stated; (5) provides an appropriate statement in context of having or not having convincing evidence in favor of H_a based on reported answers. Partially correct if the response satisfies 2 – 4 of the components OR if the communication is weak. Incorrect otherwise. Note: component 3 does not have to be labeled as the P-value but should not be incorrectly labeled.

Practice Exam 2 Scoring Worksheet

Section I: Multiple Choice

_______________ × 1.2500 = ___________

Number Correct (out of 40) — Weighted Section I Score (Do not round)

Section II: Free Response

Question 1 _________ × 1.8750 = __________
(out of 4) (Do not round)

Question 2 _________ × 1.8750 = __________
(out of 4) (Do not round)

Question 3 _________ × 1.8750 = __________
(out of 4) (Do not round)

Question 4 _________ × 1.8750 = __________
(out of 4) (Do not round)

Question 5 _________ × 1.8750 = __________
(out of 4) (Do not round)

Question 6 _________ × 3.1250 = __________
(out of 4) (Do not round)

Composite Score

___________ + ___________ = _______________

Weighted Section I Score + Weighted Section II Score = Composite Score (Round to nearest whole number)

AP Score Conversion Chart

Composite Score Range	AP Score
73-100	5
59-72	4
44-58	3
32-43	2
0-31	1

Textbook Mapping
Free Response Practice Exam 2

FRQ Part	Learning Objective	Skill	CED Unit	Textbook Section	SKILL #
1a	UNC-1.G	2.B	1	1C	2
1b	VAR-2.B	3.A	1	1E	3
1c	UNC-1.K	4.B	1	1D	4
1d	UNC-1.L	2.B	1	1D	1
2a	VAR-3.C	1.C	3	3C	1
2b	VAR-3.C	1.C	3	3C	1
2c	VAR-3.B	1.B	3	3C	1
3a-i	VAR-4.A	3.A	4	4B	3
3a-ii	VAR-4.E	3.A	4	4B	3
3a-iii	VAR-4.E	3.A	4	4B	3
3b	UNC-3.E	3.A	4	4F	3
3c	UNC-3.F	3.B	4	4F	3
4a	VAR-6.D, VAR-6.E, VAR-6.F, VAR-6.G, DAT-3.B	1.F, 1.E, 4.C, 3.E, 4.E	6	6D	1, 3, 4
4b	UNC-5.A, UNC-5.D	1.B, 4.B	6	6D	1, 4
5a	VAR-4.D	3.A	4	4C	4
5b-i	UNC-1.P	2.D	2	2A	2
5b-ii	UNC-1.P	2.D	2	2A	2
5c-i	VAR-8.J	1.E	8	8B	1
5c-ii	VAR-8.I	1.F	8	8B	1
6a	VAR-7.C	1.F	7	7B	1
6b	VAR-7.D	4.C	7	7B	4
6c		2.C	7		2
6d		3.E	7		3
6e		4.E	7		4

[Please feel free to cut out the flashcards in the following pages to use for review!]

Individuals & Variables	What is a resistant measure?
Categorical Variables Quantitative Variables (discrete vs. continuous)	Mean vs. Median
Distribution	Standard Deviation (Variance)
Describing/Comparing Distributions of Quantitative Data	Interpret Standard Deviation

• A resistant measure is not affected much by extreme values. • Resistant measures include: median, *IQR*, Q_1, Q_3 • Non-resistant measures include: mean, SD, range, correlation, slope and *y* intercept of least-squares regression line	• An individual is an object described in a set of data. Individuals can be people, animals, or things. • A variable is a characteristic that can take different values for different individuals. • In a typical data table, individuals are in the rows and variables are in the columns.
In general, • Skewed Left (Mean < Median) • Skewed Right (Mean > Median) • Roughly Symmetric (Mean ≈ Median)	• Categorical variables take values that are labels which place individuals into groups (categories). • Quantitative variables take number values that are quantities—counts or measurements. • Quantitative variables are *discrete* if their possible values have gaps between them and *continuous* if they can take any value in an interval on a number line.
Sample SD = $s_x = \sqrt{\frac{\sum (x_i - \bar{x})^2}{n-1}}$ (Variance = s_x^2)	• The distribution of a variable tells us what values the variable takes and how often it takes each value. • For categorical variables, display a distribution with a bar chart. • For quantitative variables, display a distribution with a dotplot, stemplot, histogram, or boxplot.
• Standard Deviation measures variability by giving the "typical" distance that the values in the data set are from the mean. • "The heights of students at our school typically vary from the mean height by about 3.1 inches."	Discuss the following, **in context** (using variable name) with **comparison phrases** (e.g., greater than) for center and variability • **Shape** – Skewed Left, Skewed Right, Roughly Symmetric; Unimodal (single-peaked), Bimodal (double-peaked), Approximately Uniform • **Outliers** – State whether there appears to be any outliers, gaps, and/or clusters. • **Center** – Mean or Median • **Variability** – Range, *IQR*, or Standard Deviation

Parameter vs. Statistic	Five Number Summary & Boxplot
Outlier Rules	Empirical Rule
Percentiles	Standard Normal Distribution
Interpret a standardized score (z-score)	Normal Distributions: Finding Area

• The five-number summary of a distribution of quantitative data consists of the minimum, the first quartile Q_1, the median, the third quartile Q_3, and the maximum. • A boxplot is a visual representation of the five-number summary.	• A parameter is a number that describes some characteristic of a population, such as μ, σ, or p. • A statistic is a number that describes some characteristic of a sample, such as $\bar{x}$, s_x, or $\hat{p}$. • Statistics are used to estimate parameters.
If a distribution of data is approximately normal, then, • About 68% of the values will be within 1 SD of the mean • About 95% of the values will be within 2 SD of the mean • About 99.7% of the values will be within 3 SD of the mean	Outliers $< Q_1 - 1.5(IQR)$; Outliers $> Q_3 + 1.5(IQR)$ where $IQR = Q_3 - Q_1$ OR Outliers < mean – 2(SD); Outliers > mean + 2(SD)
• The normal distribution with mean = 0 and SD = 1. • Table A includes areas under the Standard Normal Distribution.	• The pth percentile of a distribution is the value with p% of the observations less than or equal to it. • For example, a student who scores at the 90th percentile on a test scored the same or better than 90% of the other test takers.
1. Draw a normal distribution with horizontal axis labeled/scaled with mean/SD, boundary labeled, and area of interest shaded. **2. Perform calculations—show your work** i. Standardize each boundary value (calculate a z-score) and use technology (mean = 0, SD = 1) or Table A. ii. Use technology without standardizing: normalcdf(lower, upper, mean, SD). Label inputs!	$z = \dfrac{\text{value} - \text{mean}}{\text{standard deviation}}$ • A standardized score (z-score) describes how many standard deviations a value falls from the mean of the distribution and in what direction. • “Jessica’s test score was 2.3 standard deviations *less than* the mean ($z = -2.3$).”

Transforming Data / Effect of Changing Units	Normal Distributions: Finding Boundaries
Normal Distribution	Marginal, Joint, & Conditional Relative Frequency
Explanatory vs. Response Variables	Making Predictions / Extrapolation
Association	Interpret Slope and y-intercept

1. Draw a normal distribution with horizontal axis labeled/scaled with mean/SD, area of interest shaded and labeled, and unknown boundary marked. **2. Perform calculations—show your work** i. Use technology (mean = 0, SD = 1) or Table A to find the value of z with the appropriate area to the left of the boundary, then unstandardize. ii. Use technology without standardizing: invnorm(area, mean, SD). Label inputs!	• Adding "a" to every member of a data set adds "a" to the measures of center, but does not change the measures of variability or the shape. • Multiplying every member of a data set by a positive constant "b" multiplies the measures of center by "b" and multiplies most measures of variability by "b," but does not change the shape.
• **Marginal:** The proportion or percentage of individuals in a two-way table that have a specific value for one categorical variable. • **Joint:** The proportion or percentage of individuals in a two-way table that have a specific value for one categorical variable *and* a specific value for another categorical variable. **Conditional:** The proportion or percentage of individuals that have a specific value for one categorical variable *among* a group of individuals that share the same value of another categorical variable (the condition).	• A normal distribution is described by a symmetric, single-peaked, mound-shaped curve called a normal curve. Any normal distribution is completely specified by two parameters: its mean μ and standard deviation σ. 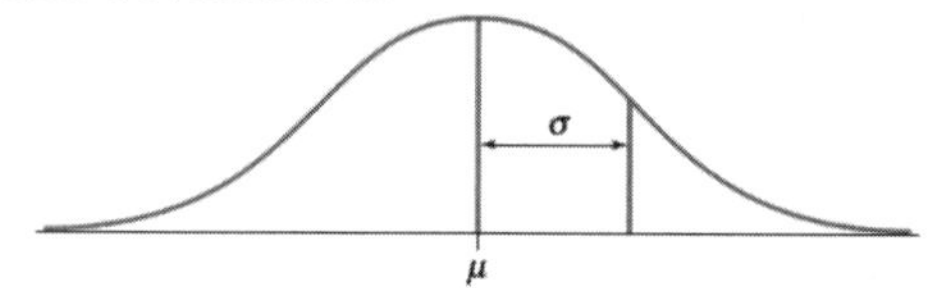
• When using a regression line in the form $\hat{y} = a + bx$, $\hat{y}$ is the "estimated" or "predicted" y-value for a given x-value. • Extrapolation is the use of a regression line for prediction outside the interval of x-values used to obtain the line. The further we extrapolate, the less reliable the predictions become.	• A response variable measures an outcome of a study. • An explanatory variable may help predict or explain changes in a response variable.
• Slope: The amount (and direction) the predicted value of y changes when x increases by one unit. • "The predicted cost of a car decreases by about $1285 for each additional year." • y-intercept: The predicted value of y when x is 0. • "The predicted cost of a car is about $23,450 when it is $x = 0$ years old."	• Two variables have an association if knowing the value of one variable helps to predict the value of the other variable. • If knowing the value of one variable does *not* help predict the value of the other variable, there is *no association* between the two variables (the two variables are *independent*).

Describing the Association in a Scatterplot	Calculate and Interpret a Residual
Interpret *r*	Reading Computer Output for Regression
Least-Squares Regression Line	Interpreting a Residual Plot
Interpret the standard deviation of residuals *s*	Population, Census, Sample

Residual = $y - \hat{y}$ (Actual y – Predicted y)

- A residual measures the difference between the actual y value and the y value that is predicted by the LSRL and gives the direction of the difference.

"The car cost $1500 more than the price predicted by the LSRL with x = years."

Discuss these, **in context** (both variable names):

- **Direction:** positive, negative, no association
- **Unusual features:** clusters, outliers, high leverage points
- **Form:** linear, nonlinear
- **Strength:** strong, moderate, weak

"There is a moderate, positive, linear association between height and weight for HS students with no unusual features."

Using foot length (x) to predict height (y):

Predictor	Coef	SE Coef	T	P
Constant	103.4100	19.5000	5.30	0.000
Foot length	2.7469	0.7833	3.51	0.004

S = 7.95126   R-Sq = 48.6%   R-Sq(adj) = 44.7

y intercept = 103.41   slope = 2.7469

SD of residuals = 7.95126 $r^2 = 0.486$

For *linear* relationships between x and y, correlation measures **strength** and gives the **direction**.

- r is always between –1 and 1
- Close to zero = very weak
- Close to 1 or –1 = strong
- Exactly 1 or –1 = perfectly straight line
- Positive r = positive association
- Negative r = negative association

- A residual plot is a scatterplot of residuals on the vertical axis and the x-values (or predicted y-values) on the horizontal axis.
- If there is a leftover curved pattern in the residual plot, consider using a regression model with a different form.

If the residual plot shows only random scatter, the regression model is appropriate.

- The least-squares regression line is the line that makes the sum of the squared residuals as small as possible.
- $\hat{y} = a + bx$, where $\hat{y}$ is the predicted value of y for a given value of x, a is the y intercept, and b is the slope.

- The population in a statistical study is the entire group of individuals we want information about.
- A census collects data from every individual in the population.
- A sample is a subset of individuals in the population from which we collect data.

- s measures the size of a typical residual. That is, s measures typical distance between the actual y-values and the predicted y-values
- "The cost of a car typically varies by about $2375 from the price predicted by the LSRL with x = years."

• An experiment deliberately imposes treatments on experimental units. Well-designed experiments allow cause-and-effect conclusions. • An observational study observes individuals and measures variables of interest but does not attempt to influence the responses. A sample survey is one type of observational study. • Observational studies that examine existing data for a sample of individuals are called *retrospective*. Observational studies that track individuals into the future are called *prospective.*	• r^2 measures the proportion (or percentage) of the variability in the response variable that is accounted for (or explained by) by the explanatory variable in the linear model. • "48% of the variability in the cost of a car is accounted for by the LSRL with x = years."
• An SRS (simple random sample) of size n is a sample chosen in such a way that every group of n individuals in the population has an equal chance to be selected as the sample. • To select an SRS using technology, label each individual in the population with a distinct label from 1 to N, use a random number generator to obtain n *different* integers from 1 to N, and choose the corresponding individuals.	• Points with *high leverage* have much larger or much smaller x-values than the other points. • An *outlier* is a point that does not follow the pattern of the data and has a large residual. • An *influential point* is any point that, if removed, substantially changes the slope, y intercept, correlation, coefficient of determination r^2, or standard deviation of the residuals s. High-leverage points and outliers can both be influential.
• **Label.** Give each member of the population a numerical label with the *same number of digits*. • **Randomize.** Read consecutive groups of digits of the appropriate length from left to right across a line in table. Ignore any group of digits that wasn't used as a label or that duplicates a label already in the sample. Continue until you have chosen n different labels.	• Nonlinear associations between two quantitative variables can sometimes be changed into linear associations by transforming one or both variables using logarithms, square roots, etc. • Once we transform the data to achieve linearity, we can fit a least-squares regression line to the transformed data and make predictions. • If the y variable was transformed, remember to back transform when calculating predicted values.
• Split population into homogeneous (similar) groups (called strata) based on anticipated response, select an SRS from each stratum, and combine the SRSs to form the overall sample. • When strata are chosen properly, a stratified random sample will produce a better estimate (less variable/more precise) than an SRS of the same size.	When choosing between different models to describe a relationship between two quantitative variables: • Choose the model whose residual plot has the most random scatter. • If there is more than one model with a randomly scattered residual plot, choose the model with the largest coefficient of determination, r^2.

Interpret the coefficient of determination r^2	Experiment vs. Observational Study
High-Leverage, Outliers, and Influential Points in Regression	SRS How to Select an SRS
Transformations to Achieve Linearity	Using a Random Digit Table to Select an SRS
Choosing a Model	Stratified Random Sampling

Cluster Sampling	Placebo Effect
Systematic Random Sampling	Control Groups & Blinding
Bias	Random Assignment and Completely Randomized Designs
Confounding	Replication and Control

The placebo effect describes the fact that some subjects in an experiment will respond favorably to any treatment, even an inactive treatment.	• Split the population into groups (often based on location) called clusters, randomly select clusters, and include each member of the selected clusters in the sample. • When clusters are chosen properly, cluster sampling is more efficient than simple random sampling. It is ideal if the members of each cluster are heterogeneous with respect to the variable being measured.
• A control group is used to provide a baseline for comparing the effects of other treatments. A control group may be given an inactive treatment (placebo), an active treatment, or no treatment. • When the subjects don't know which treatment they are receiving, and the people interacting with the subjects and measuring the response variable also don't know, the study is *double-blind*. If only one group doesn't know, the study is *single-blind.*	• Systematic random sampling selects a sample from an ordered arrangement of the population by randomly selecting one of the first k individuals and choosing every kth individual thereafter. • For example, choosing at random one of the first 10 names on a list and then selecting every 10th name thereafter.
• The purpose of random assignment is to create groups of experimental units that are roughly equivalent at the beginning of the experiment. This allows for cause-and-effect conclusions. • If treatments are assigned to experimental units completely at random (i.e., with no blocking), the result is a *completely randomized design*.	• The design of a statistical study shows bias if it is very likely to underestimate or very likely to overestimate the value you want to know. • Convenience samples, voluntary response samples, undercoverage, nonresponse, and response bias (e.g., wording of questions, lack of anonymity) can result in bias.
• Replication is giving each treatment to enough experimental units so that any differences in the effects of the treatments can be distinguished from chance differences between the groups. • Control is keeping other variables constant for all experimental units, especially variables that are likely to affect the response variable. Control helps avoid confounding and reduces variability in the response variable.	• Confounding occurs when two variables are associated in such a way that their effects on a response variable cannot be distinguished from each other. • If you are asked to identify a possible confounding variable in a given setting, you are expected to explain how the variable you choose (1) is associated with the explanatory variable and (2) is associated with the response variable.

Experimental Units, Factors, Levels, Treatments	Blocking and Matched Pairs Designs
Scope of Inference: Generalizing to a Larger Population	Probability Model, Sample Space, & Event
Scope of Inference: Cause and Effect	Probability Rules
Interpreting Probability	Conditional Probability

• Prior to random assignment, divide experimental units into groups (blocks) of experimental units that you expect to respond similarly. Then, randomly assign treatments within blocks. Blocking helps avoid confounding and accounts for the variability in the response variable due to the blocking variable. • A matched pairs design uses blocks of size 2 (using two similar units or giving both treatments to each unit in random order).	• An experimental unit is the object to which a treatment is randomly assigned. When the experimental units are human beings, they are often called subjects. • A factor is an explanatory variable that is manipulated and may cause a change in the response variable. The different values of a factor are called levels. All combinations of levels form the treatments.
• A probability model is a description of a random process that consists of two parts: a list of all possible outcomes and the probability of each outcome. • The list of all possible outcomes is called the sample space. • An event is a subset of the possible outcomes from the sample space of a random process.	• We can generalize the results of a study to a larger population if we *randomly select* from that population. • However, be aware of sampling variability—the fact that different samples of the same size from the same population will produce different estimates.
• General Addition Rule ("or" means add): $P(\text{A or B}) = P(A \cup B) = P(A) + P(B) - P(A \cap B)$ • General Multiplication Rule ("and" means mult.) $P(\text{A and B}) = P(A \cap B) = P(A) \cdot P(B \mid A)$ • Complement Rule: $P(A^C) = 1 - P(A)$	• Cause-and-effect conclusions are possible if we *randomly assign* treatments to experimental units in an experiment and find a *statistically significant* difference. • When an observed difference in responses between the groups in an experiment is so large that it is unlikely to be explained by chance variation in the random assignment, the results are called statistically significant.
• Probability that one event occurs given that another event is already known to have occurred. $P(\text{A given B}) = P(A \mid B) = \frac{P(A \cap B)}{P(B)} = \frac{P(\text{both})}{P(\text{given})}$	• The probability of any outcome of a random process is a number between 0 and 1 that describes the proportion of times the outcome would occur in a very long series of trials. • "If I were to flip a coin many, many times, I would expect to get heads in about 50% of the flips."

Law of Large Numbers	Two Events are Mutually Exclusive (Disjoint) if…
Conducting a simulation	Two Events are Independent If…
Random Variable & Probability Distribution	Combining Random Variables
Discrete vs. Continuous Random Variables	Binomial Setting and Random Variable

- Events A and B are mutually exclusive (disjoint) if they share no outcomes. That is,

$$P(\text{A and B}) = P(\text{A} \cap \text{B}) = 0$$

- The Law of Large Numbers says that if we observe more and more trials of any random process, the observed proportion of times that an event occurs approaches its probability.

- Events A and B are independent if knowing that Event A has occurred (or has not occurred) doesn't change the probability that Event B occurs.

$$P(\text{B}) = P(\text{B} \mid \text{A}) = P(\text{B} \mid \text{A}^C)$$

or

$$P(\text{A}) = P(\text{A} \mid \text{B}) = P(\text{A} \mid \text{B}^C)$$

or

$$P(\text{A and B}) = P(\text{A}) \cdot P(\text{B})$$

- Describe how to set up and use a random process to perform one trial (repetition) of the simulation. Identify what you will record at the end of each trial.
- Perform many trials of the simulation.
- Use the results of your simulation to answer the question of interest.

$$\mu_{X+Y} = \mu_X + \mu_Y \text{ and } \mu_{X-Y} = \mu_X - \mu_Y$$

- If X and Y are independent (knowing the value of X doesn't help predict the value of Y), $\sigma_{X+Y} = \sqrt{\sigma_X^2 + \sigma_Y^2}$ and $\sigma_{X-Y} = \sqrt{\sigma_X^2 + \sigma_Y^2}$

- If X and Y are normally distributed, then $X + Y$ and $X - Y$ are also normally distributed.

- A random variable takes numerical values that describe the outcomes of a random process.

- The probability distribution of a random variable gives its possible values and their probabilities.

- **B**inary? Each trial can be classified as success/failure
- **I**ndependent? Trials are independent (check 10% condition when sampling without replacement)
- **N**umber? Number of trials (n) is fixed in advance
- **S**ame probability of success? The probability of success (p) is the same for each trial.

- A discrete random variable X takes a countable set of possible values with gaps between them on a number line. Display its probability distribution with a table or histogram.

- A continuous random variable can take any value in an interval on the number line. Display its probability distribution with a density curve.

Calculate/Interpret the Mean (Expected Value) of a RV	Calculating Binomial Probabilities
Calculate/Interpret the Standard Deviation of a RV	Describing a Binomial Distribution
Linear Transformation of a Random Variable	Geometric Setting and Random Variable
Calculating Geometric Probabilities	Accurate vs. Precise

1) Define the variable, state the distribution and parameters, and identify the values of interest. 2) Perform calculations: • $P(X=x)=\binom{n}{x}(p)^x(1-p)^{n-x}$ = Binompdf(n, p, x) • $P(X \le x)$ = Binomcdf(n, p, x) • Remember to label inputs in calculator (n, p,	• Mean (Expected Value) of a Discrete RV: $\mu_X = E(X) = \sum x_i P(x_i)$ • The mean (expected value) is the average value of a RV after many, many trials of a random process. "If I play the game many, many times, I would lose about \$0.05 per game, on average."
• Mean: $\mu_X = E(X) = np$ • Standard Deviation: $\sigma_X = \sqrt{np(1-p)}$ • Shape: Approximately normal if $np \ge 10$ and $n(1-p) \ge 10$	• Standard Deviation of a Discrete RV: $\sigma_X = \sqrt{\sum (x_i - \mu_x)^2 P(x_i)}$ • The standard deviation measures how much the values of a RV typically vary from the mean in many, many trials of a random process. • "If I play the game many times, the amount I win typically varies from the mean by about \$0.15."
• Arises when we perform independent trials of the same random process and record the number of trials it takes to get one success. On each trial, the probability p of success must be the same. • X = number of trials needed to achieve one success (includes the success trial)	• If $Y = a + bX$ $\mu_Y = a + b\mu_X$ $\sigma_Y = \|b\|\sigma_X$ • SD not affected by adding constant. • Shape does not change (unless b is negative)
A statistic/estimator is **accurate** if it is *unbiased* (its sampling distribution is *centered* in the right place). A statistic/estimator is **precise** if it has little *variability* (its sampling distribution isn't very *spread* out).	1) Define the variable, state the distribution and parameter, and identify the values of interest. 2) Perform calculations: • $P(X=x) = (1-p)^{x-1}p$ = Geometpdf(p, x) • $P(X \le x)$ = Geometcdf(p, x) Remember to label inputs in calculator (p, x)!

Describing a Geometric Distribution	Describe the Sampling Distribution of $\hat{p}$
Density Curve	Describe the Sampling Distribution of $\hat{p}_1 - \hat{p}_2$
What is a Sampling Distribution?	Describe the Sampling Distribution of $\bar{x}$
Unbiased Estimator	Describe the Sampling Distribution of $\bar{x}_1 - \bar{x}_2$

• Center: $\mu_{\hat{p}} = p$ • Variability: $\sigma_{\hat{p}} = \sqrt{\frac{p(1-p)}{n}}$ if $n < 0.10N$ when sampling without replacement (*10% condition*) • Shape: Approximately normal if $np \geq 10$ and $n(1-p) \geq 10$ (*Large Counts condition*)	• Mean: $\mu_X = E(X) = \frac{1}{p}$ • Standard Deviation: $\sigma_X = \frac{\sqrt{1-p}}{p}$ • Shape: Skewed to the right with a peak at $X = 1$
• Center: $\mu_{\hat{p}_1-\hat{p}_2} = p_1 - p_2$ • Variability: $\sigma_{\hat{p}_1-\hat{p}_2} = \sqrt{\frac{p_1(1-p_1)}{n_1} + \frac{p_2(1-p_2)}{n_2}}$ if $n_1 < 0.10N_1$ and $n_2 < 0.10N_2$ when sampling without replacement (*10% condition*) • Shape: Approximately normal if n_1p_1, $n_1(1 - p_1)$, n_2p_2, $n_2(1 - p_2)$ all ≥ 10 (*Large Counts condition*)	• A density curve models the distribution of a continuous random variable with a curve that is always on or above the horizontal axis and has area exactly 1 underneath it. • The area under the density curve and above any specified interval of values on the horizontal axis gives the probability that the random variable falls within that interval.
• Center: $\mu_{\bar{x}} = \mu$ • Variability: $\sigma_{\bar{x}} = \frac{\sigma}{\sqrt{n}}$ if $n < 0.10N$ when sampling without replacement (*10% condition*) • Shape: Approximately normal if population distribution is approximately normal or if $n \geq 30$	• The distribution of values taken by a statistic in all possible samples of the same size from the same population. It describes the possible values of a statistic and how likely these values are. • Contrast with the distribution of the population and the distribution of a sample.
• Center: $\mu_{\bar{x}_1-\bar{x}_2} = \mu_1 - \mu_2$ • Variability: $\sigma_{\bar{x}_1-\bar{x}_2} = \sqrt{\frac{\sigma_1^2}{n_1} + \frac{\sigma_2^2}{n_2}}$ if $n_1 < 0.10N_1$ and $n_2 < 0.10N_2$ when sampling without replacement (*10% condition*) • Shape: Approximately normal if, for each sample, population distribution is approximately normal or $n \geq 30$	• A statistic used to estimate a parameter is an unbiased estimator if the mean of its sampling distribution is equal to the value of the parameter being estimated. • In other words, the sampling distribution of the statistic is centered in the right place.

Central Limit Theorem (CLT)	Margin of Error
Point Estimate vs. Interval Estimate	Critical Value
Interpreting a Confidence Interval	Standard Error of a Statistic
Interpreting a Confidence Level (Meaning of 95% Confidence)	One-Sample z Interval for p

• The margin of error of an estimate describes how far, at most, we expect the point estimate to vary from the population parameter. The margin of error decreases when: • The sample size increases (multiple the sample size by 4 to cut the margin of error in half) • The confidence level decreases	If the population distribution is non-normal, the sampling distribution of the sample mean $\overline{x}$ will become more and more normal as *n* increases.
• The critical value is a multiplier that makes a confidence interval wide enough to have the stated capture rate. • z^* for confidence intervals involving proportions • t^* for confidence intervals involving means and slope	• Point estimate: The single-value best guess for the value of a population parameter. • For example, $\hat{p} = 0.63$ is a point estimate for the population proportion p. • Interval Estimate = Confidence Interval = interval of plausible values for a parameter based on sample data.
• The standard error of a statistic estimates how much the statistic typically varies from the value of the population parameter. • The standard *error* of a statistic uses data from a sample to estimate the standard *deviation* of the statistic. • Formulas for standard deviations and standard errors are included on the formula sheet.	"We are ___% confident that the interval from ___ to ___ captures the ____ [parameter in context]."
$\hat{p} \pm z^* \sqrt{\frac{\hat{p}(1-\hat{p})}{n}}$ Calculator: 1-PropZInt (*X* must be an integer)	"If we were to select many random samples of the same size from the same population and construct a *C*% confidence interval using each sample, about *C*% of the intervals would capture the [parameter in context]."

Inference for Proportions (Conditions)	<u>4-Step Process</u> Confidence Intervals
Finding the Sample Size (CI for p)	<u>4-Step Process</u> Significance Tests
Null Hypothesis & Alternative Hypothesis	Carrying out a Two-Sided Test from a Confidence Interval
Interpret a P-value	Type I Error & Type II Error

• STATE: State the parameter you want to estimate, and the confidence level. • PLAN: Identify the appropriate inference method and check the conditions. • DO: If the conditions are met, perform calculations. • CONCLUDE: Interpret your interval in the context of the problem.	• **Random:** Data from a random sample (2-sample: independent random samples or randomized experiment) ○ **10%:** For each sample, $n < 0.10N$ when sampling w/o replacement **Large Counts:** For each sample, $np \geq 10$ and $n(1-p) \geq 10$ (use $\hat{p}$ for confidence intervals, use p_0 for 1-sample test, use $\hat{p}_C$ for 2-sample test)
• STATE: State the hypotheses, parameter(s), and significance level. • PLAN: Identify the appropriate inference method and check the conditions. • DO: If the conditions are met, perform calculations. Calculate the test statistic and find the *P*-value. • CONCLUDE: Make a conclusion about the hypotheses in the context of the problem.	$z^*\sqrt{\frac{\hat{p}(1-\hat{p})}{n}} \leq ME$ where ME = desired margin of error If a value of $\hat{p}$ is not given, use $\hat{p} = 0.5$
• $\alpha = 1 -$ confidence level • If the null hypothesis value is in the interval, then it is a plausible value and H_0 should not be rejected. • If the null hypothesis value is not in the interval, then it is not a plausible value and H_0 should be rejected.	• The claim that we weigh evidence against in a significance test is called the null hypothesis (H_0). • The claim that we are trying to find evidence for is the alternative hypothesis (H_a). ○ The alternative hypothesis is one-sided if it states that a parameter is *greater than* the null value or if it states that the parameter is *less than* the null value. ○ The alternative hypothesis is two-sided if it states that the parameter is *different from* the null value (it could be either greater than or less than).
• **Type I Error:** Finding convincing evidence that H_a is true, when in reality H_a isn't true. (Rejecting H_0 when H_0 is true). ○ *P*(Type I error) = α • **Type II Error:** Not finding convincing evidence that H_a is true, when in reality H_a is true. (Failing **II** reject H_0 when H_a is true).	• The *P*-value is the probability of getting evidence for the alternative hypothesis H_a as strong as or stronger than the observed evidence when the null hypothesis H_0 is true. • "Assuming that the mean speed of all cars is 25 mph, there is a 0.001 probability of getting a sample mean of 37 mph or more by chance alone."

Conclusions to a Significance Test	Interpret Power
One-Sample z Test for p	Power Will Be Greater When…
Two-Sample z Interval for $p_1 - p_2$	Paired t Interval for μ_{Diff}
Two-Sample z Test for $p_1 - p_2$	One-Sample t Test for μ

• Probability of finding convincing evidence that H_a is true, when a specific alternative value of the parameter is true. In other words, the probability of avoiding a Type II error. • "Assuming that the new drug is 10 points more effective, there is a 0.89 probability of finding convincing evidence that the mean response to the new drug is greater than the mean response to the old drug."	• Because P-value $\leq \alpha$, we reject H_0. There is convincing evidence for [H_a in context]. • Because P-value $> \alpha$, we fail to reject H_0. There is not convincing evidence for [H_a in context]. α = significance level (e.g., 0.01, 0.05, 0.10)
1. The **sample size** is greater. 2. The **significance level** (α) is greater. 3. The **effect size** is greater. Effect size is the difference between the truth and the hypothesized value. 4. Better **data collection** methods are used to account for sources of variability (e.g., controlling other variables, blocking, stratifying)	H_0: p = hypothesized value = p_0 $z = \frac{\hat{p} - p_0}{\sqrt{\frac{p_0(1-p_0)}{n}}}$ Calculator: 1-PropZTest (X must be an integer)
$\bar{x}_{\text{Diff}} \pm t^* \frac{s_{\text{Diff}}}{\sqrt{n_{\text{Diff}}}}$ df = $n_{\text{Diff}} - 1$ Calculator: TInterval	$(\hat{p}_1 - \hat{p}_2) \pm z^* \sqrt{\frac{\hat{p}_1(1-\hat{p}_1)}{n_1} + \frac{\hat{p}_2(1-\hat{p}_2)}{n_2}}$ Calculator: 2-PropZInt (X_1 and X_2 must be integers)
H_0: μ = hypothesized value = μ_0 $t = \frac{\bar{x} - \mu_0}{\frac{s_x}{\sqrt{n}}}$ df = $n - 1$ Calculator: T-Test	H_0: $p_1 - p_2 = 0$ $z = \frac{(\hat{p}_1 - \hat{p}_2) - 0}{\sqrt{\hat{p}_C(1-\hat{p}_C)\left(\frac{1}{n_1} + \frac{1}{n_2}\right)}}$ where $\hat{p}_C = \frac{X_1 + X_2}{n_1 + n_2}$ Calculator: 2-PropZTest (X_1 and X_2 must be integers)

What is a t distribution?	Paired t Test for μ_{Diff}
Inference for Means (Conditions)	Two-Sample t Interval for $\mu_1 - \mu_2$
One-Sample t Interval for μ	Two-Sample t Test for $\mu_1 - \mu_2$
Two-Sample t vs. Paired t	Chi-Square Test for Homogeneity

H_0: $\mu_{diff} = 0$ $t = \frac{\overline{x}_{\text{Diff}} - 0}{\frac{s_{\text{Diff}}}{\sqrt{n_{\text{Diff}}}}}$ df = $n_{\text{Diff}} - 1$ Calculator: T-Test	• A t distribution is described by a symmetric, single-peaked, bell-shaped density curve centered at 0. Any t distribution is completely specified by its *degrees of freedom* (df). • t distributions are more variable (have more area in the tails) than the standard normal distribution. As the df increases, the t distributions approach the standard normal distribution.
$(\overline{x}_1 - \overline{x}_2) \pm t^* \sqrt{\frac{s_1^2}{n_1} + \frac{s_2^2}{n_2}}$ df = from technology Calculator: 2-SampTInt ("No" to pooling)	• **Random:** Data from a random sample (paired: random sample or randomized experiment, 2-sample: independent random samples or randomized experiment) ○ **10%:** For each sample, $n < 0.10N$ when sampling w/o replacement • **Normal/Large Sample:** For each sample, the population distribution is approximately normal or sample size is large ($n \geq 30$). If $n < 30$, graph sample data and verify no strong skewness or outliers. Include graph!
H_0: $\mu_1 - \mu_2 = 0$ (usually) $t = \frac{(\overline{x}_1 - \overline{x}_2) - 0}{\sqrt{\frac{s_1^2}{n_1} + \frac{s_2^2}{n_2}}}$ df = from technology Calculator: 2-SampTTest ("No" to pooling)	$\overline{x} \pm t^* \frac{s_x}{\sqrt{n}}$ df = $n - 1$ Calculator: TInterval
• H_0: The distribution of a categorical variable is the same for two or more populations/treatments. • Expected count: $\frac{(\text{row total})(\text{column total})}{\text{table total}}$ • $\chi^2 = \sum \frac{(\text{observed} - \text{expected})^2}{\text{expected}}$ • df = (# rows – 1)(# columns – 1) Calculator: χ^2-Test (uses matrices)	Two-sample: • "Difference in means" • Independent random samples or completely randomized experiment Paired: • "Mean difference" • Data are paired (two observations of the same variable made on the same individual or two very similar individuals)

Chi-Square Tests (Conditions)	Chi-Square Tests: Homogeneity vs. Independence
What is a χ^2 distribution?	Inference for Regression (Conditions)
Chi-Square Test for Goodness of Fit	t Interval for the Slope
Chi-Square Test for Independence	t Test for the Slope

- Both use data in a two-way table.
- Calculations (Do step) are the same for both tests.
- Homogeneity: Data from independent random samples or multiple treatment groups (all experiments should be analyzed with a test for homogeneity). One set of totals known in advance.
- Independence: Data from one random sample. Only table total known in advance.

- **Random:** Data from a random sample (test for homogeneity: independent random samples or randomized experiment).
  - **10%:** $n < 0.10N$ for each sample when sampling w/o replacement.
- **Large Counts:** All *expected* counts ≥ 5.

- **Linear:** Association between variables is linear.
- **Normal:** For each value of x, the distribution of y is approximatley normal.
- **Equal SD** For each value of x, the SD of y is the same.
- **Random:** Data from a random sample or randomized experiment.
  - **10%:** When sampling without replacement, $n < 0.10N$.

- A chi-square distribution is defined by a density curve that takes only nonnegative values and is skewed to the right.
- As df increases, the chi-square distributions become more variable, less skewed, and centered at a larger value (mean = df).
- The chi-square test statistic measures how different the observed counts are from the expected counts, relative to expected counts.

$$b \pm t^* s_b \qquad \text{df} = n - 2$$

Calculator: LinRegTInt

Predictor	Coef	SE Coef	T	P
Constant	103.4100	19.5000	5.30	0.000
Foot length	2.7469	0.7833	3.51	0.004

S = 7.95126 R-Sq = 48.6% R-Sq(adj) = 44.7%

$2.7469 \pm t^*(0.7833)$

- H_0: The distribution of a categorical variable in a population is the same as hypothesized distribution
- Expected Counts: Sample size times hypothesized proportion in each category.
- $\chi^2 = \sum \frac{(\text{observed} - \text{expected})^2}{\text{expected}}$
- df = # of categories – 1
- Calculator: χ^2 GOF-Test (uses lists)

$$t = \frac{b - \text{hypothesized slope}}{s_b} \qquad \text{df} = n - 2$$

Calculator: LinRegTTest

Predictor	Coef	SE Coef	T	P
Constant	103.4100	19.5000	5.30	0.000
Foot length	2.7469	0.7833	3.51	0.004

S = 7.95126 R-Sq = 48.6% R-Sq(adj) = 44.7%

$t = \frac{2.7469 - 0}{0.7833} = 3.51$; Two-sided P-value = 0.004

- H_0: There is no association between two categorical variables in one population.
- Expected count: $\frac{(\text{row total})(\text{column total})}{\text{table total}}$
- $\chi^2 = \sum \frac{(\text{observed} - \text{expected})^2}{\text{expected}}$
- df = (# rows – 1)(# columns – 1)
- Calculator: χ^2-Test (uses matrices)

NOTES

NOTES

NOTES

NOTES

NOTES

NOTES